PHYSIOLOGY, GROWTH AND DEVELOPMENT
OF PLANTS IN CULTURE

Physiology, Growth and Development of Plants in Culture

Edited by

P.J. LUMSDEN, J.R. NICHOLAS and W.J. DAVIES

SPRINGER SCIENCE+BUSINESS MEDIA, B.V.

Library of Congress Cataloging-in-Publication Data

Physiology, growth, and development of plants in culture / edited by
  P.J. Lumsden and J.R. Nicholas and W.J. Davies.
      p.   cm.
  Includes index.
  ISBN 978-0-7923-2516-1   ISBN 978-94-011-0790-7 (eBook)
  DOI 10.1007/978-94-011-0790-7
    1. Plant propagation--In-vitro--Congresses.   2. Plant physiology-
  -Congresses.   3. Crops--Physiology--Congresses.   I. Lumsden, P. J.
  II. Nicholas, J. R. (Julian R.)   III. Davies, W. J.
  SB123.6.P46   1994
  631.5'3--dc20   93-20900

ISBN 978-0-7923-2516-1

Printed on acid-free paper

Contents

2. Applications

Morphogenesis and regeneration

Rooting and acclimatization

Rejuvenation

Conclusions

Editorial

Introduction

Over recent years, micropropagation has passed through various stages in its development as a genuine industrial plant production process. Like many new technologies, progress has not been as rapid as many expected and, even now, relatively few crops are produced commercially. There are two obvious reasons why progress has been limited: the biology is not properly understood and the economics restrict application.

The following description is still frequently used to describe micro-propagation. 'Small pieces of plant are removed from a selected parent and sterilised to remove contaminants and diseases. They are then placed in sterile culture vessels containing a gel based medium that contains all things necessary for ideal growth (including carbohydrates, organic and inorganic nutrients, and growth regulators). The cultures are placed in a growth room with controlled temperature, irradiance and daylength so that, in its ideal culture, the plant is capable of unrestricted (exponential) growth'. Unfortunately, for the majority of plant species ideal culture conditions have not been achieved; either growth rates are low and / or the quality of the material is poor. Therefore the range of plants being produced is limited. Until recently, relatively little work had been done to find out why the culture is not ideal for plant growth. However, during the past decade, tissue culture companies and others have invested considerable effort to reduce the empirical nature of the production process. There have been many attempts to reduce variability in the rooting process, too often the source of unexpected and costly plant losses. Subsequently serious attention has been given to the control of contamination in culture and the control of plant water relations during planting. Many groups have been involved in work on metabolic and nutritional constraints to plant growth and quality. Advances continue to be made in the regeneration and propagation of recalcitrant species. These and other projects made up the programme of the conference 'Physiology, Growth and Development of Plants and Cells in Culture', held at Lancaster University in September 1992, sponsored by Neo Plants Ltd, IAPTC, BSPGR and the SEB. One purpose of the conference was to shed some light on the limitations to growth and development of *in vitro* cultured material, and to indicate directions in which future work should go in order to enhance and expand its use in plant production.

There are two reasons why people work on micropropagation related subjects. The first is because the culture provides a good experimental platform. For this, the objective is a greater understanding of biological processes; the work does not need to have any obvious application. Rooting of woody plants in particular is very difficult to investigate 'ex vitrum'; during a year the changes in light, temperature and daylength make replication of experimental treatments impossible. The use of *in vitro* culture allows variables, which are otherwise uncontrollable, to be standardised. The other reason for working in culture is to develop and improve commercially applicable technologies. There is a danger that these two justifications for work can become confused. It often happens that institutional research workers try to justify their approach to the commercial world. This doesn't always work because the commercial world is not always consulted first, and work then proceeds in the wrong direction. Similarly, in trying to produce commercially justifiable projects there is too great an emphasis on achieving a result rather than understanding the mechanism. When this happens, work tends to be empirical. Industrial customers demand very high standards of reliability and quality. It is not always possible to provide these if the method of production has been developed empirically; if anything goes wrong there is no reliable way of correcting them. It is therefore important to know **why** things work as well as to know **that** they work, but the two approaches need to relate closely to each other. Thus a second aim of the conference was to introduce subject specialists to micropropagators; beginning new discussions on the constraints and potentials affecting the commercialisation of micropropagation.

The *in vitro* environment

There has been much debate in recent years on the question of whether the established culture methods, involving sugar and agar, should continue to be used. There are many who argue that plants would grow better in a high light, low carbohydrate system. The major justification for this argument comes from two directions; the culture exposes the plant to unnaturally high humidity and suppresses the need and opportunity for photosynthesis. To overcome these two biologically damaging side effects, the recommendation is to remove them by taking away the sugar in the medium and increasing the light and CO_2. This subject is referred to by Chaves (this volume) and more extensively in the review by Buddendorf-Joosten and Woltering. It has also been suggested (Firn, this volume), that growth regulators and nutrients should be provided continuously, rather than with each monthly subculture.

There is much sense in these ideas; it is certainly not wise to create an artificial environment and expect to be able to force subjugated plants of all genera to grow in it. Much better to tailor the environment to suit the biological requirements of the plant. There are, however, many distinct advantages to the 'traditional' culture system. In industry, the overriding demand from the market is for reliability (being: the ability to produce a good quality product on time). The culture, containing sugar and growing under fluorescent light of known irradiance, often grows economically and reliably. Drastically altering the culture may cause us to lose the advantages we have without satisfactorily replacing them.

An alternative to removing sugar and other established components of the culture system is to make more effort to understand how plants interact with the *in vitro* environment. The culture might then be modified to the benefit of the plant, without losing the advantages of the system. Once the biological process is understood it will not take long for the industry to find ways of adapting. There are good examples of this in the chapters by Welander (carbohydrates), Lees (light), Cassells and Roche, Santamaria and Davies (water and gas exchange). This debate will continue and, when a change to the culture system is justified, it will have to be made quickly and effectively by companies who want to remain competitive.

Contents of this volume

The line of thought that it is as important to know **why** things work as to know **that** they work was followed in constructing the programme for the meeting. Firstly, specialists working on fundamental aspects of plant physiology and biochemistry with relevance to tissue culture were invited. Micropropagators tend to be generalists, being reasonably knowledgeable about many subjects, rather than being expert in one; too often work has been done by micropropagators without realising that others already know what they are trying to find out. This was certainly true when we began to work on contamination. It is also true of light, photosynthesis and nutrition; it may also be true of water relations. Secondly, invited speakers from micropropagation were chosen because they are amongst the best workers in the field and because they have at some time worked on problems associated with the subject given to them.

The book resulting from the conference is entitled 'Physiology, Growth and Development of Plants in Culture' simply because there were so few contributions which dealt only with cell culture. We hope that it will be seen as a standard text as much as simply the proceedings of the conference. It has been divided into two main sections, the first dealing with aspects of the *in vitro* environment — light, nutrients, water, gas, and the second with applied aspects of the culture process — morphogenesis, acclimation, rejuvenation, contamination. Within each section there are contributions from the invited speakers, which give a general background to the particular area, plus contributions from groups actively working in the particular area. Specialist contributions have been written as mini reviews, while offered papers and shorter papers of poster contributions are presented as experimental papers with methods, results etc. Where possible the two types of contribution have been grouped to give a coherent view of a particular area, a valuable contrast of pure and applied presentations on most subjects that we hope will provide the basis for productive work in the future. As Lorenz and Leslie wrote in their book 'The Financial Times on Management', '(successful organisations)....rely, at almost all levels, on teams of people from different specialist disciplines, rather than on either individuals or single departments'.

Julian Nicholas

Environmental constraints to photosynthesis in *ex vitro* plants

M.M. CHAVES

Dep. Botânica e Engenharia Biológica, Instituto Superior de Agronomia, Tapada da Ajuda, 1399 Lisboa Codex, Portugal

1. Introduction

Plants transferred to a different environment may become more susceptible to various stresses because they have not developed adequate patterns of resource allocation and evolved the morphological and physiological features required by the new 'demands'. This is the case for micropropagated plants which often do not survive transfer from *in vitro* culture to the greenhouse or the field [56], where high irradiances and low air humidity become stressful to the young plants just starting to become autotrophic. Understanding mechanisms of stress physiology as well as the plant's potential to acclimate to new environments is of particular importance if we are to predict and improve performance and survival of plants during the process of acclimation, or acclimatization. The latter is the horticultural terminology for acclimation, meaning the guided process of adjustment of the plants to a new environment [20].

Most stress responses are not fully expressed in isolated systems but occur as an integrated response of the whole plant [28]. However, many regulatory systems are located in leaves, and may change as a function of leaf development or with environmental changes [24]. Leaves control the so-called 'source strength', and therefore carbon assimilation by the whole plant, and are partly responsible for carbon allocation to the different parts of the plant.

In developing leaves the components of the photosynthetic apparatus have to be synthesised and assembled into functional units in order to allow leaves to transduce light energy into chemical energy and become autotrophic [3]. In the particular case of plantlets or shoots that are growing *in vitro*, carbon and nitrogen allocation to the different components of the photosynthetic apparatus follow the pattern of shade leaves. Moreover some species exhibit *in vitro* stomatal structure and functioning which are different from those of greenhouse or field-grown plants (see also chapters by Santamaria & Davies, and Mansfield) [46] and which increase the fragility of micropropagated plants when removed from culture.

After transfer from culture, changes in functioning and structure may occur in leaves developed before the transfer (e.g. stomatal response to humidity [10])

1

P.J. Lumsden, J.R. Nicholas and W.J. Davies (eds.), Physiology, Growth and Development of Plants in Culture, 1–18, 1994.

but as a rule modifications take place in leaves formed after the plants are removed from culture [29, 31].

As far as photosynthesis is concerned differences in photosynthetic rates among various species *in vitro* appear in the literature (see references in [56]). These discrepancies have been attributed either to real differences between genotypes or to the methodologies used to assess photosynthetic rates. According to Grout [34] some species (e.g. strawberry and cauliflower) never develop *in vitro* leaf photosynthetic capability whereas others (e.g. *Dieffenbachia*) produce leaves that are photosynthetically competent and will eventually adapt to autotrophic conditions.

In this paper a brief description of carbon balance in developing and shaded leaves will be given, followed by the analysis of photosynthetic response to the three major environmental stresses that micropropagated plants will encounter when displaced from *in vitro* to greenhouse or field conditions – high light, high temperature and water deficits.

2. Photosynthesis in developing leaves

In most dicotyledon species the major increases in cell number occur during lamina unfolding, 90% of all cells in a leaf being produced by divisions during this period (Fig.1 and [22]). Variation in final area of a leaf is associated with differences in numbers rather than size of epidermal cells. On the other hand,

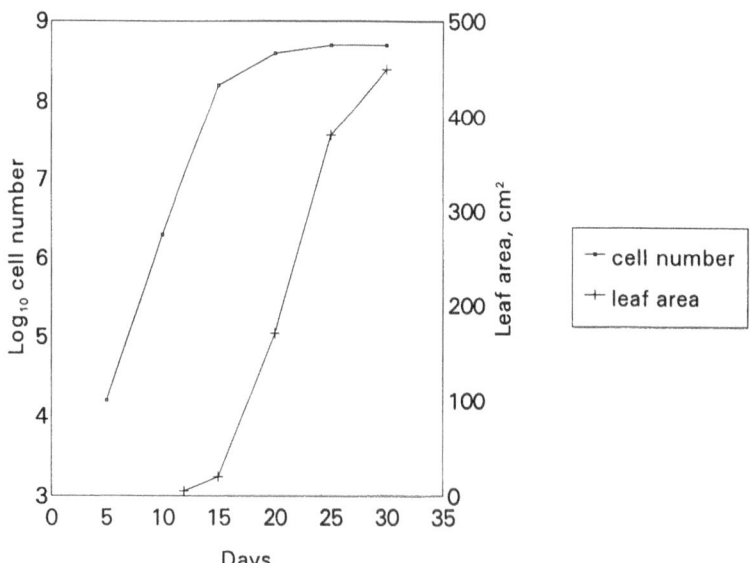

Fig. 1. Time-course of leaf growth in area and in cell number at node 2 on cucumber plants held at 4.4 MJ m^{-2} of irradiance (adapted from [42] and [49]).

whereas cell number dictates the potential size of a leaf, expression of that potential in terms of area and thickness depends on cell enlargement. Supply of photoassimilate is thus a stronger limitation to division than to enlargement of leaf cells [42].

An exponential increase in the import by leaf primordia is observed with time, the phase of rapid import of assimilates corresponding to the structural development of the phloem. According to Dale [22] the development of major veins in a leaf occurs acropetally with the need for nutrition of the leaf in its early development as an importing organ, whereas vein maturation occurs from the leaf tip downwards in a process closely linked to export (see also [28]). Most data in the literature indicate that import of carbon is maximal before the leaf reaches half of its final size, but can still take place after export of assimilates has begun [22]. For export to occur, local production of organic carbon must exceed local demand. When the leaves are fully mature storage carbohydrates such as starch begin to be formed. This pool of assimilates is tightly controlled either by the source/sink relationships, the photoperiod, or the environmental stresses such as leaf water deficits, as will be further discussed.

Studies on a wide range of species have shown that net photosynthetic rates increase rapidly during leaf development, reaching a maximum before full leaf area expansion is attained (ranging from 35 to 75% of the maximum leaf area), then decrease during late developmental stages (see references in [63]). Some plant species exhibit more than one maximum in net photosynthetic rate during the leaf life span. These secondary maxima are ascribed to various events in the plant's development such as unfolding of leaves, flowering etc (e.g. *Phaeolus, Glycine* and *Capsicum*), [60]. The ontogenic patterns of net photosynthesis are usually repeated in the sucessive leaves formed in a plant. Leaves inserted in the middle of a shoot are the ones exhibiting maximal final blade sizes and photosynthetic activities. Only in mature plants are leaf insertion gradients similar to the ontogenic changes of an individual leaf [60], consequently care must be taken when extrapolating the results obtained on studies of leaf insertion gradients to leaf age responses. The quantitative expression of the ontogenetic changes on photosynthesis is dependent on environmental factors, mainly the irradiance. However plant control on leaf ontogeny seems to be stronger than environmental-induction [65].

The increase in photosynthetic activity during leaf expansion is correlated with the increase in numbers of stomata per unit leaf area and the number of chloroplasts per cell. The formation of chloroplast ultrastructure, chlorophyll accumulation and synthesis of other components of the photosynthetic apparatus proceed almost in parallel. At early stages of leaf growth, the capacity for light energy transduction seems to be non-significant as compared to the potential supply of chemical energy from respiration [3] and is in accordance with the high rates of import of assimilates in growing leaves as mentioned earlier in this chapter. As soon as chlorophyll accumulates, the leaves begin to develop photochemically competent thylakoids. On the other hand, an increase in the carboxylation activity per unit of leaf area of ribulose 1,5-bisphosphate carboxylase-oxygenase (Rubisco) has often been reported accompanying leaf development [4, 11]. This is correlated with the increase in the amount of this enzyme observed in the leaves up to full maturity. Similar increases in Rubisco activity and amount have been observed from the young leaves of the top of the shoot, followed by a decline in older leaves, as shown in mature tobacco plants (Fig. 2; see also [68]). An increase in the ratio of glucose/fructose concentrations was also observed in younger leaves of tobacco. The inhibitory effects of high concentrations of glucose on Rubisco activity reported by Krapp *et al.* [39] may partly explain the lower activities of this enzyme in younger leaves. In C_4 plants either a continuous increase in the activity of phosphoenolpyruvate carboxylase up to a maximum or an increase followed by a decrease have been reported during leaf ontogeny [30].

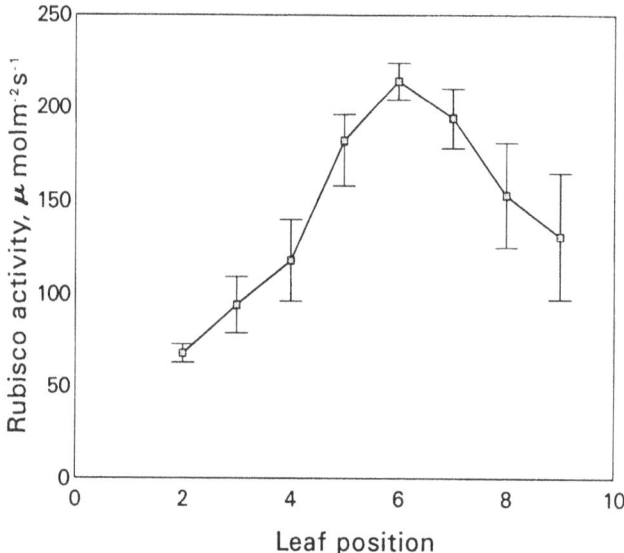

Fig. 2. Rubisco activity as a function of leaf position of middle-aged *Nicotiana tabacum* plants. Leaves are numbered from the base of the shoot. Bars represent SE of the mean. (A Krapp, MM Chaves, JS Pereira and M Stitt, unpublished).

In addition to the limitations imposed by lower stomatal conductance and lower Rubisco activity on carbon uptake by young leaves, the very low rates of net photosynthesis are also associated with the high rates of dark respiration in those leaves [41]. The increase in respiration at early stages of leaf expansion presumably reflects mitochondrial development in the leaf cells as well as high requirements for respiratory activity of imported assimilates during active growth. The gradual decline in respiratory activity after rapid expansion of the leaf is accomplished is attributed to the aquisition of photosynthetic competence of the leaf, resulting in a decreased requirement for respiratory activity. Since a young leaf contains populations of both autotrophic and heterotrophic cells the significance of change in respiration rate during development is quite complex [3].

Net carbon uptake in young leaves is also dependent on the rate of photorespiration. Presumably as a result of difficulties in accurately measuring photorespiratory rates some conflicting data on ontogenetic changes in photorespiration are apparent in the literature. According to some authors photorespiratory rates in leaves of C_3 species increase during leaf ontogeny to a maximum, obtained usually after the maximal net photosynthetic rate, and then slowly decline (see [12] for references). Other studies led to the conclusion that photorespiratory activity decreases throughout leaf ontogeny (see [3]).

In summary, changes in the carbon balance observed in a developing leaf are the result of the differential development of the metabolic pathways involved in energy transduction. High rates of respiration at the early stages of leaf growth are noteworthy. These presumably account for the metabolic demands of the growing heterotrophic tissues and are responsible for the very low net CO_2 uptake by young leaves. Ontogenetic studies of individual leaves are difficult to undertake because different parts of the same expanding leaf are often at different developmental stages, with significant consequences for analysis and sampling. For these reasons, there are still no mechanistically based models of carbon balance in a developing leaf.

3. Characteristics of leaves from shaded habitats

The light environment during growth strongly influences growth and development of plants and determines leaf morphology and physiology. When shaded, plants of the 'shade-tolerant' type grow slowly, increase leaf expansion and exhibit acclimation of the photosynthetic apparatus to the low photon flux densities (PFD). If they are 'shade-avoiding' species (or sun plants) stem growth is greatly enhanced whereas leaf expansion is reduced [23]. These types of response involve not only the effects of the photon flux density but also the quality of light, which under shade involves a decrease in the photo-morphogenetically important red: far-red ratio. The reasons for the different response of leaf area to shade in 'shade-tolerant' or in 'shade-avoiders' are unclear. According to Dale [23] one possible interpretation is that in 'shade-avoiders' leaf expansion is closely linked to the available photoassimilates whereas in 'shade-tolerant' species light quality appears to be the major control of the individual leaf dimension.

In 'shade-tolerant' species the increase in leaf expansion leads to a marked enhancement of specific leaf area — i.e. leaf area per unit leaf weight — and therefore to larger but thinner leaves [8]. As a consequence lower rates of dark respiration per unit leaf area are observed in these leaves and therefore lower light compensation points [47]. The low rates of respiration are particularly important at the whole plant level because a decrease in leaf night respiration brings about a significant increase in plant carbon gain [33].

Table 1. Average values of fresh weight per leaf area, chlorophyll content and ratio of chlorophyll a/b and soluble protein/chlorophyll in 4 or 5 species of shade and sun species (adapted from [8] and [47])

	Fresh weight per leaf area (g dm-2)	Chlorophyll a+b		Ratio chlor. a/b	Ratio soluble protein/chlor
		per fresh weight (mg g-1)	per leaf area (mg dm-2)		
Shade species	0.96±0.06	3.10±0.10	2.98±0.28	2.85±0.09	3.30±0.40
Sun species	2.50±0.19	1.94±0.13	4.68±0.25	3.5±0.15	11.08±1.13

Acclimation processes in shade leaves involve an increased investment in the light-harvesting complexes which is apparent as a higher proportion of chlorophyll *b* relative to chlorophyll *a*, a higher content of total chlorophyll expressed on a weight basis and a lower ratio of soluble protein/chlorophyll than in leaves grown under high light i.e. sun leaves (Table 1). Chloroplasts with large irregularly arranged grana stacks (as many as 100 thylakoids per grana stack) are typical of shade leaves [18]. This is presumably the way shade plants achieve the highest density of light-harvesting pigments and increase the efficiency of light capture [1]. On the other hand, shade leaves have low rates of photosynthetic electron transport which are presumably a result of diminished amounts of some of the electron carriers [8]. The content of soluble protein per unit leaf area and of Rubisco decreases in shaded leaves in comparison to leaves grown with sufficient light [7] . This low capacity for electron transport and for carboxylation enables shade leaves to decrease the maintenance costs under low PFDs. However, as a consequence of the low investment in the biochemical machinery for photosynthesis, light-saturated rates of carbon uptake are much lower in shade than in sun leaves [9].

The level of carbohydrates in a leaf and its daily turnover are very sensitive to the amount of incident light and plant growth rate. Therefore the lower contents of soluble carbohydrates and starch in shade leaves reflect the differences in photosynthetic rates between sun and shade leaves [47]. However, protein and soluble amino acids expressed on a dry matter basis may be higher in shade leaves [37].

Shade leaves exhibit lower carboxylation efficiencies than sun leaves due to lower amounts of Rubisco. Therefore plants growing in growth cabinets or greenhouses usually respond to CO_2 enrichment in the atmosphere by increasing leaf net carbon gain and water use efficiency. It seems that young, fully unfolded leaves are more responsive to CO_2 enrichment than very young or old leaves [38]. At the whole plant level, growth may be enhanced (especially the root system) and this has proved to be particularly beneficial during acclimatization of ex-*vitro* plants, especially when supplementary lighting is provided (Fig.3) [27, 44]. Furthermore, the accumulation of starch in aerial

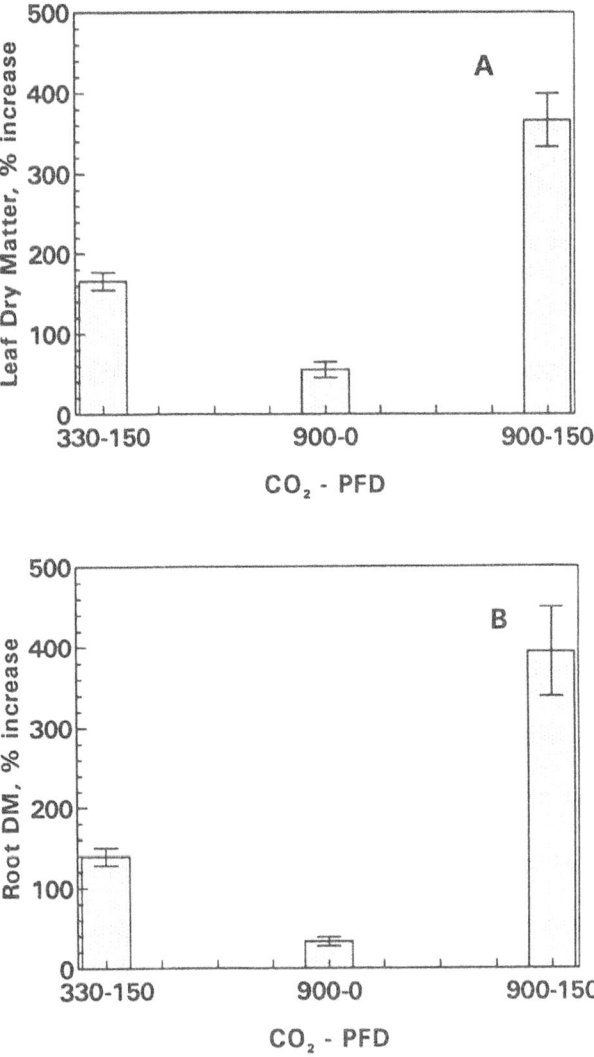

Fig. 3. Effect of the concentration of CO_2 in the air (330 and 990 ppm) and two levels of supplementary lighting (photon flux densities, PFD of 0 and 150 μmol m^{-2}s^{-1}) on leaf dry matter (A) and on root dry matter (B), expressed on percent increase as compared to the control (330 ppm CO_2 and no supplementary light), for Kent variety of strawberries. Bars represent SE of the mean (adapted from [27]).

organs of CO_2-enriched plants may favour establishment of young propagules from shoot tips [43]. However, care must be taken with this practice because some species may exhibit negative acclimation to high $[CO_2]$ in the longer term, including decreased levels of Rubisco and similar or even lower growth rates than plants growing at ambient $[CO_2]$ [16].

4. Acclimation of shade-grown plants to high irradiance: the risk of photoinhibition

Many herbaceous plants are capable of photosynthetic acclimation to a 10- to 20- fold difference in irradiance during growth, characterized by a two- to five-fold increase in the rate of light- and CO_2-saturated photosynthesis and no change in quantum yield, as shown for spinach by Osmond [53]. Shade grown plants, even though capable of acclimating to high irradiance, become very susceptible to light stress when transferred from shade, unless they undergo a process of acclimation during a step by step increase in the incident irradiance. In fact, due to the high capacity to capture light and a non-equivalent ability to use it in carbon reduction, the photosynthetic machinery may become overenergized and light stressed (for a review see [40, 55]. The response of plants to the transfer from shade to sun is difficult to predict as it will depend on a balance between photoinibitory damage, repair processes and accelerated senescence, which may be species dependent. It seems also that nitrogen availability, presumably associated with increased demand of Rubisco and repair processes will influence plant response to high irradiance stress [32].

When plants are subjected to a large excess of excitation energy, whether they are shade plants transferred to direct sunlight or sun plants whose ability to utilize excitation energy in photosynthetic carbon assimilation is impaired by some kind of stress, they exhibit a reduced quantum yield and sometimes a reduced photosynthetic capacity, i.e. respectively, the initial slope and the CO_2 saturated rate of photosynthesis as a function of irradiance (Fig.4 and [21]). The causes of this decline may be simply a 'down-regulatory' process accompanied by an increase in the non-photochemical quenching, q_{NP}, responsible for a thermal deactivation of the excitation energy [26] or a reversible inhibition of photosynthesis corresponding to a damage of the reaction centres of PSII (photoinibitory process). Photooxidation leading to leaf death may occur when damages become irreversible. In the near future, it will be possible to estimate for different species and conditions, the partitioning of excitation energy among protective (diverting energy from reaction centres), productive (transduction of energy by reaction centres and electron transport) and photoinhibitory processes (damage to the reaction centres) as outlined by Osmond and Chow [54]. In fact, to assess changes in quantum yield and changes in the various pathways of photochemical and non-photochemical energy utilization it is possible to use the chlorophyll *a* fluorescence technique. To obtain information on changes in mesophyll photosynthesis, measurements of

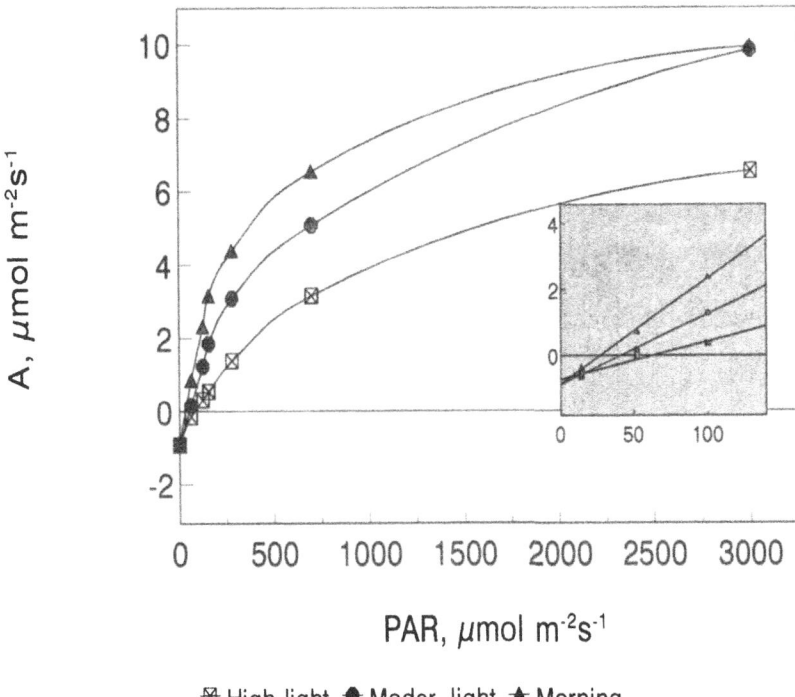

Fig. 4. CO$_2$-saturated rates of photosynthetic O$_2$ evolution at 22 °C, as a function of incident photon flux density, determined in *Vitis vinifera* L. Leaf discs were sampled in the morning (triangles) or after exposure of the leaves to a constant level of either moderate [700 μmol m^{-2}s^{-1} (circles)] or high [1350 μmol m^{-2}s^{-1} (squares)] photon flux density. Throughout the light treatment, leaf temperature (22.7±0.3 °C) and leaf-to-air water vapour pressure difference (10±2 Pa KPa^{-1}) were kept constant. All points are means of three replicates except for leaf discs sampled in the morning when n=6. The insert shows the values used to calculate the apparent quantum yield (from [21]).

CO$_2$-dependent O$_2$ evolution at saturating conditions of light and CO$_2$ can be done by using leaf-disc oxygen electrodes [13]. It is more difficult to distinguish between 'down-regulation' and photoinhibition, since there is still some undefinition regarding the term photoinhibition and an absence of a unique set of criteria to describe the damage or the repair of PSII [6, 54]. However, studies of recovery processes and their time scale may be most useful. In fact, whereas 'down-regulatory' processes are reversed very quickly (few hours), recovery from photoinhibition takes longer, (e.g. some days) because it presumably requires resynthesis and insertion of the damaged polypeptides into the PSII reaction center [54].

A good index of plant acclimation to a sunny habitat seems to be the increased amount of Rubisco [8, 53]. These increases in Rubisco can be achieved without any change in the amount of chlorophyll, as is the case of *Phaseolus* or can be accompanied by an increase in chlorophyll (e.g. *Alocasia*) [53]. It is noteworthy that differences in Rubisco are still apparent in plants grown at two different but high irradiances, as in the case of grapevines grown in the greenhouse (1000 μmol m^{-2}s^{-1} PFD) or in the field (1500–2000 μmol^{-2}s^{-1} PFD). The first group of plants showed Rubisco activities of 119±14 μmol h^{-1}mg^{-1} chlor. as compared with 198±14 μmol h^{-1}mg^{-1} chlor. in the second group [17]. Field and greenhouse plants exhibited a different sensitivity to exposure to a PFD of 2000 mol m^{-2}s^{-1} for five hours. Whereas a decrease of both the quantum yield and the photosynthetic capacity was observed in greenhouse-grown plants no significant decline was apparent in field grown plants, except in the quantum yield whenever they were subjected to water deficits [17, 21].

Some data suggest that older leaves, especially of low nitrogen plants, senesce more rapidly than younger leaves on transfer from shade to sun [32]. Under these circumstances, as is the case with water-stressed plants, senescence of older leaves make the limited nitrogen resources available for acclimation and repair of photoinhibition in younger leaves, but the capacity for remobilization of N from older leaves may vary with species.

5. High temperature stress

Young plants transferred from culture to the greenhouse or the field will probably be subjected to periods of supra-optimal temperature. Under those conditions the susceptibility to photoinhibition may increase dramatically in some plants due to a decreased ability to transduce photochemical energy by carbon metabolism.

When temperature rises net photosynthetic rate increases to a maximum, which is dependent on the species and the growth temperature regime. The slow-down in the increase in net photosynthetic rate with temperature in C$_3$ plants is due to a more than proportional increase in photorespiration as a result of a decreased solubility of CO_2 relative to O_2, leading to an enhancement of the oxygenase activity of Rubisco. At high temperature, photorespiration becomes the more important pathway of consumption of absorbed energy by leaves since in mesophyll cells photosynthesis starts to decrease as a result of a 'down-regulation' of Rubisco. Its activation state may decrease substantially with the increase in temperature as was reported by Weis and Berry [67] in cotton leaves (40% when temperature rises from 25° to 38 °C). Although the reasons for this decline in activation state of Rubisco are unclear it is possible that they are related to the depletion of the pool of Calvin cycle metabolites as suggested by Weis and Berry [67]. In fact, lower amounts of various stromal carbohydrates have been reported by Stitt and Grosse [62] and Oja *et al.* [51]. In addition to photorespiration, leaves may dissipate part of the energy absorbed by increasing the so-called 'state-transition' quenching of chlorophyll fluorescence, when leaf temperature rises. This seems to correspond to a decrease in the absorptive capacity of PSII due to the migration of core PSII away from the LHCII [64].

As already mentioned, photosynthetic response to temperature is under genetic control and is subjected to acclimation processes which take place as a function of the temperature regime during growth. Plants native from habitats where large thermal oscillations occur are usually more predisposed to acclimation [5]. However, even herbaceous species usually growing under e.g. winter-spring months in Southern Europe, may undergo photosynthetic acclimation to high temperature. This is illustrated by *Lupinus albus* plants grown at 25 °C/20 °C (day-night regime) and at 15 °C/10 °C — the first group of plants increased the thermal tolerance limit to 41 °C as compared to 38.5 °C in the second group (ML Osório and MM Chaves, unpublished). The increase in thermal resistance in this species was accompanied by a decrease of *c.* 34% in green leaf area duration, in spite of an increase of 37% in the number of leaves formed, simply because the warmer regime led to an accelerated rate of senescence (JA Passarinho, ML Osório, T Faria and MM Chaves, unpublished). Plants grown at the higher temperature also exhibited lower soluble sugars and starch contents (but a higher daily starch turnover) in the leaves than plants grown under the low temperature regime. It is interesting to note that lupins growing under natural conditions (December-May) exhibit their maximal rate of photosynthesis at 25 °C, but this temperature is already supra-optimal for growth. This illustrates the well know fact that temperature influences growth and gas exchange differently [52]. As a consequence a balance between resistance to temperature stress and best conditions for growth should be taken into account when choosing optimal growth conditions for a particular species.

As an example of a plant species which can endure much higher temperatures we may refer to grapevines. When growing under natural conditions these plants can withstand temperatures up to 47 °C without suffering irreversible damage in the photosynthetic apparatus, as revealed by the constancy in the basal fluorescence level, F_o as well as by the absence of leaf necrosis (ML Osório and MM Chaves, unpublished).

6. Water deficits

Micropropagated plants often exhibit a water imbalance upon removal from culture due to lack of stomatal control of transpiration and/or poor root performance [46, 56]. The development of leaf water deficits will have negative consequences for carbon uptake and growth, and therefore for subsequent autotrophic plant development. The nature and the extent of these effects are a function of the intensity and the duration of the stress, as well as of the genetically-determined capacity of plants to acclimate to it [13]. It is now well established that the rate of CO_2 assimilation in the leaves is depressed at moderate leaf water deficits or even before leaf water status is changed in response to a drop in air humidity [45, 59] or in soil water potential [25]. Under these circumstances stomatal control plays the most important role in controlling photosynthesis, since the photosynthetic apparatus is highly resistant to dehydration. Whenever the stress period is lengthened or water deficits occur concurrently with other stresses, such as high photon flux densities (PFDs) or high temperature, inhibitory effects of mesophyll photosynthesis may take place [15, 48], either at the enzymatic or the photochemical level. In the first case, decreased photosynthetic capacity (measured as the rate of CO_2 dependent O_2 evolution at saturating conditions of CO_2 and irradiance) usually occurs at relative water content (RWC) below 70%, which in many plants corresponds to a severe wilting of the leaves. These inhibitory effects are still reversible down to 30 to 40% of RWC as have been reported by Kaiser [36] and Chaves [13]. In the second case quantum efficiency of PSII is also affected because energy absorbed exceeds consumption by the various pathways of energy deactivation. This is shown, for example, by the decrease in F_m/F_o observed in water-stressed grapevine plants subjected simultaneously to high light [57]. When, in addition to high irradiance, water-stressed plants are subjected to temperatures above their optimum for photosynthesis, decreases in mesophyll photosynthesis may be observed at higher leaf RWC. This is the case of *Lupinus albus* plants subjected to 35 °C whose leaf photosynthetic capacity declines at RWC around 80% whereas at the optimal temperature of photosynthesis in lupins, 25 °C, this decrease is only detectable at *c.* 60 to 70% RWC (Fig. 5).

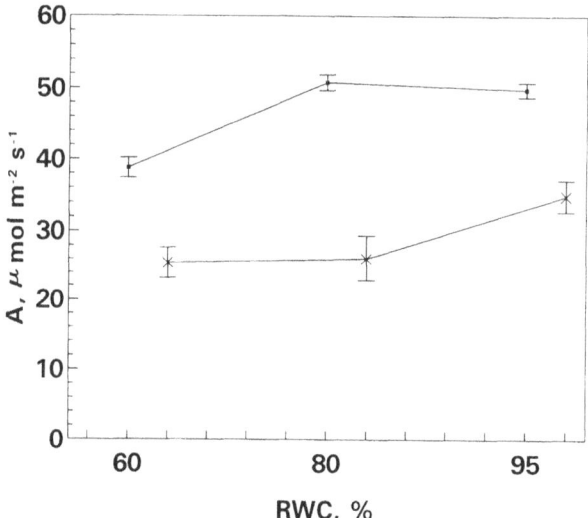

Fig. 5. Dependence of the rate of CO_2-dependent O_2 evolution (A) on the relative water content (RWC) of lupin leaf discs measured at 25 °C (squares) or 35 °C (crosses). The measurements were done at saturating irradiance and CO_2. Each value is the average of 8 to 16 measurements; bars indicate standard error of the mean (from [17]).

Shade-type plants are particularly susceptible to the combined effects of multiple environmental stresses due to their large antenna size leading to a high light interception. Experimental support for this has been given by the studies of Cleland and Melis [19] showing that a barley mutant lacking chlorophyll a/b light harvesting complex, was more resistant to photoinhibition than the wild type. In addition, shade plants seem to possess a lower capacity to repair damages caused by photoinhibition than sun plants [2].

Changes in leaf carbohydrate metabolism seem to be a very early and sensitive response to water deficits, and may exert a positive influence in plant water status [13, 57], contributing to osmoregulation [50]. Regulatory mechanisms are still unclear, but presumably involve changes both at the enzymatic [69] and the genetic levels, namely in gene repression processes by carbohydrates [61]. Therefore, it is not surprising that these types of responses are very strongly species-dependent [50, 57].

Among the main changes occurring in carbohydrate metabolism in dehydrating leaves are increases in the partitioning between sucrose and newly synthesised starch, increase or maintenance in the amount of sucrose and a dramatic decrease in the pool of starch. The latter is presumably a result of starch degradation beginning at a higher endogenous level in water-stressed than in well-watered leaves [69]. Changes in the activity and/or activation of the enzyme sucrose phosphate synthase (SPS) as a result of dehydration, have been reported in the literature and although there are some conflicting data, increased activation of SPS appears to be associated with the increased partitioning of carbon towards sucrose [69]. On the other hand, due to a low sink demand, leaves of dehydrated plants do not export assimilates and therefore are able to maintain or increase the amounts of soluble sugars, even under very low rates of carbon uptake. In spite of growth inhibition, typical effects of sink limitation on carbon metabolism, expressed by an accumulation of starch and a decline in sucrose synthesis as observed e.g. under nitrogen limiting conditions, do not occur in leaves water-stressed plants as mentioned above (see Fig. 6). However, whether down-regulatory effects on Rubisco activity and/or amount reported under several conditions [35, 66] may be due to long-term inhibitory effects of increased sugar content in dehydrated-leaves, as shown under other circumstances by Krapp *et al.* [39] remains to be tested. It is noteworthy that, even if feedback inhibition of photosynthesis does occur in water-stressed leaves it will not be the main cause of diminished carbon uptake, primarily limited by stomata. Recent work by Stitt [63] on the regulation of carbon metabolism in plants has given evidence that partitioning of carbon at the critical branching point of sucrose/starch synthesis is prone to fine modulation by various internal and external factors, making prediction of plant metabolic responses very difficult.

When discussing photosynthesis and growth under water stress conditions it is important to note that younger leaves in the canopy, mainly in herbaceous plants, are normally more resistant to water deficits than older leaves. In lupins, we observed a smaller decrease in the photosynthetic capacity in the youngest leaves of the shoots (a decrease of 22% relative to the values in controls), as well as the maintenance of Rubisco amount, whereas in older leaves the decrease in photosynthetic capacity was of the order of 40% and Rubisco amount was

Effects of water deficits

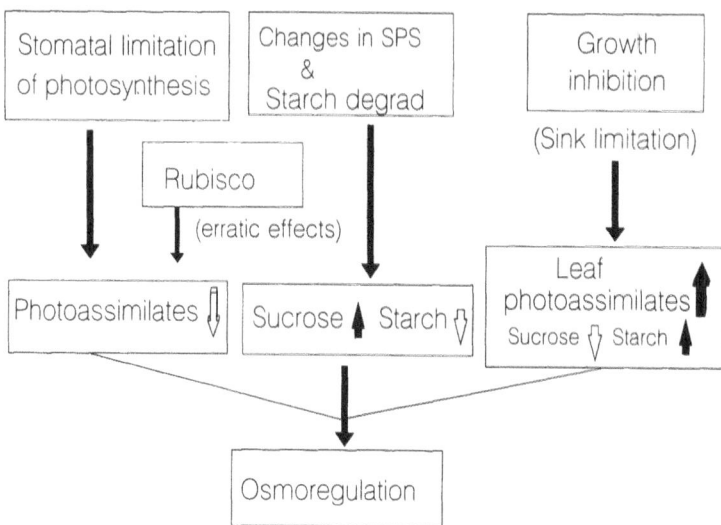

Fig. 6. Schematic presentation of the effects of water deficits on photoassimilates. These effects are mediated via effects on stomata, growth and cytosol + chloroplastic enzymes. Note the opposite effects on sucrose and starch synthesis by sink limitation and the changes in SPS and starch degrading enzymes. The final balance is in favour of sucrose at the expense of starch content, with beneficial effects of the osmoregulatory type.

reduced by 26% [14, 16]). The increased tolerance to dehydration in younger leaves may be of particular relevance in plants where a severe reduction in the size of leaf canopy occurs as a result of shedding of older leaves, because it allows a quick recovery after rehydration [13]. This is the case with lupins where a 50% reduction in leaf area was observed following a 15 days period of drought [58].

7. Conclusions

It is important to consider which characteristics may contribute to an improvement of plant performance during acclimatization. Plants removed from culture have to overcome two major limitations — heterotrophy, by producing more assimilates than importing them, and shade characteristics, essentially by increasing their Rubisco content in order to cope with an increased energy input. Limitation by Rubisco is particularly important at high irradiances and at the low intercellular CO_2 partial pressure occurring in water- or high temperature-stressed plants.

Limitations to carbon uptake due to transient water deficits may be overcome by increasing humidity or the $[CO_2]$ in the atmosphere. An enriched CO_2 atmosphere will improve water use efficiency and increase net carbon uptake in the low Rubisco leaves, leading to increased reserve accumulation in young plants. However, one has to be aware of possible feedback inhibitory effects of elevated $[CO_2]$ on photosynthesis in the long term, especially when plants are sink-limited e.g. due to poor root development.

Protection against photoinhibition may be achieved by preventing additional stress effects as incident irradiance is gradually increased and by optimizing nitrogen nutrition, which seems to decrease chloroplast susceptibility to light damage.

Acknowledgements

I am grateful to J.S. Pereira for critical reading of the manuscript.

References

1. Anderson JM, Goodchild DJ and Boardman NK (1973) Composition of the photosystems and chloroplast of the extreme shade plants. Biochim Biophys Acta 325:573–585
2. Aro E-M, Tyystjärvi E and Kettunen R (1992) Photoinhibition and degradation of the D1 protein in high- and low-light grown pumpkin leaves. Photos Research 34:104
3. Baker NR (1985) Energy transduction during leaf growth. In: NR Baker, WJ Davies and CK Ong, eds. Control of Leaf Growth, pp 115–133. Cambridge: Cambridge Univ Press
4. Baker NR and Hardwick K (1973) Biochemical and physiological aspects of leaf development in cocoa (*Theobroma cacao*). I. Development of chlorophyll and photosynthetic activity. New Phytol 72:1315–1324
5. Berry J and Björkman O (1980) Photosynthetic response and adaptation to temperature in higher plants. Ann Rev Plant Physiol 31:491–543
6. Björkman O (1987) High-irradiance stress in higher plants and interaction with other stress factors. In: J Biggins, ed. Progress in Photosynthesis Research, vol 4, pp 11–18. Dordrecht: Martinus Nijhoff
7. Blenkinsop PG and Dale JE (1974) The effects of shade treatment and light intensity on ribulose-1,5-diphosphate carboxylase activity and fraction I protein level in the first leaf of barley. J Exp Bot 25:899–912
8. Boardman NK (1977) Comparative photosynthesis of sun and shade plants. Ann Rev Plant Physiol 28:355–377
9. Bowes G, Ogren WL and Hageman RH (1972) Light saturation, photosynthesis rate, RuBPcarboxylase activity, and specific leaf weight in soybeans grown under different light intensities. Crop Sci 12:77–79
10. Brainerd KE and Fuchigami LH (1981) Acclimatization of aseptically cultured apple plants to low relative humidity. J Amer Soc Hort Sci 106:515–518
11. Callow JA (1974) Ribosomal RNA, fraction I protein synthesis and ribulose diphosphate carboxylase activity in developing and senescing leaves of cucumber. New Phytol 73:13–20
12. Catský J and Tichá I (1985) Photorespiration during leaf ontogeny. In: Z Sesták, ed. Photosynthesis during Leaf Development, pp 250–262. Dordrecht: Dr W Junk
13. Chaves MM (1991) Effects of water deficits on carbon assimilation. J Exp Bot 42:1–16
14. Chaves MM, Correia MJ, David MM, Osório J, Osório ML and Pereira JS (1990) Changes in photosynthetic carbon metabolism of *Lupinus albus* induced by slowly imposed water deficits. Physiol Plant 79 (2, Part 2): A132
15. Chaves MM, Osório ML, Osório J and Pereira JS (1992) The photosynthetic response of

Lupinus albus to high temperature is dependent on irradiance and leaf water status. Photosynthetica 27:521–528

16. Chaves MM and Pereira JS (1992) Water stress, CO_2 and climate change. J Exp Bot 45:1131–1139

17. Chaves MM, Wendler R, Osório ML, Osório J, Rodrigues ML, Faria T, Leegood R, Stitt M and Pereira JS (1992) Down-regulation of photosynthesis under soil water deficits in grapevines. J Exp Bot 43, suppl, 35

18. Chow WS, Qian L, Goodchild DD and Anderson JM (1988) Photosynthetic acclimation of *Alocasia macrorrhiza* (L) G Don to growth irradiance: structure, function and composition of chloroplasts. Australian J Plant Physiol 15:107–122

19. Cleland RE and Melis A (1987) Probing the events of photoinhibition by altering electron-transport activity and light-harvesting capacity in thylakoids. Plant Cell Environ 10:747–752

20. Conover RE and Poole RT (1984) Acclimatization of indoor foliage plants. Hort Rev 6:119–154

21. Correia MJ, Chaves MM and Pereira JS (1990) Afternoon depression in photosynthesis — evidence for a high light stress effect in grapevine leaves. J Exp Bot 41:417–426

22. Dale JE (1985) The carbon relations of the developing leaf. In: NR Baker, WJ Davies and CK Ong, eds. Control of Leaf Growth, pp 135–153. Cambridge: Cambridge Univ Press

23. Dale JE (1988) The control of leaf expansion. Ann Rev Plant Physiol Plant Mol Biol 39:267–295

24. Dale JE and Milthorpe FL, eds. The Growth and Functioning of Leaves. Cambridge: Cambridge Univ Press

25. Davies WJ and Zhang J (1991) Root signals and the regulation of growth and development of plants in drying soil. Ann Rev Plant Physiol Plant Mol Biol 42:55–76

26. Demmig B and Winter K (1988) Light response of CO_2 assimilation, reduction state of Q, and radiationless energy dissipation in intact leaves. Aust J Plant Physiol 15:151–162

27. Desjardins Y, Gosselin A and Yelle S (1987) Acclimatization of *ex vitro* strawberry plantlets in CO_2-enriched environments and supplementary lighting. J Amer Soc Hort Sci 112:846–851

28. Dickson RE and Isebrands JG (1991) Leaves as regulators of stress response. In: HA Mooney, WE Winner and EJ Pell, eds. Response of Plants to Multiple Stresses, pp 3–34. New York: Academic Press

29. Donnelly DJ, Vidaver WE and Lee KY (1985) The anatomy of tissue cultured red raspberry prior to and after transfer to soil. Plant Cell Tiss Org Cult 4:43–50

30. Downton J and Slatyer RO (1972) Variation in levels of some leaf enzymes. Planta 96:1–12

31. Fabbri A, Sutter E and Dunston SK (1986) Anatomical changes in persistent leaves of tissue-cultured strawberry plants after removal from culture. Sci Hort 28:331–337

32. Ferrar PJ and Osmond CB (1986) Nitrogen supply as a factor influencing photoinhibition and photosynthetic acclimation after transfer of shade-grown *Solanum dulcamara* to bright light. Planta 168:563–570

33. Givnish TJ (1988) Adaptation to sun and shade: a whole plant perspective. Aust J Plant Physiol 15: 63–92

34. Grout BWW (1988) Photosynthesis of regenerated plantlets *in vitro*, and the stresses of transplanting. Acta Hort 230:129–135

35. Jones HG (1973) Moderate-term water stress and associated changes in some photosynthetic parameters in cotton. New Phytol 72:1095–1104

36. Kaiser W (1987) Effects of water deficit on photosynthetic capacity. Physiol Plant 71:142–149

37. Kausch W and Haas W (1965) Chemische Unterschiede zwischen Sonnen- und Schattenblättern der Blutbreche (*Fagus sylvatica* L cv. Atropunicea). Naturwissenschaften 52:214–215

38. Kelly DW, Hickleton PR and Reekie EG (1991) Photosynthetic response of geranium to elevated CO_2 as affected by leaf age and time of CO_2 exposure. Can J Bot 69:2482–2488

39. Krapp A, Quick WP and Stitt M (1991) Ribulose-1,5-bisphosphate carboxylase-oxygenase, other Calvin-cycle enzymes, and chlorophyll decrease when glucose is supplied to mature spinach leaves via the transpiration stream. Planta 186:58–69

40. Krause GH (1988) Photoinhibition of photosynthesis. An evaluation of damaging and protective mechanisms. Physiol Plant 74:566–574

41. Kriedemann PE (1968) Photosynthesis in vine leaves as a function of light intensity, temperature and leaf age. Vitis 7:213–220
42. Kriedemann PE (1986) Stomatal and photosynthetic limitations to leaf growth. Aust J Plant Physiol 13:15–31
43. Kriedemann PE, Sward RJ and Downton WJS (1976) Vine response to carbon dioxide enrichment during heat therapy. Aust J Plant Physiol 3:605–618
44. Lakso An, Reisch BI, Mortensen J and Roberts MH (1986) Carbon dioxide enrichment for stimulation of growth of *in vitro*-propagated grapevines after transfer from culture. J Amer Soc Hort Sci 111:634–638
45. Lange OL, Lösch R, Schulze E-D and Kappen L (1971) Responses of stomata to changes in humidity. Planta 100:76–86
46. Lee N, Wetstein HY and Sommer HE (1988) Quantum flux density effects on the anatomy and surface morphology of *in vitro*- and *in vivo*- developed sweetgum leaves. J Amer Soc Hort Sci 113:167–171
47. Lichtenthaler HK (1985) Differences in morphology and chemical composition of leaves grown at different light intensities and qualities. In: NR Baker, WJ Davies and CK Ong, eds. Control of Leaf Growth, pp 201–221. Cambridge: Cambridge Univ Press
48. Ludlow MM (1987) Light stress at high temperature. In: DJ Kyle, CB Osmond and CJ Arntzen, eds. Photoinhibition, pp 89–109. Amsterdam: Elsevier
49. Milthorpe FL and Newton P (1963) Studies on the expansion of the leaf surface. III. Influence of radiation on cell division and leaf expansion. J Exp Bot 14:483–495
50. Morgan JM (1984) Osmoregulation and water stress in higher plants. Ann Rev Plant Physiol 35: 299–319
51. Oja VM, Rasulov BH and Laisk AH (1988) An analysis of the temperature dependence of photosynthesis considering kinetics of RuP_2 carboxylase and the pool of RuP_2 in intact leaves. Aust J Plant Physiol 15:737–748
52. Ong CK and Baker CK (1985) Temperature and leaf growth. In: NR Baker, WJ Davies and CK Ong, eds. Control of Leaf Growth, pp 175–200. Cambridge: Cambridge University Press
53. Osmond CB (1987) Photosynthesis and carbon economy of plants. New Phytol 106 (suppl):161–175
54. Osmond CB and Chow WS (1988) Ecology of photosynthesis in the sun and shade: summary and prognostications. Australian J Plant Physiol 15:1–9
55. Powles SB (1984) Photoinhibition of photosynthesis induced by visible light. Ann Rev Plant Physiol 35: 15–44
56. Preece JE and Sutter EG (1991) Acclimatization of micropropagated plants to the greenhouse and the field. In: PC Debergh and RH Zimmerman, eds. Micropropagation, pp 71–93. The Netherlands: Kluwer Academic Publishers
57. Quick WP, Chaves MM, Wendler R, David M, Rodrigues ML, Passarinho JA, Pereira JS, Adcock MD, Leegood RC and Stitt M (1992) The effect of water stress on photosynthetic carbon metabolism in four species grown under field conditions. Plant Cell & Environ 15:25–35
58. Ramalho JC and Chaves MM (1992) Drought effects on plant water relations and carbon gain in two lines of *Lupinus albus* L. Eur J Agron 1:271–280
59. Schulze E-D (1986) Carbon dioxide and water vapour exchange in response to drought in the atmosphere and in the soil. Ann Rev Plant Physiol 37:247–74
60. Sesták Z, Tichá I, Catsky, J., Solárová J, Pospisilová J and Hodánová D (1985) Integration of photosynthetic characteristics during leaf development. In: Z Sesták, ed. Photosynthesis during leaf development, pp 263–286
61. Sheen J (1990) Metabolic repression of transcription in higher plants. The Plant Cell 2:1027–1038
62. Stitt M and Grosse H (1988) Interactions between sucrose synthesis and CO_2 fixation. IV. Temperature-dependent adjustment of the relation between sucrose synthesis and CO_2 fixation. J Plant Physiol 133:392–400
63. Stitt M and Quick WP (1989) Photosynthetic carbon partitioning: its regulation and possibilities for manipulation. Physiol Plant 77:633–641

64. Sundby C and Anderson B (1985) Temperature-induced reversible migration along the thylakoid membrane of photosystem II regulates its association with LHC II. FEBS Lett 191:24–28
65. Tichá I, Catsky J, Hodánová D, Pospísilová J., Kase M and Sesták Z (1985) Gas exchange and dry matter accumulation during leaf development. In: Z Sesták, ed. Photosynthesis during Leaf Development, pp 157–216. Dordrecht: Dr W Junk
66. Vu JVC and Yelenosky G (1988) Water deficit and associated changes in some photosynthetic parameters in leaves of 'Valencia', orange (*Citrus sinensis* (L.) Osbeck.) Plant Physiol 88:375–378
67. Weis E and Berry JA (1988) Plants and high temperature stress. In: SP Long and FI Woodward, eds. Plants and Temperature, pp 329–346. Cambridge: The Company of Biologists
68. Zima J and Sesták Z (1985) Carbon fixation pathways, their enzymes and products during leaf ontogeny. In: Z Sesták, ed. Photosynthesis during Leaf Development, pp 145–156. Dordrecht: Dr W Junk
69. Zrenner R and Stitt M (1991) Comparison of the effect of rapidly and gradually developing water-stress on carbohydrate metabolism in spinach leaves. Plant Cell & Environ 14:939–946

Photomorphogenesis and plant development

DAPHNE VINCE-PRUE

Department of Botany, University of Reading, UK

1. Introduction

Studies of the effects of light on morphology and development in regenerating plant tissues have been carried out in rather few instances. This is surprising in view of the multiplicity of responses known to be influenced by light and also because micropropagation is one of the few commercial horticultural enterprises in which plants are grown entirely in artificial light. When natural daylight is completely replaced, the spectrum of the artificial light source becomes particularly important and yet, in a majority of cases, this is largely ignored in commercial micropropagation installations.

2. Light and lamps

All artificial light sources that are currently available differ from the spectrum of daylight in ways which have implications for photomorphogenesis, i.e., for the effects of light on plant morphology and development. Light energy vibrates at a particular range of frequencies with wavelengths between approximately 400 and 700 nm. Beyond these limits lie the ultra-violet and infra-red parts of the spectrum which are not visible to the human eye. Wavelengths shorter than about 280 nm in the ultra-violet are highly injurious to plant and animal cells. Wavelengths longer than approximately 1000 nm in the infra-red have a strong heating effect.

In order to effect a response, light must first be absorbed by a photoreceptor pigment. For higher plant photomorphogenesis, the two important pigments are phytochrome and a blue/UV-A absorbing photoreceptor (which is sometimes called cryptochrome). The blue/UV-A pigment absorbs light only in the blue (400–500 nm) and near ultra-violet (UV-A, 320–400 nm) regions of the spectrum. Phytochrome, in contrast, absorbs mainly in the red (600–700 nm) and far-red (700–760 nm) regions and, because of the characteristics of the pigment, it is primarily the ratio of red to far-red radiation rather than the absolute amount of red or far-red light which is the important factor. This ratio has been defined [19] as:

P.J. Lumsden, J.R. Nicholas and W.J. Davies (eds.), Physiology, Growth and Development of Plants in Culture, 19–30, 1994.

$$R : FR = \frac{\text{photon fluence rate in 10nm band centred on 660 nm}}{\text{photon fluence rate in 10nm band centred on 730 nm}}$$

Table 1. The effect of the ratio of red:far-red light on the long-day induced bulbing of onion. Plants received a main light period of 14.5 h under the various types of fluorescent lamps. A photoperiod of 19 h was obtained by giving 2.25 h of low-intensity light from tungsten-filament lamps both before and after the mainlight period [1]

Type of fluorescent lamp	Bulbing ratio (76 days after sowing)	R:FR ratio of lamp
Colour 29, Warm white	1.92	22.7
Colour 34, White de luxe	3.75	3.7
Colour 37	3.82	3.6
De-luxe Warm white	4.04	3.6

Natural sunlight is rich in blue and has a remarkably constant R:FR ratio of approximately 1.1 [19]. The majority of micropropagation facilities utilise tubular fluorescent lamps as the only light source. These have considerable advantages because they have a low surface temperature and emit relatively little heating infra-red radiation; consequently, they can be positioned quite close to the plant bench. Because the source is extended, the light distribution is relatively uniform. Various types of 'white' fluorescent lamp are often considered to be interchangeable for plant growth. However, they vary in both their R:FR ratio (which is higher than in sunlight, Table 1) and in their blue content. All of them differ markedly from the spectrum of sunlight. Different types of 'white' lamp have been found to have profoundly different photomorphogenetic effects and they are certainly not all equivalent for plant development and differentiation. For example, there were differences in the bulbing response of onion depending on the R:FR ratio of the particular type of 'white' lamp used (Table 1). The low blue content of certain fluorescent lamps may also be important for some responses. All fluorescent lamps emit some radiation in the UV-A range, which is not without morphogenetic effect [13].

A number of other lamp types are utilised in horticultural lighting applications [30] but not generally for micropropagation. It is worthwhile drawing attention to two of these, namely the tungsten-filament (incandescent) and the low-pressure sodium (LPS) lamps. The tungsten-filament lamp emits little blue light, has a large output in red and is particularly rich in far-red. The R:FR ratio is approximately 0.7 [19] and a major use has been to supplement the content of far-red light, especially for photoperiodic responses. A problem with this lamp as an artificial light source for plants is its high radiant heat output, which imposes additional loads on the cooling system. Used alone, it is also morphogenetically undesirable and leads to excessively elongated plants. The LPS lamp is mainly of interest experimentally; it is an extended monochromatic source with a light output entirely at 589 nm. There is no blue light and the absence of far-red radiation means that, as far as phytochrome responses are concerned, the lamp is equivalent to pure red light. It is, therefore, a useful lamp for trying to separate out photomorphogenetic effects that are specifically dependent on blue light [22].

3. Plant responses to light

Higher green plants are entirely dependent on photosynthesis and thus on light for survival. Since the location of any individual terrestrial plant is fixed, it is necessary that individuals are able to make adjustments to their physiology and/or morphology to enable them to adapt to the prevailing light environment. Such 'photomorphogenesis' is an essential property of practically all stages of the ontogeny of higher plants and many of the morphological and physiological features that are often regarded as being constitutive in a normal green plant are in fact epigenetic adaptations to the light environment.

Photosynthesis requires the input of considerable amounts of light energy and $1.25-2.50$ MJ m^{-2} d^{-1} is necessary to achieve reasonably satisfactory growth [30]; this is equivalent to $30-60$ W m^{-2} for a 12 h day. In contrast, the photomorphogenetic systems require only small amounts of light. For example, the low-energy reaction of phytochrome operates within the range of $0.1-1980$ J m^{-2} [7], which is about 10 W m^{-2} for 200 s. Even the so-called *high irradiance response* to blue light is saturated at an irradiance of 4.0 W m^{-2} of broad band blue light (for the inhibition of hypocotyl elongation in lettuce), although this must be given over several hours in order to be effective. Thus, it is the spectral balance of the lamp rather than its overall light output which is likely to be important for photomorphogenetic effects.

Table 2. Light signals for morphogenetic responses of higher plants

Signal	Type of response	Photoreceptor
Light versus dark (emergence into light)	De-etiolation of seedlings	Phytochrome; Blue/UV-A
Wavelength (R:FR ratio)	Adaptation to shade conditions	Phytochrome
Irradiance	Adaptation to shade conditions	Blue/UV-A (Phytochrome?)
Daily duration of light	Seasonal responses; flowering, dormancy, etc.	Phytochrome

By considering their information value to the plant under natural conditions, four major categories of photomorphogenetic signal can be deduced (Table 2). Plants can respond to the presence or absence of light; this may control germination (buried seeds) and, for seedlings emerging into the light, a whole range of morphogenetic responses such as greening, hook unfolding, leaf initiation, leaf expansion and stem elongation (which are often collectively called de-etiolation). For older plants growing in the light, an important factor controlling development is whether or not plants are shaded by other plants. Shade light from a leaf canopy differs from sunlight in two major respects; the total amount of light is reduced (especially in red and blue) and, because far-red light is poorly absorbed by chlorophyll, the ratio of R:FR is much reduced compared with sunlight [19]. Thus, under natural conditions, the perception of shade light could be achieved either through sensing its R:FR ratio or by the ability to respond to the change in irradiance; in practice, both systems appear to operate and may control different components of the overall response [3, 6, 16, 19]. Depending on the species, this response may be avoidance by increasing stem elongation to enable plants to outgrow their competitors and reach a better light climate. This is often accompanied by reduced lateral bud outgrowth in order to direct assimilates preferentially to elongation. Such a strategy would, however, be of no value to plants of the woodland floor. These must tolerate shade and can make biochemical and morphological adjustments to enable them to increase their photosynthetic capacity under such conditions; for example, by producing thinner, larger leaves to increase their light-capturing surface and by increasing the amount of light-harvesting chlorophyll. Finally, the daily duration of light is an important factor controlling many aspects of seasonal development.

3.1 The light-sensing pigments

Despite the variety of signals responded to (Table 2), only two light-sensing pigments — phytochrome and the blue/UV-A absorbing pigment, crypto-chrome — appear to be involved in the control of photomorphogenetic and photoperiodic responses of higher plants [13]. Cryptochrome has not yet been chemically identified. It is thought to be a flavoprotein, although variations in the action spectra for responses to blue light indicate that several somewhat different blue/UV-A photoreceptors may be involved [18]. Because responses to blue light are usually strongly dependent on irradiance [22], several workers have suggested that, under natural conditions, cryptochrome could be important for shade perception and adaptation [3]. It has also been shown that, at least in some cases, blue light absorbed by cryptochrome may increase the sensitivity of the plant to red light absorbed by phytochrome [13]. Such interactions make it harder to assess the specific function of the blue/UV-A photoreceptor, especially as phytochrome itself absorbs weakly in the blue region of the spectrum.

Whatever its role under natural conditions, plants growing in artificial light lacking blue sometimes show dramatic changes in their morphology. For example, leaf rolling and excessive stem elongation were observed in some varieties of lettuce when grown entirely under LPS lamps, which have no blue-light content; the addition of only small amounts of blue restored normal development [22]. In soybean plants, development under LPS light was similar in many respects to that of shaded plants and relatively more growth was partitioned to leaves [3]. Although it is not possible to give guidelines about the amount of blue light needed for normal morphogenesis and development (which in any case appears to vary between plants [30] and may be affected by the R:FR ratio of the light), the blue content of 'white' lamps used for plant tissue culture systems should not be ignored. For example, when red (+FR) – biased sources were compared with blue-biased ones, different growth patterns wereobtained; in general, the blue-biased sources produced shorter plants with smaller thicker leaves than those under red-biased light [34]. In contrast to cryptochrome, a good deal is known about the photoreceptor phytochrome [8, 11, 23]. First isolated in 1966, phytochrome was shown to be a protein attached to a light-absorbing chromophore. It is now known that there is more than one type of phytochrome protein and that these have different characteristics, particularly with respect to their stability within the plant. It is also suggested that the different phytochromes may have different physiological functions [29]. All types of phytochrome, however, show essentially the same photochemical properties, especially the red/far-red reversibility of the pigment. With successive short exposures to R and FR light, the physiological response depends on which treatment is given before the plant is returned to darkness. Because the response depends on the final exposure to light, it is evident that the controlling light-absorbing pigment must be photochromic, i.e., that it exists in two isomeric forms which are converted from one to the other by light in the following reaction:

synthesis in   red light

  darkness ———→ Pr $\xrightleftharpoons{}$ Pfr ———→ observed response

                      far-red-light

Thus, the form Pr absorbs red light and is converted to Pfr, which is the physiologically active molecule. The Pfr form, in contrast absorbs far-red light and is converted back to inactive Pr. A very large number of responses to light have been shown to be under the control of the red/far-red reversible reaction of phytochrome; these include many morphogenetic events such as leaf initiation, apical dominance, leaf expansion and stem elongation. Red/far-red reversible control has also been found to operate in light-regulated biochemical reactions, such as the synthesis of a number of enzymes. It is now well established that phytochrome regulates the expression of many genes and that both up-regulation (i.e., expression is increased by light) and down-regulation (expression is decreased) occur [11]. One of the most important examples of down-regulation is the phytochrome gene itself. Phytochrome has also been shown to control a number of membrane-mediated processes.

Although R/FR reversible responses have been mostly studied in young seedlings during de-etiolation, it can be shown also to operate in older plants growing in the light by giving them a brief exposure to R or FR light immediately before transfer to darkness. Such *end-of-day* R/FR reversible responses have been found to have a profound effect on many aspects of growth; for example, end-of-day FR (which would reduce the amount of Pfr) increases apical dominance and prevents lateral bud outgrowth and also markedly increases stem elongation [26].

The photoreversible nature of phytochrome means that the ratio of R:FR is an important feature of the light environment, because this will affect the amount of Pfr present in the plant. The relationship between the %Pfr established and the R:FR ratio is, however, not linear [19]. The estimated (epidermal) %Pfr is especially sensitive to R:FR in the range experienced in vegetation shade (R:FR = 1.1–0.05) and decreasing R:FR below that present in sunlight (1.1) has a proportionately greater effect on the %Pfr than increasing R:FR in the range above this. At R:FR ratios above about 1.5, the %Pfr becomes very insensitive to changes in R:FR. Consequently, fluorescent lamps which contain little FR and so have a much higher R:FR ratio than sunlight (Table 1) are likely to have smaller morphogenetic effects (compared with sunlight) than lighting treatments (such as tungsten-filament lamps) with a R:FR ratio below 1.0. Nevertheless, plants growing entirely under fluorescent lamps often differ morphogenetically from their counterparts growing in natural sunlight: they are often more compact and bushier under fluorescent light and have shorter stems.

Even though the reactions controlling photomorphogenesis are known, it is not yet possible to predict the behaviour of plants growing under spectral qualities that differ from sunlight. The most important components of the light source appear to be:
1) the R:FR ratio (the reversible reaction of phytochrome);
2) the red photon flux (an irradiance-dependent reaction of phytochrome [9]; and
3) the blue photon flux (the irradiance-dependent reaction of cryptochrome).

However, attempts to correlate the morphogenetic responses of *Vigna* with the spectral quality of the lamp type under which they were grown showed that they were not simply related to the R:FR ratio, nor to the content of blue or FR light [21]. Moreover, different species, and even cultivars, show considerable variation in their response when grown in different spectral qualities of artifical light, especially with respect to the requirement for blue light for 'normal' morphogenesis [30].

3.2 Photoperiodism

The duration of the daily light period controls many responses of interest for micropropagation, including flowering, bud dormancy, and other aspects of seasonal development. Duration perception normally occurs in the leaves and usually some leaf expansion must occur before plants become sensitive to daylength, although very young leaves are effective in certain cases [24]. In the majority of flowering plants, phytochrome appears to be the only light-sensitive pigment involved in photoperiodism, although the controlling reactions may be quite complex [32, 33]. For example, at least three separate phytochrome-dependent reactions have been distinguished in the photoperiodic control of flowering in *Pharbitis nil*. Usually, however, these reactions can only be detected by using treatments which would never be given in normal growing practice. In simple light-dark cycles, such as occur naturally, the operation of the photoperiodic control mechanism is fairly straightforward, although there are important differences between long-day and short-day plants.

The best known effect of daylength on plants is the regulation of flowering time. Some plants flower only in short days, or flowering is accelerated by them (SDP), others in long days (LDP), while some are indifferent to daylength (day-neutral plants). The main factor controlling flowering in SDP is the duration of the night, which under natural conditions is set by the times of sunset and sunrise. Because the controlling factor is the duration of darkness, a brief exposure to light in the long night can mimic the effect of a long day and suppress flowering in many SDP. In some of the most studied SDP, such as *Xanthium strumarium*, it has been shown that light is only required to convert phytochrome to the active Pfr form [24, 27]. Red light is, therefore, the most effective and only a few seconds of red light is sufficient to suppress flowering in *Xanthium*; however, if this red light is immediately followed by far-red, flowering is restored. Thus, the night-break control of flowering in SDP has been shown to be a typical R/FR reversible reaction of phytochrome, with the active form, Pfr, being inhibitory to flowering. Thus, for many SDP, a long-day effect can be obtained by giving a short day (of say 8–10 h) with a night-break of red light given during the long night. An important point is that the night-break is only effective when given at a particular time in the dark period. There is some difference of opinion as to the precise reason for this but the most probable explanation is that a time-measuring system is set in motion at dusk and reaches a light-sensitive phase after a certain number of hours have elapsed in darkness [27, 32]. This time-measuring system is probably a *circadian rhythm*.

Flowering in LDP requires (or is accelerated by) exposure to long photoperiods. In some cases, flowering can also be induced in short days if a brief night break sufficient to convert Pr to Pfr is given during the long night [27]. As in SDP, this night-break effect is R/FR reversible. Such experiments have established that flowering in LDP requires the presence of Pfr at certain times in the daily cycle, in contrast to SDP where flowering is inhibited by a night-break which forms Pfr. However, many LDP require a much longer exposure of several hours in order to promote flowering [27]. This type of response is sometimes called *light-dominant*, to emphasis the requirement for long daily exposures to light and because flowering seems not to be primarily controlled by the duration of darkness as in SDP (*dark-dominant*). Although light dominant and dark-dominant patterns of photoperiodic behaviour (Table 3) more or less correlate with the classification of plants as LDP and SDP respectively, there are exceptions. For example, the inhibition of flowering in the SDP, strawberry, has the characteristics of a light-dominant response [31] while the promotion of flowering in the LDP, *Fuchsia hybrida* requires only a short red night-break and has the characteristics of a dark-dominant response [25]. A particular feature of light-dominant responses is that light containing a mixture of R and FR wavelengths is more effective than red light alone and in most cases optimum flowering is obtained with a R:FR ratio close to, or below that found in sunlight [27]. A second characteristic is that the requirement for FR changes during the course of the day. For example, a 16 h photoperiod consisting of 8h R followed by 8h sunlight (R:FR = 1.1) has been found to be an effective long day for light-dominant plants. The reverse sequence, on the other hand, has little effect (Table 3). This changing response may be associated with a circadian rhythm of sensitivity to FR light [4, 27].

Table 3. Differences between dark-dominant and light-dominant patterns of photoperiodic response

Photoperiod treatment	Dark-dominant plants	Light-dominant plants
SD + short night-break	LDR	SDR*
LD (R + FR)	LDR	LDR
LD (no FR)	LDR	SDR*
Red extension before 8 h sunlight (R+FR)	LDR	LDR
Red extension after 8 h sunlight	LDR	SDR*

LDR: Plants respond as if receiving a long day = long-day response
SDR: Plants respond as if receiving a short-day = short-day response
(* in some cases a weak long-day response is obtained)

In micropropagation practice, plants are usually grown under fluorescent lamps for 16 to 18 h per day. For the majority of SDP, this regime will prevent flowering provided that a sufficiently high %Pfr is established by the light source. All of the commonly used 'white' fluorescent lamps would be effective since R:FR ratios of 1.0 and above have been found adequate to prevent flowering in a number of short-day plants [31]. However, for most LDP (and for light-dominant SDP such as strawberry), the R:FR ratio of most white fluorescent lamps is too high for optimum response and, in many cases, these plants will show only short-day behaviour under these conditions, even though the photoperiod is 16 h or more. A supplementation with FR light to lower the R:FR ratio is necessary in order to achieve a long-day response, although this lowered R:FR ratio may be required for only part of the day (Table 3).

Other photoperiodically controlled responses also show light-dominant and dark-dominant behaviour. Bulbing in onion has already been mentioned as an example of a response which requires long photoperiods but does not occur when the R:FR ratio of the light is too high (Table 1). Similarly, in the SD-induced bud dormancy of woody species, both light-dominant and dark dominant forms of control have been identified [28]. When manipulating photoperiod artificially, therefore, it is essential to consider the R:FR ratio of the lamps and the response characteristics of the plant in question, as well as the duration and timing of the daily light treatment.

4. Photomorphogenesis and micropropagation

Based on the results obtained with intact plants, it seems likely that manipulation of the light environment is potentially advantageous at many stages during micropropagation. A few randomly chosen examples where positive effects have been obtained experimentally illustrate this point.

Regeneration from callus has been shown to be influenced by light but the optimal conditions vary with species. For example, greater shoot regeneration occurred in near UV/blue light in tobacco callus [17] but red light was most effective in *Actinidia* [15]. In *Actinidia*, the response was dependent on the red photon flux, indicating the possible participation of the R irradiance-dependent reaction of phytochrome [9]. The amount of shoot proliferation in shoot explants has also been shown to be influenced by light conditions. For example, in *Rhododendron* cv Dopey, shoot production was doubled when explants were cultured at 9 W m^{-2} compared with 18 W m^{-2}, a level of irradiance commonly used in growth rooms [12]. A quite different approach has been taken with shoot tips of two peach rootstocks [14] where explants were grown on proliferation media for 6–8 weeks under unusual light/dark cycles. The number of shoots (as well as root length and leaf area) was greater in cycles of 4h light (L) / 2 h dark (D) and (to a lesser extent) in 8L / 4D than in more normal cycles of 16L / 8D. It is not clear that this is a true photoperiodic effect, although proliferation in both azalea [5] and apple [35] was greater in shorter (16L / 8D) than in longer (24L) photoperiods.

The induction of adventitious root formation is also a critical stage in micropropagation. This has been found to be influenced by light in several different ways. Direct effects of light on *in vitro* rooting has been reported for microcuttings of pear [2]. In the absence of auxin, 60% rooting occurred under low intensity R light but none in FR. An intermediate level of rooting occurred in the dark; however, there was no root emergence in cuttings exposed to FR (to reduce Pfr) before being placed in the dark, indicating that the rooting response is controlled by the reversible reaction of phytochrome. Indirect effects of light on rooting have also been recorded. In many woody species, cuttings from adult plants are difficult to root; culture in the presence of cytokinin can effectively rejuvenate shoots and enhance rootability, although it can take several months before this occurs. This time was decreased in explants of *Rhododendron* cv. Dopey when plants were cultured at a reduced level of irradiance[12]. Low light levels are also known to cause rejuvenation or to prolong the juvenile phase in intact plants of some species.

The subsequent field behaviour of micropropagated plants is extremely important and several photomorphogenetic effects have been observed here. For example, light versus dark during bulbil formation in hyacinth resulted in earlier leaf emergence and partially prevented the early onset of bulb dormancy which can be a problem in micropropagated plants. Dormancy of *Lilium longiflorum* can also be overcome by R light given for 16 h per day during the 2 weeks before planting [20].

Modification of the photoperiod during micropropagation has been shown to affect the field performance in strawberry, where there is evidence that plants can be manipulated towards enhanced stolon production by exposing them to long photoperiods. Several cultivars of strawberry are SDP for flower initiation. Stolon production is, however, enhanced in LD and plantlets of SDcultivars produced markedly more stolons when micropropagated in a 16 h photoperiod than in 12 h days [10]. There was, however, no effect in day-neutral cultivars. At least for floral initiation, strawberry behaves as a light-dominant plant requiring both R and FR light in order to elicit a long-day response and suppress flowering (Table 3); the spectral quality of the light is also likely to be important when manipulating the photoperiod towards stolon production.

5. Conclusions

There is now an enormous amount of information on photomorphogenesis in intact plants and a considerable understanding of both the controlling mechanisms and the processes controlled. The wavelength distribution, irradiance, and duration of light all have profound morphogenetic effects, some of which may be highly relevant for the practice of micropropagation. In intact plants, different morphogenetic processes are controlled by light in a variety of ways, some of which have been outlined here. Moreover, different species and cultivars do not necessarily respond to light in the same way. Despite its powerful morphogenetic effects, light has been given relatively little attention in micropropagation practice. A 16–18 h photoperiod under some kind of 'white' fluorescent light is often the 'standard' lighting treatment and, in most cases, there is little evidence to show whether, or not, this is the optimum light environment for any particular developmental stage of the plant in question.

References

1. Austin RB (1972) Bulb formation in onions as affected by photoperiod and spectral quality of light. J Hort Sci 47:193–504
2. Baraldi R, Bertazza G and Prediery S (1991) Effects of light quality on *in vitro* rooting of pear microcuttings. Abstract: Photomorphogenesis in Plants, Beltsville Symposium 16, p 29
3. Britz SJ and Sagar JC (1990) Photomorphogenesis and photoassimilation in soybean and sorghum grown under broad spectrum or blue-deficient light-sources. Plant Physiol 94:448–454
4. Deitzer GF, Hayes R and Jabben M (1979) Kinetics and time dependence of the effect of far-red light on the photoperiodic induction of flowering in Wintex barley. Plant Physiol 64:1015–1021
5. Economou AS and Read PE (1986) Influence of light duration and irradiance on micropropagation of a hardy deciduous azalea. J Amer Soc Hort Sci 111:146–149
6. Frankland B (1986) Perception of light quantity. In: RE Kendrick and GHM Kronenberg, eds. Photomorphogenesis in Plants, pp 219–235. Dordrecht: Martinus Nijhoff
7. Furuya M (1968) Biochemistry and physiology of phytochrome. Prog Phytochem 1:347–405
8. Furuya M (1989) Molecular properties and biogenesis of phytochrome I and II. Adv Biophys 25:133–167

9. Jabben M and Holmes GE (1983) Phytochrome in light-grown plants. Encyclopedia of Plant Physiol NS 16B:704–722
10. Jones OP, Waller, BJ, Hadlow WCC and Beech MG (1986) Modification of micropropagation procedures to control field performance. Rep E Malling Res Sta 1985, p 95
11. Jordan BR, Partis MD and Thomas B (1986) The biology and molecular biology of phytochrome. Oxford Surveys of Plant Molecular and Cell Biology 3:315–362
12. Marks TR (1986) Micropropagation of hardy ornamental nursery stock. In: PG Alderson and WM Dullforce, eds. Micropropagation in Horticulture, pp 71–83. Proceedings Inst of Hort Symposium, University of Nottingham
13. Mohr H (1986) Coaction between pigment systems. In: RE Kendrick and GHM Kronenberg, eds. Photomorphogenesis in Plants, pp 547–564 Dordrecht: Martinus Nijhoff
14. Morini S, Fortuna P, Sciutti and Muleo R (1990) Effect of different light-dark cycles on growth of fruit tree shoots cultured *in vitro*. Adv Hort Sci 4:163–166
15. Muleo P and Morini S (1990) Effect of light quality on regeneration from callus of *Actinidia deliciosa*. Acta Hort 280:155–158
16. Pausch RC, Britz SJ and Mulchi CL (1991) Growth and photosynthesis of soybean (*Glycine max* (L.) Merr.) in simulated vegetation shade: influence of the ratio of red to far-red radiation. Plant Cell Environ 14:647–656
17. Selbert M, Wetherbee PJ and Job DD (1975) The effects of light intensity and spectral quality on growth and shoot initiation in tobacco callus. Plant Physiol 56:130–139
18. Senger H and Schmidt W (1986) Diversity of photoreceptors. In: RE Kendrick and GHM Kronenberg, eds. Photomorphogenesis in Plants, pp 137–158. Dordrecht: Martinus Nijhoff
19. Smith H (1982) Light quality, photoperception, and plant strategy. Ann Rev Plant Physiol 33:481–518
20. Stimart DP, Ascher PD and Wilkins HF (1982) Overcoming dormancy in *Lilium longiflorum* bulblets produced in tissue culture. J Amer Soc Hort Sci 107:1004–1007
21. Summerfield RJ, Minchin FR, Roberts EH and Wien HC (1977) Photomorphogenetic effects on cowpea; a comparison of controlled environment and field-grown plants. Plant SciLett 8:355–361
22. Thomas B and Dickinson HG (1979) Evidence for two photoreceptors controlling growth in de-etiolated seedlings. Planta 146:545–550
23. Thomas B and Johnson CB (1991) Phytochrome Properties and Biological Action. Berlin-Heidelberg: Springer-Verlag
24. Vince-Prue D (1975) Photoperiodism in Plants. Maidenhead: McGraw Hill
25. Vince-Prue D (1976) Phytochrome and photoperiodism. In: H Smith, ed. Light and Plant Development, pp 347–369. London: Butterworths
26. Vince-Prue D (1977) Photocontrol of stem elongation in light-grown plants of *Fuchsia hybrida*. Planta 133:149–156
27. Vince-Prue D (1983) Photomorphogenesis and flowering. Encyclopedia of Plant Physiol NS 16B:457–490
28. Vince-Prue D (1984) Contrasting types of photoperiodic response in the control of dormancy. Plant Cell Environ 7:507–513
29. Vince-Prue D (1992) Phytochrome: a remarkable light sensor in plants. Plantsman 13:203–214
30. Vince-Prue D and Canham AE (1983) Horticultural significance of photomorphogenesis. Encyclopedia of Plant Physiol NS 16B:518–544
31. Vince-Prue D and Guttridge CG (1973) Floral initiation in strawberry; spectral evidence for the regulation of flowering by long-day inhibition. Planta 110:165–172
32. Vince-Prue D and Lumsden PJ (1987) Inductive events in the leaves: time measurement and photoperception in the short-day plant, *Pharbitis nil*. In: JG Atherton, ed. Manipulation of Flowering, pp 255–268. London: Butterworths
33. Vince-Prue D and Takimoto A (1987) Roles of phytochrome in photoperiodic floral induction. In: M Furuya, ed. Phytochrome and Photoregulation in Plants. Tokyo: Academic Press
34. Warrington IJ and Mitchell KJ (1976) The influence of blue- and red-biased light spectra on the growth and development of plants. Agric Meteorol 16:247–262

35. Yae RW, Zimmerman RH and Fordham I (1987) Influence of photoperiod, apical meristem and explant orientation on axillary shoot proliferation of apple cultivars *in vitro*. J Amer Sci Hort Sci 112:588–592

Effects of the light environment on photosynthesis and growth *in vitro*

ROBERT P. LEES

Neo Plants Ltd, 197 Kirkham Rd, Freckleton, Preston, Lancs. PR4 1HU, UK

1. Introduction

The transfer of plants from culture to soil is a stage at which many losses occur during micropropagation. One reason for this is the poor development of the plants *in vitro*, another may be poor photosynthetic performance.

Due to economic factors the irradiance within most culture rooms is maintained at approximately 60 μmol m^{-2} s^{-1} (approximately 5000 lux). This light level is clearly very much lower than normal daylight. In order to facilitate rapid growth *in vitro* a carbohydrate (normally sucrose) is incorporated into the micropropagation medium. This may lead to problems of microbial contamination and may also lead to problems during the transfer of plants from culture to soil.

Much current research is concentrating on developing a carbohydrate-free or autotrophic system of plant production which would involve increased irradiance and supplementary CO_2 being supplied to cultured material [2]. The research detailed here is not an attempt to develop such a system, but an attempt to understand the effects of different irradiances on plant growth *in vitro* using the system of plant production currently in common use. The aims of this work were to determine whether growth room irradiance (between 20 and 80 μmol m^{-2} s^{-1}) affected photosynthetic capability, plant morphology and survival of plants after transfer to the nursery. Three different plant species, *Clematis, Hosta* and *Daphne* were investigated.

2. Materials and methods

Clematis multiplication medium contained 'MS' salts [6], 30 g/l sucrose and 7.5 g/l agar. Rooting medium contained half strength 'Woody' salts [5], 0.5 mg/l naphthyleneacetic acid (NAA), 7.5 g/l agar, and 20 g/l sucrose. The multiplication medium used for *Hosta* contained 'MS' salts, 0.8 mg/l Benzyl adenine (BA), 0.1 mg/l NAA, 30 g/l sucrose and 7.5 g/l agar. Rooting medium contained half strength 'MS' salts, 20 g/l sucrose and 7.5 g/l agar. *Daphne*

31

P.J. Lumsden, J.R. Nicholas and W.J. Davies (eds.), Physiology, Growth and Development of Plants in Culture, 31–46, 1994.

multiplication medium contained 'MS' salts, 0.1 mg/l BA, 30 g/l sucrose and 7.5 g/l agar. For rooting, half strength 'Woody' salts, 0.5 mg/l NAA, 0.5 mg/l indolebutyric acid (IBA), 20 g/l glucose and 7.5 g/l agar were used. All media were adjusted prior to autoclaving to give a final pH of 5.7.

Plant density was 20 plants per culture vessel in all cases except for *Clematis* rooting cultures which were maintained at 15 plants per vessel. Each culture vessel held approximately 50 ml medium. Plants were maintained with a 16 h daylength, at a temperature of 24 °C during the day and 18 °C during the night. Plants were placed under Philips TLD 84 fluorescent tubes with a range of irradiances from 20 μmol m^{-2} s^{-1} to 80 μmol m^{-2} s^{-1}. Irradiance was measured using a Skye SKP215 quantum sensor.

For experimental work 200 plants were used for each treatment. Plants were subcultured onto multiplication medium and placed into the trial environment. After 28 days, plant quality was assessed visually (see below) as the material was subcultured onto fresh multiplication medium. This was repeated for three further subcultures. At the fourth subculture plant material was subcultured onto rooting medium. After a rooting period of 12 days for *Hosta*, 28 days for *Daphne*, and 56 days for *Clematis*, plants were transferred to soil.

During the multiplication stage various morphological characteristics were assessed visually: The number of shoots produced per original shoot at each subculture was recorded to give a multiplication rate. The number of shoots which possessed at least one chlorotic leaf after four weeks' growth *in vitro* was recorded, as was the number in which the shoot tip had become necrotic, the number which had rooted, the number which possessed at least one vitrified leaf, and the number of shoots which had become clearly elongated. Data for morphological characteristics are presented as the mean of four consecutive subcultures.

During growth on rooting medium the number of shoots from which roots had emerged was recorded each week until the plants were sent to the nursery for planting. On transfer to the nursery rooted and non-rooted plants were planted separately. The survival of both rooted and non-rooted plants was monitored.

Leaf anatomy was studied by fluorescence microscopy, using a technique described elsewhere [3]. For photosynthesis measurements a Hansatech LD2 leaf disc oxygen electrode (Hansatech Ltd, Kings Lynn, U.K.) was used to measure oxygen evolution from excised leaves. The method has been previously described [4], and provides information on both the light saturated photosynthetic capacity, and the quantum efficiency of photosynthesis [7].

3. Results

3.1 Effects of the light environment on photosynthesis in vitro

Table 1 shows the photosynthetic capacity of *Daphne* plants after 21 days growth under a range of irradiances on rooting medium *in vitro*. There was clearly an increase in light saturated photosynthetic capacity with increased growth room irradiance, but little difference in quantum efficiency. Plates 1a–1c show a decreased level of palisade development *in vitro* compared to nursery grown material. There was also increased palisade development with increased irradiance *in vitro*. These results demonstrate that *Daphne* produce 'sun type' leaves *in vitro* [1]. Increased photosynthetic development at increased irradiance may explain the increased photosynthetic capacity of leaves produced under higher light.

Table 1. Photosynthetic capability of *Daphne odora* after 21 days growth on rooting medium under different irradiances *in vitro*. Values are means (n=4) ± standard error

Growth irradiance (μmol m^{-2}s^{-1})	Psat (μmol m^{-2}s^{-1})	θ
20	1.85±0.35	0.0581±0.0040
40	2.90±0.37	0.0596±0.0047
60	4.16±0.28	0.0760±0.0049
80	4.64±0.12	0.0752±0.0020

Table 2. Nursery survival of *Daphne odora* after 87 days' growth on the nursery, following growth *in vitro* for 28 days under a range of irridiances (n = plants per treatment)

Growth irradiance (μmol m^{-2}s^{-1})	Rooted		Non-rooted		Total	
	n	Survival %	n	Survival %	n	Survival %
20	28	92.9	112	21.4	140	35.7
40	58	94.8	81	40.7	139	63.3
60	79	94.9	61	42.6	140	72.1
80	87	97.7	51	52.9	138	81.2

Nursery survival was enhanced for plants grown under higher irradiances *in vitro* (Table 2). This suggests that photosynthetic competence as plants are transferred to soil may be an important factor in determining transplant survival.

Table 3 shows the photosynthetic capacity of *Hosta* plants after 12 days' growth on rooting medium *in vitro*. Although there is an increase in light saturated capacity with increased growth room irradiance, the increase is not as dramatic as that for *Daphne* (Table 1). In the case of *Hosta*, there was no visible palisade development in mature leaves of plants produced on the nursery under daylight irradiances (Plate 2). These results suggest that *Hosta* produces 'shade type' leaves *in vitro* [1]. Nursery survival of plants from all treatments was 100%. There was no visible difference in the quality of *Hosta* plants after transfer to the nursery.

Table 3. Photosynthetic capability of *Hosta* after 12 days growth on rooting medium under different irradiances *in vitro*. Values are means (n=4) ± standard error

Growth irradiance (μmol m^{-2}s^{-1})	Psat (μmol m^{-2}s^{-1})	θ
20	4.06±0.93	0.1173±0.0063
40	6.15±1.57	0.0955±0.0059
60	5.69±1.26	0.1060±0.0066
80	6.55±0.82	0.1000±0.0049

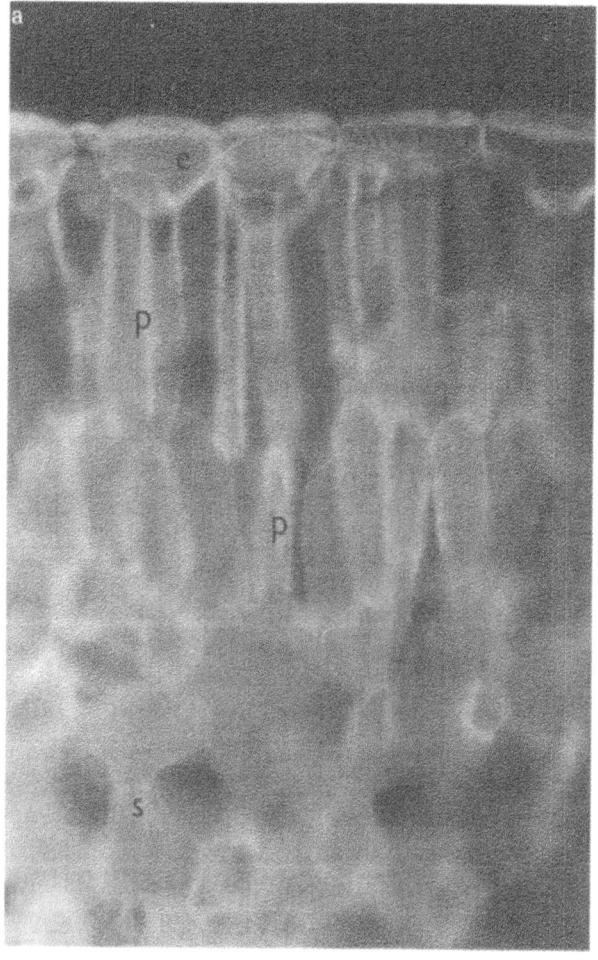

Plate 1. Leaf anatomy of *Daphne odora* plants, a) nursery grown, b) grown *in vitro* under 20 μmol m^{-2}s^{-1} and c) grown *in vitro* under 80 μmol m^{-2}s^{-1}.

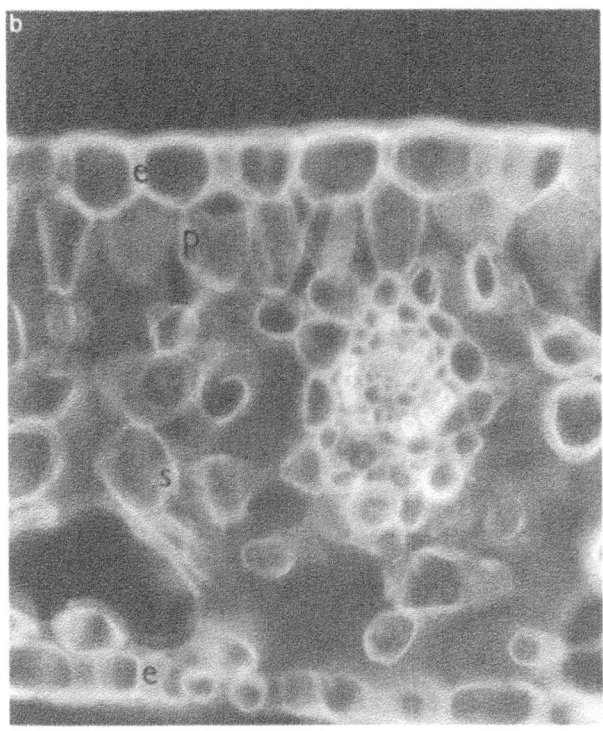

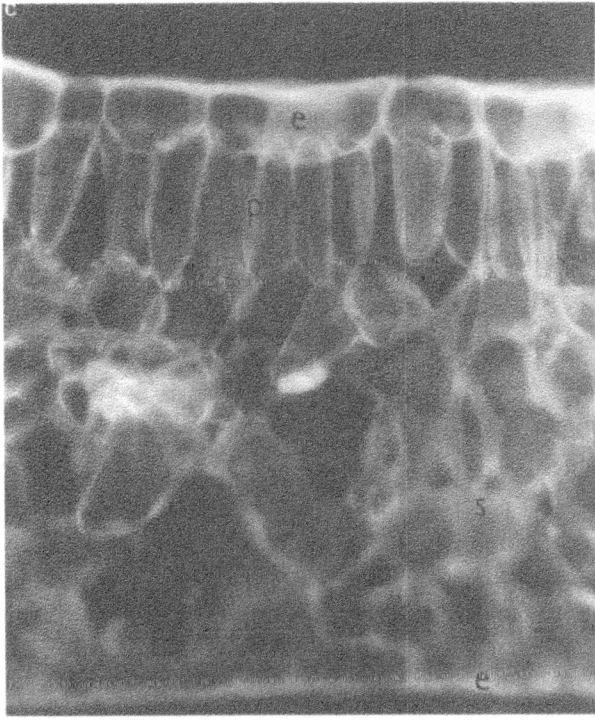

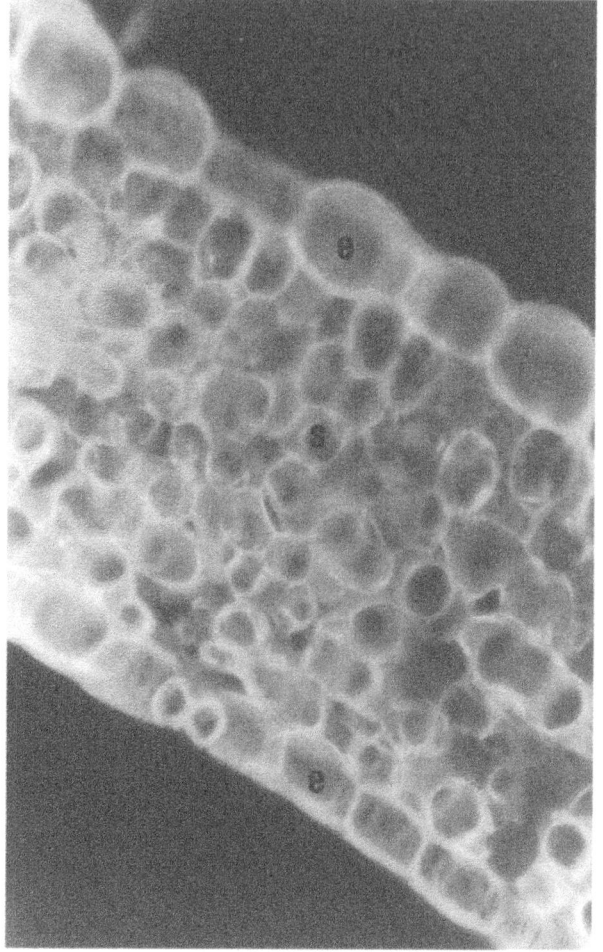

Plate 2. Leaf anatomy of nursery grown *Hosta* plants.

3.2 *Effects of the light environment on growth* in vitro

Daphne plants grown under 80 μmol m^{-2} s^{-1} (Plate 3a) were of much better quality than those grown under 20 μmol m^{-2} s^{-1} (Plate 3b), with larger leaves and thicker stems. Poor shoot quality was shown as a higher incidence of vitrification (F=7.88; df=3,35; p<0.001), and shoot tip necrosis (F=9.81; df=3,35; p<0.001). These symptoms, together with greater stem elongation (Fig. 1b) are typical shade avoidance responses, and explain the poor performance of these plants under low irradiance. The incidence of leaf chlorosis was approximately equal for all treatments. Although results show no significant effect of irradiance on multiplication rate (F=1.07; df=3,35; p<0.05), root emergence on rooting medium was significantly reduced (data not shown).

Plate 3. Daphne plants after 28 days growth on rooting medium under a) 20 μmol m^{-2}s^{-1} and b) 80 μmol m^{-2}s^{-1}.

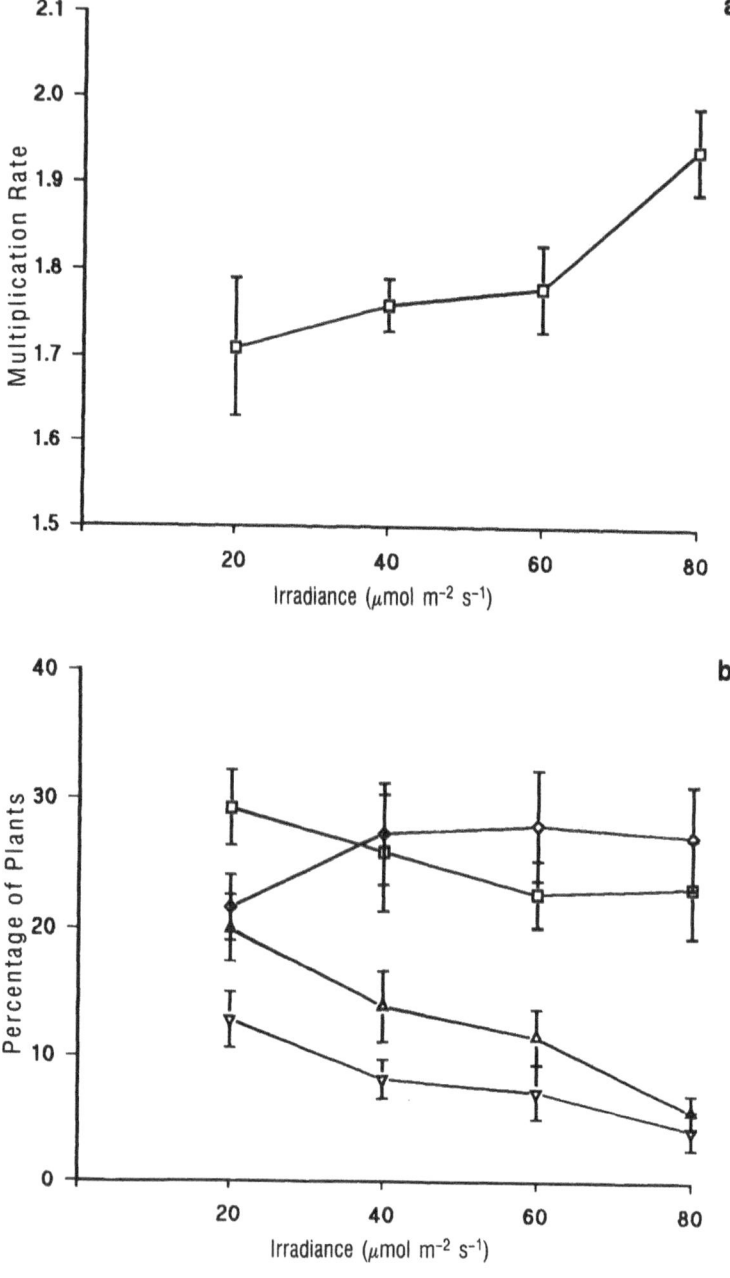

Fig. 1. Growth characteristics of *Daphne* assessed over four successive subcultures onto multiplication medium after growth under a range of irradiances *in vitro*. a) multiplication rate, b) shoot elongation. (□), shoots with vitrified leaves (Δ), shoot tip necrosis (∇), shoots with chlorotic leaves (◇). (Mean ± standard error plotted).

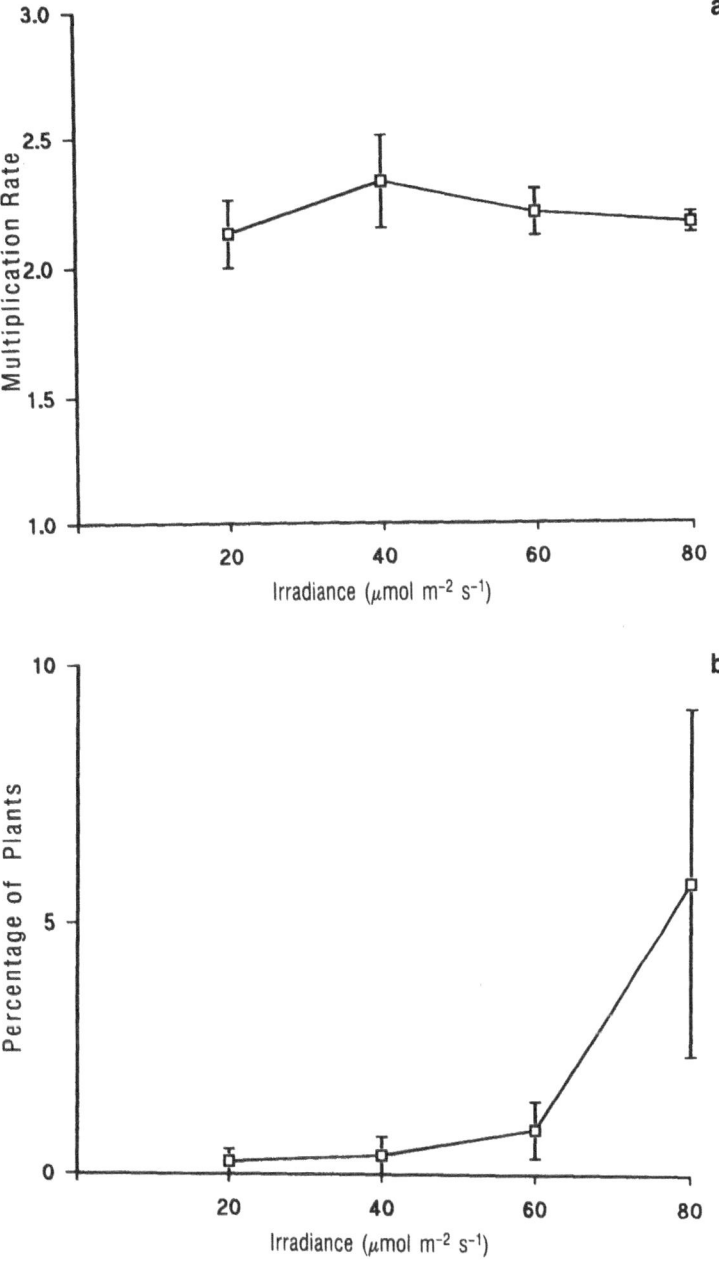

Fig. 2. Growth characteristics of *Hosta* assessed over four successive subcultures onto multiplication medium after growth under a range of irradiances *in vitro*. a) multiplication rate, b) shoots with chlorotic leaves. (Mean ± standard error plotted).

Plate 4. *Hosta* plants after 12 days growth on rooting medium under a) 20 μmol m^{-2}s^{-1} and b) 80 μmol m^{-2}s^{-1}.

Figure 2a shows that there is little effect of irradiance on the multiplication rate of *Hosta in vitro* (F=0.57; df=3,36; p>0.05). These plants are subcultured by basal division and do not exhibit any stem elongation *in vitro*. There was no evidence of poor shoot quality such as vitrification under any irradiance. There was little difference between the appearance of plants grown under 20 μmol m^{-2} s^{-1} (Plate 4a) and those grown under 80 μmol m^{-2} s^{-1} (Plate 4b), although there was a higher incidence of leaf chlorosis at higher irradiances (Fig. 2b; F=15.67; df=3,36; p<0.001), indicating a degree of photobleaching *in vitro*. *Hosta* is a typical shade tolerant plant and these results indicate that this plant also displays a shade tolerance response *in vitro*.

Figure 3a shows that multiplication rate increased with decreasing irradiance for *Clematis in vitro* (F=20.27; df=3,36; p<0.001). Shoot elongation was clearly enhanced at lower irradiances (Fig. 3b; F=8.38; df=3,36; p<0.001), and there was a greater degree of chlorosis at higher irradiance (F=10.09; df=3,36; p<0.001). Increased shoot elongation at low irradiances suggests that *Clematis* was behaving as a shade avoider *in vitro*. A greater incidence of chlorosis at higher irradiances suggests a degree of photobleaching *in vitro*. This is shown clearly by Plates 5a–5d. Plants grown under 20 μmol m^{-2} s^{-1} (Plates 5a and 5b) were of visibly better quality than plants grown under 80 μmol m^{-2} s^{-1} (Plates 5c and 5d). Shoots from 80 μmol m^{-2} s^{-1} were visibly less elongated, and had thicker stems than those from lower irradiances. Table 4 shows the photosynthetic capability of *Clematis* plants grown on rooting medium after 56 days growth *in vitro*. Both light saturated photosynthetic capacity (F=4.31; df=3,12; p<0.05), and quantum yield (F=6.22; df=3,12; p<0.01) increased as irradiance increased to 60 μmol m^{-2} s^{-1}, a result which can be explained as an increase in the development of the photosynthetic apparatus with increased irradiance (see above). Plants grown under 80 μmol m^{-2} s^{-1}, however, had a lower light saturated photosynthetic capacity and lower quantum efficiency than plants grown under lower irradiances. This suggests a direct inhibition of photosynthesis associated with growth under an irradiance greater than 60 μmol m^{-2} s^{-1}. Together with the increased incidence of leaf chlorosis at higher irradiance (Fig. 3b and Plate 5a-d), this result indicates photoinhibition of photosynthesis, and possibly some damage to the photosynthetic apparatus caused by growth at 80 μmol m^{-2} s^{-1} *in vitro*.

Table 4. Photosynthetic capability of *Clematis* on rooting medium after 56 days growth under different irradiances *in vitro*. Values represent means (n=4) ± standard error

Growth irradiance (μmol m^{-2}s^{-1})	Psat (μmol m^{-2}s^{-1})	θ
20	1.09±0.25	0.0313±0.0012
40	2.47±0.34	0.0469±0.0038
60	3.02±0.51	0.0604±0.0077
80	1.50±0.32	0.0336±0.0035

Plate 5. Clematis 'The President' after 56 days growth on rooting medium containing 20 g/l sucrose, grown *in vitro* under a) 20 μmol m^{-2}s^{-1} and b) 80 μmol m^{-2}s^{-1}.

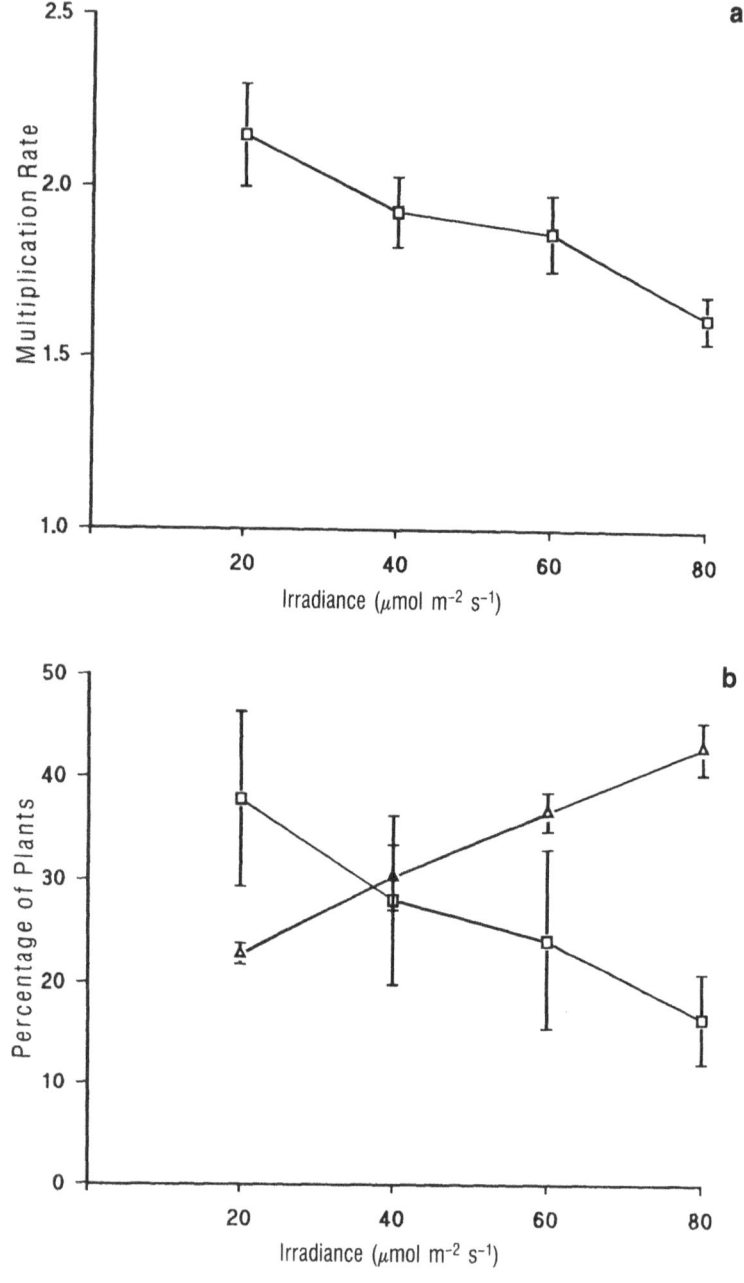

Fig. 3. Growth characteristics of *Clematis* assessed over four successive subcultures onto multiplication medium after growth under a range of irradiances *in vitro*. a) multiplication rate, b) shoot elongation (□), shoots with chlorotic leaves (Δ). (Mean ± standard error plotted).

Table 5. Nursery survival of *Clematis* after 47 days' growth on the nursery, following transfer from different growth irradiances *in vitro* (n = plants per treatment)

Growth Irradiance (μmol m^{-2}s^{-1})	Rooted		Non-rooted		Total	
	n	Survival %	n	Survival %	n	Survival %
20	39	97.4	64	87.5	103	91.3
40	42	100.0	60	81.7	102	89.2
60	34	100.0	70	92.9	104	95.2
80	42	100.0	62	69.4	104	81.7

After 56 days growth on rooting medium plants were transferred to soil. The percentage survival of plants from 80 μmol m^{-2} s^{-1} was lower than for plants transferred from lower irradiances (Table 5). This is particularly apparent for non-rooted material. This again suggests that photosynthetic competence is an important factor in determining nursery survival.

4. Discussion

These results demonstrate that normal physiological and morphological responses to shade are exhibited under the low irradiance conditions used for micropropagation. Increasing light levels *in vitro* may increase the quality of material such as *Daphne*, decrease the quality of material such as *Clematis* or have little effect on the growth of material such as *Hosta*. An understanding of the normal growth habits of plants to be micropropagated will help to optimise the light regime for each plant *in vitro*.

It is clear that there is considerable inhibition of photosynthesis in plants growing *in vitro*. This is a consequence of feedback inhibition due to the presence of medium carbohydrate, and to environmental factors such as low CO_2 levels and growth irradiance [3]. Results presented here demonstrate that an irradiance of only 80 μmol m^{-2} s^{-1} may be high enough to cause photo-inhibition of photosynthesis when feedback inhibition due to carbohydrate accumulation is so great.

Results presented in this paper demonstrate that the effects of light and medium carbohydrate combine to affect the efficiency of the micropropagation process. It may be possible to manipulate the photosynthetic capacity of plants by decreasing medium carbohydrate concentration, or by increasing growth irradiance. The effectiveness of this approach will, however depend on the natural growth response of the plant in question, and care should be taken not to increase photosynthetic ability at the expense of plant quality, and thereby reducing the efficiency of the overall process.

Acknowledgements

I would like to thank Neo Plants Ltd for financial support for this work and Sandra Wright, Julie Youngs and Birgit Funke for technical assistance.

References

1. Esau K (1967) Plant anatomy. New York: John Wiley
2. Kozai T, Oki H and Fujiwara K (1987) Effects of CO_2 enrichment and sucrose concentration under high photosynthetic photon fluxes on growth of tissue cultured *Cymbidium* plantlets during the preparation stage. In: Ducate G, Jacob M and Simeon A, eds. Symposium on Plant Micropropagation in Horticultural Industries (1987), pp 135–141. BPTC Group, Arlon, Belgium
3. Lees RP (1993) Photosynthesis and Nutrition in *in vitro* plants. PhD Thesis, University of Central Lancashire
4. Lees RP, Evans EH and Nicholas JR (1991) Photosynthesis in *Clematis* 'The President' during growth *in vitro* and subsequent *in vivo* acclimatisation. J. Exp Bot 42:605–610
5. Lloyd G and McCown B (1981) Commercially feasible micropropagation of mountain laurel *Kalmia latifolia* by use of shoot tip culture. Combined Proceedings of the International Plant Propagators Society. 30:421–427
6. Murashige T and Skoog F (1962) A revised medium for rapid growth and bioassay of tobacco tissue cultures. Physiol Plant 15:473–497
7. Walker DA (1989) The automated measurement of quantum yield and it's implications. Phil Trans Royal Soc Lon B 323: 313-326

Nutrient supply and plant growth

A. JAMES S. McDONALD

Department of Ecology and Environmental Research, Swedish University of Agricultural Sciences, P.O. Box 7072, S-750 07 Uppsala, Sweden

1. Introduction

By way of introduction, two general features of plant nutrition may be stated. Firstly, because all physiological processes in the plant are ultimately dependent upon the incorporation of one or more mineral nutrients in a form appropriate to underlying biochemistries, an increase in the size of plant organs and their correct functioning is ultimately dependent upon an appropriate availability of essential nutrients [15, 16, 21]. Secondly, the extent to which growth processes are dependent upon the current uptake of externally available nutrient will depend upon the amounts and availability of stored nutrients and the extent to which recycling occurs within the plant. Both storage and recycling are phenomena which can be of critical importance to the survival and fitness of individuals in nutrient deficient environments or under circumstances of fluctuating nutrient availability. In contrast to plantlets grown from cell culture, seeds can, for example, contain sufficient nutrients to allow considerable increase in plant size without further nutrient uptake. In micropropagated plantlets, the nutrient reserve will be minimal and current growth will predictably be largely dependent upon current availability and uptake of mineral nutrients. The first part of this paper discusses some basic relationships between nutrient supply and plant growth. The feasibility of controlling, in a quantitative manner, dry matter productivity by regulating the supply of any one mineral nutrient is emphasised. In the second part, qualitative differences in growth response dependent upon the nature of nutrient limitation are discussed. It is suggested that, by combining quantitative and qualitative aspects, plants of a predictable size, morphology, physiology and biochemistry may be grown. It is believed that this could have particular bearing not only on the commercial preparation of plantlets of a given size and appearance but also on the subsequent production and recovery of specific metabolites from plants whose micropropagation has, from the outset, been combined with appropriate nutritional practice.

P.J. Lumsden, J.R. Nicholas and W.J. Davies (eds.), Physiology, Growth and Development of Plants in Culture, 47–57, 1994.

2. Nutrient supply and control of biomass increase

The accepted dogma on plant growth response to the supply of mineral nutrients has, for the greater part, assumed the importance of nutrient concentration about the root surface to both the uptake of nutrient and subsequent plant growth. In the context of understanding mechanisms in nutrient uptake, dependence of flux on a manipulated and variable external concentration can of course yield crucial information [14, 20]. Conceptually, it is less evident that manipulation of the external concentration *per se* should have any affect on plant growth or at least on the rate of biomass increase. Indeed, it has been demonstrated that, over a wide range of external nutrient concentrations, maximum growth rate assumes a constant value for a given genotype and environment [9, 22]. These concentrations encompass the range over which nutritional studies are often made.

It may be argued that concentration alone is inappropriate to varying the time course of nutrient availability at the root surface which may be associated with a particular rate of nutrient uptake. In order to manipulate the rate of delivery (flux) to the root surface, the concentration variable must be combined with one of flow rate [9]. Thus, the flux achieved by a solution of nutrient concentration C (mol m^{-3}) and flow rate F (m^3 s^{-1}) is

$$A = FC \ (\text{mol s}^{-1}).$$

If the flux does not exceed the capacity of a root system to absorb available nutrient, then there exists the technical possibility of controlling plant growth through nutrient supply by manipulating the nutrient flux at the root surface. In practice, this can be achieved [1] although there may be considerable problems in attaining sufficient mixing of nutrient to maintain the intended concentration at the root surface equal to that being monitored in the bulk solution.

An alternative approach to maintaining nutrient flux is that which has been advocated by Ingestad, in which the external concentration variable has essentially been ignored and in which predetermined amounts of nutrient are added according to a chosen time schedule [9, 12]. As with combined concentration and flow systems, mixing of the added nutrient is essential to ensure contact with the root surface. The particular case of this approach which is most often practised is that of adding nutrients at regular intervals and at exponentially-increasing amounts [2, 8]. The implications of this procedure in the qualitative control of plant growth are potentially far-reaching and are discussed more fully in the following section. Here, it is appropriate to comment on the generality of the approach in controlling rates of biomass increase and its equivalence to approaches where extremely low external concentrations are maintained at the root surface.

A useful starting point in appreciating the Ingestad approach is to consider the relationship between growth rate and nutrient depletion under free access. In small plants, an exponential increase in biomass will occur at a rate which is a maximum for the genotype and growth environment in question. From observations on nutrient depletion, it may be calculated that nutrient absorption by the root system is also increasing at an exponential rate and that the exponent of increase is equal to that of biomass increase. In other words, the relative growth rate of the plant is equal to the relative rate of increase in nutrient uptake:

$$w_t = w_0e^{Rt} \text{ and } n_t = n_0e^{Rt}$$

where w_0 and n_0 are initial plant weight and amount of nutrient; w_t and n_t are plant weight and amount of plant nutrient at time t; R is the relative rate of increase in plant weight, equal to that of plant nutrient amount. The Ingestad contribution was to demonstrate that this free access relationship between nutrient uptake and growth was a special case of a more general phenomenon. By supplying nutrients in the proportions taken up at free access, it has been possible to choose any sub-maximal relative rate of increase in nutrient supply and experimentally control the relative rate of biomass increase. Where this is done, the relative growth rate of the plant is found to equal the relative rate of increase in nutrient supply. This is true irrespective of genotype (Fig. 1) and

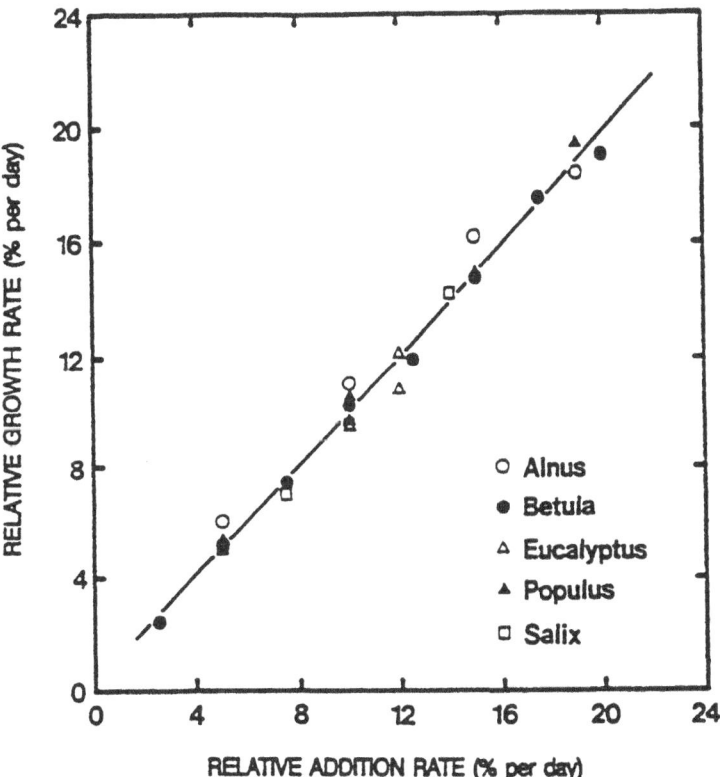

Fig. 1. The dependence of plant relative growth rate (% per day) on relative addition rate (% per day) of a complete nutrient solution. The data are from a number of experiments with different tree species [5].

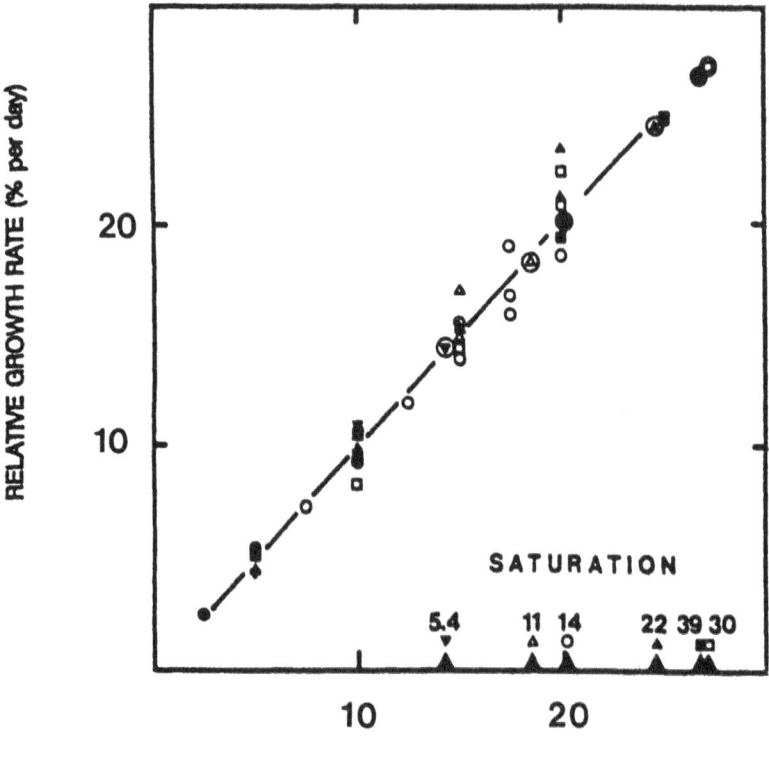

Fig. 2. The dependence of plant relative growth rate (% per day) on relative addition rate (% per day) of a complete nutrient solution. The data are for *Betula* and the saturation values indicate the relative addition rate which was associated with maximum growth at different photon flux densities (5.4, 11, 14, 22, 30 and 39 mol m^{-2} d^{-1}) [13].

growth environment (Fig. 2), both of which are instrumental in determining the maximum rate and thus the range of nutrient supply for which the relationship holds true. In other studies it has, for example, been demonstrated that the maximum growth rate of pine seedlings is less than that of birch which, in turn, is less than that of a herb such as sunflower [10, 11, 19]. From Figure 2, it can be seen that maximum growth rate in birch is greater at higher photon flux density [13]. Higher maximum values of relative growth rate have also been found at higher atmospheric concentrations of carbon dioxide [23]. Growth rate will of course eventually saturate with respect to increase in either of these variables. However, the point to note is that potentially higher growth rates can only be realised if nutrient supply is increased. In Ingestad terms, the relative rate of increase in nutrient supply must be increased to equal and maintain the higher values of plant relative growth rate.

So far, the discussion has been confined to nutrient supply in total. In practice, it might be expected that growth rate may be most limited by the supply of any one nutrient, where access to all other nutrients is essentially non-limiting, at least to biomass increase. In fact, the experimental control of relative growth rate, by the choice of relative addition rate, has been demonstrated for all nutrients which have so far been investigated: nitrogen [11]; phosphorus [3]; potassium [4]; sulphur and magnesium (T Ericsson, personal communication); iron [6]; manganese [7]; zinc and boron (A Göransson, personal communication). The repercussions of this observation for plant quality are discussed in the next section. At present, it is sufficient to note that, because relative growth rate can be accurately controlled by the relative rate of increase in any chosen limiting nutrient, an increase in plant biomass can accurately be predicted. Given a start value of biomass, an accurate prediction of plant weight at a particular point in time can be made, irrespective of genotype or growth environment for any relative rate of increase in nutrient supply, provided this does not exceed the inherent maximum. Alternatively, if a particular biomass is required at a particular point in time, an appropriate choice of relative addition rate may be made. In fact, if biomass was the only consideration, plants could be grown for longer or shorter periods of time with appropriate choice of nutrient addition rate to achieve the desired biomass at a particular point in time.

Amongst those working in plant nutrition research, there has, to some extent, been an apparent polarization of emphasis between those advocating the Ingestad approach and those persisting with the use of external concentration alone or concentration in conjunction with flow rate in the control of plant growth. Perhaps not surprisingly, elements of truth may be identified in each of the opposing views, — Ingestad in his emphasis of nutrient flux; the advocates of concentration in maintaining that sufficiently low values of external concentraton should inevitably be limiting to nutrient uptake and plant growth. At least for nitrate nutrition, these opposing emphases have apparently recently been resolved. When plants are grown at a continuous relative rate of increase in nitrate supply, equilibrium concentrations of nitrate, proportional to the relative addition rate, have been measured in the culture solution (JH Macduff, SC Jarvis, C-M Larsson and P Oscarson, personal communication). These are very low nitrate concentrations ($1-10 \times 10^{-6}$ M), being considerably less than those often used in nutritional studies. However, if these low concentrations are experimentally maintained in culture solution by monitoring nitrate concentration and continuously compensating for depletion, plant relative growth rate assumes the value for which the same equilibrium concentration may be obtained through maintenance of relative addition rate. So, it would appear that the two approaches may be reconciled at extremely low external concentrations. Whether or not the experimental maintenance of such extremely low external concentrations is an attractive practical alternative in plant culture systems will depend largely on the extent to which adequate mixing about the root surface can be ensured.

3. Nutrient supply and plant quality

The foregoing discussion has been almost entirely concerned with the control of biomass increase through regulation of nutrient supply. However, in plant culture biomass increase is only one of many possible aims. Of equal importance are the many aspects of plant quality.

These may relate to morphological and anatomical responses or to physiological and biochemical properties. The choice of limiting nutrient and the rate at which it is supplied are both likely to affect a number of morphological and physiological traits and as such, should be regarded as useful tools in the manipulation of plant quality. These are not the only nutritional varibles of interest in this context. Nutrient proportions and nutrient source (e.g. ammmonium/nitrate ratios) may both potentially have far-reaching consequences on growth response. However, the present discussion is limited to that of choice and rate of supply of nutrient.

It should be emphasised that these aspects of quality should not be confused with the experimental control of biomass increase *per se* as discussed in the previous section. It is, for example, possible to grow plants at the same relative growth rate but, by varying the choice of limiting nutrient, achieve different plant nutrient concentrations (Fig. 3). With strict nutritional control, the time course of plant biomass will be the same in both instances but the plants will differ greatly in morphology and physiology between treatments.

For example, for a given relative rate of increase in supply, a nitrogen-limited plant will have a higher net assimilation rate but a lower leaf area ratio and specific leaf area than a plant whose growth is limited by the availability of iron [6, 11]. These differences, in turn, can be attributed to differences in photosynthesis and in cell expansion rate (AJS McDonald, unpublished) [24]. The nitrogen- and iron-limited plants will also differ greatly in appearance, with the nitrogen limitation resulting in a plant with a long root system and a compact shoot comprising small leaves. On the other hand, iron limitation results in a plant with short roots and a shoot system of large leaves. Starch reserves in leaves are high in response to nitrogen limitation [17] but are negligible in response to iron limitation [6]. This may be of some importance in determining the extent to which carbon usage in expansive growth processes associated with volume increase are effectively buffered or uncoupled from current carbon uptake. This latter point might be of crucial importance to plant performance following some perturbation such as that associated with transplanting. It is not immediately obvious why one might choose to limit biomass increase through iron rather than nitrogen supply. (The fact that the leaves on an iron-limited plant are bright yellow in colour may or may not be a desirable attribute!) The point to emphasise is that both these and many other limiting nutrient possibilities exist. The nitrogen-limited and iron-limited plants will have acclimated to the limiting supply of nutrient in question and, for so long as the relative rate of increase in supply is maintained, will continue to show characteristic and predictable growth response. Which of the many

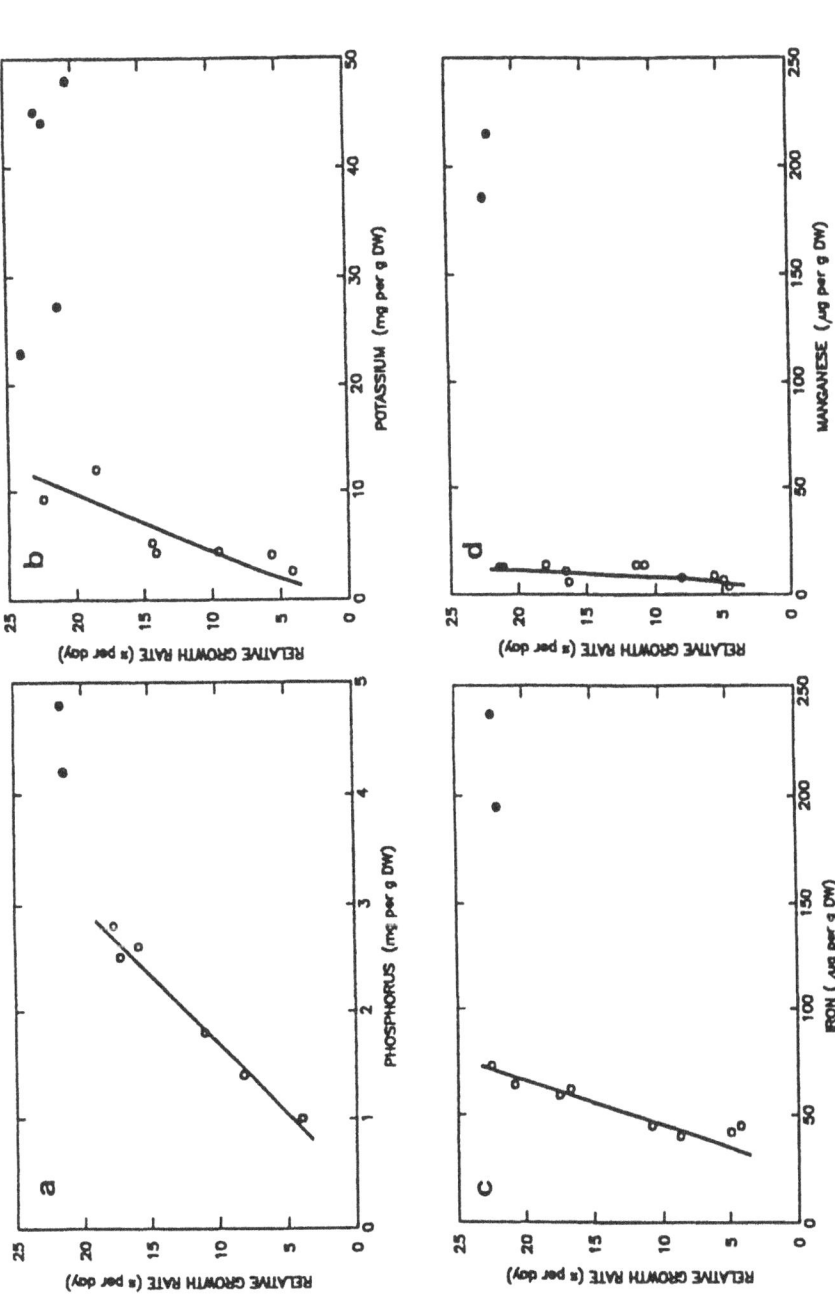

Fig. 3. The dependence of plant relative growth rate (% per day) on plant concentration of (a) phosphorus, (b) potassium, (c) iron, (d) manganese. The data are for *Betula* and, in each case, growth has been limited by the supply rate of the nutrient whose plant concentration has been measured [3, 4, 6, 7]. Open symbols: controlled exponential supply rates: filled symbols: free access to all nutrients.

possible nutrient limitations is chosen will then be a matter of aims in plant culture.

Nitrogen and iron limitation have been chosen by way of illustrating possible manipulation of plant quality through choice of nutrient. Increasing information on how some of the gross morphologies are related to both choice of limiting nutrient and supply rate is being obtained for many other nutrients. For example, two distinct categories of response with regard to partitioning of dry matter between roots and shoots have been observed (Fig. 4). In the one instance (nitrogen, phosphorus, sulphur and iron), nutrient limitation is associated with an increased allocation of dry matter to root systems. In the other category (magnesium, manganese and potassium), decreased nutrient availability results in a decreased allocation of dry matter to roots. These observations may, of course, have far-reaching consequences in an ecological context but, in the present context, should be seen primarily as providing further means of manipulating plant quality. Again, the aims in plant culture will determine the desirability of manipulating nutrient limitation in attaining one type of root system as opposed to another.

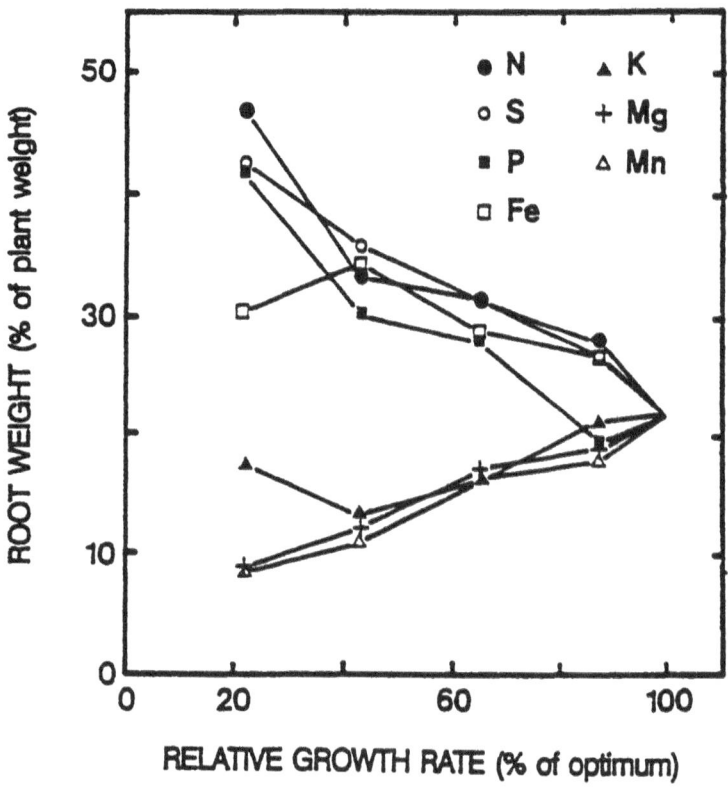

Fig. 4. The effect of controlled nutrient stress on dry matter partitioning to roots in *Betula* seedlings [5].

A number of other possibilites exist for the manipulation of plant quality, which because they interact with nutrient supply, should be mentioned. In addition to the previously mentioned nutritional variables such as nutrient proportions and source, notice should be taken of other variables associated with the growth environment.

By way of example, only one such variable will be considered, namely photon flux density. It is appropriate to emphasise that neither this variable, nor indeed any other environmental variable if held constant during plant culture, will affect the experimental control of relative growth rate through relative rate of increase in nutrient supply. Irrespective of growth environment, the prediction of plant biomass with respect to relative rate of increase in supply holds true. However, growth environment does affect the efficiency of nutrient usage in dry matter production. For example, the rate of biomass increase per amount of plant nitrogen is greater at higher photon flux density than at much lower values (cf. slopes of regressions in Fig. 5). With respect to photon flux density, relative growth rate, as determined by relative rate of increase in nitrogen supply, is

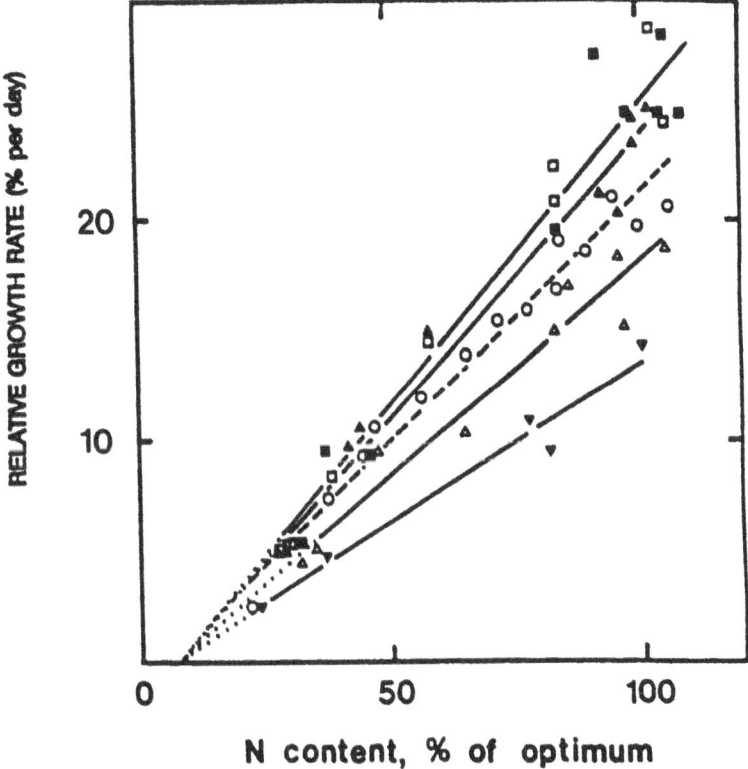

Fig. 5. The dependence of plant relative growth rate (% per day) upon plant nitrogen content (% of optimum), where nitrogen has been the most limiting nutrient to plant growth. The data are for *Betula* seedlings grown at the photon flux densities shown in Figure 1. Higher slopes indicate higher rates of biomass production per plant nitrogen [13].

associated with different morphology and physiology [13, 18]. Typically, photosynthetic rates are higher but specific leaf area is lower in plants grown at higher photon flux density. Plant chemistries may also be different. For example, leaf starch content will be greater in plants grown at higher photon flux density, despite identical relative growth rates to plants grown at lower photon flux density [17]. So, choice of growth environment, in conjunction with choice of nutrient and its supply rate, can also affect plant quality.

4. Concluding remarks: the way ahead

The potential for precision control over biomass increase through manipulation of nutrient supply has been emphasised and the appropriateness of the Ingestad approach in this regard has been noted. The generality of the relationship between relative rate of increase in supply and relative growth rate provides a potentially useful tool in manipulating supply to meet the demand for a particular biomass at a particular point in time. Both the choice of nutrient and supply rate, in conjunction with the growth environment, will affect the quality of plant material produced. Thus, plants with the same relative growth rate may have, for example, very different partitioning of dry matter between roots and shoots, different specific leaf areas or root lengths and different occurrences of plant chemistries such as carbohydrates. As these basic relationships become better understood, they provide for exciting possibilities for enhanced precision control of biomass and quality production in plant culture, over that currently practised. Micropropagation systems are particularly suitable in this regard because they allow for the possibility of manipulating quality at a very early stage in plant development. In this context, nutrient stress (by design) should not be regarded as a problem but rather as a potential means to an end. The realisation of this potential must surely constitute one of the priorities for progress in plant culture.

References

1. Edwards DG and Asher CJ (1974) The significance of solution flow rate in flowing culture experiments. Plant and Soil 41:161–175
2. Ericsson T (1981) Growth and nutrition of three Salix clones in low conductivity solutions. Physiol Plant 52:239–244
3. Ericsson T and Ingestad T (1988) Nutrition and growth of birch seedlings at varied relative phosphorus addition rates. Physiol Plant 72:227–235
4. Ericsson T and Kähr M (1993) Growth and nutrition of birch seedlings in relation to potassium supply rate. Trees 7:78–85
5. Ericsson T, Rytter L and Linder S (1992) Nutritional dynamics and requirements of short rotation forests. In: CP Mitchell, JB Ford-Robertson, T Hinckley and L Sennerby-Forsse, eds. Ecophysiology of Short Rotation Forest Crops, pp 35–65. London and New York: Elsevier Applied Science

6. Göransson A (1993) Growth and nutrition of small *Betula pendula* plants growing at different relative addition rates of iron. Trees (in press)

7. Göransson A (1993) Growth and nutrition of small *Betula pendula* plants growing at different relative addition rates of manganese. Tree Physiol (in press)

8. Ingestad T (1981) Nutrition and growth of birch and grey alder seedlings in low conductivity solutions and at varied relative rates of nutrient addition. Physiol Plant 52:454–466

9. Ingestad T (1982) Relative addition rate and external concentration; driving variables used in plant nutrition research. Plant Cell Environ 5:443–453

10. Ingestad T and Kähr M (1985) Nutrition and growth of coniferous seedlings at varied relative nitrogen addition rate. Physiol Plant 65:109–116

11. Ingestad T and Lund A-B (1979) Nitrogen stress in birch seedlings: I. Growth technique and growth. Physiol Plant 45:137–148

12. Ingestad T and Lund A-B (1986) Theory and techniques for steady-state mineral nutrition and growth of plants. Scand J For Res 1:439–453

13. Ingestad T and McDonald AJS (1989) Interaction between nitrogen and photon flux density in birch seedlings at steady-state nutrition. Physiol Plant 77:1–11

14. Marschner H (1986) Ion uptake mechanisms of individual cells and roots: short-distance transport. In: H Marschner, Mineral Nutrition of Higher Plants, pp 7–69. London: Academic Press

15. Marschner H (1986) Functions of mineral nutrients: macronutrients. In: H Marschner, Mineral Nutrition of Higher Plants, pp 195–267. London: Academic Press

16. Marschner H (1986) Functions of mineral nutrients: micronutrients. In: H Marschner, Mineral Nutrition of Higher Plants, pp 269–340. London: Academic Press

17. McDonald AJS, Ericsson A and Lohammar T (1986) Dependence of starch storage on nutrient availability and photon flux density in small birch (*Betula pendula* Roth). Plant Cell Environ 9:433–438

18. McDonald AJS, Lohammar T and Ingestad T (1992) Net assimilation rate and shoot area development in birch (Betula pendula Roth.) at different steady-state values of nutrition and photon flux density. Trees 6:1–6

19. McDonald AJS, Palmer S and Davies WJ (1993) Leaf and root extension at different nitrate availabilities in sunflower (Helianthus annuus L.). Physiol Plant (in press)

20. Mengel K and Kirkby EA (1987) Ion uptake and ionic status of plants. In: K Mengel and EA Kirkby, Principles of Plant Nutrition, 4th edition, pp 113–147. Bern: International Potash Institute

21. Mengel K and Kirkby EA (1987) Nutrition and plant growth. In: K Mengel and EA Kirkby, Principles of Plant Nutrition, 4th edition, pp 247–302. Bern: International Potash Institute

22. Olsen C (1950) The significance of concentration on the rate of ion absorption by higher plants in water culture. Physiol Plant 3:152–164

23. Pettersson R and McDonald AJS (1992) Effects of elevated carbon dioxide concentration on photosynthesis and growth of small birch plants (*Betula pendula* Roth.) at optimal nutrition. Plant Cell Environ 15:911–919

24. Terry N (1983) Limiting factors in photosynthesis. IV. Iron stress-mediated changes in light-harvesting and electron transport capacity and its effects on photosynthesis *in vitro*. Plant Physiol 71:855–860

Nutrient supply and growth of plants in culture

PIERRE DEBERGH, JAN DE RIEK and DANNY MATTHYS
Department of Plant Production, Laboratory of Horticulture, University Gent, Coupure 653, 9000 Gent, Belgium

1. Introduction

A nutrient medium for tissue culture usually consists of inorganic salts, a carbon source, some vitamins and growth regulators. In many cases a gelling agent is added. The formulation of the nutrient medium remains an important part of the development for all applications of plant tissue culture. The most used media for tissue culture is still that of Murashige and Skoog [16] or modifications from this formulation. It is now thirty years since they published their historical paper 'A revised medium for rapid growth and bioassays with tobacco tissue cultures'. It remains difficult to evaluate the impact of this paper on the development of *in vitro* culture; the medium they proposed, known by everybody as MS-medium, has proven to be satisfactory for the tissue culture of many plant species for micropropagation purposes and has undoubtedly been used more than any other for plant tissue culture work [6]. In 1981 Evans *et al.* [3] stated that the MS-medium is used in 70% of the protocols for the induction of somatic embryogenesis, and there is no reason to believe it is less for other applications of tissue culture or that it has decreased with time. It was therefore not surprising that this paper was discussed in Current Contents as one of the citation classics; it was the first plant paper which reached this status.

What developments have there been in nutrition of organs, tissues and cells in culture since that time? In the past two or three decades a large number of reports have appeared on modifications of about two dozen basic compositions [4], but more than half of them are just modifications of the MS-medium. Most of these modifications increase the complexity, and do in fact introduce unnecessary variables in comparative research. On the other hand the very important orchid markets are still using undefined media, in which potato and banana homogenates are still very popular additives.

Therefore we can conclude that (1) many if not most plant systems *in vitro* are more or less functional on a limited number of tissue culture media, and (2) for those cases where changing the composition of the basal medium was not effective, undefined components were required.

This seems to be a relative simple conclusion but it is not, since the

58

P.J. Lumsden, J.R. Nicholas and W.J. Davies (eds.), Physiology, Growth and Development of Plants in Culture, 58–68, 1994.
© 1994 *Kluwer Academic Publishers.*

relationship between the mineral components of *in vitro* media and the mineral nutrition of the explants is dynamic [19]. The overall process of mineral nutrition of plants involves a number of steps: the input of minerals; the interaction between the ions as supplied; interaction between these ions and the medium substrate; movement of the ions through the substrate to the plant surface; uptake by the plant; transport through the plant; and finally assimilation by the plant [19].

Here we intend to focus on one or two of the variables noted by Williams [19], in particular the effect of gel strength on mineral availability. We will also discuss some aspects of nutrient availability and metabolism in the double layer technique, in particular the sugar housekeeping within a double layer system, since after adding the double layer on top of the solid layer drastic quantitative and qualitative changes in carbon metabolism occur.

2. Availability of the medium ingredients

Several factors have been implicated as causes of variability in plant tissue cultures. Factors most often quoted are related to the plant material (genotype, source of the tissue, time in culture). Factors which should be amenable to careful control, such as the media components and the environmental conditions, continue to be a sometimes unsuspected source of variability [17].

As most tissues and organs are grown on defined media, the chemical nature and amounts of all components added to the medium are known. However, different abiotic factors and interactions can determine their availability (qualitatively or quantitatively). Moreover, when considering the nutrient supply in the medium we must remember that the plant *in vitro*, just as *in vivo*, does not have access to the entire volume of the medium.

It has been shown that a precipitate may form in MS-medium which may not be noticed if agar is added. The determined composition of the precipitate is mainly phosphate and iron. The composition of the medium thus changes, and after 7 days up to 50% of iron and 13 % of phosphate can be removed from solution. Precipitated elements are not necessarily unavailable to the tissues. To avoid this the pH should be reduced to 3.2 or the concentration of iron should be reduced while the EDTA level is kept the same.

Exposure of plant tissue culture media to light from fluorescent bulbs changed the growth regulating properties of the media. The light caused nutrient medium-dependent photosensitized degradation of the phytohormone indole-3-acetic acid and other media components. Formaldehyde is formed from EDTA and the medium becomes deficient in iron as it becomes unchelated. The use of appropriately filtered light (eliminate wavelengths lower than 450 nm) when culturing plant material can eliminate unnecessary variability by stabilizing the culture media composition [7, 8, 17].

3. Physicochemical characterisitics of a solidified medium

The physicochemical characteristics of a gellified medium are not constant, and may affect the diffusion of ions through the gel. We used an Instron-penetrometer to follow the hardness and the force to shear a gel (Fig. 1). There is a linear increase in hardness during the first 9 h after setting of the gel, followed by a transition phase of more or less 27 h, and an equilibrium phase (during which the gel strength can increase depending on the water loss) (Fig. 2). The force needed to shear a gel followed an analogous course, but the compressibility was not influenced by the gel age (Fig.2).

The ingredients of a culture medium can have a dramatic effect on the physicochemical characteristics. This is illustrated with a few examples.
1. In media with fructose, the Maillard reaction will take place and this results in a lowered pH, with a lowered hardness as a consequence [9].
2. For equimolar concentrations of sugars, the hardness decreased in the following order: sucrose = 1/2 (fructose + glucose) > fructose > glucose. The hardness also decreased with increasing concentrations of sugars. The compressibility was not dependent on the type of sugar, but increased with increasing sugar concentrations.
3. The water quality (distilled vs. demineralized) can be responsible for differences of up to 30% in hardness and force to shear the gel. Differences in compressibility were not significant.
4. The absence of macro elements halved the hardness and the force to shear the

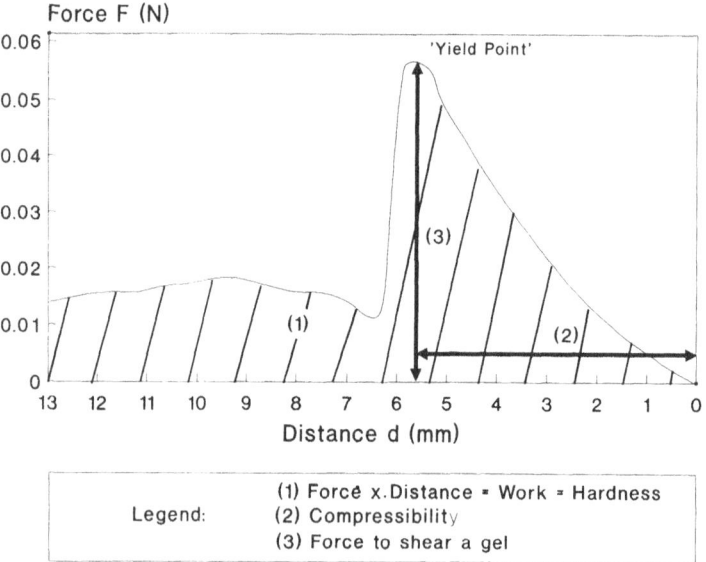

Fig. 1. Illustration of different physical characteristics of an agarized culture medium obtained with an Intron penetrometer.

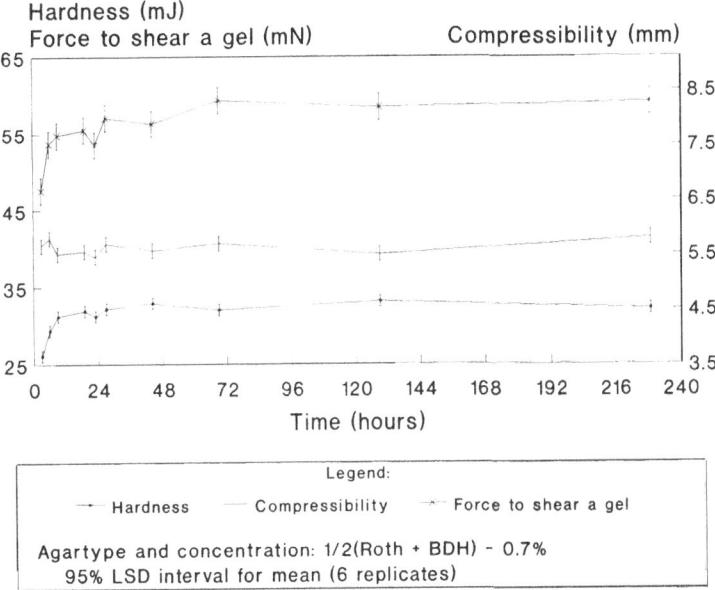

Fig. 2. Changes in the hardness, the force to shear a gel and the compressibility of an agarized medium as a function of time.

gel. On the other hand the compressibility increased with increasing concentrations of macro elements.

Not only the ingredients, but also the way a nutrient medium is prepared can have rather important consequences on the physicochemical properties of the nutrient gel.

1. Stirring the medium caused a significant decrease in hardness when agar was added at 50 °C. However, when the agar was added at boiling temperature there was an increase of the hardness when stirred, and a decrease when not stirred, compared to addition at 50 °C.

2. Lengthening the boiling period (4 min instead of 2) caused a significant increase in hardness of the medium, whether mineral elements were present or not. This effect cannot be attributed to evaporation of water, as precautions were taken to minimize these losses.

These examples clearly demonstrate how abiotic factors determine the physical and chemical characteristics of a medium, and may therefore interfere in the nutritional status of the system.

4. Delivery systems

Nutrients can be provided in cultures as a nutrient gel, a solution, a mist or a combination of those. Most often liquid media lead to higher proliferation rates, but this is quite often accompanied by the occurrence of hyperhydricity

[2], a major physiological disorder occurring in tissue cultures. Proliferation on solid media is usually much lower but the intrinsic quality is better and it is easier to control the process of bud formation (lower frequency of adventitious buds for identical hormonal conditions).

There is a growing interest in liquid culture systems but so far their use remains limited to specific categories of plants, such as geophytes [20], or to plants which are able to regenerate from homogenates [10] or from nodules [14]. However, for many plants, belonging to very different groups (woody, herbaceous, bulbs, corms, conifers) interesting results have been obtained with a nutrient delivery system, called a *double layer system*.

4.1 Double layer system

The double layer technique consists of applying a liquid medium on top of cultures established on an agar medium. The liquid medium can be applied either at the beginning of the culture or after the cultures have developed on agar medium for a few weeks. Adding liquid medium can be used to favour propagation [15, 18] or to induce elongation and rooting [11, 12].

Since these original publications the system has been used by different authors, for different plants and in most cases it has proven to be an efficient system, for different reasons: the propagation ratio is increased, the labour input is decreased, or a more homogeneous reaction is obtained, and vigour of the cultures is improved. The system also opens new possibilities for automation [1] by maintenance of cutting hedges *in vitro*. Depending on the plant species and the stage, hyperhydricity can be a problem, but this can be overcome by controlling the water retention capacity of the container.

However, the reasons for the efficiency of the double layer system are not well understood, and different hypotheses can be formulated.

1. The availability of nutrients. The water potential of normal gelled MS-medium is approximately -0.7 MPa, which is much lower than in a soil under normal conditions. Therefore the mobility of water, and accordingly of the nutrients in solution, is reduced. By adding a liquid layer on top of the gelled medium, the availability of the nutrients can be improved, especially in those cases where the liquid medium is applied directly after inoculating the explants.

   The situation is somewhat different when the liquid medium is provided at the end of a normal stage II period. Indeed, by that time the major part of the mineral elements has been taken up by the plants [12]. Nevertheless, the effect of the double layer is more than of providing nutrients, as is clearly shown by the data of Table 1. Adding the double layer is more effective than transplanting to a fresh medium of the same composition. Moreover, in some cases significantly improved results can be obtained by just adding water as liquid medium in stage III [13].

2. A liquid layer on top of a gelled medium can modify the gas exchange, e.g. by avoiding the formation of an (impermeable) film on top of the medium,

or by creating an anaerobic zone at the base of the shoots planted in the agar medium.

3. In solid media toxic metabolites can accumulate at the base of the explants; with a liquid layer on top these products can either be diluted or they diffuse in the medium, which can avoid toxic concentrations.

Table 1. Fresh weight (FW), shoot characteristics and regrowth of *Cordyline terminalis* shoots 4 weeks after the start of the stage III-treatment. (1) subculture on agar stage III-medium, (2) adding 20 ml of stage III-medium on top of the final stage II-culture. Non-rooted shoots were transplanted to the greenhouse (Maene, 1985)

Treatment	Average fresh FW per container (g)	Average number of shoots ≥ 2,5 cm	Average weight of shoot at transplantation time	Average weight of the plantlets 4 weeks after transplanting (g)
(1)	23,2±2,3	22,4±5,5	0,139±0,013	0,174
(2)	28,0±1,6	51,4±6,9	0,191±0,022	0,269

4.1.1 Carbon metabolism and translocation in a double layer system

With *Rosa multiflora* 'Montsé' as a test plant we followed the evolution of different sugars during stage II and during stage III, applied as a double layer system (both media contain 3% sucrose, and for stage III also charcoal 0.3%). Rather different curves were obtained (Fig.3). During stage II there was an almost linear decrease of sucrose (84.3 mg/jar/day), while the concentrations of glucose and fructose increased (resp. 23.6 and 26.2 mg/jar/day). From these values it can be calculated that there was an overall uptake of sugars of 34.5 mg/jar/day.

When a liquid stage III medium was applied, the evolution was completely different. Within 1 week most of the sucrose had disappeared from the medium, while the glucose and fructose concentrations increased. Over the rest of the culture period the glucose and fructose concentration gradually decreased.

We tested the invertase activity in the medium, at different stages during the culture period and no activity could be detected. Therefore one must consider the activity of cell wall bound invertases or other pathways with combined sucrose uptake and subsequent release of glucose and fructose.

The addition of liquid stage III medium as a double layer after 4 weeks resulted in an acceleration of the dry matter increase (Fig. 4). In this experiment radioactive sugar was added to the culture medium. The decrease in radioactivity of the culture medium was used to calculate the sugar uptake; this is presented in Figure 5, as well as the amount of sugar remaining in the medium.

To investigate the pathway of carbon metabolism, experiments were run with pulse treatments (40 h), using selectively C^{14-}labelled glucose at C1 or C6. Afterwards the plant material was extracted with ethanol and the distribution of the label was followed in both the ethanol soluble and insoluble fractions. From

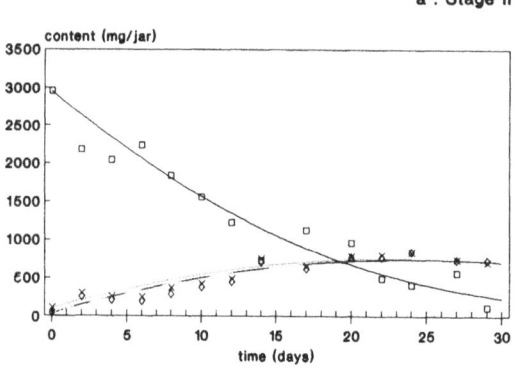

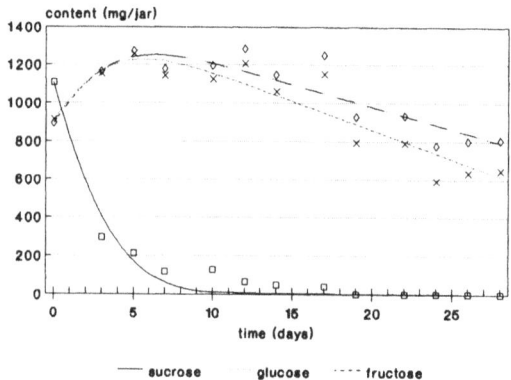

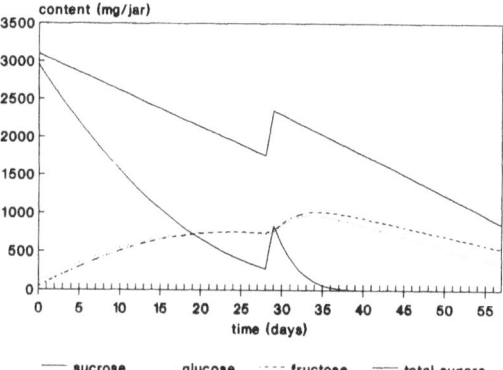

Fig. 3. Rosa multiflora 'Montsé' cultures at stage II for 4 weeks, after which a liquid stage III medium was delivered on top of the established stage II culture.

a) Changes in concentration of different sugars during stage II, b) Changes in sugar concentration after addition of stage III medium, c) combined curves from a) and b).

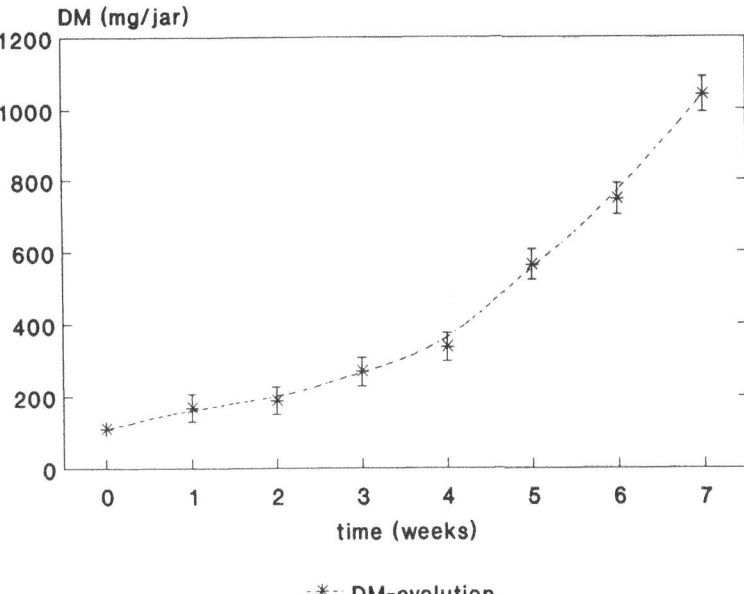

Fig. 4. Rosa multiflora 'Montsé' cultures at stage II for 4 weeks, after which a liquid stage III medium was delivered on top of the established stage II culture. Changes in dry matter (DM) of shoots.

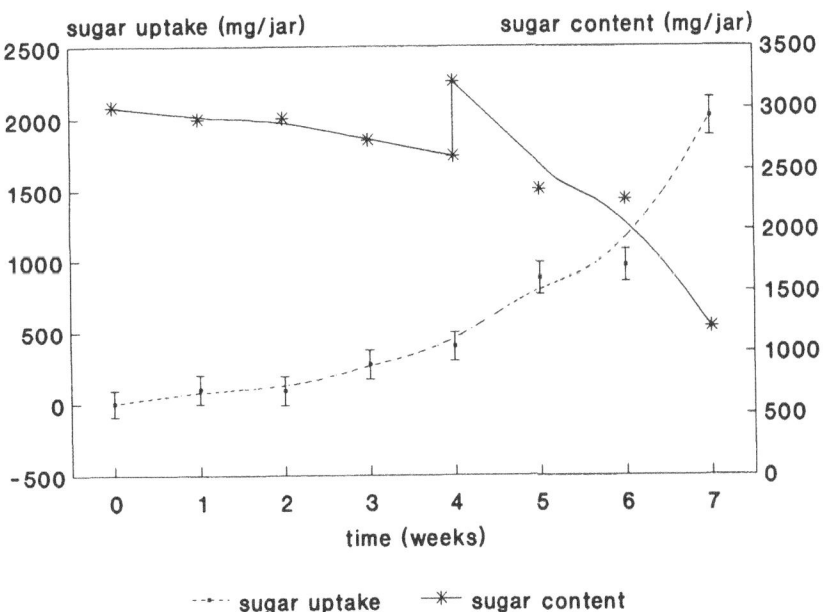

Fig. 5. Rosa multiflora 'Montsé' cultures at stage II for 4 weeks, after which a liquid stage III medium was delivered on top of the established stage II culture. Sugar uptake and decrease of the sugar content of the medium over the total culture period.

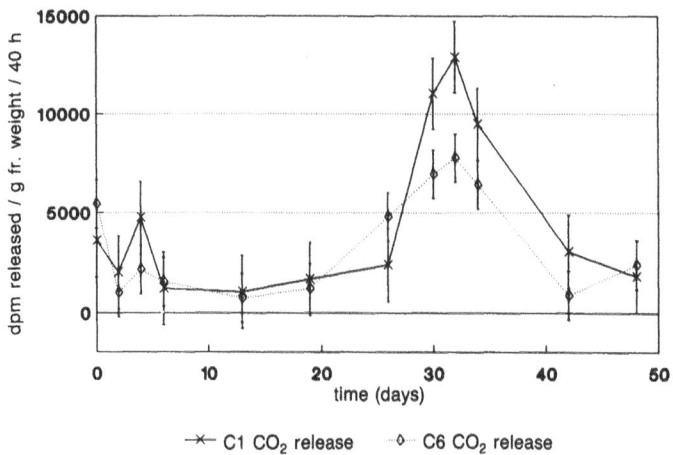

Fig. 6. Rosa multiflora 'Montsé' cultures at stage II for 4 weeks, after which a liquid stage III medium was delivered on top of the established stage II culture.
Carbon metabolism after pulse treatment (40 h) with selectively C^{14}-labelled glucose, at C1 and C6. Afterwards the plant material was extracted with ethanol and the distribution of the label was followed in both the ethanol soluble and insoluble fraction. From the ethanol soluble fraction the relative parts of the label detected was sucrose, glucose-fructose and other compounds were determined by TLC.

the ethanol soluble fraction the relative parts of label detected as sucrose, glucose-fructose and other compounds were determined by TLC (thin layer chromatography). Data are presented in Figures 6, 7, 8.

From Figure 6 it is clear that after adding the stage III medium (day 28), there was an increase in respiration, and the release of $C1-CO_2$ was significantly higher than the release of $C6-CO_2$. This is an indication that more carbon is metabolized through the pentose phosphate shunt (providing intermediates for new synthetic reactions) than through the Krebs cycle.

From Figures 7 and 8 it is evident that, after 40h a considerable part of the glucose had already accumulated in the ethanol insoluble fraction (cell wall components and all proteins).

During stage III an increasing part of the label was recovered in the ethanol soluble fraction, both for C1 and C6 label. The major part was recovered from the sucrose fraction. As the 40 h incubation took place on medium without sucrose, sucrose must have been synthesized from the glucose which had been taken up, suggesting that translocation of sugar becomes more important as shoots elongate.

These metabolic studies clearly support the idea that extra sucrose should be supplied when a liquid stage III medium is provided.

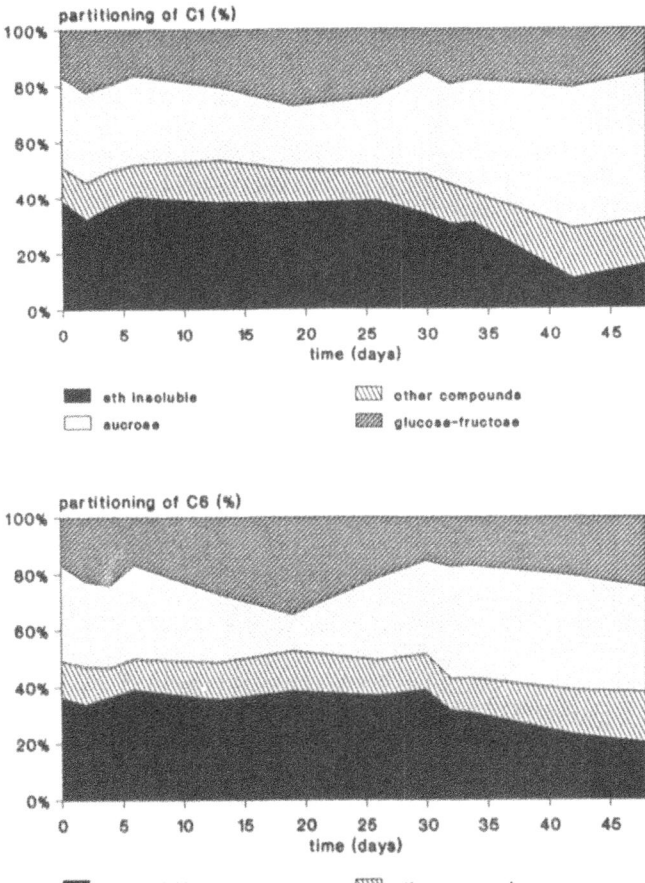

Fig. 7. Rosa multiflora 'Montsé' cultures at stage II for 4 weeks, after which a liquid stage III medium was delivered on top of the established stage II culture.
Carbon metabolism after pulse treatment (40 h) with selectively C^{14}-labelled glucose, at C1 and C6. Afterwards the plant material was extracted with ethanol and the distribution of the label was followed in both the ethanol soluble and insoluble fraction. From the ethanol soluble fraction the relative parts of the label detected was sucrose, glucose-fructose and other compounds were determined by TLC.

5. Conclusions

It is impossible to discuss the nutrient supply of tissue culture systems, without considering the chemical, physical, physicochemical and biological factors involved. Actually, most of these factors are not sufficient to allow interpretation of the results. Therefore it remains very difficult to compare results from the same laboratory and of course between different ones.

References

1. Aitken-Christie J and Jones C (1987) Towards automation: Radiata pine shoot hedges *in vitro*. Plant Cell Tiss Org Cult 8:185–196
2. Debergh P, Aitken-Christie J, Cohen D, Grout B, von Arnold S, Zimmerman R and Ziv M (1992) Reconsideration of the term 'vitrification' as used in micropropagation. Plant Cell Tiss Org Cult 30:135–140
3. Evans DA, Sharp WR and Flick CE (1981) Growth and behavior of cell cultures: embryogenesis and organogenesis. In: Thorpe TA, ed. Plant tissue culture: methods and applications in agriculture, pp 45-114. New York: Academic press
4. Gamborg OL (1991) Media preparation. In: K Lindsey, ed. 'Plant tissue culture manual', A1:1–24, Kluwer Academic Publishers
5. Gamborg OL, Murashige T, Thorpe TA and Vasil IK (1976) Plant tissue culture media. In Vitro 12:473–478
6. George EF and Sherrington PD (1984) Plant propagation by tissue culture. Exegetics Ltd Basingstoke pp 709
7. Hangarter RP and Stasinopoulos TC (1991a) Repression of plant tissue culture growth by light is caused by photochemical change in the culture medium. Plant Sci 79:253–257
8. Hangarter RP and Stasinopoulos TC (1991b) Effect of Fe-catalyzed photooxidation of EDTA on root growth in plant culture media. Plant Physiol 96:843– 847
9. Hsiao KC and Bornman CH (1991) Further studies on autoclaved-induced toxicity in tissue culture media: gauging sugar breakdown by spectrophotometry. Physiol Plant 82:261–265
10. Janssens J and Sepelie M (1989) *In vitro* multiplication of *Blechnum* and *Peleae rotundifolia* (Forst)Hookl by homogenisation. Sci Hort 38:161–164
11. Maene LJ (1985) Optimalisering van de overgang van weefselteelt plantjes naar *in vivo* omstandigheden. PhD-thesis, State University Gent, pp 221
12. Maene LJ and Debergh PC (1985) Liquid medium additions to established tissue cultures to improve elongation and rooting in vivo. Plant Cell Tiss Org Cult 5:23–33
13. Maene LJ and Debergh PC (1986) Problems related to *in vivo* rooting of *in vitro* propagated shoots. In: Schäfer-Menuhr A (Ed) Advances in agricultural biotechnology – *In vitro* techniques, Propagation and Long term storage pp, 59–72. Martinus Nijhoff/Dr. W Junk Publishers
14. McCown BH, Zeldin EL, Pinkalla HA and Dedolph RR (1987) Nodule culture: a developmental pathway with high potential for regeneration, automated propagation and plant metabolite production from woody plants. In: Hanover JW and Keathley DE (Eds) Genetic manipulation of woody plants, pp. 146–166. New York: Plenum Press
15. Molnar GY (1987) A new patented method for mass propagation of shoot cultures. Acta Hort 212:125–130
16. Murashige T and Skoog F (1962) A revised medium for rapid growth and bioassays with tobacco tissue cultures. Physiol Plant 15:473–497
17. Stasinopoulos TC and Hangarter RP (1990) Preventing photochemistry in culture media by long-pass light filters alters growth of cultured tissues. Plant Physiol 93:1365–1369
18. Viseur J (1987) Micropropagation of pear, *Pyrus communis* L., in a double-phase culture medium. Acta Hort 212:117–124
19. Williams RR (1992) Towards a model of mineral nutrition *in vitro*. In: Kurata K, Kozai T (Eds) Transplant Production Systems, pp 213–229. Dordrecht: Kluwer Acad Publ
20. Ziv M (1991) Morphogenetic patterns of plants micropropagated in liquid medium in shaken flasks or large scale bioreactor cultures. Israel J Bot 40:145–153

The effect of plant density and macronutrient nutrition on *Iris* shoot cultures

S. PRYCE[1,2], P.J. LUMSDEN[1], F. BERGER[3], J.R. NICHOLAS[2] and C. LEIFERT[4]*

[1] *Department of Applied Biology, University of Central Lancashire, Preston PR1 2HE, UK*
[2] *Neo Plants Ltd., Freckleton, Lancashire PR4 1HU, UK*
[3] *Institute für Pflanzenbau und Tierhygiene in den Tropen und Subtropen, Universität Göttingen, 3400 Göttingen, Germany*
[4] *Department of Plant & Soil Science, University of Aberdeen, Aberdeen AB9 2UE, UK*
* *Corresponding author*

1. Introduction

Micropropagation of *Iris* is usually carried out to produce plants free of bacterial soft rot (caused by *Erwinia carotovora*) or for the initial multiplication of new varieties [1]. However, due to the higher cost of micropropagation, large scale commercial propagation is still mainly carried out by traditional methods.

We have therefore investigated the effect of macro-nutrient nutrition on *Iris* growth rates *in vitro* in an attempt to increase the productivity of micro-propagation.

2. Materials and methods

Iris germanica L. '5 Star Admiral' were initiated as described previously [2] and grown on modified Murashige and Skoog's medium [5] containing 3.6 μM BA and 0.54 μM NAA. Plants used in experiments had been grown *in vitro* for longer than 1 year and were free of detectable microbial contaminants [4]. During the experiments plants were sterility tested at every subculture and cultures which became contaminated were discarded since contaminants were shown to affect growth rates of plants [3]. Unless stated otherwise, 10 shoots were grown in irradiated plastic (clear polystyrene K-resin) containers (6.5 cm high, 8 cm diameter) and subcultured onto fresh media at 4-week intervals.

The multiplication rate was determined by assessing the number of shoots with a similar size per tub after 4 weeks in the growth room. The growth efficiency at different plant densities was calculated as: ((No. of plants/culture vessel) x multiplication rate) − No. of plants/culture vessel = the number of new shoots formed per culture vessel. The fresh weight gain was determined by weighing the culture vessels before and after the 4 week growth period with and without plants and dividing the final weight of plants by the weight of plants initially planted on the medium. The nutrient analysis of media was carried out as described earlier [6].

To determine the effect of increased nitrogen, phosphate or potassium

69

P.J. Lumsden, J.R. Nicholas and W.J. Davies (eds.), Physiology, Growth and Development of Plants in Culture, 69–71, 1994.

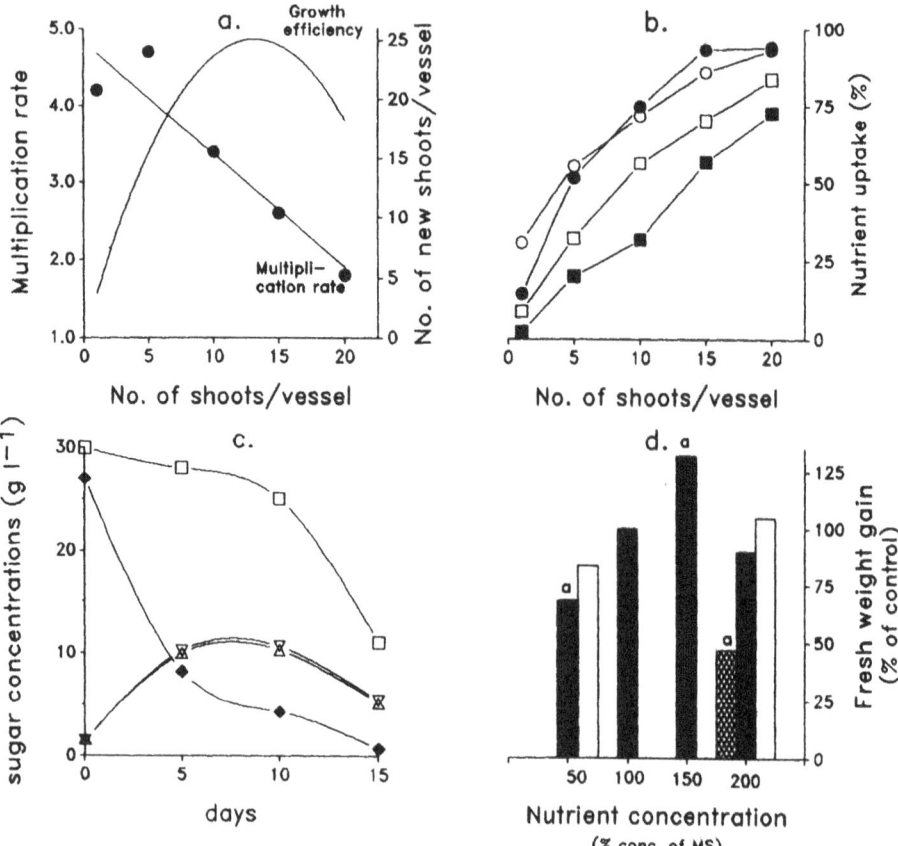

Fig. 1. Effect of plant density on the multiplication rate (a), nutrient uptake from the medium (b), sugar concentration in the medium (c), and effect of potassium, phosphate and nitrogen concentration in MS medium on fresh weight gain (d) of *Iris* shoot cultures. Nutrient uptake from the medium (b): ■, potassium; ○, phosphate; ●, nitrate; □, total sugar (mean is of 10 determinations). Sugar concentration in the medium (c): □, total sugar; ◆, sucrose; △, glucose; ▽, fructose. Fresh weight gain at different nutrient concentrations (d) : ■, nitrogen; □, phosphate; ▨, potassium (the letter a above the bars indicates that results were significantly different to the control p=0.05)

concentrations on the growth of *Iris*, 50 shoots were assessed per treatment for 4 (for phosphate trial) or 2 (for nitrogen trial) subsequent subcultures. Results from different subculture dates were pooled and tested by analysis of variance using the epistat software.

3. Results and discussion

Like many other plant species [1] *Iris* showed decreased multiplication rates with increased plant density (Fig. 1a). When the growth efficiency was calculated from the first order regression curve it was found to be maximal at 13 shoots/culture vessel. This, and the finding that *Iris* growing at higher plant densities took up nearly all the phosphate, nitrate and sugar from the medium (Fig. 1b & c) suggested that plant growth at higher plant densities is limited by the macronutrient supply in MS medium.

However, when *Iris* were grown on MS medium (10 shoots per culture vessel) with nitrogen increased by 50% there was only a small increase in fesh weight (Fig. 1d). Doubling the phosphate in the same conditions resulted in a culture fresh weight gain similar to control plants. Increasing the potassium concentration resulted in a significant reduction in the fresh weight gain (Fig. 1d). These results indicate that normal phosphate and potassium concentrations were sufficient, and that nitrate was the first nutrient to become limiting.

Acknowledgements

We would like to thank NEO PLANTS Ltd. for financing this study and for their technical assistance.

References

1. Leifert C, Pryce S, Lumsden PJ and Murphy KP (1991) Nutrition of *in vitro* cultured plants. In: KH Goulding, ed. Horticultural Exploitation of Recent Biological Developments, pp. 43–57. Preston: Lancashire Polytechnic Publication Service
2. Leifert C, Pryce S, Lumsden PJ and Waites WM (1992) Effect of medium pH on different plant species growing *in vitro*. Plant Cell Tiss Org Cult 30:171–179.
3. Leifert C, Ritchie J and Waites WM (1991) Contaminants of plant tissue and cell cultures. World J Microbiol Biotechnol 7:452–469
4. Leifert C, Waites WM, Camotta H and Nicholas JR (1989) *Lactobacillus plantarum*; a deleterious contaminant of plant tissue cultures. L Appl Bacteriol 67:363–370
5. Murashige T and Skoog F (1962) A revised medium for rapid growth and bio-assays with tobacco tissue cultures. Physiol Plant 15:473–497
6. Pryce S, Lumsden PJ, Berger F and Leifert C (1993) Effect of plant density and macronutrient nutrition on *Delphinium* shoot cultures. J Hort Sci 68:807–813

Growth and mineral nutrition in micropropagated delphinium during a subculture period

KENNETH P. MURPHY[1,2], JULIAN R. NICHOLAS[1],
CARLO LEIFERT[3] and PETER J. LUMSDEN[2]

[1] Neo Plants Ltd., 197 Kirkham Road, Freckleton, Lancashire, PR4 1HU, UK
[2] Department of Applied Biology, University of Central Lancashire, Corporation Street, Preston, PR1 2HE, UK
[3] Department of Plant & Soil Science, University of Aberdeen, Aberdeen AB9 2UE, UK

1. Introduction

Previous work with delphinium in culture has indicated that there is significant depletion of minerals in the growing media during the course of a 28 day subculture period [2]. This raises the question of whether the decline in mineral nutrients affects the growth of the plants. If at some point in the subculture period the mineral nutrient availability to the plants becomes growth limiting then it may be possible to adjust the quantity or combination of minerals being supplied in the media to alleviate the problem and thus improve the growth rate or quality of the plants, or even extend the subculture period.

Delphinium grown *in vitro* suffer a decline in vigour when subjected to long subculture periods (in excess of 35 days). This is often manifested as yellowing of the leaves leading to necrosis and reduced growth. Leaf chlorosis often portends mineral deficiencies especially of elements such as iron, manganese, magnesium and zinc. To determine whether the symptoms seen in culture are due to a mineral deficiency the first stage is to determine what concentrations of the elements are present in the tissues and to see how the growth of the plants affects these concentrations. Thus to examine whether changes in the concentrations of mineral elements were correlated with growth or quality of the plants the concentrations of a number of essential mineral nutrient elements were determined concurrently with measurements of growth rate over a six week period.

2. Materials and methods

Four hundred and fifty genotypically identical delphinium (variety Princess Caroline) were subcultured into thirty culture vessels (15 plants per vessel) each containing 50g of Murashige and Skoog media [4]. At weekly intervals for six weeks five vessels were selected at random and harvested. The fresh weight gain

P.J. Lumsden, J.R. Nicholas and W.J. Davies (eds.), Physiology, Growth and Development of Plants in Culture, 72–76, 1994.

of the plants from each vessel and the weight loss of the media were measured on a per vessel basis. The plants were then dried and digested using nitric and perchloric acids [3] and the digest analyzed for calcium, iron, magnesium, manganese, sodium and zinc using an atomic absorption spectrophotometer (Instrument Laboratory 151). Protocols for obtaining optimum response and minimizing interference problems with the atomic absorption measurements, had been developed previously for each element. An air/acetylene flame was used for all determinations.

3. Results and discussion

3.1 Growth

Plant weight generally increased each week during the six week subculture period (Fig. 1), though relative growth rate [1] decreased. The weight gain per week levelled off at the 4 week point both for dry and fresh weight. This appears to justify the common use of a 4 week subculture period.

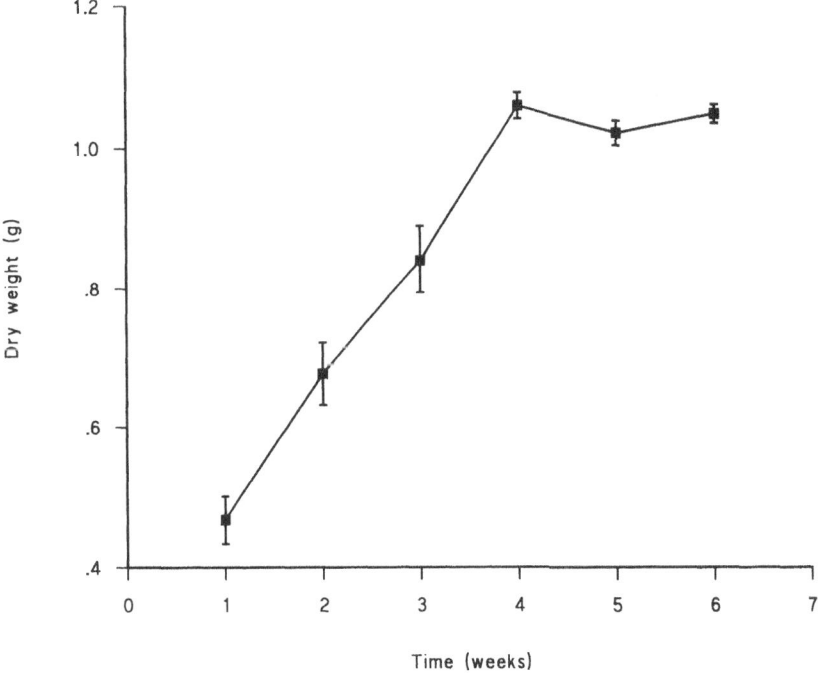

Fig. 1. Dry weight gain of delphinium *in vitro* per week for six weeks following subculturing. 95% confidence limits are shown. N = 5 culture vessels with 15 plants per vessel.

3.2 Changes in concentration of elements

The concentrations of magnesium, manganese and zinc in the plants changed in a consistent pattern. The concentration declined for 3 or 4 weeks before increasing again towards the end of the six week period (Fig. 2). The tissue concentration of iron also declined but did not increase after the initial drop and was significantly lower after six weeks than at the start of the subculture period (Fig. 2). The concentration of sodium, which is unlikely to be an essential mineral for delphinium, was especially high at about 200 fold that regarded as adequate. A high concentration of sodium in the agar or the use of sodium hydroxide for setting media pH may be the cause of this high sodium concentration. Calcium was the only element which was found to increase in concentration in the plant tissue during the subculture period (data not shown). The macronutrients nitrogen, phosphorous and potassium were not measured in this study but have been reported to decline in concentration in culture media [6]. Phosphate concentration has been shown to be depleted by a large percentage, up to 70% in some reports (Postma pers. comm.).

Magnesium was the only element found to be at a concentration which may be considered deficient [5]. A possible reason for this is competition for uptake from other cations such as calcium, manganese and especially potassium.

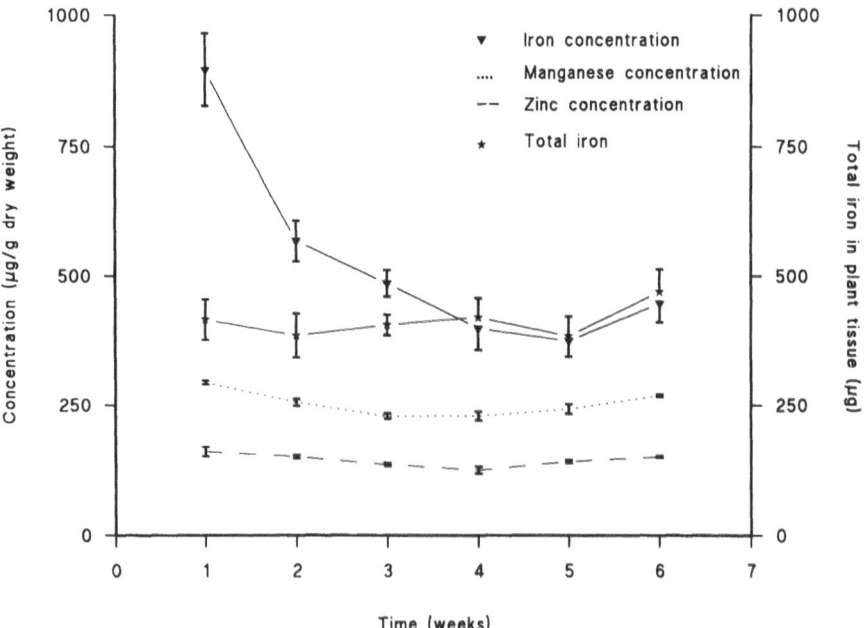

Fig. 2. Concentrations of manganese, iron and zinc and the total quantity of iron present in delphinium plants for six weeks following subculturing. 95% confidence limits are shown. N = 5 culture vessels with 15 plants per vessel.

3.3 Growth rate and uptake of minerals

The initial decline in tissue concentration of some minerals, followed by an increase later in the culture period may be due to the initial high growth rate of the plants. This would have the effect of 'diluting' the mineral content of the plants if the growth was faster than the mineral uptake. For calcium however the uptake must be faster than the growth rate since the concentration increased. This implies that water uptake is occurring because calcium uptake is not an active process and depends mainly on water uptake. This would be despite the high humidity in the culture vessels which is thought to severely reduce transpiration in *in vitro* plants [7]. One explanation is that as the culture ages the media dries out by evaporation to the headspace and subsequently out of the vessel; this would have the effect of concentrating the media relative to the plant leading to increased uptake by diffusion. Alternatively transpiration may still be occurring (due to the slight difference in temperature between the plant material and the gas in the vessel) at a sufficient rate to carry calcium into the tissues.

The concentration of iron declined continuously and the total mass of iron within the plants remained almost constant (Fig. 2) regardless of substantial growth, showing that there was no net uptake of iron. Despite this apparent lack of iron uptake the iron concentration in the plant tissue is in a range that is deemed adequate for many species [5]. However the decline in the relative growth rate was strongly and significantly correlated with the decline in the tissue concentrations of iron and zinc. Though this does not prove that iron and zinc are directly responsible for the decline in growth it does indicate that they may be important to the vigour of delphinium in culture.

The low concentration of magnesium and the decline in concentration of iron in the plants suggest that mineral nutrients may become growth limiting for delphinium not perhaps during a single subculture period but possibly over the course of a series of subcultures. Although delphinium grow on Murashige and Skoog medium this is not necessarily the optimum medium for each subculture. Williams [8] has pointed out that plants *in vitro* may show better growth or quality if the basic mineral nutrient constituents of the media they are cultured on are varied rather than always kept constant. The reason for this is that varying the constituent mineral salts in culture media from subculture to subculture may neutralize any long term deterioration induced by continuous growth on a specific media recipe, for example by increasing the iron or magnesium uptake intermittently in delphinium cultures. The implications of this are that manipulating the mineral constituents of media may improve delphinium growth in culture.

Acknowledgements

The authors wish to thank Myerscough College for the provision of analytical facilities and Neo Plants Ltd. for technical and financial support.

References

1. Hunt R (1981) Plant Growth Analysis. London: Edward Arnold
2. Lumsden PJ, Pryce S and Leifert C (1990) Effect of mineral nutrition on growth and multiplication of *in vitro* cultured plants. In: HJJ Nijkamp, LHW van der Plas and J van Aartrijk, eds. Progress in Plant Cellular and Molecular Biology, pp 108–114. Dordrecht: Kluwer Academic Publishers
3. Ministry of Agriculture Fisheries and Food (1986) The Analysis of Agricultural Materials, 3rd edition. London: HMSO
4. Murashige T and Skoog F (1962) A revised medium for rapid growth and bio-assays with tobacco tissue cultures. Physiol Plant 15:473–497
5. Pearcy RW, Ehleringer JR, Mooney HA and Rundel PW (1990) Plant physiological ecology: field methods and instrumentation. London: Chapman and Hall
6. Pryce S, Lumsden PJ, Berger S and Leifert C (1993) Effect of plant density and macronutrient nutrition on *Delphinium* shoot cultures. J Hort Sci 68:807–813
7. Ritchie GA, Short KC and Davey MR (1991) *In vitro* acclimatization of Chrysanthemum and sugar beet plantlets by treatment with paclobutrazol and exposure to reduced humidity. J Ex Bot 42:1557–1563
8. Williams RR (1992) Towards a model of mineral nutrition *in vitro*. In: K Kurata and T Kozai, eds. Transplant Production Systems, pp 213–229. Dordrecht: Kluwer Academic Publishers

The effect of nitrogen sources and initial pH of the media with or without buffer on *in vitro* rooting of jackfruit (*Artocarpus heterophyllus* Lam.)

B.N. SATHYANARAYANA and JENNET BLAKE

Unit for Advanced Propagation Systems, Department of Horticulture, University of London (Wye College) Wye, Ashford, Kent TN25 5AH, UK

1. Introduction

The influence of nitrogen on growth and morphogenesis of tissue cultures has been well established. Availability of nitrogen and the form in which it is presented are of great importance. Nitrate has generally been considered to be the most essential form of nitrogen for the culture of tissues. However in many cases culturing tissues on nitrate alone as a source of nitrogen has not been very successful [9, 15, 16]. It has been found that a reduced nitrogen source such as ammonium is essential for the optimal use of nitrate ion by tissue cultures. A number of studies have shown that growth of tissue cultures is possible on a medium containing ammonium as the sole nitrogen source provided that a Kreb's cycle dicarboxylic acid is also present in the medium [7]. Tobacco cells were found capable of growing on ammonium alone, provided citrate, malate and pyruvate were added to the culture medium [3, 6]. Later Dougall and Verma [4] demonstrated that suspension cultures of wild-carrot could be grown with ammonium as a sole nitrogen source even in the absence of any exogenous Kreb's cycle acid. However a combination of nitrate and ammonium is still considered to be an ideal source of nitrogen for tissue cultures [10, 15]. George and Sherrington [8] suggest that as most tissue culture media have not been buffered, the concentration of ammonium and nitrate ions have probably been adopted more to obtain practical pH control than because of the requirement of plant tissues for one form of nitrogen or another. Uptake of ammonium and nitrate forms of nitrogen has been reported to be correlated with the pH of the medium. Uptake of nitrate requires an acid pH, but the medium drifts towards alkalinity [1]. In contrast, ammonium uptake shifts the medium pH towards acidity [4, 5, 7, 18] and consequently further uptake of ammonium is inhibited. Therefore it is evident that in an unbuffered medium efficient nitrogen uptake depends on the presence of both ions. Most of the studies on the effects of different nitrogen forms and related pH factors have been concentrated on cell culture and embryogenesis studies using unbuffered media. No definitive studies regarding these factors have been reported as far as *in vitro* rooting is concerned. Hence this study was

P.J. Lumsden, J.R. Nicholas and W.J. Davies (eds.), Physiology, Growth and Development of Plants in Culture, 77–82, 1994.

undertaken to investigate the effect of sources of nitrogen and associated pH changes on *in vitro* rooting.

2. Materials and methods

In vitro grown shoot cultures of jackfruit were grown on semi-solid medium containing half-strength MS medium [12] with 2% sucrose, 5 x 10^{-6} M of BAP (6-benzylaminopurine) and 0.6% Agar as used by Rahman and Blake (1988). Medium pH was adjusted to 5.7 before autoclaving. The stock of shoot cultures was maintained by subculturing at intervals of 6–7 weeks.

For the rooting experiment, 10-week-old microcuttings of 2.5–3.5cm length were selected. Basal leaves were removed leaving the two top leaves (or three in cases where the leaf sizes were relatively small). The microcuttings were grouped by size so that they could be distributed uniformly between the treatments. They were transferred to liquid medium (supported by Declon blue polyurethane mesh) containing half-strength MS macro- and micro-nutrients, organic compounds (except glycine), 2% sucrose and IBA at 10^{-6} M. Variations in the nitrogen supply were made by omitting NH_4^+ and/or NO_3^- and this necessitated manipulations of ammonium nitrate (NH_4NO_3) and potassium nitrate (KNO_3) in the MS medium. Compounds such as ammonium citrate (($NH_4)_3C_6H_5O_7$), potassium citrate ($K_3C_6H_5O_7$), ammonium chloride (NH_4Cl) potassium chloride (KCl) and sodium nitrate ($NaNO_3$) were used to balance the nutrient composition of media during manipulation of nitrogen source. MES (2-(N-morpholino)ethanesulfonic acid), an effective and non-toxic organic buffer for our system (as found from the results of a preliminary experiment) at a concentration of 4 g/litre was used. pH was adjusted as necessary with $1M$ NaOH or HCl before autoclaving. Plant cultures were incubated in growth rooms at 25 °C temperature with a 16 h photoperiod. Light was provided by fluorescent lamps (1.83m C84 TDL/Philips) giving an irradiance of 55–62 μmol m^{-2}s^{-1}. Analysis of Variance (maximum-likelihood) of the data was obtained by following categorical data procedure (CATMOD) of SAS Statistical Package (SAS Institute Inc., Cary, NC, USA).

3. Results

Nitrogen sources, initial pH of the medium and presence or absence of buffer showed significant effects on rooting of jackfruit microcuttings *in vitro* (Table 1). Nitrogen in the combined form of ammonium and nitrate gave a higher percentage of rooting than either ammonium or nitrate supplied as a sole source of nitrogen. However, when ammonium citrate was used to supply ammonium as a sole source of nitrogen, rooting was significantly higher compared with ammonium chloride and was comparable to rooting obtained with ammonium and nitrate combination. Rooting was significantly lower at an initial pH of 4.5

as compared to either 5.7 or 6.5 (Table 1). Rooting was higher at pH 6.5 than 5.7. Rooting was also significantly improved by the addition of buffer (Table 1).

Table 1. *In vitro* rooting of jackfruit microcuttings: effect of nitrogen sources and initial pH of the media with or without MES buffer; per cent rooting at 30 days. Figures are means of 9 replicates. ± values are SE of means

Nitrogen source	Buffer ±	Initial pH			Average effect of nitrogen source
		4.5	5.7	6.5	
+ Ammonium	–	44±16	88±12	100	
+ Nitrate	+	67±16	89±10	88±12	79±6
+ Nitrate	–	0	33±16	44±16	
	+	67±16	33±16	33±16	35±7
+ Ammonium (citrate)	–	38±11	89±10	89±10	
	+	29±17	89±10	89±10	70±6
+ Ammonium (chloride)	–	0	22±14	13±12	
	+	0	38±17	89±10	27±6
Average effect of buffer	– 47±5				
	+ 59±5				
Average effect of initial pH		31±6	60±5	68±3	

Effects of nitrogen source on rooting showed a significant interaction with pH and buffer (Table 1). Rooting with the combination of ammonium and nitrate, though similar at higher initial levels of pH of 5.7 and 6.5, was slower at an initial pH of 4.5 and a considerable number of microcuttings which failed to root initially developed callus at the base and eventually rooted.

As seen with the ammonium and nitrate combination, irrespective of buffering status of the media, rooting with ammonium alone supplied through ammonium citrate was also poor at an initial pH of 4.5 compared with the higher levels of initial pH (Table 1). However the percentage of rooting with initial pH levels of 5.7 and 6.5 was high and comparable to the rooting obtained with the combined source of nitrogen.

When ammonium was supplied as the sole source of nitrogen through ammonium chloride, rooting was inhibited especially at pH 4.5 whether or not the medium was buffered (Table 1). Irrespective of presence of the buffer at the initial pH of 5.7, rooting did not differ significantly and was at a low level. At the initial pH of 6.5, in presence of the buffer rooting was significantly higher.

Nitrate as a sole source of nitrogen was not beneficial for rooting. However, the results observed in this experiment have shown that nitrate could induce over 60% rooting at the low initial pH of 4.5 provided the medium was buffered

(Table 1). At higher initial pH levels rooting was significantly lower and did not differ significantly with the presence or absence of the buffer.

4. Discussion

Differential uptake of ammonium and nitrate ions are known to be influenced by pH and efficient utilization of nitrogen depends on the presence of both these ions. Apart from its role as a reduced source of nitrogen, the ammonium ion also effectively buffers plant nutrient media in the presence of nitrate and so enhances nitrate uptake [8]. In the present study when a combination of ammonium and nitrate as a source of nitrogen was used, rooting occurred, even in the absence of buffer irrespective of initial pH levels ranging from 4.5 to 6.5 although rooting was slower at pH 4.5.

Ammonium citrate as compared to ammonium chloride when supplied as the sole source of nitrogen promoted good rooting. Citrate has a buffering capacity and monitoring the changes in pH of the medium containing ammonium citrate (as a sole source of nitrogen) showed that the pH remained relatively stable even without MES buffer [14]. This effect was also demonstrated here in that the extra buffering effect of the added MES did not further increase rooting. Thus the good percentage rooting obtained at the initial pH of 6.5 with ammonium citrate may be largely attributed to the role of citrate in buffering the medium. The high percentage of rooting obtained at pH 5.7 may not be attributable solely to the buffering capacity of the citrate since low rooting was observed at pH 5.7 with ammonium chloride even when the media was buffered. A similar effect was observed at pH 4.5. Thus it appears that citrate apart from its role in buffering the medium may be playing some other role. Citrate has been shown to play a part in assimilating ammonium to amino acids [2]. Addition of citrate, pyruvate and malate have been shown to sustain the growth of tobacco cells on ammonium alone [3, 6]. In a recent study citric acid added as a supplement at the root initiation and elongation stage has been shown to increase the percentage rooting in shoots of apple, possibly by decreasing peroxidase activity [17].

Total inhibition of rooting with ammonium chloride at pH 4.5 and a low percentage at pH 5.7 irrespective of the presence of MES buffer, could be due to the reduced uptake of ammonium in acidic conditions. It could be seen from the results of the present experiment that rooting was significantly higher at a higher pH of 6.5 when the medium was buffered. Monitoring the changes in pH of the media containing ammonium chloride (as sole source of nitrogen) has shown that the pH drifted down rapidly to 5.2 by 10 days and 4.5 by 30 days whether initially adjusted to 6.5 or 5.7 [14]. This result could be explained from the reported findings [11] that ammonium utilization rate decreased with decreasing pH and increased with increasing pH. In contrast to ammonium, nitrate uptake is higher at an acid pH [1] which was confirmed by the relatively high percentage of rooting obtained with nitrate as the sole source of nitrogen at an initial pH of 4.5 when buffered.

The results have shown the importance of a medium containing both ammonium and nitrate sources of nitrogen for good rooting of jackfruit. It was possible to obtain good rooting with ammonium alone as the source of nitrogen at initial pH levels of 5.7 and 6.5 provided it was supplied through ammonium citrate rather than ammonium chloride. However when the medium was buffered with MES, a high percentage of rooting was obtained with ammonium chloride at the highest pH of 6.5. Nitrate in general as the sole source of nitrogen, was not beneficial to root growth.

Acknowledgements

BNS acknowledges the award of Academic Staff Scholarship by the Commonwealth Scholarship Commission in the United Kingdom and the University of Agricultural Sciences, Bangalore, India, for deputation to higher studies.

References

1. Bayely JM, King J and Gamborg OL (1972) The ability of amino compounds and conditioned medium to alleviate the reduced nitrogen requirements of soybean cells grown in suspension cultures. Planta 105:25–31
2. Beevers L and Hageman RH (1980) Nitrate and nitrate reduction. In: PK Stumpf and EE Conn, eds. The Biochemistry of plants. vol.5, Amino Acids and Derivatives, ed. BJ Miflin, pp 115–168. New York: Academic press
3. Behrend J and Mateles RI (1975) Nitrogen metabolism in plant cell suspension cultures. II. Role of organic acids during growth on ammonia. Plant Physiol 58:510–512
4. Dougall DK and Verma DC (1978) Growth and embryo formation in wild carrot suspension cultures with ammonium ion as a sole nitrogen source. In Vitro 14:180–182
5. Fowler MW, Watson R and Lyons I (1982) Substrate utilization, carbon and nitrogen, by suspension cultured plant cells pp. 225–8. In proceedings of 5th congress of Plant Tissue and Cell culture ed. Fujiwara A, The Japanese Association for Plant Tissue Culture, Tokyo
6. Gamborg OL (1970) The effects of amino acids and ammonium on the growth of plant cells in suspension culture. Plant Physiol 45:372–375
7. Gamborg OL and Shyluk JP (1970). The culture of plant cells with ammonium salts as the sole nitrogen source. Plant Physiol 45:598–600
8. George EF and Sherrington PD (1984) Plant Propagation by Tissue Culture. Hants, England: Exegetics Ltd
9. Halperin W and Wetherell DF (1965) Ammonium requirement for embryogenesis *in vitro*. Nature 295:519–520
10. Kirby EG, Leustek T and Lee MS (1987) Nitrogen nutrition. In: JM Bonga and DJ Durzan, eds. Cell and Tissue Culture in Forestry I. General Principles and Biotechnology, pp. 67–88. Dordrecht: Martinus Nijhoff
11. Martin SM and Rose D (1976) Growth of plant cells (Ipomoea) suspension cultures at controlled pH levels. Can J Bot 54:1264–1270
12. Murashige T and Skoog F (1962) A revised medium for rapid growth and bioassays with tobacco tissue cultures. Physiol Plant 15:473–97
13. Rahman MA and Blake J (1988) The effects of medium composition and culture conditions on *in vitro* rooting and *ex vitro* establishment of jackfruit (*Artocarpus heterophyllus* Lam.). Plant Cell Tiss Org Cult 13:189–200

14. Sathyanarayana BN and Blake J (Unpublished)
15. Selby C and Harvey BMR (1990) The influence of composition of the basal medium on the growth and morphogenesis of sitka spruce (*Picea sitchensis*) tissues. Ann Bot 65:395–407
16. Smith DL and Krikorian AD (1990) pH control of carrot somatic embryogenesis. In: HJJ Nijkamp, LHW Van Der Plas and J Van Aartrijk, eds. Progress in Plant Cellular and Molecular Biology. Proceedings of the VIIth International Congress on Plant Tissue and Cell culture, pp 449–453. Dordrecht: Kluwer Academic Publishers
17. Standardi A and Fausto R (1990) Effects of some antioxidants on *in vitro* rooting of apple shoots. Hort Sci 25:1435–1436
18. Wetherell DF and Dougall DK (1976) Sources of nitrogen on growth and embryogenesis in cultured wild carrot tissue. Physiol Plant 37:97–103

Carbon compounds and their influence on *in vitro* growth and organogenesis

MARGARETA WELANDER and NATHALIE PAWLICKI
*The Swedish University of Agricultural Sciences, Department of Horticulture, S-230 53 Alnarp,
Sweden*

1. Introduction

The aim of this article is to discuss *in vitro* growth and organogenesis in relation
to transport and metabolism of carbon compounds. Carbohydrates play a
prominent part in the nutrition and the structure of a plant. Photosynthesis
provides the carbon from CO_2 and the potential energy whereas respiration
makes the energy available. The synthesis of carbohydrates occurs mainly in the
leaves whereas the photosynthetic products are used in nongreen cells often
remote from the leaves. Many carbon compounds are synthesized in the leaves
but very few are present in the phloem sap. Although sucrose is the main sugar
in the translocation stream in most plants several other carbon compounds have
been identified [36, 18]. It is also known that the ability to metabolize different
types of sugars differ within the plant kingdom [28, 1, 17]. Translocation and
partition of assimilated carbon are regulated in response to both leaf and plant
ontogeny as well as environmental conditions [16, 30]. The composition of the
phloem sap seems to correspond to the substances being translocated [24, 13,
20]. However when analyzing the phloem sap not only the collection technique
is of importance but also the detection and quantification techniques. It has
been shown that *in vitro* cultures of various plant species differ in uptake and
utilization of different carbon sources which in its turn affect growth and
development [33]. This article deals with the following areas. 1) Transport of
carbohydrates; 2) Analysis of the phloem sap; 3) Assimlate allocation and
partitioning; 4) *In vitro* growth as influenced by the carbon source; 5)
Organogenesis as influenced by the carbon source.

2. Transport of carbohydrates

The list of sugars and sugar alcohols published by Zimmerman and Ziegler [36]
is taxonomically interesting as there is a striking grouping of sugars within given
families. For example in Malus and Prunus within the family *Rosaceae*, sorbitol
(D-glucitol) is the primary product of photosynthesis and the major trans-

*P.J. Lumsden, J.R. Nicholas and W.J. Davies (eds.), Physiology, Growth and Development of Plants
in Culture*, 83–93, 1994.

location substance [32]. In *Fraxinus americana* L. and *Syringa vulgaris* L. both belonging to the family *Oleaceae*, the sugar alcohol mannitol is the main carbon compound in the translocation stream together with verbascose, stachyose, and raffinose [28, 29]. Raffinose, stachyose and verbascose are nonreducing sugars containing sucrose bound to one, two or three galactose molecules. No hexoses are found in the sieve tube exudate and all transport sugars are nonreducing [14]. One explanation might be that reducing sugars, having an exposed aldehyde or ketone group are more reactive and thus unprotected during translocation.

3. Analysis of phloem sap

The collection technique of the phloem sap is important in order to avoid contamination from cells adjacent to the sieve tube. Several methods have been used to collect phloem exudate. These involve making incisions in the bark of trees, allowing cut stems, petioles or inflorescence stalk to exude or by using the aphid stylet technique. However most speices yield only limited quantity of phloem sap from the cut surface. King and Zeevart [15] showed that addition of EDTA, EGTA and citric acid into a solution bathing the cut surface could greatly increase exudation by reducing callose formation on the sieve plate. Not only the collection technique of the phloem sap is important but also the methods for detection and quantification. For example in some routine methods for the analysis of carbohydrates in plant extracts, the sugar alcohols (polyols) may be very easily overlooked. For example on paper chromatograms polyols have similar mobilities to their corresponding sugars in most of the commonly used solvents and they do not react with most of the used detection reagents. There is no single method which is universally applicable for quantitative and qualitative analysis of all monosaccharides. The method of choice will depend on accuracy required and resources available. There are two techniques available for the quantitative analysis of mixtures of monosaccharides, high performance liquid chromatography and gas-liquid chromatography. If only a qualitative analysis is needed, either paper or thin-layer chromatography may be used. Single monosaccharides may be identified by means of mass spectrometry, infra-red spectroscopy or proton or carbon-13 nuclear magnetic resonance, and quantified by use of specific colorimetic or enzymatic assays [4].

4. Assimilate allocation and partitioning

It is also well known that within the leaves, partition of assimilated carbon and export are regulated in response to both leaf and plant ontogeny and environmental conditions. A leaf is transformed from a heterotrophic importing organ to an autotrophic exporting organ. Autoradiographs show that the export of photosynthate begins at the tip and progress towards the base [30].

During transition the tip exports sugar while the base imports. Import stops at the same developmental stage in both albino and green leaves indicating that it is not only accumulation of sufficient photosynthate in the sieve element but also other changes must occur to drive translocation [31]. One hypothesis is that the unloading pathway is blocked in mature leaves thus accumulating enough sugar for export to be initiated. The control of partitioning must also involve regulation of enzyme activity so that transported substances are protected. In the source tissue and the translocation stream one might expect low levels of degrading enzymes whereas enzymes for synthesis would be in the sink tissue. Several attempts have been made to demonstrate such a relationship in species translocating mainly sucrose [21, 10, 11]. Generally the capacity to synthesize sucrose increases in the source leaf. However the results are difficult to interpret because at the same time there are changes in enzymes such as invertase and sucrose synthase normally associated with utilization [22]. Control may involve inhibition or sequestration in a cellular compartment away from sucrose.

To avoid problems in sucrose metabolism, sink-source interconversions have been studied in other species where sucrose is not the major translocated carbon compound. Loesher *et al.* [17] suggest that sorbitol metabolism in apple could be associated with mechanisms regulating partitioning of source and sink activity. In green leaves sorbitol (D-glucitol) is synthesized via a NADPH-dependant glucose(aldolase)-6-phosphate reductase (A6PR), whereas in non green tissue sorbitol is degraded via a NAD-dependant sorbitol dehydrogenase (SDH). In leaves undergoing the transition from sink to source, SDH activity reached a minimum as A6PR peaked. These changes were related to increases in leaf carbohydrate levels, especially sorbitol, and to increases in rates of photosynthesis. There are also differences in the ability of young and mature leaves to metabolize sorbitol and in the distribution of sorbitol enzymes in leaves at transitional developmental stages.

In those species using mannitol as a respiratory substrate it has been shown that metabolic breakdown of mannitol can occur via mannitol dehydrogenase. In celery a key enzyme, mannose-6-phosphate reductase (M6PR), used in mannitol synthesis, has been isolated and located within the cytoplasm [25]. The distribution of M6PR is a function of leaf age and plant part [9].

5. Carbon source and *in vitro* shoot growth

Mannitol and sucrose are found in the phloem sap in the Genus *Syringa* [36]. Welander *et al.* [33] showed that in *Syringa*, mannitol and sucrose at 88 mM were the best carbon compounds to promote growth *in vitro* (Fig. 1). Sorbitol, which is not found in either the leaves or the translocation stream was not able to sustain growth; in *Syringa* only one or two axillary shoots were formed per explant (Table 1). To obtain a satisfactory multiplication rate both shoot tips and nodal segments had to be used. For this reason it is important to obtain long shoots with nodes 'easy to separate'. Both mannitol and sucrose promoted

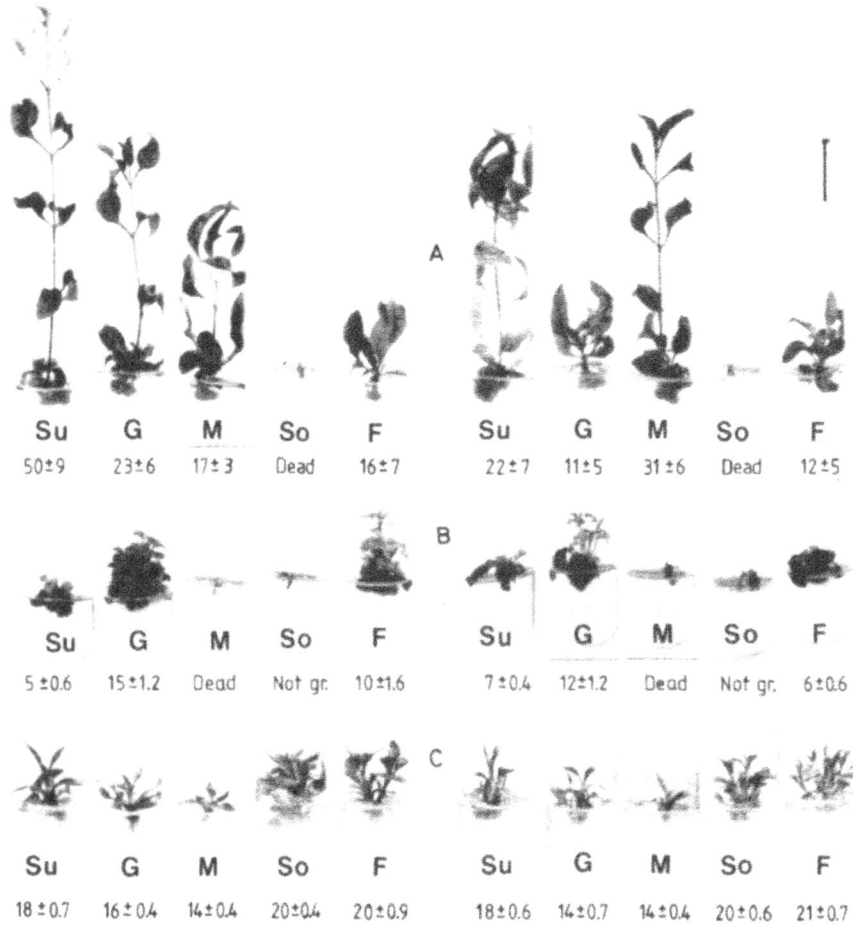

Fig. 1. Morphology of shoots derived from shoot tip explants of *Syringa* (A), *Alnus* (B) and M9 (C), cultured on different carbon sources. SU = sucrose, M = mannitol, So = sorbitol and F = fructose. The values below each illustration denote length of shoots ≥ 5mm ± SE. Scale bar = 10 mm. The concentration of the carbon sources is 88 mM in photos on the left half of the figure and 175 mM on the right (from 33).

shoot elongation. In *Malus*, sorbitol is the main translocation substance [17]. Sorbitol has been demonstrated to be an effective carbon source *in vitro* for apple tissue [5]. Welander *et al.* [33] showed that shoot multiplication in the recalcitrant apple rootstock M9 could be improved by adding sorbitol in the multiplication medium (Table 1 and Fig. 1). In *Alnus*, sucrose is the main sugar transported in the phloem. *Alnus* explants were not able to grow on either sorbitol or mannitol (Table 1 and Fig. 1). The inability of mannitol to sustain growth in *Alnus* has been shown earlier [27]. In our experiments glucose was superior to all other carbohydrates in promoting shoot multiplication. However Trembley and Lalonde [27] found that sucrose was optimum for *Alnus glutinosa* whereas all other *Alnus* species grew better on glucose. From these experiments it can be concluded that sugar compounds normally found in the sieve-tube exudate can be used as an indicator of a suitable *in vitro* carbon source. In tissue culture it is likely that different carbon sources present in the medium will be readily and directly transported by diffusion across the cut stem surfaces. Hence reducing sugars as glucose and fructose, not normally transported in the sieve-tubes can be taken up and metabolized. However sugars normally translocated in the phloem are probably more protected and can reach the site of metabolism without being degraded. Species deficient of a certain carbon compound in the leaves or in the translocation stream probably lack the enzymes capable to utilize it.

Table 1. Effects of five carbohydrates at two concentrations on the number of shoots produced per shoot tip from *Syringa vulgaris, Alnus glutinosa* and *Malus pumila* M9. ± SE is given for each mean value

			Carbohydrates				
Variate	Species	Conc mM	Sucrose	Glucose	Fructose	Mannitol	Sorbitol
No. of shoots	S. vulgaris	88	1.2 ±0.1	1.0±0	1.0±0	1.3±0.2	dead
		175	1.0±0	1.1±0.1	1.0±0	1.3±0.2	dead
No. of shoots	Alnus	88	1.8±0.1	4.8±0.4	3.5±0.3	dead	1.2±0.2
		175	1.3±0.1	3.4±0.4	1.0±1.0	dead	0.4±0.1
No. of shoots	M9	88	3.0±0.2	2.9±0.2	3.3±0.2	1.0±0	5.0±0.2
		175	2.0±0.2	3.3±0.2	4.2±0.2	1.0±0	5.9±0.3

6. Carbon source and organogenesis

6.1 Adventitious shoot formation

In experiments with adventitious shoot regeneration from leaf explants of the apple rootstocks Ottawa 3 and M9 it was shown that optimum shoot regeneration occured between 3% and 4% sucrose in the medium [34]. Higher concentrations caused red pigmentation which could be a sign of osmotic stress. Not only the carbohydrate concentration is important for organogenesis but also the carbon source. This has been demonstrated in our experiments with the apple rootstock M9 cultivar Jork in which leaf segments were exposed to different carbon sources (Fig. 2 and Fig. 4). Both the percentage of leaf segments forming adventitious shoots and the number of shoots per segment increased in the presence of sorbitol at 220 mM in comparison with the other carbon sources tested. Interestingly the positive response to sorbitol was enhanced when the

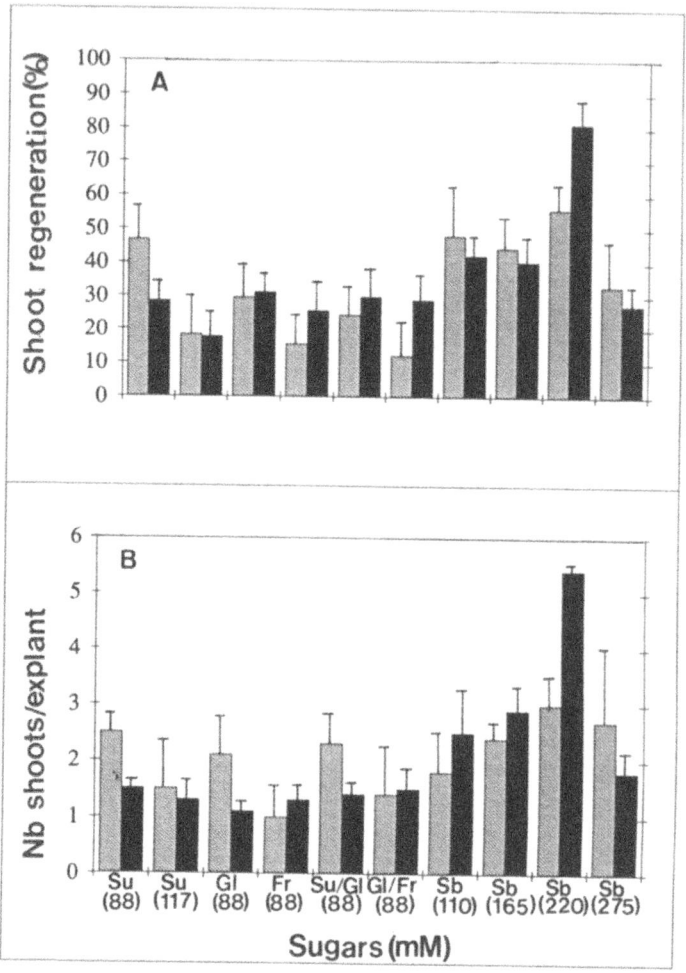

Fig. 2. Histogram showing the influence of different sugars at various levels and combinations on shoot regeneration of apple. A. Percent explants forming shoots from normal (□) and pretreated (■) mother shoots. B. Mean number of shoots per explant from normal (□) and preteated (■) mother shoots. Su= sucrose, Gl=glucose, Fr=fructose, Sb=sorbitol, Su/Gl and Gl/Fr 44/44 mM of each. Vertical bars denote ±SE. Pretreatment of mother shoots concisted of two weeks exposure to complete dark and 4°C during the shoot multiplication phase prior to the start of the regeneration experiments.

mother shoots were pretreated with a dark and cold period prior to regeneration. A combination of two sugars as Su/Gl and Gl/Fr at the same molarity as the sugars alone always resulted in a regeneration capacity similar to the lowest value obtained for the individual sugars. Sucrose is composed of glucose and fructose but a combination of glucose and fructose at the same molarity as sucrose never resulted in the same regeneration capacity. This is true for both shoot and root regeneration.

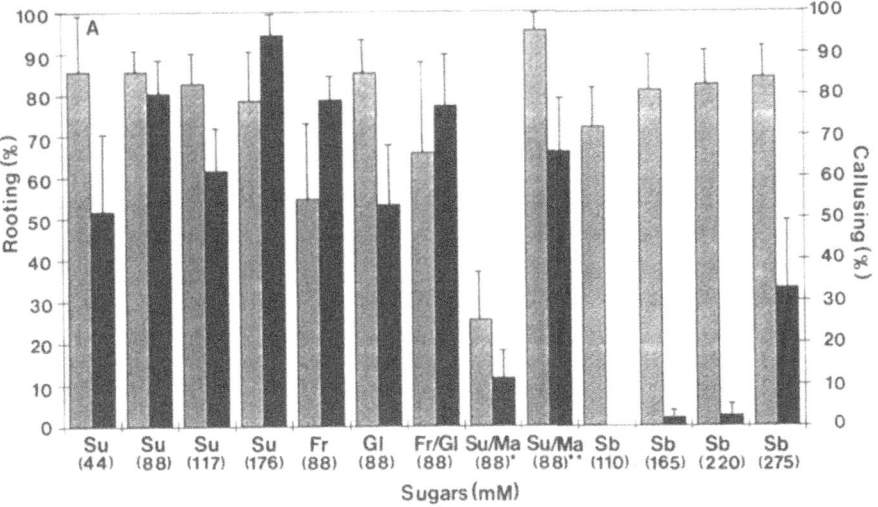

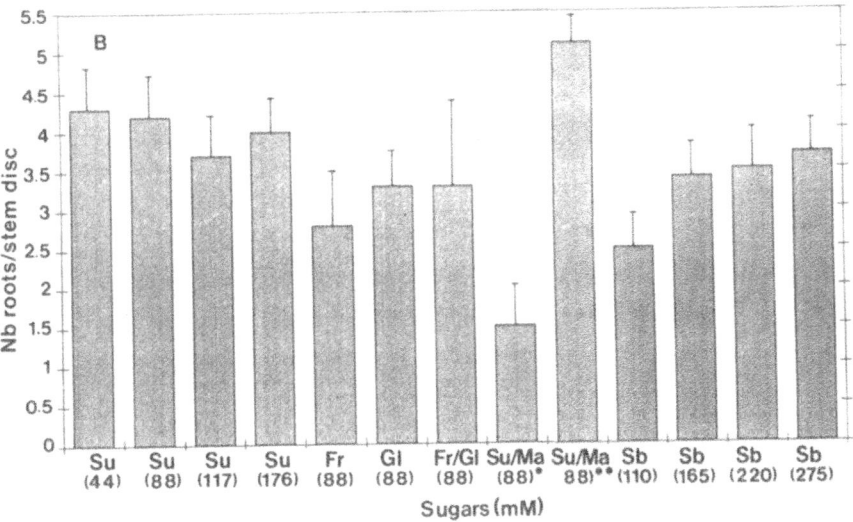

Fig. 3. Histogram showing the influence of different sugars at various levels and combinations on root regeneration and callus formation of apple. A. Percent explants forming roots (□) and callus (■). B. Mean number of roots per stem disc. Vertical bars denote ±SE. Ma= mannitol, Su/Ma* 30/58 mM, Su/Ma** 58/30 mM, Fr/Gl 44/44 mM, other abbreviations as in Figure 2.

Previous studies have indicated the importance of carbohydrate metabolism in the organ initiation process. Organ initiation is associated with utilization of accumulated starch and free sugars of the medium. For example during shoot formation in tobacco callus, increased rates of respiration were observed and

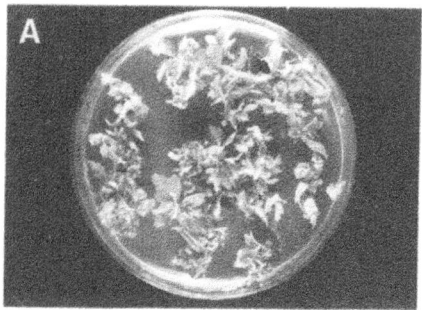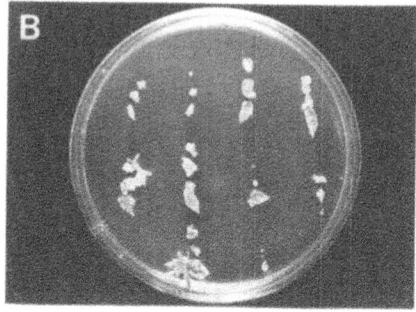

Fig. 4. Petri dishes showing shoot regeneration of apple in the presence of sorbitol at 220 mM (A) and sucrose at 117 mM (B). Explants were taken from pretreated mother shoots (see Fig.2).

differences in enzyme activities could be correlated with accumulation and utilization of starch [26]. Activation of the glycolysis along with the pentose phosphate pathways and enhanced glucose oxidation support the hypothesis that organ initiation has a high requirement for reducing power (NADPH)[12]. However it appears that carbohydrates have two roles. One is as a carbon and energy source and the other as an osmotic agent. It has been shown in tobacco callus that the number of shoots formed increased with increasing sucrose content in the medium up to about 3% (w/v) after which there was a decrease in the number of shoots formed. The inhibition of shoot formation above 3% sucrose could be duplicated by substituting mannitol for sucrose on a molar basis. Mannitol can be taken up by tobacco cells but not be metabolized. Therefore the inhibition was considered to be of an osmotic nature [3]. The osmotic requirement as well as type and concentration of sugar for shoot regeneration from leaf pieces of *Solanum melongena* L. has also been demonstrated [19].

In experiments with Convolvulus [7] it was shown that if the carbon compounds sorbitol and ribose were used in the medium, shoot formation from leaf explants was inhibited. However if the sorbitol was present only during the first 8 days of culture, shoot formation was reduced but not inhibited. Explants grown on ribose medium were completly inhibited unless the explants were transferred to a ribose free medium at day 5. One explanation for the different responses to the two carbon sources could be that sorbitol can be taken up but is not metabolized and thus can interfere by acting as an osmoticum whereas ribose can be metabolized but instead interferes with production of reducing equivalents thus being detrimental to shoot formation.

6.2 Adventitious root formation

The effect of different types and concentrations of sugars alone and in combination on root regeneration was also investigated (Fig. 3). Stem discs about 1 mm thick were cut from the base of *in vitro* produced apple shoots (M9

Jork) and placed on a Petri dish containing the rooting medium as described earlier [35]. Sorbitol at different concentrations did not enhance root regeneration compared to other sugars but reduced callus formation. However, as for shoot regeneration, the lowest root regeneration capacity was obtained with fructose. Sucrose concentrations above 44 mM did not increase root regeneration. When mannitol (30 mM) was added together with sucrose (58 mM) to a concentration corresponding to 88 mM sucrose both rooting percentage and number of roots per disc increased significantly. It seems that the optimum concentration of sucrose is between 44 and 58 mM. When sucrose and mannitol were added at the same molarity but in the reverse ratio the rooting response decreased.

7. Conclusions

In this paper we have discussed the metabolism and transport of carbon compounds in association with *in vitro* growth and organogenesis. It is obvious that the substances being translocated in the phloem and the ability to metabolize certain carbon compounds differ between plant species and that there is a striking grouping within given families. Interestingly it appears that the main sugar compounds being translocated can be used to promote growth *in vitro* as well as organogenesis. Those sugars that are not metabolized can be used to alter the osmotic conditions in the tissue which also seem to influence adventitious shoot and root formation.

References

 1. Bieleski DP (1977) Accumulation of sorbitol and glucose by leaf slices of *Rosaceae*. Australian J Plant Physiol 4:11–24
 2. Bieleski DP (1982) Sugar alcohols. In: FA Loewus and W Tanner, eds. Encyclopedia of Plant Physiology, New Series, vol. 13A, pp 158–192. Heidelberg: Springer-Verlag
 3. Brown DCW, David Leung DWM and Thorpe TA (1979) Osmotic requirement for shoot formation in tobacco callus. Physiol Plant 46:36–41
 4. Chaplin MF (1986) Monosacharides. In: MF Chaplin and JF Kennedy, eds. Carbohydrate Analysis, a Practical Approach, pp 1–36. Oxford, Washington DC: IRL Press
 5. Chong C and Taper CD (1971) Daily variation of sorbitol and related carbohydrates in *Malus* leaves. Can J Bot 49:173–177
 6. Chong C and Taper CD (1972) *Malus* tissue cultures.I. Sorbitol (D-glucitol) as a carbon source for callus initiation and growth. Can J Bot 50:1399–1404
 7. Christansen ML (1987) Causal events in morphogenesis. In: CE Green, DA Somers, WP Hackett, DD Biesboer, eds. Plant tissue and cell culture 1987. New York: Alan R. Liss Inc
 8. Coffin R, Taper CD and Chong C (1976) Sorbitol and sucrose as carbon source for callus culture of some species of the *Rosaceae*. Can J Bot 54:547–51
 9. Davis JM and Loesher WH (1990) [^{14}C]-Assimilate translocation in the light and dark in celery (*Apium graveolens*) leaves of different ages. Physiol Plant 79:656–662
10. Giaquinta RT (1978) Source and sink leaf metabolism in relation to phloem translocation. Plant Physiol 61:380–385

11. Giaquinta RT (1983) Phloem loading of sucrose. Ann Rev Plant Physiol 34:347–387
12. Haissig BH (1982) Activity of some glycolytic and pentose phosphate pathway enzymes during the development of adventitious roots. Physiol Plant 55:261–272
13. Hall SM and Baker DA (1972) The chemical composition of *Ricinus* phloem exudate. Planta 106:131–140
14. Humphreys TE (1988) Phloem transport-with emphasis on loading and unloading. In: DA Baker and JL Hall eds. Solute Transport in Plant Cells and Tissues, pp 305–345. Longman Scientific and Technical, UK
15. King RW and Zeevaart JAD (1974) Enhancement of phloem exudation from cut petioles by chelating agents. Plant Physiol 53:96–103
16. King RW (1976) Implication for plant growth of the transport of regulatory compounds in phloem and xylem. In : IF Wardlaw and JB Passioura, eds. Transport and Transfer Processes in Plants, pp 415–31. London, New York: Academic Press
17. Loescher WH, Marlow GC and Kennedy RA (1982) Sorbitol metabolism and sink-source interconversion in developing apple leaves. Plant Physiol 70:335–3915.
18. Loescher WH, Fellman JK, Fox TC, Davis JM, Redgwell RJ and Kennedy RA (1985) Other carbohydrates as translocated carbon sources: Acyclic polyols and photosyntetic carbon metabolism. In: RL Heath and J Preiss, eds. Regulation of Carbon Partitioning in Photosyntetic Tissue, pp 309–332. Rockville MD: American Society of Plant Physiologists
19. Mukherjee SK, Rathinasabapathi and Gupta N (1991) Low sugar and osmotic requirements for shoot regeneration from leaf pieces of *Solanum melongena* L. Plant Cell Tiss Org Cult 25:13–16
20. Pate JS (1976) Nutrients and metabolites of fluids recovered from xylem and phloem: significance in relation to long distance transport in plants. In: JF Wardlaw and JB Passioura, eds. Transport and Transfer Processes in Plants, pp 251–81. London and New York: Academic Press
21. Pollock CJ (1976) Changes in the activity of sucrose-synthesizing enzymes in developing leaves of *Lolium temulentum*. Plant Sci Lett 7:27–31
22. Pollock CJ and Lloyd EJ (1977) The distribution of acid invertase in developing leaves of *Lolium temulentum*. Planta 133:197–200
23. Pua EC and Chong C (1984) Regulation *in vitro* shoot and root regeneration in 'Macspur' apple by sorbitol (D-glucitol) and related carbon sources. J Amer Soc Hort Sci 110:705–09
24. Richardson PT and Baker DA (1982) The chemical composition of cucurbit vascular exudates. J Exp Bot 33:1239–47
25. Rumpho ME, Edwards GE and Loescher WH (1983) Pathway for photosynthetic carbon flow to mannitol in celery leaves. Plant Physiol 73: 869–873
26. Thorpe TA (1978) Physiological and biochemical aspects of organogenesis *in vitro*. In: TA Thorpe ed. Frontiers of Plant Tissue Culture, pp 49–58. Calgary, Canada: University of Calgary Printing Services
27. Tremblay FM and Lalonde M (1984) Requirements for *in vitro* propagation of seven nitrogen-fixing *Alnus* species. Plant Cell Tiss Org Cult 3:189–199
28. Trip P, Krotkov G and Nelson CD (1964) Metabolism of mannitol in higher plants. Am J Bot 51:828–835
29. Trip P, Nelson CD and Krotkov G (1965) Selective and preferential translocation of C^{14} – labeled sugars in white ash and lilac. Plant Physiol 40:740–747
30. Turgeon R and Webb JA (1973) Leaf development and phloem transport in *Cucurbita pepo*: Transition from import to export. Planta 113:269–281
31. Turgeon R (1984) Termination of nutrient import and development of vein loading capacity in albino tobacco leaves. Plant Physiol 76:45–48
32. Wallaart RAM (1980) Distribution of sorbitol in *Rosaceae*. Phytochem 19:2603–2610
33. Welander M, Welander NT and Brackman AS (1989) Regulation of *in vitro* shoot multiplication in *Syringa*, *Alnus* and *Malus* by different carbon sources. J Hort Sci 64:361–366
34. Welander M and Maheswaran G (1992). Shoot regeneration from leaf explants of dwarfing apple rootstocks. J Plant Physiol 140:223–228
35. Welander M and Pawlicki N (1992). A model system for studying root regeneration in woody species. Acta Hort 336:225–230

36. Zimmerman MH and Ziegler H (1975) List of sugars and sugar alcohols in sieve-tube exudates. In: A Pirson and MH Zimmerman, eds. Encyclopedia of Plant Physiology. New series vol.I, pp 480–503. Berlin, Heidelberg, New York: Springer-Verlag

The effect of carbohydrate source and pH on *in vitro* growth of *Vitis vinifera* cultivars Black Hamburg and Alvarino

SEÁN MAC AN tSAOIR[1] and VASSILIS DAMVOGLOU[2]
[1] *Department of Plant Science, Queens University, Belfast and Dept. of Agriculture for Northern Ireland*
[2] *Technical Education Institute, Heraklion, Crete, Greece*

Introduction

The grape is the most widely grown fruit crop in the world and research into tissue culture of *Vitis* species has been ongoing since the 1940's [10]. At first grapes proved recalcitrant but successful micropropagation of the cultivar 'Sylvaner Reisling' was reported in 1978 [8]. Micropropagation of grape vines has been used to produce pathogen free stock [1] and also to produce large numbers of plants when a new cultivar is being introduced to the industry [7]. There is little doubt that tissue culture will contribute to the future genetic improvement of grape vines [11].

Carbon source [13] and pH [9] have both been shown to affect plant growth in culture. While sucrose is the most common carbohydrate source used in plant tissue culture, some plants can metabolise other sugars more efficiently [5]. The pH of the medium will affect whether salts will remain in a soluble form, the uptake of nutrients and growth regulators and the gelling efficiency of agar [6]. This paper reports the effects of carbohydrate source on *in vitro* growth of *Vitis vinifera* cultivars Black Hamburg and Alvarino and the effect of pH on Alvarino.

2. Materials and methods

In vitro cultures of Alvarino were obtained from Enrique Ferro (C.S.I.C., Santiago de Compostela, Spain) while cultures of Black Hamburg were established by apical meristem culture. The cultures were grown in MS [11] with BAP (1mg/l), zeatin (5mg/l), sucrose (30g/l), agar (Oxoid difco 6g/l) and pH 5.5 [3]. The media was dispensed into 35mm glass test-tubes (20 ml/tube), covered with plastic caps and autoclaved. The culture conditions were a constant 23 °C 16h light under either Grolux Thorn 65/80W or Gelbweiss Osram L65/80W/23 lights (there were no significant interactions between the type of lights and explant growth). The explants were subcultured every four weeks. The range of carbohydrates used was glucose, sucrose, sorbitol, fructose (all 30 g/l) and

94

P.J. Lumsden, J.R. Nicholas and W.J. Davies (eds.), Physiology, Growth and Development of Plants in Culture, 94–97, 1994.

fructose/glucose mix (15g of each/l) and the explants were destructively harvested after eight weeks. The pH range investigated was from 5.0 to 7.5; pH was adjusted by addition of either 1M NaOH or 1M HCl as appropriate before autoclaving. The pH of the medium was measured after one culture passage as described by Jona and Webb [8]. All results were subjected to an analysis of variance (the probability values and estimation of standard errors are shown where appropriate). In the pH experiment there were 14 replicates of two node segments per pH treatment. Three of these replicates had a $2mm^3$ piece of callus at their base while the other eleven had no callus. This unequal division was taken into account in the analysis.

3. Results and discussion

3.1 Carbohydrate

The plants were distructively harvested after two culture passages and the results are shown in Table 1. Both cultivars grew well on sucrose and poorly on sorbitol. However, the importance of testing other sugars is illustrated by the fact that while Alvarino grew best on sucrose, Black Hamburg grew best on fructose. The carbohydrate response could have been caused by effects on water potential or metabolism/uptake differences. In this experiment water potential was not measured but in experiments where carbohydrates and water potential were compared, the source of carbohydrate was found to be the important factor [13].

Table 1. The effect of carbohydrate on Total Dry Weight (mg) after eight weeks in culture

Carbohydrate	Black Hamburg	Alvarino
Sucrose	340	140
Glucose	320	50
Sorbitol	150	20
Fructose	520	50
Fructose/Glucose	360	40

Treatment	Significance	
	E.S.E.	P
Sugar	35	0.004
Cultivar	22	<0.001
Sugar x Cultivar	50	0.013

3.2 pH

The results for shoot fresh weight (Table 2) show that increasing pH significantly reduced shoot growth and the presence of callus also reduced shoot growth. Similar results were obtained for total fresh weight (shoots plus callus, data not shown).

Table 2. The effect of pH on shoot fresh weight (mg) of Alvarino after four weeks in culture

| Callus (±) | pH | | | | |
	5.0	5.5	6.0	7.0	7.5
+	30	14	15	16	14
—	79	75	15	9	13

| Treatment | Significance | |
	E.S.E.	P
pH	8	<0.001
Callus	8	0.025

Plant cells can maintain a relatively constant cytoplasmic pH (pHc) across an external pH range of 4 to 9 (±0.1 pHc per external pH unit) [2]. Figure 1 shows that plant cells can also modify the external pH in the same manner. The explants raised or lowered the pH, depending on which end of the scale they were growing. The pH of the media (without plants-controls) was not measured until the end of the culture period. However the reduction in pH in the controls has been observed before [9]. The significant modification of the external pH by the explants could result from differences in nitrate and ammonia uptake. At low pH nitrate is taken up generating excess OH^- (which secreted into the medium will raise the pH) whereas at higher pH's NH_4^+ is metabolised generating excess H^+ (lowering pH of the media) [2]. This would suggest that plant growth would result regardless of initial pH since MS medium contains both N sources. However as Table 2 shows, good shoot growth occurs only at the lower end of the pH range. It is possible that an energy dependent proton pump system is in operation at the higher pH draining energy supplies which would otherwise be utilised for shoot growth. The explants with callus were not able to raise the pH to the same extent as those without callus, but were able to lower the pH to a greater degree perhaps indicating that the callus could metabolise nitrogen in the NH_4^+ form in preference to the nitrate form.

In conclusion it is important to determine which carbohydrate source is optimal for a particular plant. While all explants could raise or lower the pH of the media, explants which had no callus showed a much more uniform manipulation of the pH and produced better growth.

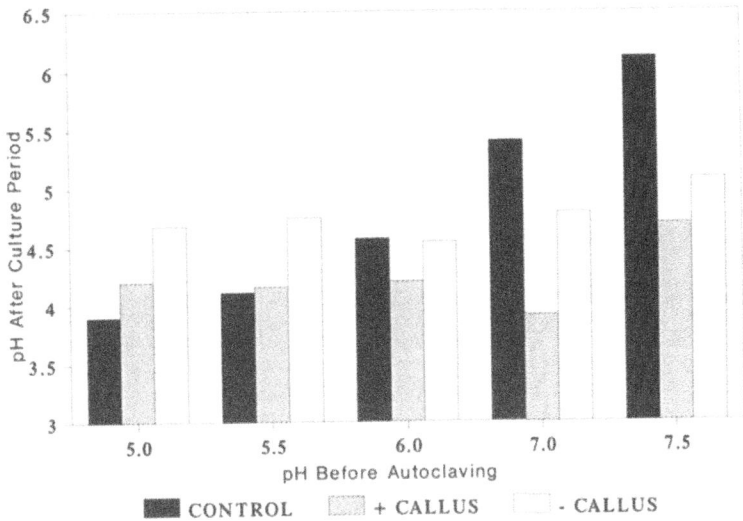

Fig. 1. Changes in media pH in the presence of explants with or without callus after 4 weeks of culture. Solid bars, media alone; heavy shading, explants with callus; light shading, explants without callus.

References

1. Barlass M (1987) Elimination of stem pitting and corky bark diseases from grapevines by fragmented shoot apex culture. Ann Appl Biol 110:653–656
2. Caponetti JD, Hall GC and Farmer RE, (1971) *In vitro* growth of black cherry callus: effects of medium, environment and clone. Bot Gaz 132:313–318
3. Felle H, (1988) Short-term pH regulation in plants. Physiol Plant 74:583–591
4. Ferro E, Gonzales, JL and Ballester A, (1990) Aplicacion de tecnicas *in vitro* al saneamiento de clones seleccionados de *Vitis vinifera* cv. Alvarino Invest Agr Prod Veg 5:383–397
5. George, EF., Puttock DJM and George, HJ (1987) Plant culture media Vol 1, Exergetics Ltd. Westbury pp 227–296
6. George,EF and Sherrington,PD (1984) Plant Propagation by Tissue Culture. Handbook and Directory of Commercial Laboratories. Exergetics Ltd. pp 119–201
7. Gray DJ and Klein CM (1989) *In vitro* micropropagation and plant establishment of 'Blanc du Bois' grape. Proc Fla State Hort Soc 102:221–223
8. Jona R and Webb KJ (1978) Callus and axillary bud culture of *Vitis vinifera* 'Sylvaner Reisling'. Sci Hort 9:55–60
9. Leifert C, Pryce S, Lumsden PJ and Waites WM (1992) Effect of medium acidity on growth and rooting of different plant species growing *in vitro*. Plant Cell Tiss Org Cult 30:171–179
10. Morel G (1944) Sur le development de tissues de vigne cultives *in vitro*. C.R.Soc Biol Paris 138:62
11. Mullins MG (1990) Tissue culture and the genetic improvement of grapevines: A review. Acta Hort 280:11–22
12. Murashige T and Skoog F (1962) A revised medium for rapid growth and bio assays with tobacco tissue cultures. Physiol Plant 15:473–497
13. Seingre D, O'Rourke J, Gavillet S and Moncousin Ch (1992) Influence of gelling agent and carbon source on the *in vitro* proliferation rate of apple rootstock EM IX. Acta Hort 289:151–155
14. Staikidou I (1992) M.Sc. Thesis Narcissus Bulbil formation *in vitro*. Queens University of Belfast

Evidence for the interactive involvement of carbohydrates in the control of differentiation of *in vitro* explants

STEPHEN MILLAM

Crop Genetics Department, Scottish Crop Research Institute, Invergowrie, Dundee DD2 5DA, UK

Abbreviations: 2,4-D = 2,4-Dichlorophenoxyacetic acid

1. Introduction

The process of regeneration in plant cells is a key biological process, and the foundation on which all plant tissue culture techniques are based. Despite extensive studies the basis of plant regeneration is mainly unclear, for instance there are many reports of genotype-specificity in plant regeneration protocols. Overcoming such specificity, resulting in the application of protocols to a wider range of genotypes or species, is an elemental objective for many applied plant genetic investigations. Plant cell and tissue culture techniques facilitate studies on the response of single, and interactive factors affecting the differentiation process in plants.

The effects of plant growth regulators have been the primary focus in studies of plant development *in vitro*, and such methods underpin many of the regeneration systems currently in use in both experimental and commercial applications of plant tissue culture. Most systems involve exposing explants to defined levels of growth regulators for the entire culture period. There have been reports on the induction of tissues into a developmental response by exposures to cytokinin or auxin, [4, 1] of shorter time periods than normally used, i.e. for the initiation period of the culture only. In addition, the role of carbohydrates in the differentiation process has been the focus of recent investigation in a range of plant tissue culture studies, including explant [9, 8] anther culture [5] and somatic embryogenetic systems [2, 11]. Many tissue culture protocols are designed for a specific morphogenic response system of a one-step nature, whereby a standard culture medium is supplemented with a carbohydrate source and growth regulator component empirically derived for that species and objective. It has been suggested, however, that the mode of regeneration may be critically dependent on such factors as the key time of exposure to growth regulator, or alteration and application of carbohydrate source in the media. Consequently, the use of more highly defined systems could be of great benefit in studies on the development and differentiation of tissue explants into

P.J. Lumsden, J.R. Nicholas and W.J. Davies (eds.), Physiology, Growth and Development of Plants in Culture, 98–103, 1994.
© 1994 *Kluwer Academic Publishers.*

organised structures. The findings from such investigations would also be of use in the establishment of more broader-based regeneration systems. The rationale behind this present study was therefore to develop a strategy for the quantification of the effect of a simple factor, in this case the exposure to auxin, and to allow clear observations of the effect of one additional factor (in this case carbohydrate source) on the model protocol.

2. Materials and methods

2.1 Plant material

Seeds of flax cv. Antares were surface sterilised using a three-step method as previously described [9]. The seeds were then placed onto 9 cm Petri dishes containing Murashige and Skoog [10] medium solidified using 0.8% Difco Bitek agar, and incubated at 22 °C in the dark for six days. Following this germination period, the central sections of the hypocotyls were aseptically excised into 5 mm explants, randomised and plated onto 9 cm Petri dishes (ten explants per plate) of media.

2.2 Culture media

Basal media described above was supplemented with either 0, 5, 50 or 100 μM 2,4-D. Each auxin treatment was duplicated with one set of replicates having sucrose at 50 mM as the carbohydrate source and the other having 50 mM maltose. In all cases the sugars and auxin were added to the medium following filter-sterilisation through 0.2 μM Nalgene filters.

2.3 Growth conditions

The explants were exposed to the 2,4-D treatments for periods of 1, 3, 7, 10 and 21 days under conditions of 22.0 + / − 2.0 °C, 16 h photoperiod of photon fluence density 140 μmol m^2 sec^{-1}. The explants were removed from the 2,4-D medium after the period of treatment and replated onto the appropriate basal medium under the same conditions.

2.4 Assessment

The developmental response was assessed after a total of 21 days culture. The number of shoots and the incidence of roots and callus arising from each explant was recorded and the mode of differentiation (e.g. from callus, epidermis or cut ends) noted.

3. Results

Shoot formation occurred from the control explants at similar rates on both carbohydrate sources (Fig. 1). In the 2,4-D treated explants different patterns of development were recorded. Low-level, short-time auxin exposure (5 μM for 1 day) significantly enhanced shoot formation over the control figure but 3 days exposure and longer repressed shoot formation to either zero or below that of the control. Conversely, in both the 100 μM auxin treatments no shoot formation was recorded in the zero exposure, but 1 day exposure resulted in mean shoot numbers of 1.4 and 1.6 respectively in the sucrose and maltose treatments. Rooting was inhibited in all but the lowest auxin treatments but was at a much higher rate in the sucrose treatment. The data for shoot and root

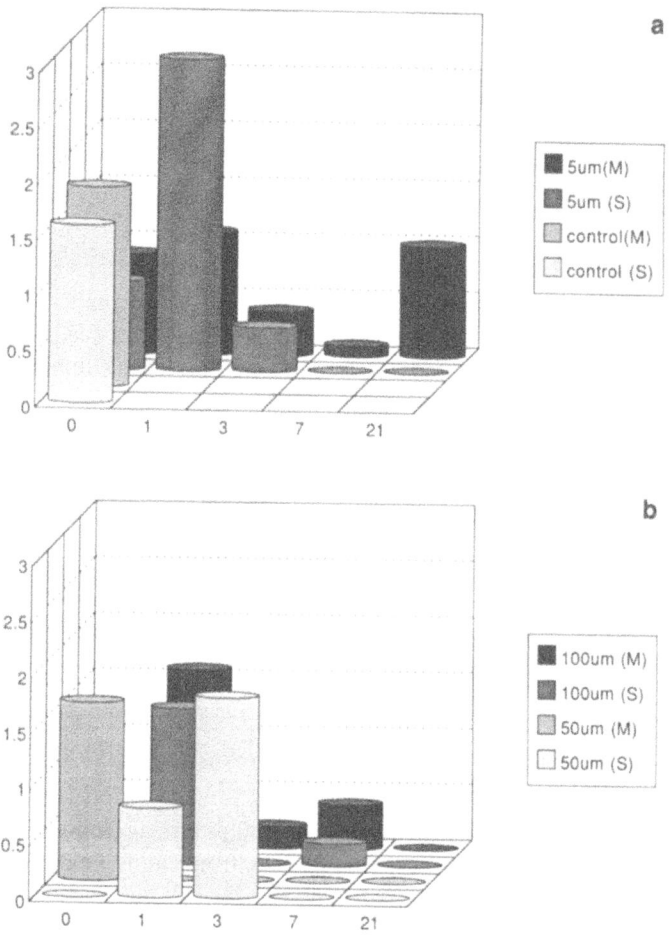

Fig. 1. Representation of shoot formation following exposure to 2,4-D at 0, 5 μM (a) and 50 and 100 μM (b) for 0,1,3,7,10 and 14 days. (M = maltose treatment, S = sucrose).

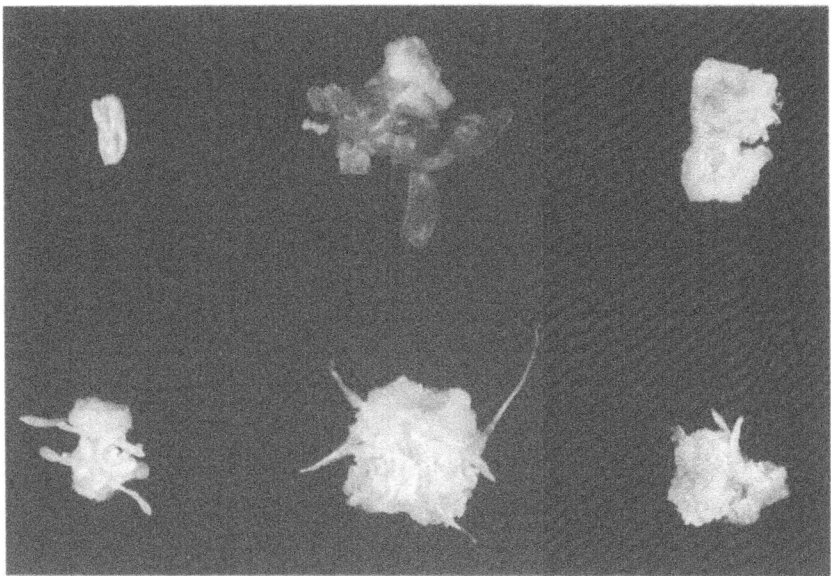

Plate 1. Explants exposed to (from top left, left to right) 0,1,3 and 7,10,14 (bottom left, left to right) days exposure to 5μM 2,4-D with sucrose as the carbohydrate source.

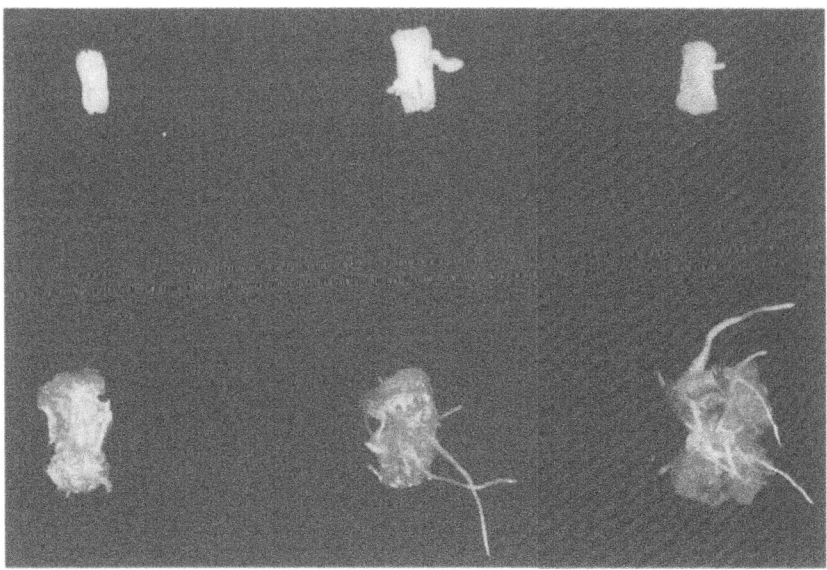

Plate 2. Explants exposed to (from top left, left to right) 0,1,3 and 7,10,14 (bottom left, left to right) days exposure to 5 μM 2,4-D with maltose as the carbohydrate source.

production for the control and 5 μM auxin treatments were analysed statistically using analysis of variance and generalised linear model. This showed that there were significant (p <0.001) differences in shoot production rates on sucrose and maltose at these levels, and in comparison with the control. The effect of sucrose and a 1 day exposure to 2,4-D resulted in significantly more shoots than any other treatment. There was also a significant interaction between carbohydrate and time period for this early exposure. No such patterns were found in the higher auxin treatments. The mode of regeneration (see Plates 1 and 2) was similar in both carbohydrate treatments with shoots arising from the epidermis rather than the cut ends of the hypocotyls in the 0, 1 and 3 day exposures, and rooting (of a spontaneous nature) from cut ends, epidermal surfaces and, where present, callus. The callus arose from cut ends initially but in the longer exposure treatments spontaneous callus arose and generally the callus from the maltose treatments was much greener in colour.

4. Discussion

From the results presented it is apparent that only a short exposure to auxin is required to elicit an enhanced regenerative response. Furthermore, by increasing the exposure period, clear differences in the form of morphogenic response occur. Moreover, exposure to auxin for at least 3 days was required for significant root and callus formation, but only a 1 or 3 day exposure for significant shoot formation. The level of 2,4-D found to be most effective in our research was comparatively low. In other studies it was found that early exposure to high levels (100 μM) of 2,4-D resulted in the initiation of a developmental response and that the reduction of the auxin resulted in the induction of this response [3]. It has been suggested [4] that the minimum concentration and duration of 2,4-D treatment required for inductive effect is genotype specific. Our previous work, employing a standard regeneration protocol, has shown genotypic effects in the response of flax cultures to growth regulators [7] but preliminary studies (in progress) indicate that the response seen in our present study may be observed in a broader range of cultivars than expected. Carbohydrates have previously been demonstrated to be an important factor in processes controlling regeneration [1]. The competence of the carbohydrate may be related to the developmental stage [2] and the statistically significant interaction between the carbohydrate and auxin at the early stages of culture found in our studies implies a control mechanism role for carbohydrates. This indicates that the very early stages of induction into the regeneration process, prior to the initiation of morphogenesis is a critical area for more detailed investigation. The role of peroxidases in the early stages of culture and differentiation of flax explants *in vitro* has been studied previously [6] and findings, particularly related to the importance of peroxidases in IAA-oxidase activity during early culture would appear to have relevance to the further elucidation of the findings presented here. The involved and interacting underlying mechanisms of regeneration of plant cells or explants back into intact plants are not fully understood, but the precise mechanism of involvement of the factors analyzed in this paper may provide the basis for further applied biochemical and molecular investigations. The flax hypocotyl regeneration system described in this paper seems to be an excellent model system for such research.

Acknowledgements

This work was supported by the Scottish Office Agriculture and Fisheries Department. The author would like to thank Diane Davidson for her excellent technical assistance.

References

1. Antonelli M and Druart Ph (1990) The use of a brief 2,4-D treatment to induce leaf regeneration on *Prunus canescens*. Acta Hort 280:45–50
2. Babbar SB and Gupta SC (1986) Effect of carbon source on *Datura metel* microspore embryogenesis and the growth of callus raised from microspore-derived embryos. Biochem Physiol Pflanzen 181:331–338
3. Chee RP and Cantcliffe DJ (1988) Selective enhancement of *Ipomea batatas* Poir. embryogenic and non-embryogenic callus growth and production of embryos in liquid culture. Plant Cell Tiss Org Cult 15:149–156
4. Dudits DL, Bogre L and Gyorgyey J (1991) Molecular and cellular approaches to the analysis of plant embryo development from somatic cells *in vitro*. J Cell Sci 99:475–484
5. Finnie SJ, Powell W and Dyer AF (1989) The effect of carbohydrate composition and concentration on anther culture response in Barley (*Hordeum vulgare* L.). Plant Breeding 103:110–118
6. McDougall G, Davidson D and Millam, S (1992) Alterations in surface-associated peroxidases during callus development and shoot formation of explants of *Linum usitatissimum*. J Plant Physiol 140:193–200
7. Millam S and Davidson D (1993) The role of carbohydrates and growth regulators in the induction of morphogenesis *in vitro*. Proc. 2nd FAO Regional Workshop on Flax, Brno, Czechoslovakia (in press)
8. Millam S and Hodgson VJ (1991) *In vitro* response of thin-layer floral internode sections of *Brassica oleracea* L. to a range of sucrose and maltose levels. J Plant Physiol 138:620–621
9. Millam S, Davidson D and Powell W (1992) The use of flax as a model system for studies of organogenesis *in vitro*: effect of carbohydrate. Plant Cell Tiss Org Cult 28:163–166
10. Murashige T and Skoog F (1962) A revised medium for rapid growth and bioassays with tobacco tissue cultures. Physiol Plant 15:473–497
11. Strickland SG, Nicholl JW, McCall CM and Stuart DA (1987) Effect of carbohydrates on alfafa somatic embryogenesis. Plant Sci 48:113–121

Factors affecting cell expansion; hydroponic roots as a model system

JEREMY PRITCHARD

Ysgol Gwyddorau Biolegol, Coleg Prifysgol Gogledd Cymru, Bangor, Gwynedd, LL57 2UW, Wales

1. Introduction

The complexity and variety of plant form is ultimately the result of cell expansion. These same basic processes occur in plant cells in culture and in callus tissue. An understanding of the behaviour of cells in culture will benefit from an understanding of the mechanism of cell expansion and the factors which influence it. In addition considering the basic process of cell expansion may help to understand the changes which occur from single non-differentiated culture cells of callus tissue to whole plants.

The scope of this paper is limited to the control of cell expansion and does not consider cell division, differentiation or the control of cell shape. To discuss the factors which control cell expansion we can consider a hypothetical 'ideal' cell. Actually this is not unlike a unicellular algae or a single cell in cell-culture.

Plant cells consist of a central vacuole bounded by the tonoplast enclosed by the plasmalemma, the cytoplasm is contained within these two membranes. Outside the plasmalemma is the cell wall. In order to expand a cell must increase in volume which means that water has to enter the cell. In addition the cell wall must expand to accept this increase in volume. Cell expansion is therefore influenced by the properties of the cell wall and the ability of the cell to take up water [5, 27].

The third major factor with a role in controlling cell expansion is the hydrostatic pressure within the cell [21]. Turgor pressure is generated by the opposition of the cell wall to water drawn into the cell by osmosis across the semi-permeable plasma membrane. In common with other molecules water moves spontaneously down its chemical potential gradient.

The chemical activity of water is reduced by the accumulation of solutes within the cell; the osmotic pressure of the cell sap increases. (The osmotic pressure generated by a solute dissolved in a solvent is equivalent to the hydrostatic pressure that would be needed to prevent the osmotically driven flow of a solvent molecule from a pure solvent across an ideal semi-permeable membrane). Osmotic pressure and turgor pressure can be related in an equation describing the water potential:

P.J. Lumsden, J.R. Nicholas and W.J. Davies (eds.), Physiology, Growth and Development of Plants in Culture, 104–119, 1994.

$$\Psi_w = P - \Pi + mgh \tag{1}$$

This equation links turgor and osmotic pressure and allows the direction of water movement to be predicted. Generally, except for tall trees the mgh term can be discounted. Water will move from a high to a low water potential. Lowering the hydrostatic pressure in a cell will result in a lower value of Ψ_w resulting in a stimulation of water flow into the cell. In an ideal cell at equilibrium the water potential gradient between the inside and the outside of the cell is zero and turgor pressure is balanced by an equivalent osmotic pressure and no net water flow occurs. The turgor in the cell is determined by the difference between the osmotic pressure inside (Π_i) and that immediately outside the cell (Π_o);

$$P = \Pi_i - \Pi_o \tag{2}$$

Any loosening of the cell wall will result in a decrease in P thus making Π_w more negative so that water moves in and the cell expands. A more detailed treatment of the thermodynamics of water movement is given by Nobel [14].

1.1 Growth with restricted water supply

The turgor pressure, osmotic pressure and water flow characteristic of a cell can be linked in the equation;

$$r = L(\delta\Delta\Pi - P) \tag{3}$$

Where L = volumetric hydraulic conductance, δ = solute reflection coefficient and $\Delta\Pi$ is the osmotic pressure difference between the protoplast and its immediate environment.

Measurements of single cell values of the hydraulic conductivity indicated that water flow into single cells is probably not limiting to growth [11, 17]. However a tissue or organ is rarely a simple sum of its parts; the physiology and anatomy of an organ may result in the path way of water flow from its source, perhaps in the xylem, to the expanding tissue being limiting. In this case turgor pressure will not be determined by equation 2; since the system is not in equilibrium, the water potential of the cell will be less than zero (see equation 1). Such water potentials are termed growth induced water potentials (GIWP) and there has been much debate about their magnitude in growing tissues [1, 4]. In practice, whether they are large enough to be detected depends on the availability of water and the length and resistance of the pathway from the water source to the growing tissue. It is important to note that even in growing tissues that have no apparent GIWP, Ψ_w must be slightly lower in the growing zone to allow water entry into expanding cells [24].

1.2 Growth with an unrestricted water supply

Single cells in culture and a collection of cells in a root growing in hydroponics probably approximate to the behaviour of an 'ideal' cell. Thus the hydraulic conductivity of single cells is generally not low enough to make water entry into the cell the largest resistance.

If water is not limiting to growth then only P and the properties of the cell walls influence expansion rate;

$$r = \phi(P - Y) \tag{4}$$

where r = growth rate, ϕ = wall extensibility, P = turgor pressure and Y = yield threshold. ϕ and Y describe the physical properties of the cells wall. In this formulation, first described by Lockhart [12] and popularised by Ray and co-workers [21] expansion is a function of the extensibility of the cell wall once the minimum turgor required for growth is exceeded.

The relationship between P and r described by equation 3 has been demonstrated experimentally in leaves of *Betula* and *Acer* [26], in pea roots growing in soil [9] and in roots of wheat and maize growing in hydroponics [16, 19]. Figure 1 shows the growth of whole maize roots as a function of the turgor pressure within the growing cells. Clearly a change in any of the parameters P, Y or ϕ (or a combination of the three) can result in an alteration in growth rate.

1.3 Growth and turgor pressure

In our work we have attempted to examine the effects of these three parameters on cell extension in hydroponically grown roots and to show how their interrelationship is affected by environmental factors.

We have used the pressure probe [10] to measure the turgor pressure of single cells. Growth was measured as increase in length (time-lapse photography allowed high resolution measurement). Changes in cell wall properties Y and ϕ could be measured by constructing plots of growth vs turgor pressure [16, 19]. The physical state of the cell walls could also be followed by simply comparing growth and turgor pressure; if two cells have the same growth rate but different turgor pressures (the driving force for extension) then the cell with the lower turgor must have more extensible or looser walls. Conversely if two cells have different growth rates but the same turgor pressure the cell with the higher growth rate will have the more extensible walls.

These methods follow cell wall properties *in vivo*. In addition the stress/strain relations of a tissue can be measured using an Instron type tensiometer [3, 29]. The spatial resolution of this technique is not good, at least 5 mm length of tissue is needed and it is impossible to use for single cells, however it can prove useful in providing comparative information on tissue rheology.

2. Treatments which change growth rate

2.1 Developmental changes in unstressed roots

2.1.1 Growth
The growth rates of Figure 1 were for the whole root and the turgor pressure was an average for the whole growing zone. The estimates of ϕ and Y therefore relate to all the growing cells. Measurement of the *profile* of growth indicated a wide variation, (Fig. 2a); growth rate accelerated from near zero at the apex to its maximum at around 4–5 mm from the root tip then decelerated so that growth had ceased at 10–12 mm from the root tip.

2.1.2 Turgor pressure
Turgor pressure was constant along the apical 12 mm of the root despite the wide variations in growth measured over the same region (Fig. 2b). Turgor was also constant in underlying cells since no differences in turgor were measured radially across the root (Fig. 2c).

Since turgor was constant but growth varied, the rate of solute import into the expanding cells must be tightly regulated [5]. The delivery of solutes to cells at the centre of the growing zone, where cells will almost double in volume every hour, far exceeds the solute import to cells at 10 mm from the tip, where expansion has almost ceased.

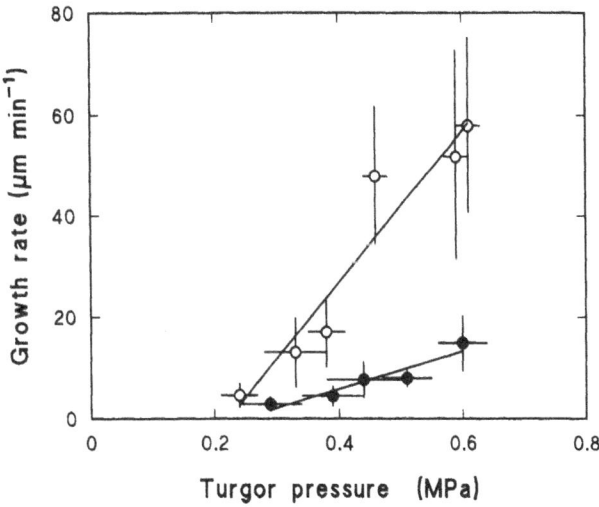

Fig. 1. Growth rate of entire maize root as a function of cell turgor pressure. Roots were bathed in a series of mannitol solutions at 30 °C (○) and 15 °C (●) and growth and turgor measured (modified from Pritchard, Barlow, Adam, & Tomos 1990). Growth is the mean of 8–17 ±s.d. measurements, turgor is mean of <30±s.d.

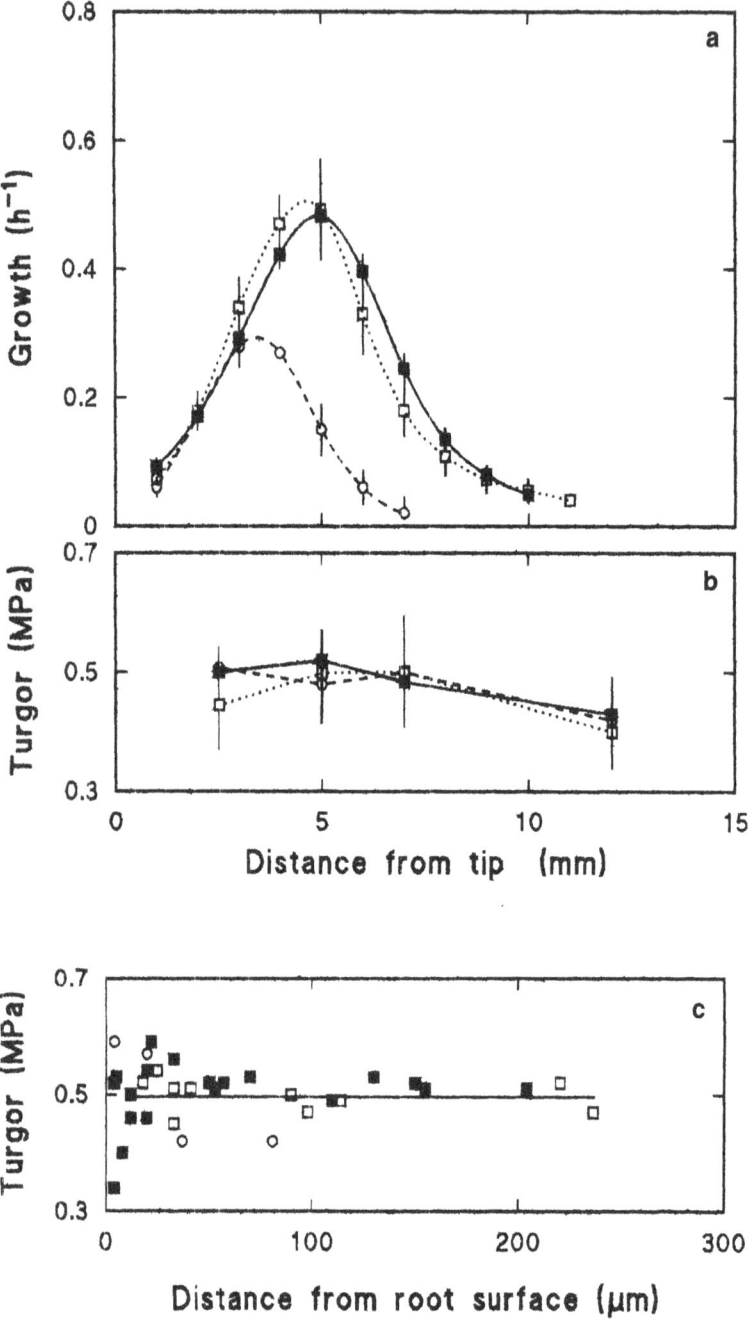

Fig. 2. Growth rate and turgor pressure of individual cells over the apical 12 mm of maize roots. Unstressed roots (■), after 24 h in 200 mol.m^{-3} mannitol (0.48 MPa) (□), after 24 h in 400 mol.m^{-3} mannitol (0.96 MPa) (○). (a) Local longitudinal growth rate n >15±sd; (b) Turgor pressure of epidermal and surface cortical cells measured longitudinally along the root tip n = 5–25 ± s.d. (c) Turgor pressure measured across the root radius at 7 mm from the root tip, each point is a single measurement.

2.1.3 Osmotic pressure

Vacuolar osmotic pressure was also constant along the apical 10 mm of unstressed roots. However it exceeded the turgor pressure by about 0.19 MPa over the whole growth profile (Fig. 3). Thus cell Ψ was −0.19 MPa while the water potential in the control medium (0.5 mol.m^{-3} CaCl$_2$) was only 0.004 MPa. A similar situation has been shown for root cells of maize, tomato, wheat and rice grown in nutrient solutions of different water potential (J. Pritchard unpublished data). The step in water potential between cell and the bathing medium requires explanation. A number of possible reasons for this may be proposed.

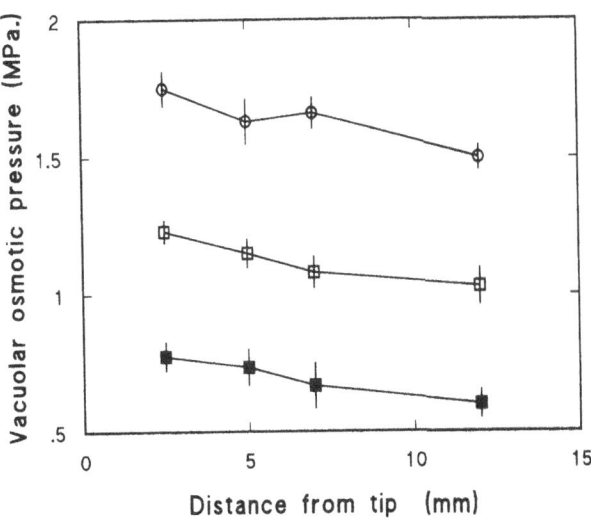

Fig. 3. Osmotic pressure profiles, individual cells over the apical 12 mm of maize roots. Unstressed roots (■), after 24 h in 200 mol.m^{-3} mannitol (0.48 MPa) (□), after 24 h in 400 mol.m^{-3} mannitol (0.96 MPa) (○). n = >12 ±s.d.).

1. Cell-wall hydrostatic pressure could be lower than atmospheric due to transpiration tension. This seems unlikely since transpiration rate was minimal in these plants. In addition the root cell-walls would be expected to be in hydrostatic equilibrium with the bathing solution surrounding them.
2. If the entry of water into the cells were limiting growth, turgor would be less than the potential maximum (growth-induced water potential [24]). This would imply a greater discrepancy between turgor and osmotic pressure in the region where growth is most rapid which (within experimental limits) is not observed.
3. The average reflection coefficient δ of the cell solutes to the plasma membranes of the root cells may be significantly less than one (a value of 0.75 would be consistent with the observations).
4. The osmotic pressure of the cell wall immediately adjacent to the plasma-membrane is not zero, i.e. there is an un-stirred layer within the wall

(equation 2). This could be stabilised by a flow of water across the wall and into the cell. If such a flow were driven by growth the effect would be greatest in the fastest growing region, which has been precluded above. Water flow due to transpiration stream across the cortex is a second source of such flow. This would contradict the view that the expanding-zone of the root is relatively isolated from the transpiration steam [22].

2.1.4 Cell wall properties

Since the driving force for growth P remains constant along the growth profile the properties of the cell wall must vary along the root [20; equation 4]. Specifically the longitudinal cell walls must loosen (become more extensible) as growth accelerates and then tighten, slowing cell expansion during the decelerating phase. Thus developmental changes in both parameters correlate with the decrease in growth rate 5–10 mm from the root tip.

2.2 Osmotic stress

2.2.1 Short term osmotic stress

Following 20 min of mannitol treatment, growth of the cells in the accelerating region was unchanged in comparison to the control roots despite having a turgor 0.3 MPa lower (Fig. 4). Since the driving force for extension was unchanged, the wall rheological properties (Y and/or) ϕ must have altered within the period of the experiment (20 min) to loosen the wall and so maintain growth rate in the root apex. Thus there are turgor sensitive changes in the cell wall parameters. In marked contrast this wall-loosening does not occur in cells that are about to cease expansion at 5–10 mm from the root tip.

2.2.2 Long term osmotic stress

In the above example, short term immersion in mannitol reduces the turgor pressure (as Π_o increased, equation 2). In addition to the rapidly-induced changes in wall properties, cells reduce Π_i by accumulating solutes in order to recover P. Turgor adjustment occurred over a period of some hours, but the rate of recovery is dependant on the stage of development of the cells. Recovery was faster at 2.5 mm from the root tip (Fig. 5a) than at 7 mm from the tip (Fig. 5b).

Following adjustment, turgor was similar to that of unstressed roots (Fig. 2a). However, local growth rate did not recover to unstressed rates but was reduced over the growth profile (Fig. 2b). An asymmetric reduction in growth following water stress was previously noted by Sharp and co-workers [23]. The reduction was greater towards the base of the growing region indicating that stress-induced hardening increases with development. The hardening of the cell wall following osmotic stress was confirmed by measuring the plastic and elastic extensibility using a tensiometer [29] on the basal (5–10 mm) region of the growing zone following 24 h of osmotic stress.

At higher levels of stress, osmotic adjustment may not be able to maintain turgor pressure at the control level so that turgor may become growth-limiting.

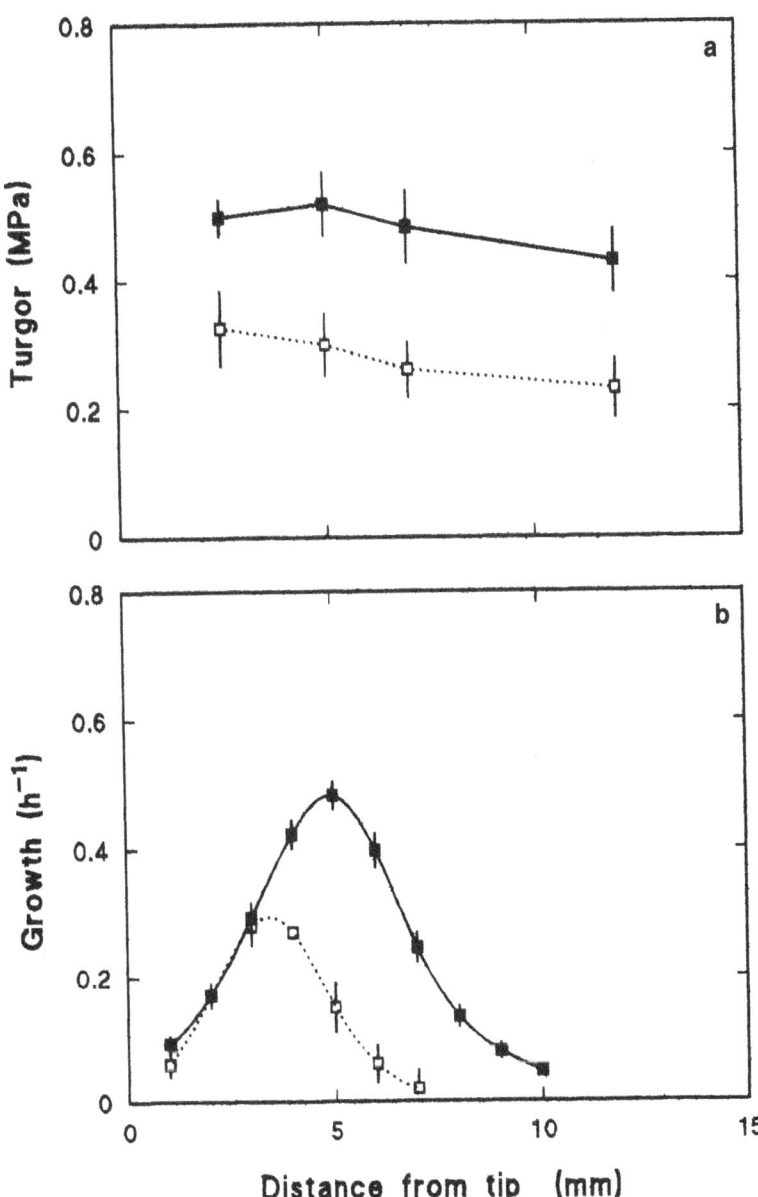

Fig. 4. Local growth and turgor pressure profiles 10–0 min before (■) and 10–20 min after (□) transfer from 0 mol.m^{-3} (0 MPa) to 200 mol.m^{-3} mannitol (0.48 MPa). (a) Turgor pressure, n = 5–15±s.d; (b) Local growth, n = 15±s.d.

In a study of the effect of osmotic stress on wheat roots, turgor was maintained in the root apex but was reduced in the more mature regions as the external concentration of mannitol was increased [20]. At the highest mannitol concentration (400 mol.m^{-3} or 0.96 MPa) turgor pressure was reduced in even

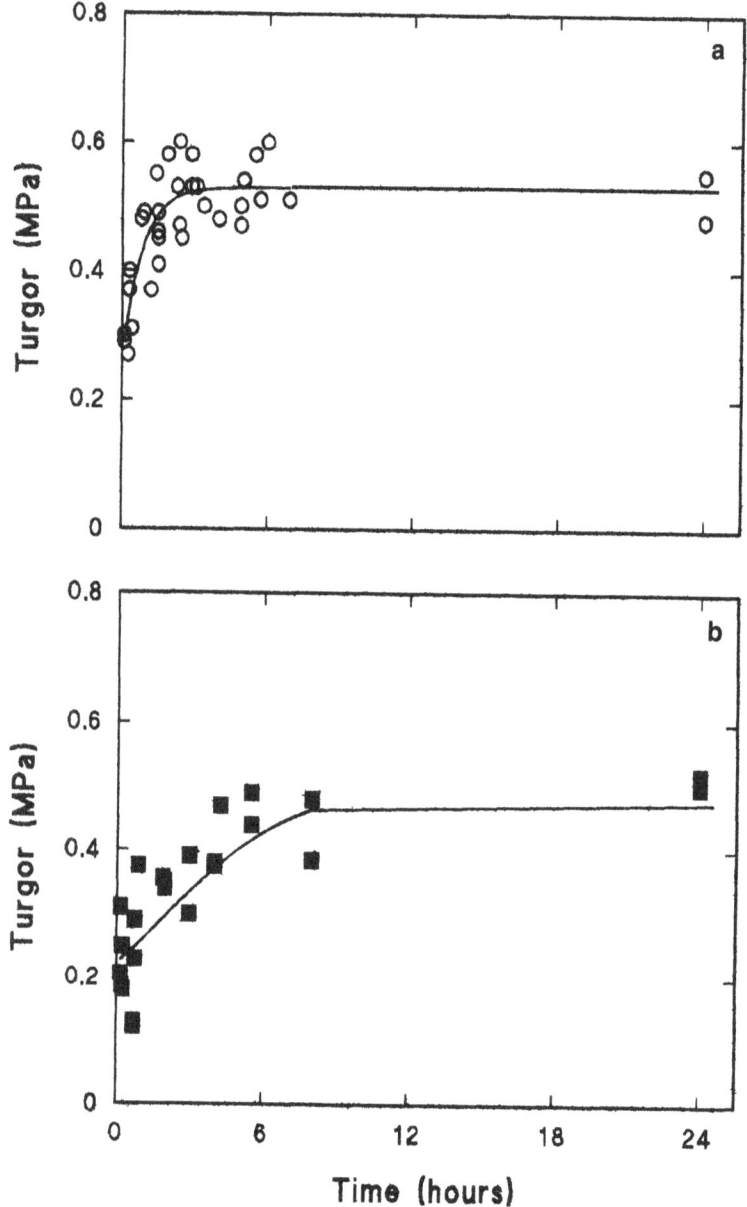

Fig. 5. Time course of turgor recovery following immersion in 200 mol.m^{-3} mannitol, each point is a single determination. (a) Cortical cells 2.5 mm from the root tip; (b) 7 mm from the root tip. (■) 10 μm from the root surface, (□) 204 μm from the root surface.

the regions of the growing zone. This implied that the apical region preferentially undergoes osmotic adjustment. The basal regions of the growing zone only receive solute for adjustment if the apical regions have sufficient.

In the present study the difference of −0.19 MPa between the water potential of cell and medium remained the same following the mannitol-induced water stress, since the cell osmotic pressure rose by the same amount as the osmotic pressure of the medium (Fig. 3). This confirmed that full osmotic adjustment of the protoplast occurred. Thus moderate water stress results in transient changes in turgor but in the long term, expansion is reduced by stress-induced changes in cell wall properties. At higher levels of stress, growth may be reduced by an inability to maintain turgor.

2.3 Temperature

Changes in temperature can have marked effects on plant growth. Root growth rate is severely reduced at 5 °C in comparison with 20 °C (16; Fig. 1). This large decrease in growth is not caused by a change in turgor pressure, indeed turgor pressure was increased over the period of reduced growth [16]. Once again it appeared that cell walls hardened, and this was confirmed by measurements of 'Instron' extensibility of the apical region. Tissue plasticity was decreased following the start of the low temperature treatment (Table 1). In addition, the *in vivo* wall rheological parameters of and Y of the whole root were measured following a decrease in root temperature from 30 °C to 15 °C. The reduction in root growth was caused by a decrease in wall extensibility $\varnothing$ and not by an increase in Y (Fig. 1).

Table 1. Growth, turgor pressure and cell wall properties of the extension zone of wheat roots in the presence and absence of potassium. Data from [17]

Treatment	Growth rate (mm 24h^{-1})	Turgor Pressure (MPa)	Plastic Extensibility (% extension 8g^{-1} load)
0.5 mol m^{-3} CaSO$_4$	32±.6	0.63±.02	4.7±0.3
0.5 mol.m^{-3} CaSO$_4$ plus 10 mol.m^{-3} KCL	20±0.2	0.62±.01	1.3±.08

2.4 Ionic effects

The nutrient solution which bathes the growing tissue is of obvious importance for continued growth of the cells. It provides solutes for turgor maintenance and its osmotic pressure is also of crucial importance (see above). However there is an additional and direct effect of some inorganic ions on the properties of the cell wall which have important consequences for expansion.

The effect of H$^+$ ions on growth is well characterised (e.g. 3,6). However other inorganic ions can have a pronounced effect on cell expansion. Wheat roots in hydroponics have maximal elongation rate in solutions containing only 0.5 mol.m^{-3} CaCl$_2$. Addition of low concentrations of K$^+$ salts to the growth medium reduced root growth rate [17]. For example, roots in 0.5 mol.m^{-3} CaCl$_2$ grew 32 mm in 24 h whereas those in 10 mmol.m^{-3} KCl grew at only 20 mm over the same period (Table 1). Turgor pressure was unchanged in the growing region suggesting that cell expansion was restricted by changes in the cell wall properties. Again this could be confirmed by measurement of the plastic extensibility of the apical region (Table 1). Potassium was effective in reducing root growth at concentrations greater than 0.1 mol.m^{-3}. Sulphate also reduced root growth by a similar effect on the cell wall properties of the growing cells. Na$^+$, Ca^{2+}, Cl$^-$ and NO$_3^-$ ions had no discernable effect on the growth rate of the roots or the cell wall properties [17].

These ion-induced changes in the cell wall properties appear to be specific to a species. For example, osmotically insignificant concentrations of potassium chloride did not effect the extension of rice roots but Mg^{2+} and NH$_4^+$ at concentration above 0.01 mol.m^{-3} and NO$_3^-$ above 1 mol.m^{-3} reduced extension, presumably by a similar effect on cell wall properties (although this was not measured directly). These specific ion effects on the cell wall may have an eco-physiological role, the faster growth rates obtained in nutrient deficient conditions may allowing the root to explore a greater soil volume.

2.5 Excised roots

The plant cell in culture is an example of a growing system stripped down to its essentials. We attempted to grow wheat roots in culture to enable a tighter experimental control over the environment of the expanding cells [18]. However, it proved difficult to re-establish the growth rate of intact roots in the excised system illustrating the subtlety of the requirements for proper cell expansion. Isolating the tissue and providing solutes for sufficient turgor generation does not guarantee normal extension. When root growing zones were excised from the plant, growth declined by 80% within an hour. The slowing of growth was accompanied by a decrease in tissue plasticity (measured as a decrease in extensibility using a tensiometer) once again indicating that changes in cell wall properties are responsible for growth cessation. Following 24 h incubation in various nutrient solutions, elongation of the roots was severely reduced yet turgor pressure was unchanged (Table 2) and in some cases increased above that of intact roots (data not shown). The hardening of the cell wall could not be loosened by addition of a variety of supplements to the incubation medium such as IAA, gibberellin, cytokinin, ABA and various pH solutions. It is clear that the plant is more than a simple aggregate of cells and

Table 2. Growth, turgor pressure and cell wall properties of the apical region of intact and excised wheat roots. Data from [18]

Treatment	Growth rate (mm 24h^{-1})	Turgor pressure (MPa)	Plastic Extensibility (% extension 8g^{-1} load)
Intact root	28	0.68±0.02	2.5±0.1
Excised root	0.1	0.62±0.02	0.9±0.1

care must be taken with data from experiments using excised tissue and plant cell cultures if extrapolation to the behaviour of whole plants is required.

3. Cell wall biochemistry

Alterations in cell growth are largely caused by changes in cell wall properties and not by variation in turgor pressure. These changes must have a basis in the structure of the wall. The cell wall consists of a network of cellulose microfibrils embedded in a pectin / hemicellulose matrix. The microfibrils are inelastic in the longitudinal direction so the orientation of the microfibril is crucial in determining the properties of the cell wall; cells tend to expand at 90 ° to the predominant microfibril orientation [15]. Microfibrils must separate as a consequence of expansion so that any increase in cross-linkage between them will restrict this extension and harden the cell wall. Xyloglucan is a component of the hemicellulose fraction of the cell wall and may be able to form cross-links between microfibrils [25, 8] . Another cross-link may be formed by the linkage of two ferulic acid residues in the wall to form di-ferulic acid cross-links. Obviously, any enzyme which can form or break these cross-linkages will have an effect on the cell wall properties.

3.1 Xyloglucan endoglycotransferase a cross-linking enzyme in the cell wall

One possible wall-crosslink may be formed by xyloglucan hemicelluloses tethering adjacent cellulose microfibrils [25]. Their length of 150–1500 nm could readily span the gap between microfibrils (about 30 nm) with each end hydrogen-bonded onto the surface of adjacent cellulose surfaces. Fry and colleagues]8] have recently characterised a xyloglucan-endotransglycosylase (XET) activity from plant walls that can cut such xyloglucan molecules. If that cross-link were load-bearing, XET activity could result in wall loosening. XET could also reform these bonds and may therefore be a control point for cell wall properties.

We measured the activity of XET along the root profile using the technique described by Fry [8] in collaboration with Drs. Fry & Hetherington at Edinburgh University (Fig. 6a). Enzymic activity increased along unstressed roots away from the tip to a maximum of 37.8×10^{-3} cpm mg dw^{-1} in the 3rd mm. It decreased proximal to this so that in the 12th mm activity was 14.5×10^{-3} cpm mg dw^{-1}. Note that maximum growth was at 4.5 mm from the tip. The increase and decrease in enzyme activity precedes acceleration and deceleration of growth in a way that suggests cause and effect. Significant enzyme activity, however, remains in fully mature tissue in which growth has stopped so that the correlation is better with wall loosening than tightening.

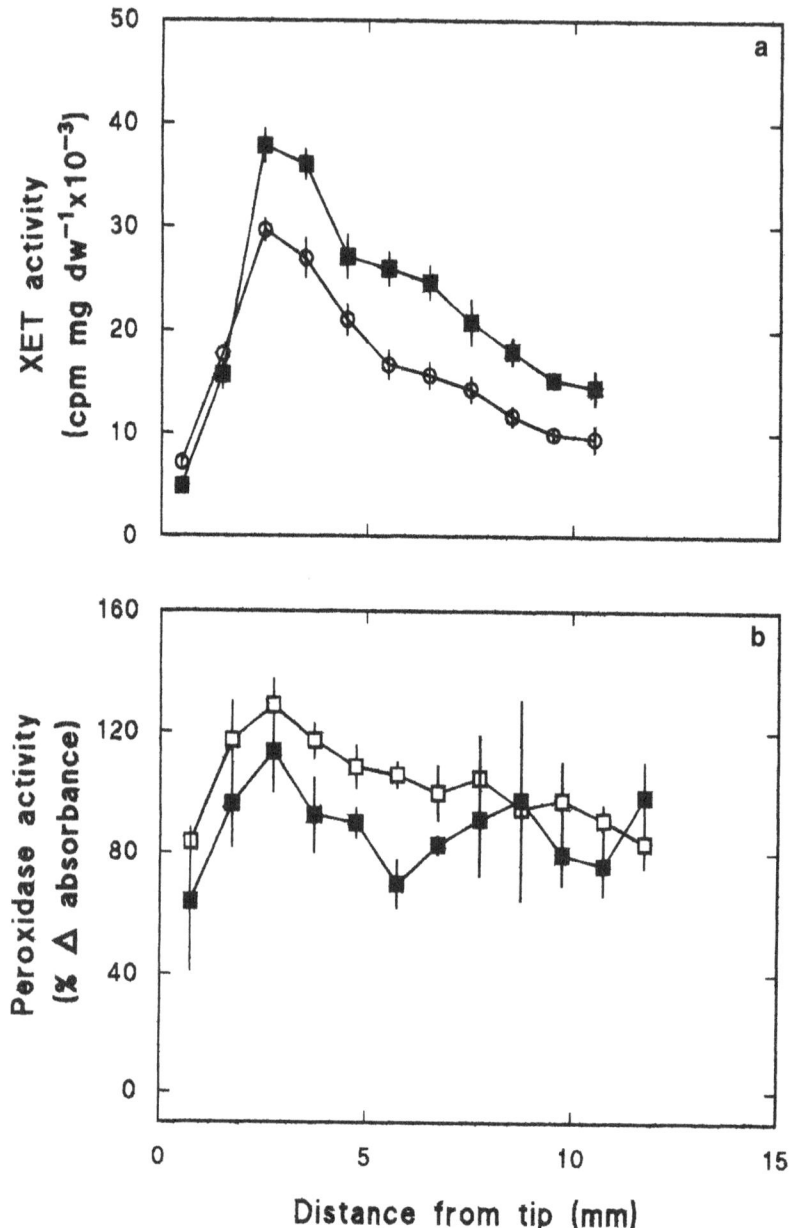

Fig. 6. Profile of wall enzyme activities. 6a; Xyloglucan endoglycosyltransferase activity along the apical 12 mm. Each point represents 4 independent measurements. Unstressed roots (■), 24 h after immersion in 400 mol.m^{-3} mannitol (0.96 MPa) (○). 6b; Peroxidase activity exactable with 1M NaCl along the apical 12 mm in unstressed (■) and roots grown in 400 mol.m^{-3} mannitol (0.96 MPa) for 24 h (○).

3.2 XET activity following 24 h osmotic stress

Following 0.96 MPa of stress the XET activity was decreased over the root apex (Fig. 6a). The decrease in this putative cell-wall enzyme therefore correlates with the stress-induced hardening of the cell wall.

3.3 Peroxidase

Peroxidase enzymes may increase cross-linkage between cell-wall components [7]. Changes in peroxidase levels might, therefore, be involved in hardening the cell wall during development as has recently been demonstrated in leaf tissue [13]. The level of enzyme activity (sequentially extracted from the wall by weak buffer, high salt and digestion of cellulose) was unchanged along the growing region, showing no correlation with the hardening of the cell wall as growth ceased (Fig. 6b). However, peroxidase activity *did* increase over the whole growth profile in roots that had been exposed to 400 mol.m^{-3} (0.96 MPa) mannitol (Fig. 10) suggesting that the increase in cell-wall peroxidase may be responsible for hardening the wall following stress.

Thus it remains unclear which factor hardens the wall in the proximal region of both stressed and unstressed roots, since an increase in peroxidase was measured in this region. It is not a prerequisite for a putative wall hardening agent that it increases as growth ceases. Peroxidase activity may be responsible for the wall hardening during development if it catalyses cross-links which gradually accumulate during development. In addition there are a large number of peroxide iso-enzymes present in plants, any one of which may form cross-links between wall components. Any relevant developmental changes in this iso-enzyme might be masked by the large background level of peroxidase activity.

A decrease in XET activity cannot be the sole cause since non-growing tissue still has significant XET activity. Perhaps during development the wall structure becomes resistant to XET. Alternatively a different wall cross-link may predominate, so that the xyloglucan cross-linkages are no longer load bearing.

4. Conclusions

The changes in wall properties discussed here must have a basis in the structure and/or biochemistry of the wall. Using the examples of changes in wall properties caused by development and by water stress it appeared that cross-linking enzymes such as XET or wall peroxidase may have a role in controlling wall extensibility. Further work localising enzymes to different cell types with antibodies and better characterisation of the substrates and products of the enzymes (xyloglucan and ferulic acid for example) will help to further clarify the control of cell extension by the wall.

Turgor pressure appears to be tightly regulated, being maintained constant despite variations in external water potential and the huge differences in dilution rate imposed by growth. The changes in solute flux that this implies deserves further investigation.

The implications of this work for cell culture and micro-propagation are, in the main, obvious. The temperature, nutrient composition and osmotic pressure of the nutrient solution are usually optimised by the experimenter. This study describes some biochemical and physiological results of manipulation of these parameters.

It is important to bear in mind that there are a number of obvious differences between cells in culture and even a simple growing system such as a root, apart from simple physical effects [28]. Only the outer surface of the epidermis is in contact with the bathing medium, whereas in culture every cell is fully exposed to the medium. In addition, each cultured cell gets its carbon from the medium but a cell embedded in a tissue maybe at the end of a complex delivery route. A cell in culture is unable to modify the huge extended apoplast of the bathing medium, so that care must be taken to provide the correct nutrients for optimising cell wall expansion. Indeed the degree of expansion may well be manipulated by quite subtle changes in the ion concentration of the bathing solution. For micro-propagated plants, root/shoot ratios and root branching may be similarly under the investigators control.

Water stress has clear effects on intact plants but its effect on cultured cells may be less obvious, moderate stress will increase cell osmotic pressure and/or alter wall biochemistry, more extreme stress could result in loss of turgor an cessation of expansion. Whatever the differences between culture cells and intact plants, it is clear both can throw light onto the study of expansion. It is to be hoped that what is known about cell expansion will prove useful to those working on cultured cells.

Acknowledgements

I would like to thank Dr. A.D. Tomos for his encouragement and continuing input of ideas throughout this work.

References

1. Boyer JS (1985) Water transport. Ann Rev Plant Physiol 36:473–516
2. Cleland RE (1983) The capacity for acid induced wall loosening as a factor in the control of *Avena* coleoptile cell elongation. J Exp Bot 34:676–680
3. Cleland RE (1984) The Instron as a measure of immediate past wall extensibility. Planta 160:514–520
4. Cosgrove DJ (1984) Hydraulic aspects of plant growth. Whats new in Plant Physiol 15:5–8
5. Cosgrove DJ (1986) Biophysical control of plant cell growth. Ann Rev Plant Physiol 37:377–405
6. Edwards KL and Scott TK (1974) Rapid growth responses of corn root segments: Effect of pH on elongation. Planta 119:27–37

7. Fry SC (1987) Formation of isodityrosine by peroxidase isoenzymes. J Exp Bot 38:853–862

8. Fry SC, Smith RC, Renwick KF, Martin DJ, Hodge SK and Matthews KJ (1992) Xyloglucan endotransglycosylase, a new wall-loosening enzyme-activity from plants. Biochem J 282: 821–828

9. Grecean EL and Oh JS (1972) The physics of root growth. Nature 235:24

10. Hüsken D, Zimmermann U and Steudle E (1978) Pressure probe technique for measuring water relations of cells in higher plants. Plant Physiol 61:158–163

11. Jones H, Leigh RA Wyn Jones RG and Tomos AD (1988) The integration of whole root and cellular hydraulic conductivities in cereal roots. Planta 174:1–7

12. Lockhart JA (1965) An analysis of irreversible plant cell elongation J Therol Biol 8:264–275

13. MacAdam JW, Nelson CJ and Sharp RE (1992) Peroxidase activity in the leaf elongation zone of tall fescue. I. Spatial distribution of ionically bound peroxidase activity in genotypes differing in length of the elongating zone. Plant Physiol 99:872–878

14. Nobel PS (1974) Introduction to Biophysical Plant Physiology. San Fransisco: Freeman

15. Preston RD (1974) The Physical Biology of Plant Cell Walls. London: Chapman and Hall

16. Pritchard J, Adam JS, Barlow PW and Tomos AD (1990) Biophysics of the inhibition of root growth by low temperature. Plant Physiol 93:222–230

17. Pritchard J, Tomos AD and Wyn Jones RG (1987) Control of wheat root extension growth I effects of ions on growth rate, wall rheology and cell water relations. J Exp Bot 38:948–959

18. Pritchard J, Wyn-Jones RG and Tomos AD (1988) Control of wheat root growth; the effects of excision on root growth, wall rheology and root anatomy. J Exp Bot 176:399–405

19. Pritchard J, Wyn Jones RG and Tomos AD (1990) Measurement of yield threshold and cell wall extensibility of intact wheat roots under different ionic, osmotic and temperature treatments. J Exp Bot 412:669–675

20. Pritchard J, Wyn-Jones RG and Tomos AD (1991) Turgor, growth and rheological gradients of wheat roots following osmotic stress. J Exp Bot 42:1043–1049

21. Ray PM, Green PB and Cleland RE (1972) Role of turgor in plant growth. Nature 239:163–164

22. Sanderson J, Whitbread FC and Clarkson DT (1988) Persistent xylem cross-walls reduce the axial hydraulic conductivity in the apical 20 V of barley seminal root axes:Implications for the driving force for water movement. Plant Cell Physiol 11:247–256

23. Sharp RE, Silk WK and Hsiao TC (1988) Growth of the primary maize root at low water potentials I spatial distribution of expansive growth. Plant Physiol 87:50–57

24. Silk WK and Wagner KK (1980) Growth sustaining water potential distributions in the primary corn root. Plant Physiol 66:859–863

25. Smith RC and Fry SC (1991) Endotransglycosylation of xyloglucans in plant cell suspension cultures. Biochem J 279:529–535

26. Taylor G and Davies WJ (1986) Yield turgor of growing leaves of *Betula* and *Acer*. New Phytol 104:347–353

27. Tomos AD (1985) Physical limitations of leaf cell expansion In: NR Baker, WJ Davies and CK Ong, eds. Control of leaf growth, pp 1–33. SEB seminar series, Cambridge: Cambridge University Press

28. Tomos AD, Malone M and Pritchard J (1989) The biophysics of differential growth. Environ Exp Bot 29:7–24

29. Van Volkenburgh E, Hunt S and Davies WJ (1983) A simple instrument for measuring cell wall extensibility. Ann Bot 51:669–672

Some aspects of stomatal physiology relevant to plants cultured *in vitro*

T.A. MANSFIELD

Division of Biological Sciences, Institute of Environmental and Biological Sciences, Lancaster University, Lancaster LA1 4YQ, UK

1. Introduction

Stomatal pores allow, and by their opening and closing movements regulate, most of the gaseous exchange between the leaves of higher plants and the atmosphere. They achieve a delicate balance between two of the fundamental requirements of plants: the acquisition of CO_2 for photosynthesis and prevention of excessive loss of water by transpiration. Most mesophytic plants contain only a small amount of water compared with the potential rate of transpiration on a dry sunny day. This means that a breakdown in stomatal regulation of water loss, even for a time as short as a few minutes, can cause severe damage or death.

The necessary degree of control of gaseous exchange is achieved by a very complex array of responses of stomata both to environmental factors and to the physiological state of the plants. The stomata can justifiably be looked upon as miniature sense organs. In the terminology used for animals we might say that they possess the senses of sight, smell, touch and taste: they react to light, to the gaseous composition of the atmosphere, to mechanical stimuli such as the vibrations caused by wind, and to chemicals that indicate the presence of other organisms on the leaf surface [6]. Furthermore they can adjust their pattern of behaviour in response to intercellular signalling mechanisms that reflect the physiological status of different parts of the plant.

There are many different ways in which abnormalities of stomatal functioning can be induced. They can be conveniently categorised as (1) disturbances in the normal mechanical relations between cells in the epidermis, and (2) changes in factors which determine the turgor of guard cells.

2. Mechanical disturbances

Stomatal opening is achieved mainly by an increase in turgor of the guard cells. It is easy to construct a working model that displays realistic changes in dimensions of the pore as the pressure is increased within two inflatable

120

P.J. Lumsden, J.R. Nicholas and W.J. Davies (eds.), Physiology, Growth and Development of Plants in Culture, 120–131, 1994.
© 1994 *Kluwer Academic Publishers.*

structures representing the guard cells (Plate 1). In this model the radial construction of the cell walls is simulated, and also the heavy thickening along the walls that border the stomatal pore. If either of these features is improperly developed the opening/closing mechanism is unlikely to function correctly. Guard cells that are enlarged in one or more dimensions are usually the outcome of a failure to develop the normal cell wall structure during development. A disturbed appearance of the guard cells can, however, occur even when they themselves are normal in structure. The turgor of the surrounding cells in the epidermis (in some cases these are specialised subsidiary cells) is important, and if a cell on one side of the stomatal apparatus loses turgor the unilateral drop in pressure causes the appearance seen in Plate 2. Distortion of this kind is often seen on epidermis that has been peeled from a leaf for examination under the microscope. It is difficult to obtain epidermal peels without causing damage to some cells. It is, however, unlikely that stomatal malfunctioning on intact leaves will often occur in this manner.

If the development of leaves produces stomata of abnormal structure that cannot function mechanically like the model in Plate 1, there is little point in exploring further for deficiencies in physiological processes. There are, however, many ways in which stomata that are fully capable of functioning normally can be disabled or changed profoundly in their responses.

3. Physiological disturbances

It is now well established that the turgor changes of guard cells are mainly determined by the uptake and extrusion of potassium ions, sometimes accompanied by anions such as chloride [20]. The plasma membranes of the guard cells contain ATP-fuelled proton pumps, and the primary event in stomatal opening appears to be a transport of protons from inside the guard cells to the apoplastic space surrounding them. This produces an electrical potential between the interior and exterior of the cells (the interior becoming more negative) which represents a driving force for the inward movement of cations. Potassium is the main ion to respond, probably because of its abundance in the apoplast and because of the presence of suitable channels in the membranes. The proton pump may be regarded as the initiator of the sequence of processes leading to changes in guard cell turgor, and it should perhaps be the first one to consider when we recognise any tendency for stomata to display abnormal opening. The toxin known as fusicoccin, which is produced by the fungus *Fusicoccum amygdali* causes stomata to open to such an extent that leaves wilt and die. It is known to cause a massive efflux of protons across the plasma membranes of many cell types [11], and when it does this in guard cells they accumulate large amounts of potassium and become totally unresponsive to factors that normally cause them to lose turgor, e.g. darkness. Fusicoccin thus leads to uncontrolled stomatal opening and all their regulatory responses are overridden. Similar disturbances, though smaller in

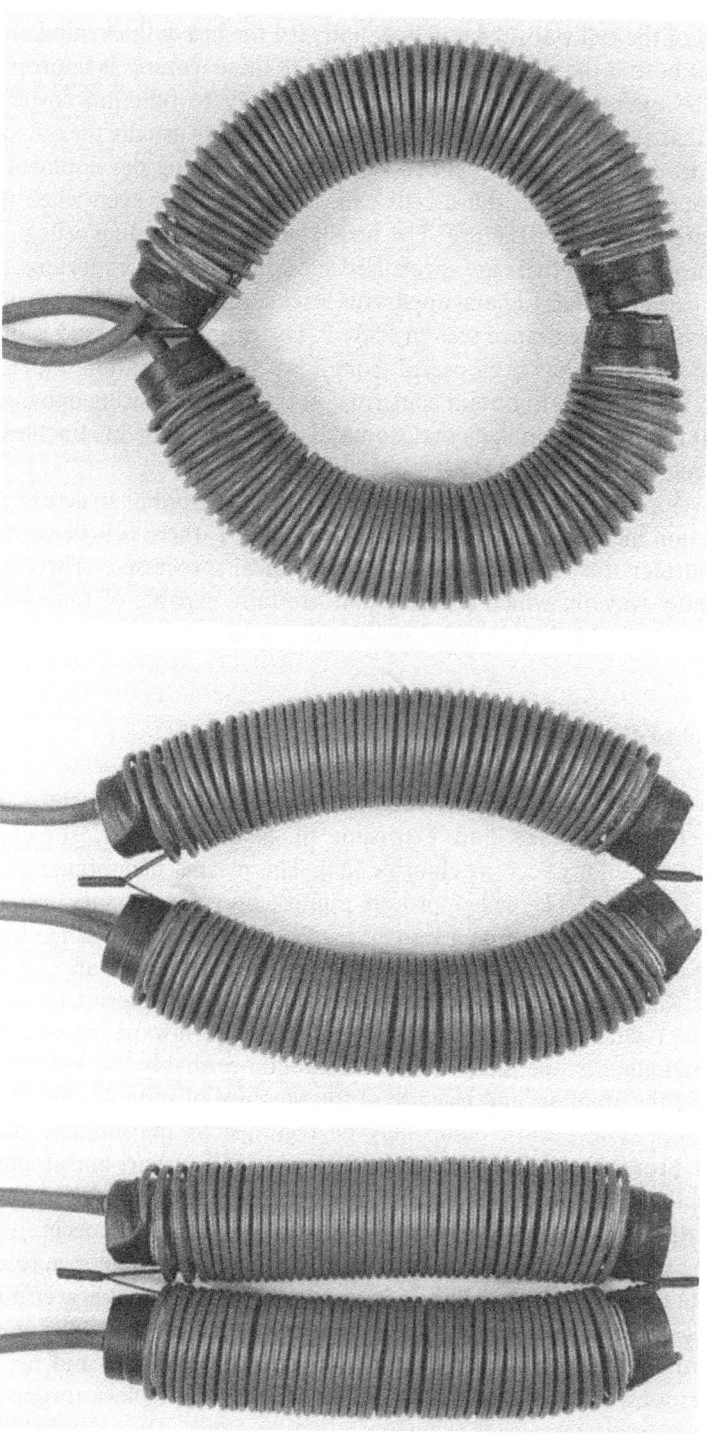

Plate 1. A model of the stomatal apparatus constructed in the author's laboratory by A.J. Travis. Radial micellation within the guard cell walls is represented by a spiral of stiff wire, and the guard cells are fixed together at their ends by a strip of flat plastic that passes through the spiral of wire adjacent to the pore. Inflation of the pneumatically operated model gives a realistic representation of the shape of the stomatal pore in the living plant.

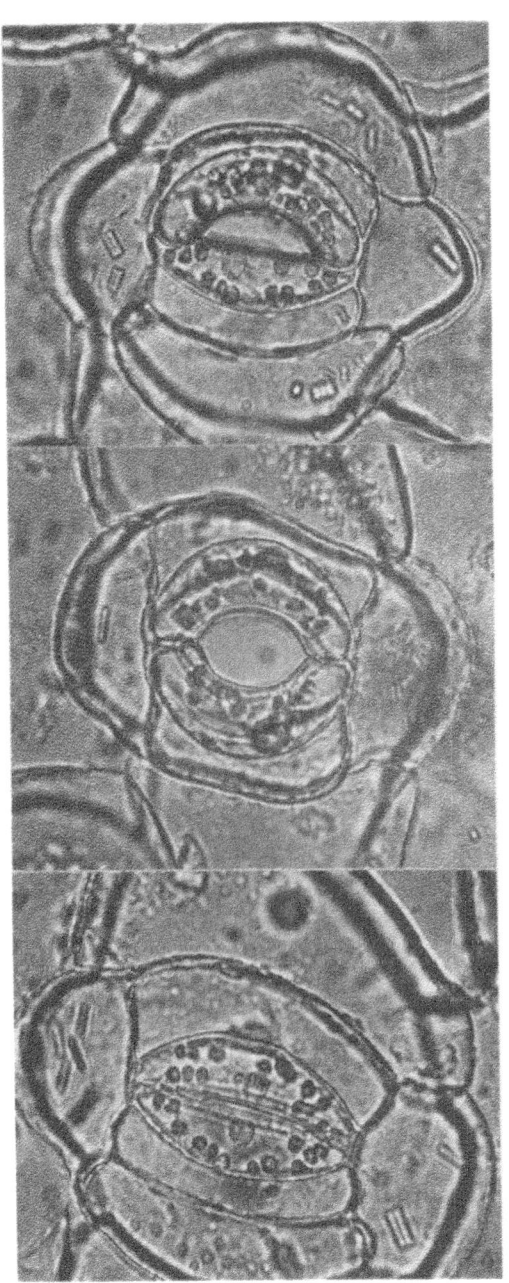

Plate 2. Left and centre: closed and open stomata of *Commelina communis* as viewed under the light microscope. Right: asymmetric pore which is probably the result of reduced turgor in the subsidiary cell adjacent to the guard cell on the right.

magnitude, have been identified as a result of the action of endogenous plant hormones.

3.1 Hormonal control of stomata

The true significance of the control of stomata by endogenous hormones only became apparent in the late 1960s and early 1970s after the discovery and identification of abscisic acid (ABA). It was found in greatly increased quantities in leaves that were allowed to wilt, and it was soon discovered that application of synthetic ABA to leaves caused substantial and prolonged stomatal closure. Thereafter for about 15 years it was generally believed that ABA operated as a 'stress hormone', being produced in the leaves of plants when they began to experience water deficit stress in the form of reduced water potential and/or loss of turgor. The ensuing stomatal closure would then reduce transpiration and create an improvement in leaf water relations [7].

More recently it has been discovered that ABA can provide control of water loss from leaves *before* they experience any measurable water stress. When a plant is growing in drying soil the roots are able to 'sense' the declining availability of water. They then pass the information to the leaves and the stomata close partially, and leaf expansion may be reduced, to improve the economy of water use. There is strong evidence that ABA is produced by the roots and that this constitutes the signal that passes, via the transpiration stream in the xylem, to the leaves. These new discoveries, which have revolutionised our thinking about the way in which plant water relations are regulated, have been described in detail by Davies and Zhang [2]. It is probably true to say that the regulatory role of ABA is essential for the survival of land plants in all but the least stressful of environmental conditions.

Mutants deficient in ABA content have been discovered in several plant species [18]. These are usually called 'wilty' mutants because they cannot maintain their turgor when they are grown in a normal environment that imposes a significant transpirational load upon the leaves. They do, however, survive and grow normally if they are provided with a regular supply of exogenous ABA. The studies of ABA-deficient mutants have also revealed abnormalities in the development of guard cells. It has been found that the stomata of some wilty tomato mutants remain open even when the guard cells are plasmolysed. Tal, Imber & Gardi [17] noted that if mutant leaves that were fully developed were treated with ABA, the stomata were incapable of closing even when all the turgor of the guard cells was lost during plasmolysis. However, if the leaves of mutants were supplied with ABA during development, the stomata were able to close in the same way as those of the normal genotype.

These observations of the development of non-functional,or partly disabled, stomata in ABA-deficient plants are likely to be important in relation to micropropagated material which has been grown in an environment free of water deficit stress. The virtual absence of ABA during leaf development may result in stomatal guard cells with wall structures that are incapable of allowing stomatal closure even when a severe water deficit is experienced.

Another matter that may be important is the interference with the action of ABA caused by other plant hormones. Indole-3-acetic acid (IAA) had not been thought to play a major part in regulating stomata until the work of Pemadasa [12]. He discovered that stomata on the adaxial (upper) surfaces of leaves which usually open less widely than those on abaxial surfaces, could be induced to function normally if they were supplied with IAA. It was found by Snaith & Mansfield [15, 16] that the dose-response curves for ABA-induced closure of stomata in epidermal strips were greatly altered by the inclusion of IAA in the incubation medium (Fig. 1). In the presence of a very high (physiologically extreme) concentration of IAA (10^{-1} mol m^{-3}) the inhibitory action of ABA was virtually abolished. It was also found that IAA had no effect on stomatal aperture when epidermis was incubated in a CO_2-free environment in the light,

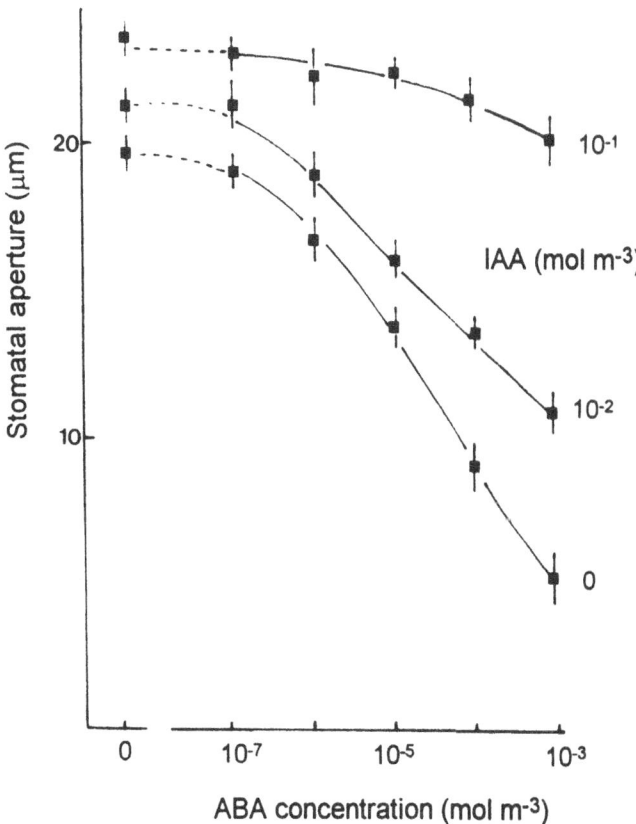

Fig. 1. The influence of IAA on the response of stomata of *Commelina communis* to ABA. A favourable concentration of KCl (100 mol m^{-3}) was used to support wide stomatal opening in the light. Abaxial epidermis was incubated for 3 h in the presence of the different concentrations of the two hormones. 95% confidence limits are shown. Redrawn from data of Snaith & Mansfield [15].

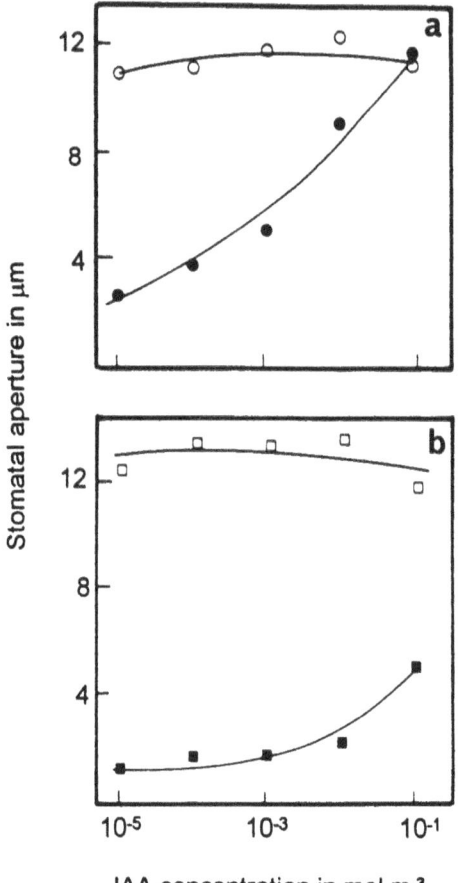

Fig. 2. (a) Stomatal opening on detached abaxial epidermis of *Commelina communis* after incubation for 3 h in light in different IAA concentrations. o = Zero CO_2, • = 700 μl l^{-1} CO_2 . Points are means of 60 measurements; (b) As in (a) but with 10^{-1} mol m^{-3} ABA in the medium. o = zero CO_2, ■ = 700 μl l^{-1} CO_2. Redrawn from Snaith & Mansfield [16].

but in the presence of CO_2, opening was strongly influenced by an increasing supply of IAA (Fig. 2a). The suppression by IAA of the inhibitory effect of CO_2 on stomata could be overcome by applying 10^{-5} mol m^{-3} ABA (Fig. 2b).

These experiments revealed a complex interplay between CO_2, IAA and ABA in determining stomatal aperture. The closure of stomata caused by increasing CO_2 concentrations is thought to be a part of an elegant system of controls by which the plant regulates its gas exchange and consumption of water [8]. It appears that this system can be disrupted by an excess of IAA in the epidermal tissue. Even if the plant forms increased amounts of ABA as it experiences water stress there may be no stomatal closing response if the concentration of IAA is high. Similarly, if wind disturbs the boundary layer and delivers more CO_2 to the CO_2-sensitive regions of the guard cells, the water-conserving closing response of the stomata will not occur in the presence of excess IAA.

A similar situation appears to occur in the presence of high concentrations of cytokinins. Ten different cytokinins, some natural and some synthetic, were found to promote stomatal opening in a survey by Jewer and Incoll [4]. In many cases, however, application of exogenous cytokinins to leaves does not have a large effect on the stomata, probably because endogenous cytokinins are present in sufficient quantities. The most dramatic effects have been found when cytokinins are applied in the presence of ABA. Blackman and Davies [1] found little response in young maize leaves to the application of zeatin or kinetin. When stomatal closure was induced by ABA, on the other hand, high concentrations of both these cytokinins could overcome the inhibitory effect (Fig. 3).

There appears to be a complex interplay between different hormones in the determination of stomatal behaviour. Changes in the relative concentrations of endogenous hormones almost certainly play a vital role in imposing a ceiling on the extent to which stomata can open. There is thus a mechanism for physiological adjustment to different conditions in the aerial and edaphic environments. When a micropropagated plant is transferred from the benign environment in which it has been raised to one which allows a normal rate of transpiration, it is challenged to adjust its hormonal levels more quickly than is rarely, if ever, likely to be required in a natural situation. It faces additional problems if it has been grown in the presence of high concentrations of auxins and/or cytokinins, because they may counteract the action of any endogenous ABA that is produced in order to close the stomata and provide protection.

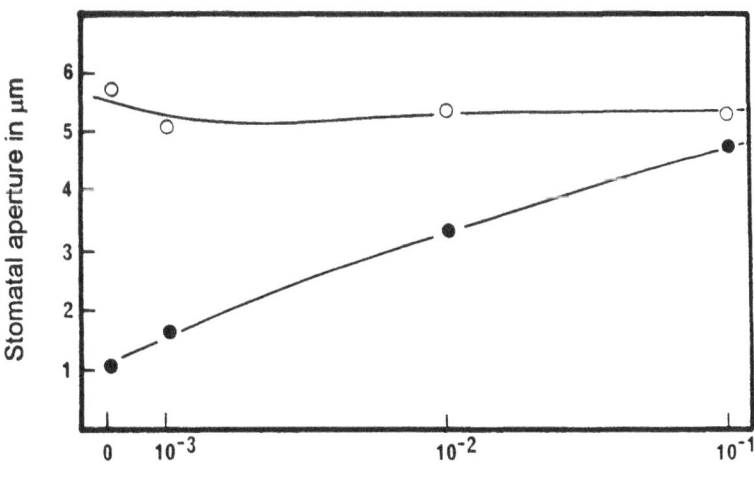

Fig. 3. Stomatal apertures on maize leaf pieces incubated on a range of kinetin concentrations with (•) or without (o) 10^{-1} mol m^{-3} ABA. Means of 60 observations. Redrawn from data of Blackman & Davies [1].

3.2 Disruption of normal ionic relationships

The movements of potassium into and out of guard cells as stomata open and close are well documented. The changes in concentration are on a scale that is much larger than in most plant tissues. In *Commelina communis*, for which more information is available than for any other species, the K^+ concentration in the guard cells of closed stomata is below 100 mol m^{-3}, and climbs to over 800 mol m^{-3} when the stomatal pore attains its full aperture [5].

There are no plasmodesmatal connections between mature guard cells and their neighbours. This means that the membrane-based processes that achieve the active transport of K^+ have to operate in the context of the concentrations of the ion in the apoplastic space between the guard cells and their neighbours. Attempts to estimate these concentrations have not been satisfactory, but evidence from experiments on stomatal behaviour in isolated epidermis suggests that the apoplastic K^+ concentration is of great importance in determining the normal responsiveness of stomata to light, CO_2, ABA and other factors. When isolated epidermis is incubated in a stirred aqueous medium it is possible to impose changes in K^+ concentration in the apoplastic space simply by altering the concentration in the medium. Figure 4 shows the effect of increasing the K^+ concentration from zero to 200 mol m^{-3} on the equilibrium stomatal aperture achieved after 3 h of incubation. The change in K^+ concentration not only increased the maximum aperture achieved, but also altered the sensitivity of the

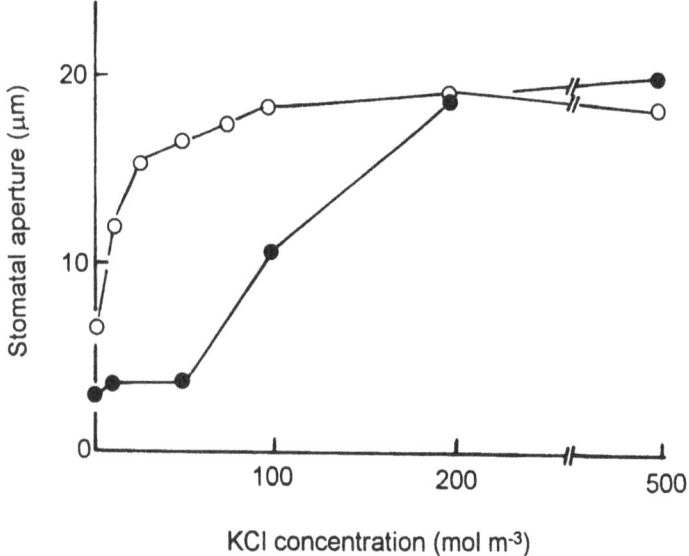

Fig. 4. Apertures of isolated stomata of *Commelina communis* incubated for 2.5 h in 0–500 mol m^{-3} KCl solutions with (•) or without (o) 10^{-1} mol m^{-3} ABA. Each point is the mean of 30 measurements. Standard errors are within the limits of the insignia. Before being transferred to the final incubation media, stomata were open to 17.0±0.3 mm. Redrawn from Wilson *et al.* [19].

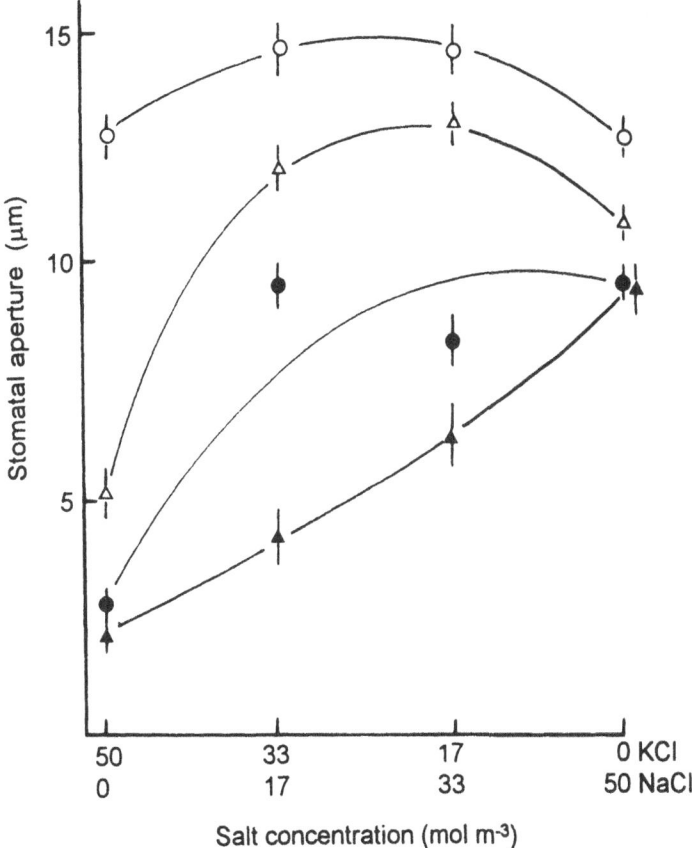

Fig. 5. The effect of different proportions of NaCl and KCl on stomatal aperture under various conditions. o = light, – CO_2; ● = dark, – CO_2; △ = light, + CO_2; ▲ = dark, + CO_2. Vertical bars represent 90% confidence limits. Redrawn from data of Jarvis & Mansfield [3].

stomata to light, CO_2 concentration and ABA. The maximum sensitivity was at 50 mol m^{-3} K$^+$ but at 200 mol m^{-3} K$^+$ and above there were no statistically significant effects of any of these agents. Figure 4 shows how the ability of stomata to respond to ABA is profoundly affected by the concentration of K$^+$ in the medium both above and below the 'optimum'. These data were obtained using 'isolated' stomata, i.e. epidermis in which all the cells apart from the guard cell pairs had been inactivated.

Stomatal guard cells of some plant species appear to be unable to discriminate between sodium and potassium ions. In the case of *Commelina communis*, Na$^+$ and K$^+$ ions are able to support stomatal opening equally well. Figure 5 shows that when the proportions of the two ions were changed but the total concentration was maintained at 50 mol m^{-3}, there was no significant change in stomatal opening. However, as the proportion of Na$^+$ ions was progressively increased the magnitude of the response to two 'closing' treatments, darkness and CO_2, decreased. Na^+ ions can accumulate in guard cells and support stomatal opening, but it appears that mechanisms for the extrusion of these ions to achieve stomatal closure may be lacking.

Taken together, these data show clearly that the ionic balance in plants cultured *in vitro* may be very important in determining the ability of stomata to respond correctly to environmental factors when the plants are transferred to a normal aerial environment. An excessive accumulation of K^+ or Na^+, may stimulate wide stomatal opening and a much reduced ability to close in response to agents such as darkness, CO_2 and ABA.

3.3 The importance of calcium ions

Although some effects of calcium on stomata have been known for a long time, it is only since 1985 that the essential role of calcium in the normal functioning of guard cells has become apparent. In 1985 there was a major step forward in understanding the mode of action of ABA, when it became clear that the hormone interacts with calcium ions as it induces stomatal closure. At that time it was already well established that calcium regulates many cellular activities in animals through its involvement in hormonal responses, and in signalling between cells. It was found that some of the drugs that block calcium channels in animal cells, and also lanthanum ions which have the same effect, can reduce or completely inhibit the response of guard cells to ABA [9].

Subsequent research has shown that concentration of calcium within the cytoplasm of the guard cells is all-important. When ABA is applied to guard cells a rapid rise in the intracellular concentration of calcium occurs within a few minutes [10, 14]. If ABA is regarded as the hormonal 'first messenger', then calcium becomes the 'second messenger' which induces a sequence of reactions in the guard cells. The first of these is probably to bind with the small protein calmodulin. The calmodulin-Ca^{2+} complex is then thought to bring about various cellular responses by selectively stimulating specific enzymes that are involved in regulating the ionic balance of the guard cells.

There is good evidence that other factors that cause stomatal closure, e.g. darkness, may also require the involvement of calcium ions [13]. An insufficient supply of calcium to a plant grown *in vitro* is likely to reduce the ability of its stomata to close in response to external and endogenous stimuli. The normal tight regulation of gas exchange to prevent excessive water loss may therefore be prevented.

References

1. Blackman PG and Davies WJ (1983) The effects of cytokinins and ABA on stomatal behaviour of maize and *Commelina*. J Exp Bot 34:1619–1626
2. Davies WJ and Zhang J (1991) Root signals and the regulation of growth and development of plants in drying soil. Ann Rev Plant Physiol & Plant Mol Biol 42:55–76

3. Jarvis RG and Mansfield TA (1980) Reduced stomatal responses to light, carbon dioxide and abscisic acid in the presence of sodium ions. Plant Cell and Environ 3:279–283

4. Jewer PC and Incoll LD (1980) Promotion of stomatal opening in the grass *Anthephora pubescens* Nees by a range of natural and synthetic cytokinins. Planta 150:218–221

5. MacRobbie EAC (1987) Ionic relations of guard cells. In: E Zeiger, GD Farquhar and IR Cowan, eds. Stomatal Function, pp 125–162. Stanford, California: Stanford University Press.

6. Mansfield TA (1986) The physiology of stomata : new insights into old problems. In: FC Steward, JF Sutcliffe, JE Dale, eds. Plant Physiology, a Treatise, volume IX, pp 155–224. New York: Academic Press

7. Mansfield, TA, Wellburn AR and Moreira TJS (1978) The role of abscisic acid and farnesol in the alleviation of water stress. Phil Trans Roy Soc Lond B284:471–482

8. Mansfield TA and Davies WJ (1985) Mechanisms for leaf control of gas exchange. BioScience 35:158–164

9. Mansfield TA, Hetherington AM and Atkinson CJ (1990) Some current aspects of stomatal physiology. Ann Rev Plant Physiol & Plant Mol Biol 41:55–75

10. McAinsh MR, Brownlee C and Hetherington AM (1990) Abscisic acid induced elevation of guard cell cytosolic Ca^{2+} precedes stomatal closure. Nature 343:186–188

11. Marrè E (1979) Fusicoccin :a tool in plant physiology. Ann Rev Plant Physiol 30:273–288

12. Pemadasa MA (1982) Differential abaxial and adaxial stomatal responses to indole – 3-acetic acid in *Commelina communis* L. New Phytol 90:209–219

13. Schwartz A (1985) Role of Ca^{2+} and EGTA on stomatal movements in *Commelina communis* L. Plant Physiol 79:1003–1005

14. Schroeder JI and Hagiwara S (1990) Repetitive increases in cytosolic Ca^{2+} of guard cells by abscisic acid activation of nonselective Ca^{2+} permeable channels. Proc Nat Acad Sci USA 87:9305–9309

15. Snaith PJ and Mansfield TA (1982) Stomatal sensitivity to abscisic acid: can it be defined? Plant Cell & Environ 5:309–311

16. Snaith PJ and Mansfield TA (1982) Control of the CO_2 responses of stomata by indol-3-ylacetic acid and abscisic acid. J Exp Bot 33:360–365

17. Tal M, Imber D and Gardi I (1974) Abnormal stomatal behaviour and hormonal imbalance in *flacca,* a wilty mutant of tomato. Effect of abscisic acid and auxin on stomatal behaviour and peroxidase activity. J Exp Bot 25:51–60

18. Taylor IB (1991) Genetics of ABA synthesis. In: WJ Davies and HG Jones, eds. Abscisic Acid: Physiology and Biochemistry, pp 23–37. Oxford: Bios Scientific Publishers

19. Wilson JA, Ogunkanmi AB and Mansfield TA (1978) Effects of external potassium supply on stomatal closure induced by abscisic acid. Plant Cell & Environ 1:199–201

20. Zeiger, E, Farquhar GD and Cowan IR (1987) Stomatal Function. Stanford, California : Stanford University Press

Effects of low light intensity and high air humidity on morphology and permeability of plant cuticles, with special respect to plants cultured *in vitro*

G. KERSTIENS

Lancaster University, Institute of Environmental and Biological Sciences, Division of Biological Sciences, Lancaster LA1 4YQ, UK

1. Introduction

It has become something of an established fact that plants cultured *in vitro* from tissue explants have less cuticular waxes than normal plants. This appears plausible, for in the virtually water vapour-saturated atmosphere of the culture vessel they don't 'need' a good water barrier on their surface, and thus wax synthesis is more or less superfluous. When plantlets are removed from the culture vessels after some weeks and exposed to the greenhouse atmosphere for further cultivation, rapid desiccation and eventually death occur frequently and cause great economic losses at this stage. It was found that with carnation plantlets the survival rate after removal from the culture vessels was 25 times higher if the plantlets had formed glaucous leaves *in vitro*, as compared to the majority of plantlets which had formed non-glaucous leaves [33]. From findings such as this, the general conclusion has been made that high cuticular water loss contributes substantially to the desiccation problem, in addition to transpiration through non-functional stomata and water uptake problems due to underdeveloped or absent roots.

It is the aim of this contribution to review the evidence for these conceptions and to ask how close they are to reality. We will ask whether an apparent lack of structural waxes on the plant surface, as seen by scanning electron microscopy (SEM), bears any relationship to the effectiveness of the cuticle as a transpiration barrier, and whether it is altogether possible with today's knowledge of cuticular function to draw any conclusions from morphological observations of the cuticle on its properties as a diffusion barrier. Since conditions of low light and very high humidity are typical for *in vitro* culture, studies on the effects of these two variables on cuticular traits of normal plants have been included to supplement the data on vessel-grown plants.

The cuticle is a membrane of very complex structure which covers all primary aerial parts of plants [2, 10]. Imbedded in a highly cross-linked and insoluble polyester network, the cutin (which may also contain more polyethylene-like domains), is a reticulum of polysaccharide fibres and a mixture of mostly

132

P.J. Lumsden, J.R. Nicholas and W.J. Davies (eds.), Physiology, Growth and Development of Plants in Culture, 132–142, 1994.
© 1994 *Kluwer Academic Publishers.*

Table 1. Effects of low light intensity and/or high air humidity on thickness, wax content and permeability of cuticular membranes.

Effect on ...	thickness				wax content				permeability[a]			
Type of treatment[b]	i	h	s	v	i	h	s	v	i	h	s	v
Species [reference]												
Brassica napus [36]							-					
Brassica oleracea												
var. *botrytis* [8]								-				
-- [35]								-				
var. *capitata* [34]								-				
var. *gemmifera* [1]					-	-						
-- [18, 19]		-			o				o			
Chrysanthemum morifolium [31]								-				
Citrus aurantium[c]		-				o				o		
-- [7, 22]						o				o		
Citrus mitis [4]			o				o					
Delphinium sp.[d]												o
Dianthus caryophyllus [18,19]		-				-				o		
-- [31]								-				
-- [37]								-				
Eucalyptus spp. [9]							-					
Ficus elastica[c]		-				-				o		
Hedera helix [30]			-			-						
Malus domestica (fruit) [15]							o					
Maranta leuconora [31]								o				
Pisum sativum [11]							-					+
-- [12]			-									
Quercus velutina [17]			-									
Spathiphyllum wallissii [31]								+				
Vitis vinifera (fruit) [23]			-				-					

a Thickness, wax content or permeability: - reduced, o not affected, + increased by treatment. b Treatments: i: reduced irradiance, h: increased air humidity, s: shading (or other combination of reduced irradiance and increased air humidity), v: *in vitro* culture with moisture saturated air as compared to seedlings grown in the culture room outside vessels [8, 31, 34], or as compared to plantlets grown in vessels with reduced air humidity [34, 35, 37]. c G Kerstiens and J Schönherr, unpublished. d G Kerstiens, unpublished.

saturated aliphatic long-chain compounds. These compounds, which can be removed from the cuticle with organic solvents, are usually referred to as 'waxes' and typically contain amongst other compounds alkanes, alcohols, fatty acids and their respective esters. The insoluble mixture of cutin and polysaccharides is called the cuticular matrix. The wax molecules are at least partly arranged in structures with a very high degree of order. These crystallites occur within the matrix (intracuticular waxes) as well as on top of it (epicuticular waxes). In the latter case they may be visible with the scanning electron microscope on the plant surface as rods, platelets, dendrites, tubules etc., and may cause a glaucous appearance of the leaf. The microscopic appearance of epicuticular waxes as 'amorphous', that is without visible crystalline structure, does not necessarily imply that those waxes are amorphous or non-crystalline in a strict physical sense, however. That is why the terms 'structural' and 'non-structural' epicuticular waxes are preferred here to 'crystalline' and 'amorphous'.

Our current understanding of how the cuticle works as a water barrier on the leaf surface is quite detailed yet incomplete (e.g. with respect to the role of the polysaccharide reticulum in transport phenomena) [16, 22, 26]. There is no indication that water traverses the cuticle by bulk flow through pores. Instead, water and any other non-ionic substance diffuse through the cuticle principally in the same way as it diffuses, for instance, through plastic film. At least in the relatively thin region towards the outer surface of the cuticle which has been found to determine the overall permeability of the membrane, each water molecule has to find its way through the polymer network and around regions which are rendered virtually inaccessible by wax crystallites. Wax removal increases water permeability by two to three orders of magnitude. In accordance with established models for synthetic membranes, it is assumed today that it is mainly the number of 'paths' around those crystallites inside the cutin network, or in other words the distribution and size of those wax crystallites, that determine the water permeability of the cuticle. As soon as the diffusing molecules reach the surface, diffusion around epicuticular wax structures is relatively easy due to the much higher diffusivity in air as compared to inside the cutin network, and thus contributes only little to the total resistance to diffusion from the epidermal cell wall to the atmosphere.

2. The evidence

The results of studies where the effects of different light intensities and/or air humidities on cuticular traits were investigated have been summarised in Table 1. Shading experiments (usually carried out without forced ventilation of the air) are shown separately because in addition to the reduction in photon fluence rate a decrease in leaf and air temperature is likely to occur, resulting in a higher relative air humidity (r.h.) and lower water vapour concentration difference between leaves and air. It is obvious from Table 1 that both a reduction in light intensity and an increase in air humidity tend to decrease the amount of wax and the thickness of the cuticular membrane.

With one exception [17], all information given in Table 1 about effects on the thickness of the cuticular membrane has been inferred from data about the weight per unit surface area of cuticles isolated enzymatically from the leaf. If the mass of the cuticle per unit area decreases, then the membrane can have become thinner, or it can have become less dense. Since it was found that the densities of cuticles from a broad variety of species were remarkably similar [25], effects on cuticle weight per area have been interpreted here as a change in membrane thickness.

An interesting case in terms of the methods applied to measure cuticular thickness is presented in a paper by Reed and Tukey [18]. They found that there was less mass of the cuticle per unit surface area in the low light treatment, which is in accordance with all the other findings, whereas the transmission electron microscopical investigation of leaf sections stained with heavy metal salts revealed thicker cuticles in the low light environment with both species tested. Assuming that the density of the cuticle hadn't changed much, one has to conclude that the ultrastructure of the cuticle had been changed, thus influencing the distribution of the stain. On the other hand, Osborn and Taylor [17] used a very similar staining technique, and they found that leaves grown in the shade had thinner cuticles.

As to the amount of wax, visual observations with the SEM and analytic determination of the amount of wax recovered from the cuticle by dissolution in organic solvents using a wide variety of methods have been pooled in Table 1. Whereas semi-quantitative SEM studies are obviously restricted to structural epicuticular waxes, there is no way to decide to what extent intracuticular and epicuticular waxes have been recovered as a result of a short dipping or a longer period of immersion of the leaf in an organic solvent. Quantitative comparisons between data from different studies using different solvents and techniques and periods of solvent applications, and even between data from a single study referring to cuticles which might differ in structure, thickness or wax composition are virtually impossible.

Riederer and Schneider [21] found that immersion of *Citrus aurantium* L. (bitter orange) leaves for 30 min in hexane yielded only ca. 15% (3 μg cm^{-2}) of the total wax content of the cuticle (20 μg cm^{-2}), and a 30 min immersion in chloroform yielded ca. 30%. Longer immersion periods resulted in the leakage of intracellular material. Total wax content was determined with cuticles isolated and thoroughly extracted after leaf immersion. It follows that dipping a leaf into an organic solvent for only a few seconds, as it is often done, is likely to grossly underestimate the total amount of wax present. It has to be stressed once more that the epicuticular waxes which are claimed to be recovered by those dipping procedures most probably play a very small part in the water barrier function of the cuticle.

It is found that *in vitro* culture usually reduces the amount of wax visible by SEM or recovered with organic solvents, when compared to plants grown at the same light level and lower air humidities (seedlings grown beside the vessels in the culture room, or tissue cultures grown inside vessels containing a desiccant). Grout and Aston [8] reported that the reduction in recovered wax from cauliflower (*Brassica oleracea* L. var. *botrytis*) seedlings grown in vessels and from micropropagated cauliflower plantlets was identical (more than 70% as compared to plants grown *ex vitro* in the culture room). Sutter and Langhans

[34] found that artificial reduction of air humidity inside the culture vessels to about 35% r.h. produced a wax load on cabbage (*Brassica oleracea* L. var. *capitata*) leaves that came close to that found on seedlings grown beside the vessels (recovered amount of wax 14.8 and 20.6 μg cm^{-2}, respectively), whereas plantlets cultured under normal *in vitro* conditions reached only 3.2 μg cm^{-2}. The absence of any effect of *in vitro* culture, however, and even an increased amount of wax as compared with material grown *ex vitro* has also been noted [31] (Table 1).

These various effects of *in vitro* culture on the amount of wax appear to follow a certain pattern. Whereas all those species whose wax cover was reduced under normal (high humidity) *in vitro* conditions naturally possess glaucous leaves (cabbage [34], carnation [31], cauliflower [8]) or high loads of non-structural waxes (chrysanthemum [31]), the amount of wax recovered from the two species which were found not to follow this trend (*Maranta leuconora*, *Spathiphyllum wallisii* L. [31]) indicated that their natural wax load was below average. It should be noted that, contrary to the expectation derived from earlier findings [33], there was no correlation between the amount of wax present on the micropropagated material, and the survival rates after transfer to the greenhouse [31]. Survival was rated according to growth, colour and vigour of the plants two weeks after the transfer. Although chrysanthemum, which had the highest amount of wax, also received the highest ratings (modest losses), carnation and *Spathiphyllum*, whose wax loads differed by a factor of about 3, had very similar ratings in the medium range, and *Maranta*, with a wax coverage similar to *Spathiphyllum*, was at the lower end of the scale (high losses) [31].

The same pattern apparently applies to studies with increased air humidity. The wax content was decreased on the two relatively wax-rich species, *Ficus elastica* Roxb. var. *decora* (rubber tree; 100% recovery: 140 μg cm^{-2} at 40% r.h.; G Kerstiens and J Schönherr, unpublished) and *Brassica oleracea* L. var. *gemmifera* (Brussels sprouts; <100% recovery: 76 μg cm^{-2} at optimum conditions) [1], whereas *Citrus aurantium* (100% recovery: 20 μg cm^{-2}) did not respond (G Kerstiens and J Schönherr, unpublished, and [22]). It should not be overlooked, however, that cuticular traits often show a high degree of variability [4, 26], and that it may be more difficult to detect changes when the absolute amounts of wax are comparatively small.

There are few studies on the effects of light and air humidity on cuticular permeability. Hunt and Baker [11] found a strong inverse correlation between the amount of wax they recovered from *Pisum sativum* L. leaves grown in a range of different environments and the amount of 1-naphthylacetic acid (NAA, a growth regulator) which penetrated the leaves during a fixed period of time. NAA uptake from an acid buffer solution (supplying the solute in an undissociated state) was higher and wax recovery was smaller when leaves were taken from an environment with lower irradiance (40 vs. 80 J m^{-2} s^{-1}) and higher relative humidity (87 vs. 70%).

Reed and Tukey [18, 19] found no effect of light intensity (145 and 440 μE m^{-2} s^{-1}) on the flux of rubidium and phosphate ions through cuticles isolated from carnation (*Dianthus caryophyllus* L.) and Brussels sprouts plants. Kerstiens and Schönherr (unpublished) and Geyer and Schönherr [7] found no effect of air humidity (50 and 90–95% r.h.) on cuticular water permeability of bitter orange and rubber tree. With all four species, cuticular thickness (mass per unit surface area) was affected by the treatments, but effects on the wax content were only found with carnation and rubber tree (Table 1).

Kerstiens (unpublished) measured cuticular water permeability of *Delphinium* sp. cv. Princess Caroline plants (courtesy of Neo plants Ltd., Freckleton, U.K.) grown for six weeks *in vitro* or for seven months *ex vitro* in a culture room. Water permeability was measured with a moisture analyser connected to a cup attached to the adaxial, astomatous leaf surface (for details see [14]) and is given as water permeance (a parameter defined in the same way as stomatal conductance) in Table 2. There was no significant difference between leaves grown *in vitro* and *ex vitro*, and the results were well within the range of values found for other species (Table 2). Cuticular water permeance of micropropagated oak material (*Quercus robur* L.; courtesy of Dr M. Welander, Alnarp, Sweden) which was susceptible to desiccation at the transfer stage (M. Welander, personal communication) was also within the range of normal plants (G Kerstiens, unpublished) (Table 2). Even the maximum values observed with micropropagated material of the two species were by no means excessively high (4.0 and $6.2 \cdot 10^{-5}$ m s^{-1}, respectively) compared with typical values for stomatal conductance (see below).

These findings are somewhat contradictory to the results obtained by Sutter [32]. Studying micropropagated sweetgum (*Liquidambar styraciflua* L.), apple (*Malus domestica* Borkh.) and cherry (*Prunus avium x pseudocerasus*) plantlets, she found values of 50, 190 and $210 \cdot 10^{-5}$ m s^{-1}, respectively, for stomata-free adaxial leaf surfaces. A second investigation by the same author with a somewhat different method resulted in an estimate for apple cuticular water permeance of $40–100 \cdot 10^{-5}$ m s^{-1} [29]. All these values are one to two orders of magnitude higher than the ones reported above. Apart from the different instruments used in the two studies, the main difference lay in the air humidity to which the leaf surface inside the cup of the respective instruments was exposed, being very low in the first investigation [cf. 14] and very high in the latter two [29, 32]. Although a certain effect of air humidity on cuticular water permeability can be expected [26], this can hardly explain the huge differences found.

It is important to note that all water permeability studies reported here were carried out in such a way that stomata-bearing leaf surfaces were excluded from the gas exchange system. It is not sufficient to assume that stomata are closed (e.g. in darkness or after excision of the leaf) and then measure whole-leaf transpiration. Due to the fact that cuticular water permeances are typically two to three orders of magnitude smaller than maximum conductances, any slight imperfection in stomatal closure will cause a high relative error in the determination of cuticular water permeance.

Table 2. Cuticular water permeance P of plants grown from tissue culture *in vitro* as compared to normal plants (results from selected studies).

Species [reference]	P (10^{-5} m s^{-1})	
Acer pseudoplatanus [a]	2.3	
Aechmea fasciata [16]	0.42	
Allium cepa [28]	1.9	
Betula pubescens [a]	1.0	
Citrus aurantium [16]	1.0	
-- [7]	0.85	*
Clivia miniata [24]	0.51	*
-- [3]	0.11	*
Coffea arabica [6]	0.22	*
Corylus avellana [a]	1.9	
Delphinium sp. (*in vitro*) [a]	2.7	
(*ex vitro*) [a]	2.0	
Fagus sylvatica [14]	3.7	
Ficus elastica [3]	0.43	*
Hedera helix [3]	0.27	*
-- [14]	0.31	
Ilex aquifolium [5]	0.80	*
-- [14]	1.1	*
Maianthemum bifolium [16]	7.7	*
Nerium oleander [3]	0.33	*
Polygonatum multiflorum [16]	7.2	*
Prunus avium [a]	0.59	
Prunus laurocerasus [14]	1.7	*
Pyrus communis [27]	2.1	*
Quercus robur (*in vitro*) [a]	3.9	
Schefflera actinophylla [3]	0.08	*
Vinca minor [16]	0.75	*

[a] G Kerstiens, unpublished. All experiments were carried out with nonstomatous adaxial leaf surfaces. Values marked '*' were obtained with cuticles isolated enzymatically. The coefficient of variation was in the range of 20--50% in all cases.

3. Conclusions and speculations

Whereas it is clear that a reduction in cuticular thickness does not lead to an increase in water permeance (see *Citrus aurantium* and *Ficus elastica* in Table 1, and [3, 26]), the results reported here do not really indicate whether high water permeability of the cuticle can result if wax production is suppressed *in vitro* or otherwise. In cases where such a reduction occurred, water permeability wasn't measured, and where water permeance was measured (and always found to be unaffected), there was no effect on wax content, or the amount of wax wasn't determined. Even the very high values for cuticular water permeance reported by Sutter [32] did not correlate with unusually low amounts of wax recovered from the leaves. What can be said, though, is that in the only studies where water permeance and wax coverage were studied with the same material, there was no correlation between the permeance of single specimens of isolated cuticle and either their total wax coverage or the amounts of single groups of wax compounds ([7, 22] and G Kerstiens and J Schönherr, unpublished). These studies included the low-wax species *Citrus aurantium*, where effects of a lack of waxes on permeability would appear to be most likely to occur.

Riederer and Schneider [22] estimated that an undisturbed molecular monolayer (molecular axis perpendicular to the leaf surface) of an 'average' wax compound would have a water permeance of only $0.2 \cdot 10^{-5}$ m s^{-1}. Such a monomolecular wax cover would correspond to only 0.4 μg wax cm^{-2}, much less than ever observed. Were all waxes typically present in the cuticle arranged in such a way as to produce the minimum permeability possible, this value would be infinitesimally small. Cuticles only very rarely reach the barrier quality of one perfect monolayer (cf. Table 2), and therefore a simple correlation between the amount of wax and cuticular permeance is hardly to be expected.

Cuticles suffer from the risk of damage by 'thermal stress', i.e. the heating up of the leaf surface. Temperatures above 45 °C were found to increase water permeability strongly and irreversibly due to changes in the molecular order of the matrix and/or waxes [25, 26]. Probably there is also a permanent mechanical strain acting on the cuticle which results from changes in leaf volume due to changes in water status, and in certain cases from leaf movements and deformations in the turbulent air. It is not unreasonable to expect that these stresses tend to disrupt crystalline wax arrangements and thus produce more diffusion 'paths'. Thus one can speculate that the water barrier in the cuticle has to be actively *maintained* in some way or the other, most likely by keeping up a supply of fresh waxes diffusing into the cuticle. On the other hand, the absence of mechanical strains acting on the cuticle inside a culture vessel once a leaf has fully expanded (no turgor changes, no wind etc.) might allow for the formation of a good water barrier even if plantlets cultured *in vitro* lack a substantial portion of their natural wax coverage.

Even if the necessity for permanent active maintenance of the water barrier should not exist, other functions which are mainly related to epicuticular waxes, such as interactions with pathogens or rendering the surface relatively unwettable [13], will be impaired by the natural change in physical wax structure [20] and the loss of waxes from the plant surface by wind and rain if there is no continuous supply of new wax material. It is a well-known fact that in many cases epicuticular waxes which had been artificially removed were replaced within days [e. g. 9, 20].

One could speculate further that since it is difficult to imagine a feedback mechanism regulating the synthesis of wax compounds according to the actual need, plants might be tuned to maintain a certain rate of wax synthesis, provided the available resources allow it. The fact that the bulk of studies involving a reduced supply of light (columns l and s in Table 1) showed a reduction in waxes could tentatively be explained by a low light-related shortage of resources (energy, reduction equivalents) for the synthesis of waxes, which are in chemical terms highly reduced and thus 'expensive' compounds. An increase in light intensity would normally be correlated with an increased risk of thermal stress for the cuticle, so an increase in the rate of wax synthesis under those circumstances may well have an adaptive advantage, too. Should a lack of cuticular waxes indeed present a problem with micropropagated material, then perhaps an increase in light supply during the *in vitro* period may prove advantageous, thus avoiding problems accompanying culture in vessels with reduced air humidity.

As far as the humidity response of wax synthesis is concerned, the picture is less straightforward. The *in vitro* 'treatment' as shown in Table 1, i.e. normal *in vitro* culture as compared to *in vitro* culture with reduced humidity or as compared to growth of seedlings in the culture room, resulted in a reduction of waxes in the relatively wax-rich species, but not so if the species had a relatively low natural wax load [31]. The same pattern holds true for the response of normal plants to increased air humidity. It appears that high humidity can suppress wax synthesis to a certain degree, but only if the natural wax production is tuned to relatively high rates anyway. A certain threshhold rate of wax synthesis appears to be upheld, perhaps enough to produce a good water barrier under favourable circumstances.

Whereas the fact that seedlings grown *ex vitro* produced more waxes than micropropagated tissue *in vitro* can have many reasons and need not necessarily be related to the lower air humidity, the really intriguing finding is the invariable increase of waxes if micropropagated material was produced under conditions of lowered air humidity [34, 35, 37]. Cuticular wax load did not increase, though, during 12 days of acclimatisation to laboratory conditions in uncapped culture vessels [32], but it did increase on persistent leaves within ten days after transplanting [8]. There is no ready explanation for all those findings, but the possibility exists that growth under conditions of high humidity might reduce the 'vigour' or 'vitality' of plants, by whatever mechanism, and that the reduction in wax synthesis is an expression of that reduced vitality rather than a direct response to the environment.

The 'vitality hypothesis' provides an alternative approach of interpreting the findings that micropropagated carnation plants which had randomly produced glaucous leaves had much higher survival rates after transfer to the greenhouse [33]. Perhaps both the improved ability to survive and the production of structural epicuticular waxes were independent results of a less disturbed physiology, which had occurred at random, and the increased production of waxes did not contribute directly to avoid desiccation and death. Nonetheless, wax production might still prove to be a valuable *indicator* of a less impaired physiology.

Acknowledgements

I am grateful to Prof. Bill Davies who suggested this paper, and to the European Environmental Research Organisation for the award of a fellowship held during the preparation of this paper.

References

1. Baker EA (1974) The influence of environment on leaf wax development in *Brassica oleracea* var. *gemmifera*. New Phytol 73:955–966
2. Baker EA (1982) Chemistry and morphology of plant epicuticular waxes. In: DF Cutler, KL Alvin and CE Price, eds. The Plant Cuticle, pp 139–165. London: Academic Press
3. Becker M, Kerstiens G and Schönherr J (1986) Water permeability of plant cuticles: permeance, diffusion and partition coefficients. Trees 1:54–60
4. Davis DG (1978) Effect of light quality and irradiance on development of Citrus mitis leaf cuticles. Bot Gaz 139:390–392
5. Garrec JP and Kerfourn C (1989) Effets de pluies acides et de l'ozone sur la perméabilité à l'eau et aux ions de cuticules isolées. Environ Exp Bot 29:215–228
6. Garrec JP and Plebin R (1986) Perméabilité au fluorure d'hydrogène (HF) des cuticules avec ou sans stomates de feuilles: influence de la présence des stomates et comparaisons avec la perméabilité à l'eau. Environ Exp Bot 26:2 99–308
7. Geyer U and Schönherr J (1990) The effect of the environment on the permeability and composition of *Citrus* leaf cuticles. I. Water permeability of isolated cuticular membranes. Planta 180:147–153
8. Grout BWW and Aston MJ (1977) Transplanting of cauliflower plants regenerated from meristem culture. I. Water loss and water transfer related to changes in leaf wax and to xylem regeneration. Hort Res 17:1–7
9. Hallam ND (1970) Growth and regeneration of waxes on the leaves of *Eucalyptus*. Planta 93:257–268
10. Holloway PJ (1982) Structure and histochemistry of plant cuticular membranes: an overview. In: DF Cutler, KL Alvin and CE Price, eds. The Plant Cuticle, pp 1–32. London: Academic Press
11. Hunt GM and Baker EA (1982) Developmental and environmental variations in plant epicuticular waxes: some effects on the penetration of naphthylacetic acid. In: DF Cutler, KL Alvin and CE Price, eds. The Plant Cuticle, pp 279–292. London: Academic Press
12. Juniper BE (1960) Growth, development, and effect of the environment on the ultra-strucure of plant surfaces. J Linn Soc (Bot) 56:413–418
13. Kerstiens G (in prep.) A review of cuticular barrier properties to water, solutes and pathogen penetration of leaves grown in polluted atmospheres. In: M Yunus and M Iqbal, eds. Plant Growth and Air Pollution
14. Kerstiens G and Lendzian KJ (1989) Interactions between ozone and plant cuticles. II. Water permeability. New Phytol 112:21–27
15. Knuth D, Neubeller J and Stösser R (1987) Vergleich der Sonnen- und Schattenseite von Apfelfrüchten. II. Lipide der Fruchtschale. Gartenbauwiss 52:140–144
16. Lendzian KJ and Kerstiens G (1991) Sorption and transport of gases and vapors in plant cuticles. Rev Environ Contam Toxicol 121:65–128
17. Osborn JM and Taylor TN (1990) Morphological and ultrastructural studies of plant cuticular membranes. I. Sun and shade leaves of Quercus velutina (Fagaceae). Bot Gaz 151:465–476
18. Reed DW and Tukey HB Jr (1982a) Light intensity and temperature effects on epicuticular wax morphology and internal cuticle ultrastructure of carnation and Brussels sprouts leaf cuticles. J Amer Soc Hort Sci 107:417–420

19. Reed DW and Tukey HB Jr (1982b) Permeability of Brussels sprouts and carnation cuticles from leaves developed in different temperatures and light intensities. In: DF Cutler, KL Alvin and CE Price, eds. The Plant Cuticle, pp 267–278. London: Academic Press

20. Riederer M (1989) The cuticle of conifers: structure, composition and transport properties. In: ED Schulze, OL Lange and R Oren, eds. Forest Decline and Air Pollution (Ecological Studies vol. 77), pp 157–192. Berlin: Springer-Verlag

21. Riederer M and Schneider G (1989) Comparative study of the composition of waxes extracted from isolated leaf cuticles and from whole leaves of *Citrus*: Evidence for selective extraction. Physiol Plant 77:373–384

22. Riederer M and Schneider G (1990) The effect of the environment on the permeability and composition of *Citrus* leaf cuticles. II. Composition of soluble cuticular lipids and correlation with transport properties. Planta 180:154–165

23. Rosenquist JK and Morrison JC (1989) Some factors affecting cuticle and wax accumulation on grape berries. Am J Enol Vitic 40:241–244

24. Schmidt HW, Mérida T and Schönherr J (1981) Water permeability and fine structure of cuticular membranes isolated enzymatically from leaves of *Clivia miniata* Reg. Z Pflanzenphysiol 105:41–51

25. Schreiber L and Schönherr J (1990) Phase transitions and thermal expansion coefficients of plant cuticles. Planta 182:186–193

26. Schönherr J (1982) Resistance of plant surfaces to water loss. In: Encyclopedia of Plant Physiology 12B, pp 153–179. Berlin: Springer-Verlag

27. Schönherr J and Lendzian KJ (1981) A simple and inexpensive method of measuring water permeability of isolated plant cuticular membranes. Z Pflanzenphysiol 102:321–327

28. Schönherr J and Mérida T (1981) Water permeability of plant cuticular membranes: the effects of humidity and temperature on the permeability of non-isolated cuticles of onion bulb scales. Plant Cell Environ 4:349–354

29. Shackel KA, Novello V and Sutter EG (1990) Stomatal function and cuticular conductance in whole tissue-cultured apple shoots. J Amer Soc Hort Sci 115:468–472

30. Skoss JD (1955) Structure and composition of plant cuticles in relation to environmental factors and permeability. Bot Gaz 117:55–72

31. Sutter EG (1985) Morphological, physical and chemical characteristics of epicuticular wax on ornamental plants regenerated *in vitro*. Ann Bot 55:321–329

32. Sutter E (1988) Stomatal and cuticular water loss from apple, cherry, and sweetgum plants after removal from in vitro culture. J Amer Soc Hort Sci 113:234–238

33. Sutter E and Langhans RW (1979) Epicuticular wax formation on carnation plantlets regenerated from shoot tip culture. J Amer Soc Hort Sci 104:493–496

34. Sutter E and Langhans RW (1982) Formation of epicuticular wax and its effect on water loss in cabbage plants regenerated from shoot-tip culture. Can J Bot 60:2896–2902

35. Wardle K, Dobbs EB and Short KC (1983) *In vitro* acclimatization of aseptically cultured plantlets to humidity. J Amer Soc Hort Sci 108:386–389

36. Whitecross MI and Armstrong DJ (1972) Environmental effects on epicuticular waxes of *Brassica napus* L. Aust J Bot 20:87–95

37. Ziv M (1986) *In vitro* hardening and acclimatization of tissue culture plants. In: LA Withers and PG Alderson, eds. Plant Tissue Culture and its Agricultural Application, pp 187–196. London: Butterworths

Vitrification in relation to stomatal deformation and malfunction in carnation leaves *in vitro*

MEIRA ZIV[1,2] and TAMAR ARIEL[1]

Department of Agricultural Botany[1] and The Otto Warburg Center for Biotechnology in Agriculture[2], The Hebrew University of Jerusalem, Rehovot, Israel

1. Introduction

Morphological and physiological disorders in leaves are often observed under culture conditions intended to promote rapid bud proliferation. High water potential in the medium and relative humidity in the culture vessel, elevated levels of sucrose and growth regulators and the presence of certain ions in unfavorable ratios, as well as low light intensity induce anomalous leaf anatomy and morphology termed 'vitrification'. Abnormal shoot morphogenesis has been termed hyperhydration, glassiness, succulence and translucence, depending on the severity of the phenomenon. These descriptions are based on visual characterizations and depend on the external morphology of the studied species [12, 24, 33]. *In vitro* plants having abnormal leaves function poorly *ex vitro* due to inadequate photosynthetic activity and high rates of water loss which lead eventualy to plant desiccation [3, 7, 9, 32, 34, 37]. The major causes for poor plant survival are underdeveloped mesophyll, poor cuticular wax formation and malfunctioning guard cells [9, 27, 34, 35]. Stomata from vitreous leaves remained open and did not respond to darkness, abscisic acid (ABA), high Ca^{2+} and hypertonic solutions — stimuli which usually cause stomatal closure [3, 22, 29, 34, 35]. Ziv *et al.* [37] reported that the defect in stomatal function resided in the guard cell walls which, unlike the protoplasts, did not contract in hypertonic solution. The mechanical properties of the guard cell walls in the stomatal apparatus are determined by the cell wall structure and composition [22]. Since both hyperhydration and guard cell malfunction result from changes in the cell walls, this research investigated the relationship between leaf hyperhydration and stomatal deformation and malfunction with the purpose of using stomatal function as a simple, more accurate parameter to evaluate vitrification.

143

P.J. Lumsden, J.R. Nicholas and W.J. Davies (eds.), Physiology, Growth and Development of Plants in Culture, 143–154, 1994.
© 1994 *Kluwer Academic Publishers.*

2. Materials and methods

Carnation (*Dianthus caryophyllus* L. cv. Ceris royallete) apical buds (3–5 mm in size) were isolated from *in vitro* 6–7 cm long plantlets. The buds were cultured in 10 x 2.2 cm test tubes, in 10 ml liquid medium supported by filter paper bridges or on the upper part of a 0.8% agar solidified medium slant as previously described [34, 36], 25 buds per treatment. Isolated buds were also cultured in 20 ml of liquid medium in 125 ml Erlenmeyer flasks on a gyratory shaker at 100 rpm. Murashige and Skoog [MS, 21] medium was supplemented with 2.7 μM of naphthalene acetic acid (NAA) and 2.3 μM of kinetin (KN). The cultures were kept under a 16 h white fluorescent light (30 μmoles m^{-2}s^{-1}) at 25±1 °C. For experiments studying the effect of NH_4NO_3 levels in the medium on leaf development, shoot apices 0.5–0.8 mm long, including two leaf primordia, were isolated from freshly harvested carnation cuttings, 25 shoots per treatment. The shoot apices were cultured in either agar or liquid medium supported by filter paper bridges. The medium consisted of MS minerals with 20, 15, 10 and 5 mM or without NH_4NO_3. Normal plants throughout all experiments had glaucous leaves resembling leaves in young greenhouse-grown carnation plants. Stomatal function was determined by the incipient plasmolysis method [37].

For scanning electron microscope (SEM) studies, leaves were soaked for 15 min in chloroform at room temperature and then gradually dehydrated in a graded series (30% to 100%) of acetone solutions. The leaves were dried in CO_2 in a critical point dryer and coated with gold in a sputter coating unit (manufactured by Polaron). Leaf scans were carried out in a Jeol 15M 35 SEM. Epidermal strips removed from the lower epidermis were placed in a drop of water and examined under a Zeiss polarization microscope to determine wall birefringence in the guard cells. Lower epidermal strips were stained for 30 min in 0.01% aniline-blue in 0.07 M K_2HPO_4 for the presence of callose in the guard cells [4]. The stained strips were examined under a Leitz fluorescence microscope equipped with a fluorescence illuminator with BG3, BG12 and 490K filters. Lower epidermal strips were stained by soaking for 30 min in 0.1% calcofluor (fluorescent brightener 28, Sigma) for cellulose content. Prior to staining the epidermal strips were soaked in 2-ethanolamine solution for 12 h at 47 °C to remove pectins and hemicelluloses [13]. Callose in the leaf tissue was determined using a quantitative fluorometric method, based on the aniline-blue staining technique, as described by Kohle *et al.* [18]. Callose was measured in 200 mg of lyophilized leaf tissue from vitreous and normal leaves. Callose aniline-blue complex was determined in a Jasco FP-550 spectrofluorometer (excitation at 400 nm, emission at 510 nm) and is expressed as μg/g DW relative to a standard curve using pachyman as a source of (1–3)β-glucan. Ethylene was measured in 1 ml samples drawn through a septic plug under aseptic conditions from liquid or agar cultures. A Carlo Elba gas-chromatograph 6000 VEGA series 2 was used for the measurements.

3. Results

Cultured carnation shoots developed either normal leaves (Plate 1a) or displayed structural deformations which varied from moderate, in translucent leaves (Plate 1b), to severe in hyperhydrous succulent leaves (Plate 1c). The deformations were manifested also in the guard cells of the stomatal apparatus. In some transluscent leaves, 40–50% of the stomata observed in isolated epidermal strips (Plate 2a) did not close in hypertonic sucrose solutions. Leaves from carnation shoots, cultured in liquid shake cultures, were hyperhydrous (Plate 1c) and contained mainly two types of deformed stomata; stomata

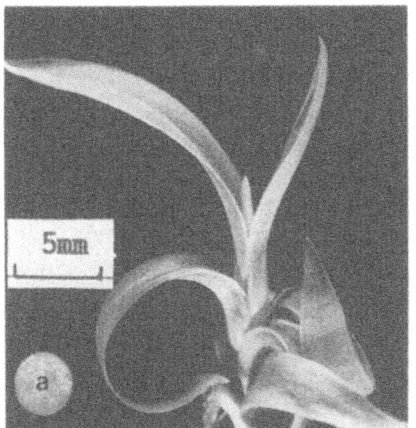
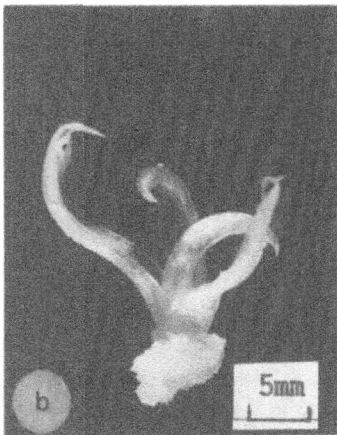
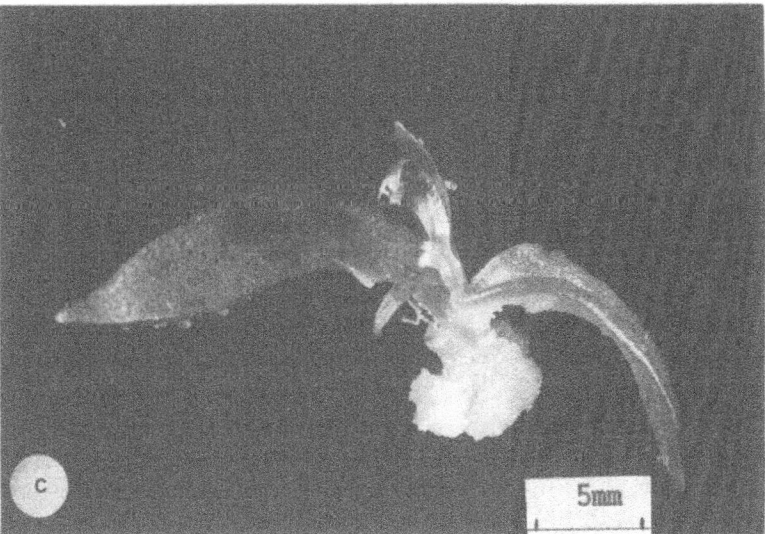

Plate 1. A normal carnation shoot from agar (a) translucent shoot from stationary liquid medium (b) and a succulent shoot from shake liquid culture (c).

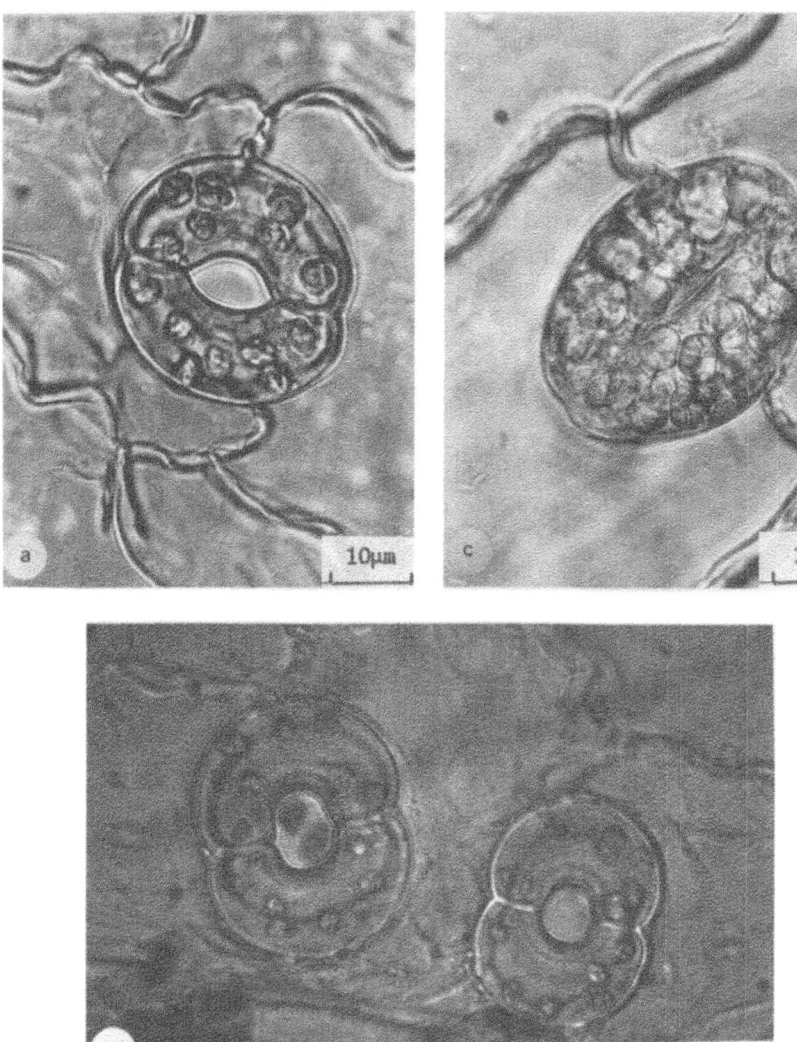

Plate 2. Stomata from translucent (a) and succulent (b, c) leaves.

consisting of thin-walled guard cells surrounding a large rounded pore, of which 95–96% remained open in 0.8 M sucrose (Plate 2b), and stomata consisting of elongated guard cells, filled with large chloroplastids, which did not open at all (Plate 2c). Severe structural deformations of the leaves were accompanied by an increase in the number of malfunctioning stomata.

A correlation between leaf and stomatal structure was also observed in carnation plants cultured in agar or liquid medium in which the NH_4NO_3 level was reduced. A decrease in NH_4NO_3 from 20 mM to 10 or 5 mM in agar solidified medium increased the number of normal plants to 82 and 96% respectively (Table 1). Replacing the NH_4NO_3, with KNO_3 serving as the sole nitrogen source in the medium also prevented vitrification in carnation leaves. In normal glaucous leaves which developed in agar medium with 10 or 5 mM or without NH_4NO_3, about 95% of the stomata closed in a hypertonic sucrose solution. In media with higher NH_4NO_3 levels a decrease in the number of the functioning stomata was found in leaves which appeared normal externally. The percent of normal glaucous plants increased from 8% in the control to about 68% and 60% in carnation shoots developed in liquid media with 5 mM or without NH_4NO_3 respectively. This increase was associated with an increase in the number of functioning stomata in the leaves, reaching 58% and 53% in plants cultured in media with 5 mM or without NH_4NO_3, respectively. Higher levels of NH_4NO_3 in liquid media increased the number of abnormal plants and the percent of malfunctioning stomata (Table 1). Variation in the final pH in the two tested types of cultures, agar and liquid, was observed after 24 days in culture. A higher pH was measured in all NH_4NO_3 levels in liquid (around 5.0–5.4), as compared to agar media (around 4.0–4.6). These differences may reflect a differential growth response and ion uptake by the shoots when cultured in liquid and agar media with the respective levels of NH_4NO_3 in the medium. Ethylene, known to affect vitrification, was sampled from shoots cultured on agar or liquid media containing 10 mM NH_4NO_3. During the first two days an initial burst with a peak of 34 and 26 nl/tube was measured in liquid and agar cultures respectively. On the fourth day ethylene evolution dropped and stabilized at about 4.5 nl/culture in agar and 10–12 nl/culture in liquid media for the rest of the culture period.

When callose presence was determined in malformed leaves a correlation between leaf and guard cell wall composition was observed. Succulent hyperhydrous leaves from liquid shake cultures contained a high level of callose

Table 1. The effect of NH_4NO_3 in agar and liquid cultures on carnation plant development, stomatal function and final pH after 24 days in culture

N (mM)		Liquid			Agar		
NH_4	NO_3	Glaucous plants (%)	Closed stomata (%)[a]	Final pH	Glaucous plants(%)	Closed stomata	Final pH
20	40[b]	8	21±3.1	5.3±0.3	60	80±6.1	4.5±0.3
15	35	30	28±4.0	5.2±0.4	73	87±8.8	4.6±0.5
10	30	46	38±3.7	5.0±0.3	82	95±10.6	4.1±0.3
5	25	68	58±6.2	5.0±0.3	96	98±7.3	4.0±0.4
0	20	60	53±8.6	5.4±0.4	95	94±8.1	4.4±0.4

[a] Stomatal response was determined in epidermal strips from the 3rd leaf.
Epidermal strips were immersed in 0.8 M sucrose and the percent of closed stomata was determined in three replicates of 25 stomata per treatment. Mean ± SE.
[b] Control — full MS mineral salts.

Table 2. Weight and callose content in leaves from greenhouse or *in vitro* carnation plants

Leaf type and culture condition	Leaf			Callose ($\mu g/g$ DW)[a]
	FW (mg)	DW		
		(mg)	% of FW	
Normal — greenhouse	261±18.1	65.2±9.8	25	86±7.2
Normal — agar	28±0.3	5.2±0.7	18	129±8.9
Transluscent — liquid	32±0.4	4.5±0.3	14	252±11.2
Succulent — liquid shake	96±10.6	10.5±1.2	11	374±10.3

[a] See Materials and methods.

which was also present, but to a lesser extent, in transluscent and normal *in vitro* leaves (Table 2). When guard cells from vitreous leaves were stained with aniline-blue and examined under a microscope (see Materials and methods), a bright fluorescence complex which formed with the callose in the cell walls, was observed in abnormal stomata (Plate 3a). The fluorescence aniline-blue complex was not observed in stomata from normal leaves in plants from agar cultures (Plate 3b). In normal and vitreous carnation leaves the structure of the cell wall, bordering the stomatal pore, was revealed in SEM observations of epidermal strips. As can be seen from Plate 4a the cell wall bordering the stomatal pore in normal guard cells is thickened, well defined and forms ridges with an elliptical pore. The cell wall bordering the stomatal pore in guard cells from vitreous leaves was thinner, and appeared damaged and torn in several places. (Plate 4b). Structural changes in the guard cell walls were evident when epidermal strips from vitreous leaves were examined under a polarizing light microscope. When cellulose microfibrils are distributed anisotropically they appear birefringent, one beam of light is refracted as two, under polarized light in the microscope. The degree of this birefringence can serve to determine the mean orientation of the microfibrils [22, 23]. Guard cells from normal leaves showed a characteristic birefringence pattern of two bright spots where the cellulose microfibrils radiate in a fan-like structure outward from the pore site (Plate 5a). The majority of stomata from vitreous leaves did not show such a birefringence pattern, but some localized bright areas which may indicate a radial or concentric arrangement of the microfibrils (Plate 5b). Additional support for the irregularity of guard cell walls in vitreous leaves came from staining epidermal strips with calcofluor. As can be seen in Plate 5c in guard cells from normal leaves, the cell wall bordering the stomatal pore appeared as a definite thick fluorescent line which was not observed in stomata from vitreous leaves (Plate 5d). These studies revealed once more that the stomatal pore was elliptical in the epidermis of normal leaves while being rounded in stomata from vitreous leaves. In addition, epidermal cells from vitreous leaves lack the undulated anticlinal walls typical of cell walls from normal leaves.

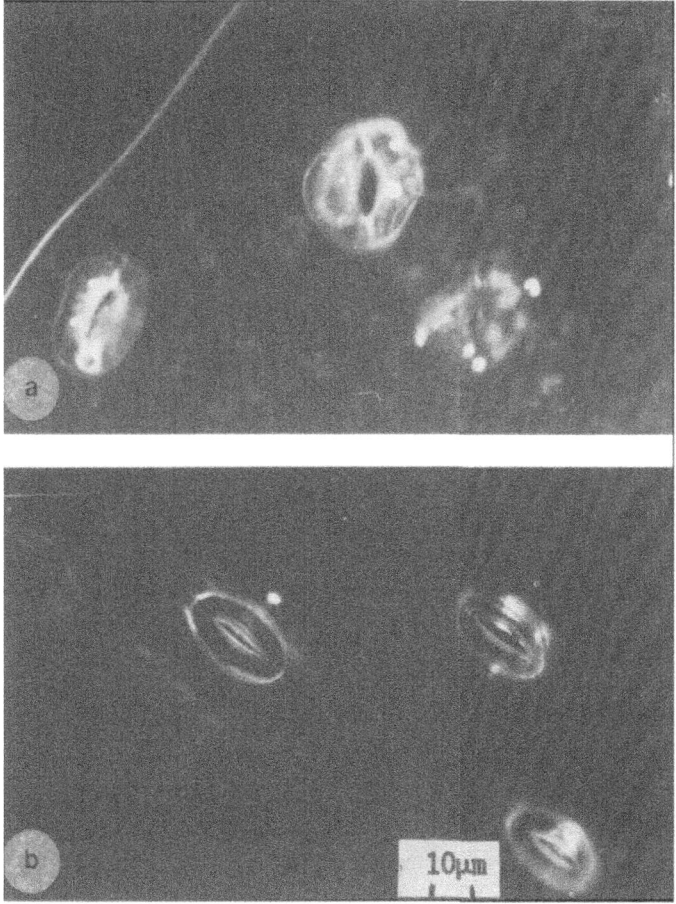

Plate 3. A guard cell from a vitreous leaf showing the fluoresence complex of callose with aniline-blue (a). In normal leaves the fluoresence complex is either missing or limited (b).

4. Discussion

The description of vitrification is often misleading and contradictory since it relies on the external hyperhydrous appearance of the leaves which varies from one species to another [12, 24, 33]. Structural deformations in vitreous leaves are attributed to various physiological changes which affect cell wall structure and composition [12, 17, 30, 33]. The changes in cell wall structure are manifested in the guard cells which fail to function and do not close in response to closing stimuli [37]. A correlation between leaf vitrescence and stomatal malfunction was found under culture conditions which induce vitrification. Vitreous carnation leaves from agar medium described as translucent [24] had a heterogenous population of normal and abnormal stomata, and only half of them closed in hypertonic solutions. Succulent leaves from liquid cultures were

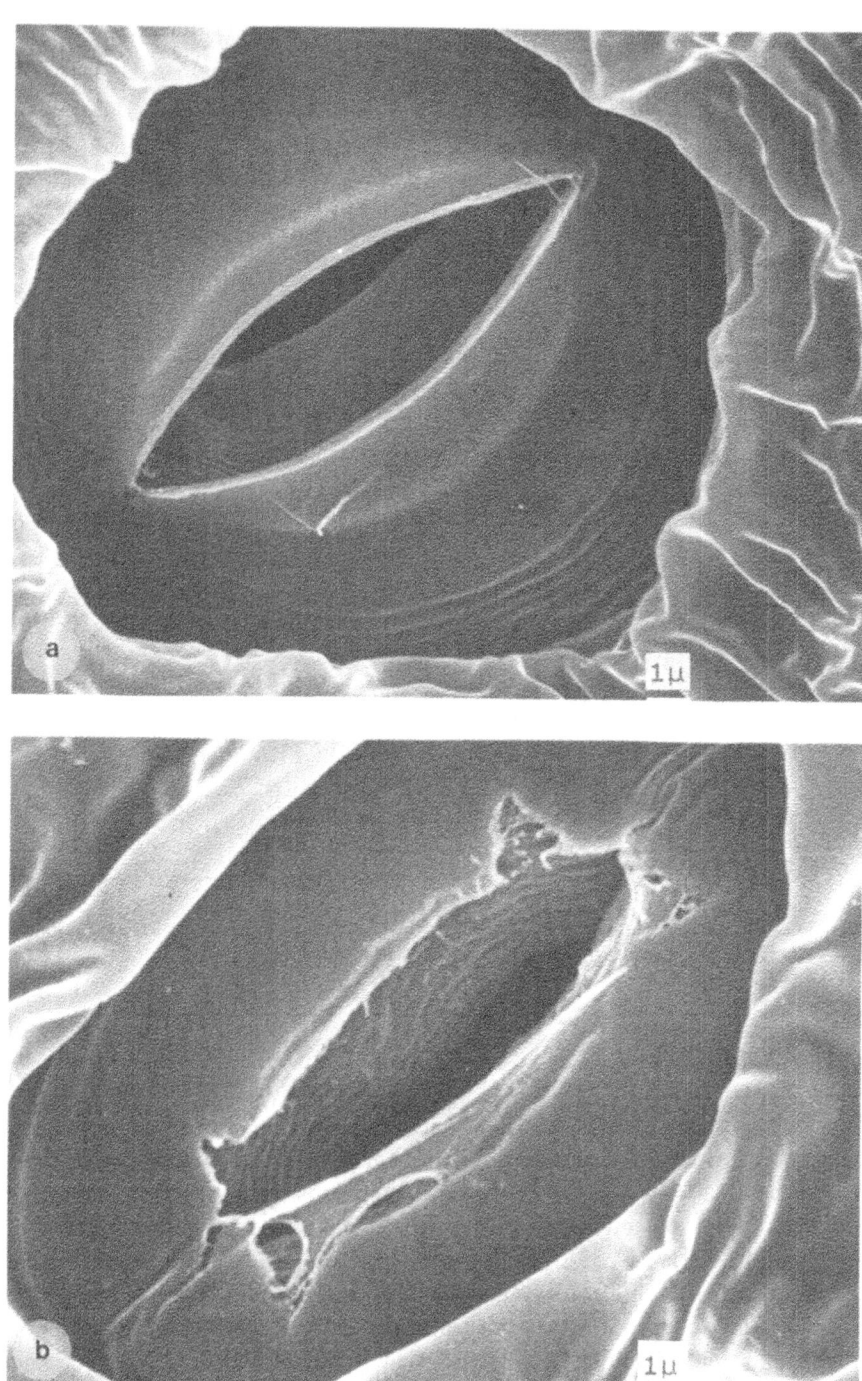

Plate 4. SEM of stomata from normal (a) and vitreous (b) carnation leaves.

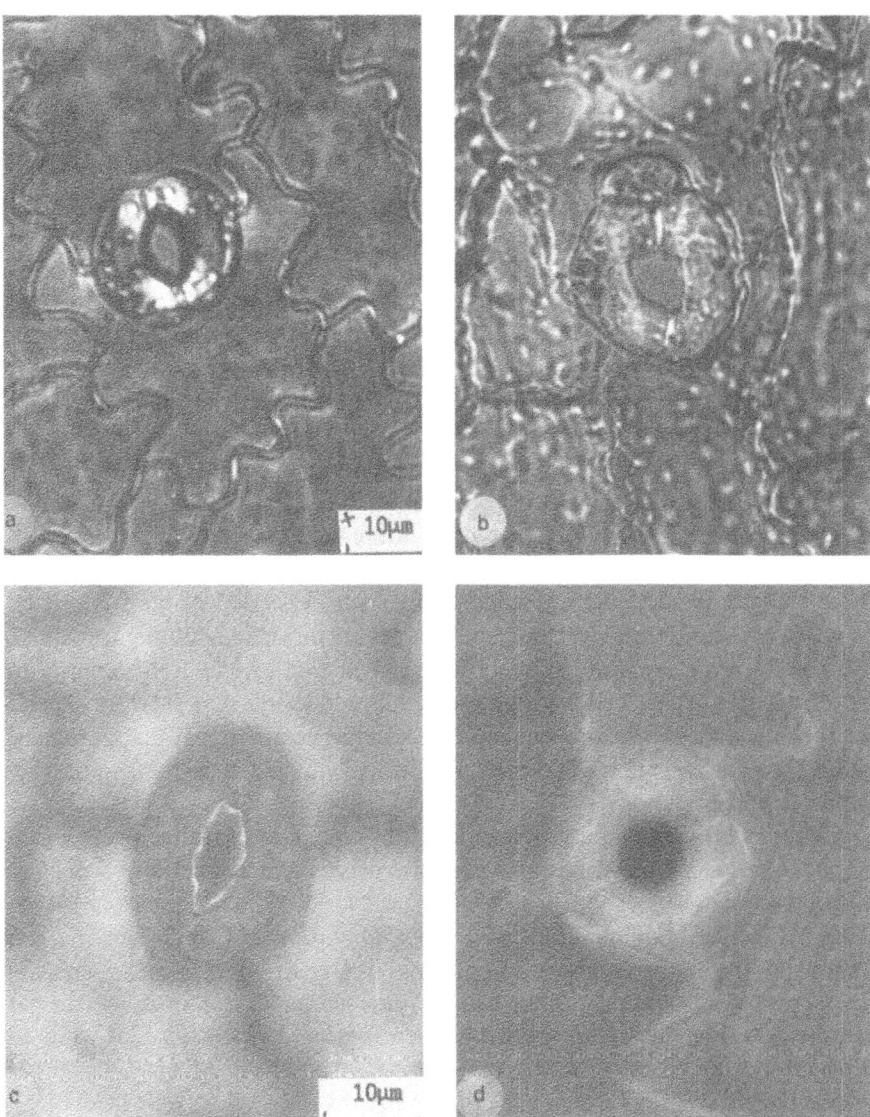

Plate 5. A stoma from a normal leaf showing birefringent images under a polarized light microscope (a) and a stoma from a vitreous leaf lacking normal microfibril orientation (b). The plane of polarizer and analyzer is indicated by the cross. Cellulose calcofluor complex fluoresence in normal (c) and vitreous (d) leaves.

severly hyperhydrated and possessed very few normal funcitoning stomata in their epidermis.

Guard cell walls have a mechanical function in stomatal movement. Their thickening, conformation and composition as well as the orientation of the cellulose microfibrils contribute to the closing mechanism of the stoma [23]. Culture conditions and factors associated with leaf hyperhydration seem to be linked to stomatal malfunction. Lowering the level of NH_4NO_3 from 20 to 10 and 5 mM increased the number of glaucous normal plants, however, not all the stomata in 'normal' appearing leaves closed in hypertonic solution. Although the data presented describes total N reduction as well as NH_4, when the levels of NH_4 were reduced and KNO_3 was supplemented, to give 60 mM N, NH_4 levels had a similar effect (Ziv unpublished). Increased levels of agar in the medium decrease vitrification [6, 36]. In semi solid media with 0.8% agar and 20 mM NH_4NO_3, the percent of normal plants was higher than in liquid media, which increased in the latter with the decrease or removal of NH_4NO_3 from the medium. The exact role of NH_4^+ in causing vitrification and in the changes in plant cell wall structure and composition *in vitro* is not clear and was reported in other species [2, 5] where it was related to changes in lignin biosynthesis.

The observation that under varying NH_4^+ levels normal appearance of carnation leaves did not always correlate with normal stomatal function, points to the role of other factors involved in the control of leaf and stomata development in addition to NH_4^+ in the culture medium, or to the stage of leaf development in the cultured shoot. Esau [11] stated that 'in a given leaf the stomata arise in succession, through a considerable period of time. In netted vein leaves the different developmental stages are mixed in a mosaic fashion so that mature and immature stomata may appear side by side'. The severity of vitrescence and guard cell deformation in cultured shoots may therfore depend on the stage of leaf development and on the culture condition at that particular stage. This can explain the presence of abnormal stomata in 'normal' appearing glaucous leaves.

Differences in the structure and function among stomata in leaves which appear normal can be related to at least two other factors in culture — pH and C_2H_4, known to effect vitrification [12, 16, 24, 33]. The differences between final pH in agar and liquid media have probably resulted from differential ion uptake by the cultured shoots. Initial rapid uptake of NH_4^+ causes a decrease in pH while NO_3^- uptake results in an increase in the pH [26]. In *Lilium* the pH dropped during the first 3 days and then increased gradually [28] while in *Cucumis* the pH changed within 48 h, and became more acid [26]. A higher pH was observed also in *Ptilotus* cultured in liquid compared to agar media [31]. The pH can affect the availability of ions such as Mg^{2+} and Ca^{2+} known to be involved in the biosynthesis of cell wall pectic substances, cellulose and callose, [8, 15]. Changes in the level of pectic substances in vitreous leaves were reported in carnation, plum and birch [19, 20, 30]. The cause of the torn damaged cell walls bordering the stomatal pore in vitreous leaves (Plate 4b) [1, 35] could be a complete and faulty division of the stomata mother cells [11]. The presence of higher levels of callose in succulent than in translucent or normal leaves *in vitro* correlated with callose presence in guard cell walls (Plate 3a). Staining techniques and microscopic examination of abnormal stomata revealed that the cell walls were low in cellulose and lacked the characteristic orientation of cellulose microfibrils (Plate 5b). In kidney shaped guard cells the major thickening of the cell walls is in the walls bordering the pore where the microfibrils are radiating away from the pore site. Guard cells exhibit a characteristic birefringence pattern consisting of two pairs of spots [23, 25] which were not observed in stomata from vitreous leaves. Faulty orientation of cellulose microfibrils affects the response of stomata to changes in turgor pressure in the guard cells, and may prevent stomatal closure. Changes in cell wall composition and structure in the mesophyll of leaves *in vitro* are related to higher levels of ethylene and relative humidity in liquid cultures [12, 16, 24]. These changes were also manifested in the walls of the guard cells and may correlate with higher levels of ethylene evolution during the first 3–4 days in culture. Partial oxygen shortage resulted in ethylene accumulation and aerenchyma formation in poorly aerated maize roots [14]. Depending on the stage of leaf development during the early culture period, changes in cell wall structure in the mesophyll, the developing stomata mother cells and the developing guard cells, can cause various degrees of vitrification and stomatal deformation. Cell wall deformation in the guard cells was found to link closely with leaf hyperhydration under culture conditions inflicting abnormal shoot morphogenesis. This relationship suggests that stomatal structure and function can provide a more accurate evaluation of vitrification, which in most reports is based on external visual symptoms of the leaves.

References

1. Ariel T (1987) The effect of culture condition on development and acclimatization of Philodendron 'Burgandi' and *Dianthus caryophyllus*, (M.Sc. Thesis, in Hebrew with an English summary), Hebrew University of Jerusalem
2. Beauchesne G (1981) Les milieux mineraux utilises en culture *in vitro* at leur incidence sur l'apparition de boutures d'aspect pathologique. C R Acad Agric Paris 67:1389–1397
3. Brainerd KE and Fuchigami LH (1982) Stomatal functioning of *in vitro* and greenhouse apple leaves in darkness, mannitol, ABA, CO_2. J Exp Bot 33:388–392
4. Currier HB (1957) Callose substance in plant cells. Am J Bot 49:478–488
5. Daguin F and Letouze R (1985) Relations entre hypolignification et etat vitreux chez *Salix babylonica* en culture *in vitro*. Role de la nutrition ammoniacale. Can J Bot 63:324–326
6. Debergh PC, Harbaoui Y and Lemeur R. (1981) Mass propagation of globe artichoke (*Cynara scolymus*). Evaluation of different hypotheses to overcome virtification with special reference to water potential. Physiol Plant 53:181–187
7. Debergh PC and Maene LJ (1981) A scheme for commercial propagation of ornamental plants by tissue culture. Sci Hort 14:335–345
8. Delmer DP (1987) Cellulose biosynthesis. Ann Rev Plant Physiol 38:259–270
9. Donnelly DJ and Vidaver WE (1984) Pigment content and gas exchange of red raspberry *in vitro* and *ex vitro*. J Amer Soc Hort Sci 109:177–181
10. Dougall DK (1980) Nutrition and metabolism. In: EJ Staba EJ, ed. Plant Tissue Culture as a Source of Biochemicals, pp 21–58. Boca Raton, Florida: CRC Press
11. Esau K (1965) Plant Anatomy. John Wiley & Sons, NY, pp 50; 163–165
12. Gaspar T, Kevers C, Debergh L, Maene L, Paques M, and Boxus P (1987) Vitrification: morphological, physiological and ecological aspects. In: JM Bonga and DJ Durzan, eds. Cell and Tissue Culture in Forestry, 1, pp. 152–166. Dordrecht: Martinus Nijhoff
13. Heslop-Harrison Y and Heslop-Harrison J (1981) The digestive glands of *Pigcuola*: structure and cytochemistry. Ann Bot 47:293–319

14. Jackson MB (1988) Aerenchyma formation in roots and leaves: regulation by oxygen shortage and ethylene, Proceedings MATO ISEP:'Signals for cell separation in plants', Turin, Italy, 46
15. Kauss H (1987) Some aspects of calcium dependent regulation in plant metabolism. Ann Rev Plant Physiol 38:47–72
16. Kevers C and Gaspar T (1985) Vitrification of carnation *in vitro*: changes in ethylene production, ACC level and capacity to covert ACC to ethylene. Plant Cell Tiss Org Cult 4:215–223
17. Kevers C, Prat R and Gaspar T (1987) Vitrification of carnation *in vitro*: Changes in cell wall mechanical properties, cellulose, and lignin content, Plant Growth Regulat 5:59–66
18. Kohle H, Jeblick W, Poten F, Blaschek W and Kauss H (1985) Chitosan-elicited callose synthesis in soybean cells as a Ca^{2+} dependent process. Plant Physiol 544–551
19. Koshuchowa S, Botcher I, Zoglauer K and Goring H (1988) Avoidance of vitrification of *in vitro* Cultured Plants. Proceedings of 6th Congress of the Federation of European Societies of Plant Physiology, Split, Yugoslavia
20. Marin JA, Gella T and Herrero M (1988) Stomatal structure and functioning as a response to environmental changes in acclimitized micropropagated *Prunus cerasus* L. Ann Bot 663–670
21. Murashige T and Skoog F (1962) A revised medium for rapid growth and bioassay with tobbacco tissue cultures. Physiol Plant 15:473–497
22. Palevitz DA (1981) The structure and development of stomatal cells. In: PG Jarvis and TA Mansfield, eds. Stomatal Physiology, pp 1–23. Cambridge: Cambridge University Press
23. Palevitz BA and Hepler PK (1976) Cellulose microfibril orientation and cell shaping in developing guard cells of Allium:The role of microtubules and ion accumulation. Planta 132:71–93
24. Paques M and Boxus P (1987) Vitrification: Review of literature. Acta Hort 212:155–166.
25. Peterson R L, Firminger MS and Dobrind LA (1975) Nature of the guard cell wall in leaf stomata of three *Ophioglossum* species. Can J Bot 58:1698–1705
26. Skirvin RM, Chu MC, Mann ML, Young H, Sullivan J and Fermanian T (1986) Stability of tissue culture medium pH as a function of autoclaving, time, and cultured plant material. Plant Cell Reports 5:292–294
27. Sutter EG and Langhans R W (1982) Formation of epicuticular wax and its effect on water loss in cabbage plants regenerated from shoot-tip culture. Ann J Bot 60:2896–2902
28. Takayama S and Misawa M (1979) Differentiation in *Lilium* bulbscales grown *in vitro*. Effect of various culture conditions. Physiol Plant 46:184–190
29. Wardle K, Quinlan A and Simpkin L (1979) Abscisic acid and the regulation of water loss in plantlets of *Brassica olerracea* L. var. Botrytis regenerated through apical meristem culture. Ann Bot 43:745–752
30. Werker E K and Leshem B (1987) Structure changes during vitrification of carnation plantlets. Ann Bot 11:19
31. Williams R R, Taji A M and Winney K A (1990) The effect of *Ptilotus* plant tissue on pH of *in vitro* medium. Plant Cell Tiss and Org Cult 22:153–158
32. Ziv M (1986) *in vitro* hardening and acclimatization of tissue culture plants. In: LA Withers and PG Alderson, eds. Plant Tissue Culture and its Agricultural Application, pp 187–196. London: Butterworths
33. Ziv M (1991) Vitrification:morphological and physiological disorders of *in vitro* plants. In: PC Debergh and R H Zimmerman, eds. Micropropagation Technology and Application, pp.45–69. Dordrecht, Boston, London: Kluwer Academic Publishers
34. Ziv M (1991) Quality of micropropagated plants — vitrification. *In vitro* Cell Dev Biol 27P:64–69
35. Ziv M and Ariel T (1988) The relationship between cell wall deformity and structural malfunction in the leaves of carnation *in vitro*. In: Proc. 11th Plant Molecular Biology, Jerusalem, 425
36. Ziv M, Meir G and Halevy A H (1983) Factors influencing the production of hardened glaucous carnation plantlets *in vitro*. Plant Cell Tiss Org Cult 2:55–65
37. Ziv M, Schwartz A and Fleminger D (1987) Malfunctioning stomata in vitreous leaves of carnation plants propagated *in vitro* — implication for hardening. Plant Sci 52:124–134

Control of water loss by delphinium plants cultured *in vitro*

JORGE M. SANTAMARIA and WILLIAM J. DAVIES
Division of Biological Sciences, I.E.B.S., Lancaster University, Lancaster, LA1 4YQ, UK

1. Introduction

In a previous paper it was shown that leaves from delphinium plants produced in culture show poor control of water loss when dehydrated [14]. This seems to be a common feature in a number of micropropagated species when transferred to soil [12, 2, 19]. This poor control of water loss in micropropagated delphinium was shown to be due to a limited capacity of stomata to close and not to a poor cuticular development [14]. One possible explanation for restricted stomatal response by cultured plants might be that they have a limited capacity to accumulate abscisic acid (ABA). Alternatively, micropropagated plants might produce ABA but their stomata exhibit reduced sensitivity to it. For a stoma to close, guard cells should be able to exclude K^+ [8] but it appears that a minimum turgor from adjacent epidermal cells is also required [3].

This paper reports on the capacity of leaves produced *in vitro* to accumulate ABA. Semi-quantitative data are presented on the ionic composition of guard cells. The turgor pressures of epidermal cells from both leaf types (*in vitro* and *ex vitro*) measured with a micro pressure probe are also presented.

2. Materials and methods

2.1 Plant material and growing conditions

Micropropagated plants from delphinium sp. cv. Princess Caroline were used. All material was non-vitrified. Leaves from plantlets produced *in vitro* (MP) were compared with leaves produced *in vitro* after acclimatization *ex vitro* for 3 months (N). MP plants were all cultured on auxin-free MS media [9] containing 7.5 g l^{-1} agar. Three different media were used:. medium 32 which contained the cytokinin, 6-benzylaminopurine (BAP) at 0.2 M and 3% sucrose (multi-plication medium); medium 191 which has no BAP with 3% glucose (rooting medium); medium 000 which is not normally used in commercial micro-propagation where plants were never exposed to BAP. Plants were grown in 300

155

P.J. Lumsden, J.R. Nicholas and W.J. Davies (eds.), Physiology, Growth and Development of Plants in Culture, 155–164, 1994.

ml vessels containing 20 plants under similar conditions to those reported in [14].

2.2 Capacity to accumulate ABA

Leaves from MP and N plants were detached and air-dehydrated under laboratory conditions. Leaves were allow to dehydrate until their fresh weight was reduced by 10% and then incubated in the dark at room temperature for 3 h. They were then freeze-dried and ABA was quantified using a radioimmunoassay [13].

2.3 Response to ABA

The stomatal response to added ABA was studied using the epidermal strip assay [10]. Carefully peeled epidermal strips from the abaxial leaf surface were incubated in a 10 mM MES (2-morpholino sulphonic acid) buffer with 50 mM KCl added at pH 6.15 (adjusted with KOH). Strips were incubated during 3 h at 25 °C under a CO_2-free air stream at ph.f.d. of 150 μmol m^{-2}s^{-1}. The MES-KCl solution was used to prepare the final solutions of various [ABA]. Twenty five randomly selected stomata were measured in each of 5 strips.

2.4 Ionic content of guard cells

Leaf pieces were mounted onto aluminium stubs with a layer of Tissue-Tek and immediately frozen by plunging them into nitrogen slush. Frozen specimens were transferred to a SEM cryo-airlock chamber avoiding contact with air. They were then coated with aluminium (50 nm thick) at a chamber vacuum pressure lower than 5 x 10 $^{-6}$ Torr. Specimens were examined on a Hexland cold stage at -160 °C using a JEOL, JSM-840A scanning electron microscope. Energy dispersive X-ray microanalysis (EDXRMA) was performed using a Link-Analytical LZ-5 detector and a Link-Analytical 860 Series 2 analyzer, connected to the SEM. The probe current used was 1×10^{-9} nA, the acceleration voltage was 15 kV, a 34 mm probe scan area at 10,000 X was used and the preset time was 200 secs. This is a semi-quantitative technique and the contents of K$^+$,Cl$^-$ and Ca^{++} are expressed as a ratio of the counts measured for the particular element and those for aluminium, as samples are coated with aluminium before being probed (e.g K net counts/Al net counts x 100). This technique eliminates possible error due to differential coating [4]] Both open and semiclosed stomata were probed from N, MP 32 and MP191 leaves.

2.5 Turgor pressures of epidermal cells

A micropressure probe [5] was used to determine the turgor of epidermal cells of both *in vitro* leaves grown on medium 191 and *ex vitro* leaves. Leaves were probed when submerged in water as *in vitro* leaves will otherwise dehydrate in 20 minutes during the process of probing.

2.6 Facilitated gas exchange in the tubs

A 0.2 μm pore size milipore filter (Whatman cellulose nitrate membrane, 25 mm diameter) was fitted over a hole on the lid of the culture vessels containing 15 plants per vessel. Stomatal characteristics of plants grown in such ventilated vessels were compared with plants grown in intact vessels. Stomatal characteristics were assessed from leaf impressions (Xantopren; Bayer) 4 weeks after plants were transferred to rooting medium 191. Twenty stomata were measured per leaf.

3. Results

3.1 Capacity to accumulate ABA

Leaves produced *in vitro* (MP) show similar capacity to accumulate ABA to that of leaves produced *ex vitro* (N) (Fig. 1).

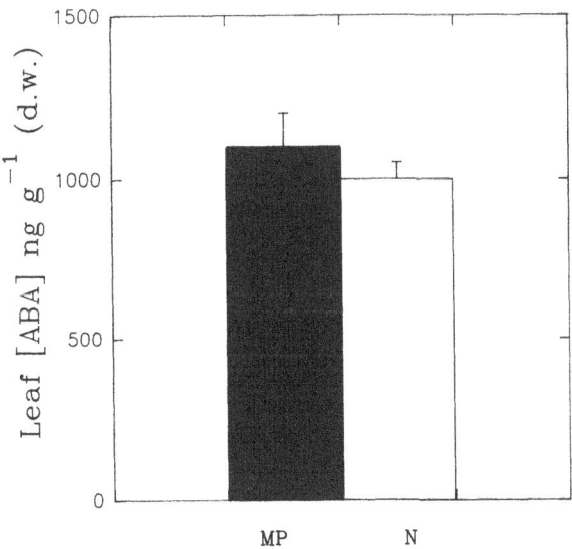

Fig. 1. Concentrations of ABA from detached leaves of N and MP plants after air-dehydration until leaves lost 10% of their fresh weight. Leaves were then incubated in the dark for 3 h. Means of 10 leaves, bars represent ± SE.

3.2 Sensitivity to ABA

Stomata from epidermal strips of MP plants growing on either rooting (191) or multiplication media [32] showed no substantial closure in response to concentrations of ABA as high as 10^{-1} mol m^{-3}, while stomata from N plants, reduced their apertures by approximately 50% when exposed to 10^{-3} mol m^{-3} ABA and closed fully when exposed to 10^{-1} mol m^{-3} ABA (Fig. 2). On the other hand, stomata from plants grown on a medium without BAP (000 medium) reduced their aperture when exposed to 10^{-3} mol m^{-3} ABA. However apertures reduced only to a limit remaining at around 20 μm at 10^{-2} and 10^{-1} mol m^{-3} ABA. The apparent inconsistency between results from plants growing in the two media lacking cytokinin (191 and 000) may be related to the fact that in the case of 191, plants were originally placed on medium 32 (with BAP) and then transferred to a BAP-free medium whereas those plants grown on 000 had never been exposed to BAP in the medium.

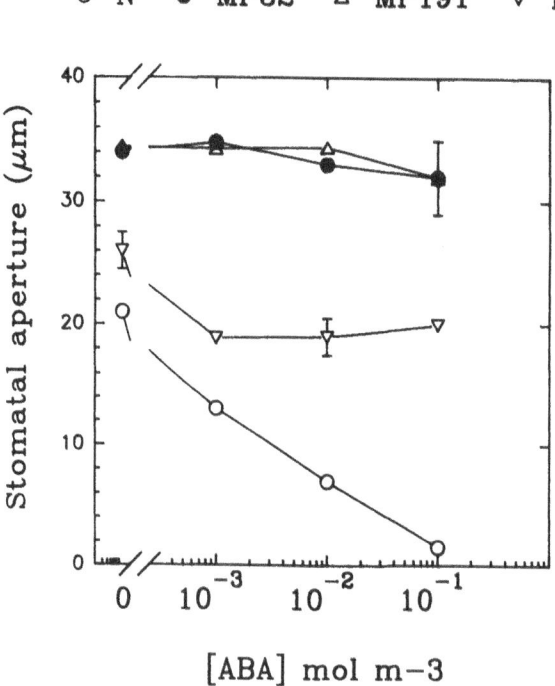

Fig. 2. Stomatal aperture from epidermal peels from both N and MP plants growing on media 32, 000 and 191 when exposed to incubating solutions with various [ABA]. Each point is the mean of 40 stomata from each of 5 epidermes, bars represent ± SE.

3.3 Ionic content of guard cells

Guard cells from both open and semiclosed stomata from MP 32, MP 191 and N leaves were probed. The relative concentrations of chloride (i.e. Cl^- counts/Al counts x 100) were similar in guard cells from open or closed stomata from all 3 treatments (Table 1). On the other hand, the relative content of potasium (K^+) in closed and semiclosed stomata was half of that in open stomata of N and MP 32 leaves. In semiclosed stomata of MP 191, however, the relative concentration of potassium did not change from that of open stomata. The calcium relative concentration of closed stomata was two times that of open stomata in N plants. Whereas in semiclosed stomata of MP32 it was 4 times that of open stomata and it was only half that of open stomata in MP 191 leaves.

Table 1. Relative ionic concentrations (counts for K^+, Cl^- and Ca^{2+}/counts Al x 100) of guard cells from both open and semiclosed stomata of N (normal) and MP (micropropagated) plants as measured by EDXRMA. Means of 25 stomata, $\pm$ SE shown in parenthesis

A) Open stomata

	N	MP	
		32	191
K^+	15.5 (5.7)	21.6 (3.1)	18.7 (1.4)
Cl^-	1.4 (0.5)	2.3 (0.2)	2.2 (0.3)
$Ca2^+$	0.5 (0.1)	0.6 (0.2)	0.03 (0.02)

B) Semiclosed stomata

	N	MP	
		32	191
K^+	7.3 (2.7)	12.4 (1.2)	18.7 (3.5)
Cl^-	1.5 (0.7)	2.4 (0.5)	2.9 (1.2)
Ca^{2+}	0.8 (0.4)	0.3 (0.1)	0.1 (0.05)

3.4 Turgor pressure of epidermal cells

Little difference was found in the turgor potential of epidermal cells from MP 191 and N leaves when probed submerged in water (Table 2).

Table 2. Turgor pressures [a] (MPa) of epidermal cells from N and MP 191 plants as measured by a micropressure probe. Means of 10 different plants per treatment, $\pm$ SE in parenthesis

N	MP 191
0.52	0.69
($\pm$ 0.04)	($\pm$ 0.11)

[a] Measured in leaves submerged in water.

3.5 Stomatal characteristics of plants from filtered and intact vessels

Stomatal characteristics from the plants grown in the intact vessels were similar to those reported in previous experiments for plants grown on medium 191 [14]. The fitting of milipore filters on the lids of the culture vessels resulted in reduced stomatal frequencies and also narrower stomatal pores, with half the apertures found in stomata from plants grown in the intact tubs (Fig. 3). No differences were found in the length of stomata grown in both treatments.

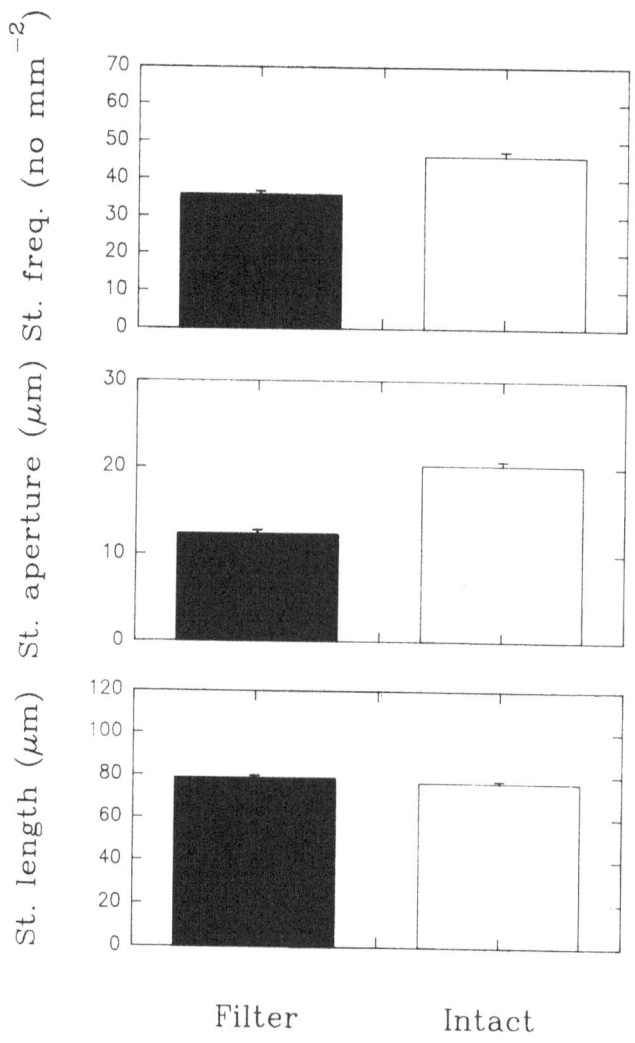

Fig. 3. Stomatal characteristics of MP 191 plants grown in intact tubs and inside vessels fitted with a milipore-filter-covered hole. Means of 40 stomata, bars show ± SE.

4. Discussion

It is clear that plants normally close their stomata in response to dehydration [18]. This has been suggested to be associated with an increased concentration of ABA in the tissues [20]. The role of ABA in promoting stomatal closure is heavily documented [7, 21]. In addition, mutants with a deficient capacity to produce ABA show a characteristic wilty appearance [16]. It was therefore expected that the poor stomatal control of water loss shown by leaves from micropropagated delphinium plants produced *in vitro* resulted from a reduced capacity to accumulate ABA. From the results shown in this paper that does not appear to be the case. Rather, the stomata are insensitive to ABA. In the long term, this insensitivity to ABA might contribute to losses of plant material when MP plants are transferred to the greenhouse since closing of stomata is necessary to to restrict water loss and mantain turgor.

This insensitivity to ABA could be related to a) lack of unaltered ABA receptors or b) the high cytokinin content in the culture medium. While the first possibility cannot be ruled out, it is less likely than the second. It has been demonstrated that cytokinins can override the ABA-induced stomatal closure of *Commelina communis* [1, 6]. The ABA insensitivity in cultured delphinium plants appears to be partly related to the cytokinin content of this medium. Stomatal apertures from plants that never experienced BAP decreased at low [ABA]. Despite this response to ABA, stomata were unable to close fully even at high [ABA]. In fact this was true for a range of other factors known to promote stomatal closure [14]. This suggests that guard cells from leaves grown *in vitro* are ionically functional but are modified structurally. All treatments which normally promote stomatal closure, caused a similar minimum aperture. Nevertheless, it cannot be ruled out that guard cells from MP delphinium plants show only a limited capacity to exclude ions. Our preliminary data using EDXRMA on guard cells showed that semiclosed stomata of MP 32 had half the relative content of $K+$ of that of open stomata, following closely what occurred in closed stomata from N leaves. However, in the case of MP 191, the $K+$ content in guard cells of semiclosed stomata was similar to that of open stomata.

For a normal stoma to close, not only do guard cells have to be able to exclude K^+ to reduce turgor [8] but also: a) the adjacent epidermal cells should have enough turgor pressure, first to prevent over-swelling of the guard cells on opening and secondly, to contribute to reduced pore size as the guard cells loses turgor at closure [3]. And b)the guard cell ventral walls should be elastic enough to allow closure after being stretched on opening [15]. It is also possible that low turgors in adjacent epidermal cells, cause stomata to open more widely than usual as they do not encounter a physical restriction. This extra stretching, may well cause an irreversible loss in the elasticity of the wall.

Our preliminary data on turgor pressures of epidermal cells from leaves grown *in vitro* (on medium 191) and *ex vitro*, do not support the possibility that the limited capacity of stomata from leaves produced in culture resulted from low pressures in the adjacent epidermal cells. Therefore, in the case of MP delphinium plants, the incapacity of stomata to close fully could result from the fact that guard cell walls are structurally altered (microfibril orientation or integrity modified).

No obvious structural guard cell wall deformation was seen in the guard cell walls of leaves grown *in vitro* using SEM (Plate 1), certainly not the extent of damage that can be seen from stomata from vitrified carnation plants (Ziv and

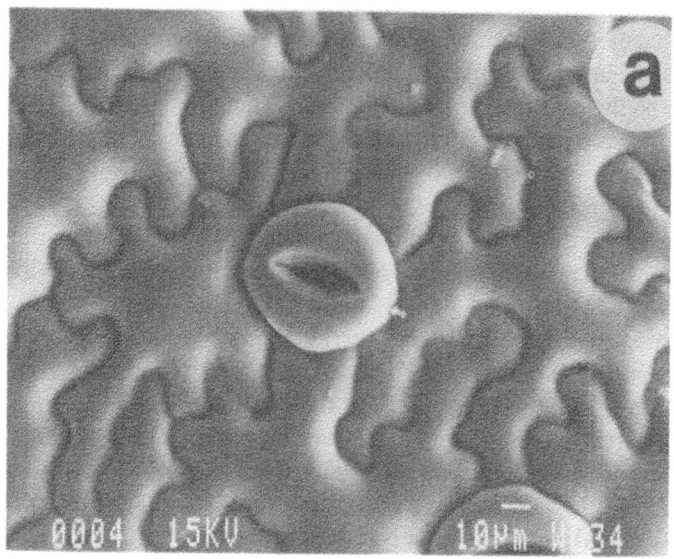

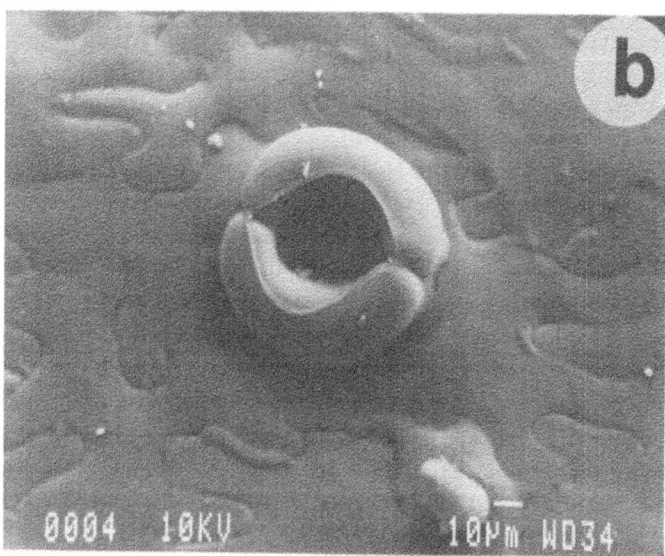

Plate 1. SEM micrographs of stomata of (a) *ex vitro*; N leaves and (b) *in vitro*; MP32 leaves at 375x magnification.

Ariel, this volume). Micropropagated delphinium plants show malfunctioning stomata without showing any visible symptoms of vitrification. This may suggest that malfunctioning stomata occur in micropropagated plants irrespective of whether they are vitrified or not. This might limit the assessment of closure mechanism as an indication of vitrification suggested by Ziv and Ariel (this volume).

It is not yet possible to draw definite conclusions on what prevents full closure of stomata from MP delphinium leaves produced in culture. However, it is interesting that by facilitating gas exchange by fitting a filter-covered hole in the lids, plants show more normal-looking stomata with narrower apertures. In fact those apertures found in plants grown in ventilated vessels were similar to those normally found in greenhouse-grown plants [14]. It is now necessary to investigate the physiological response of these stomata and to find out if plants grown in ventilated vessels show a better control of water loss.

Acknowledgements

The plant material used in this study was kindly provided by Neo-Plants LTD, Freckleton, Lancashire, U.K. Dr K Oates, U. Lancaster, took the SEM micrographs. The valuable contribution of Dr. C. Atkinson during the EDXRMA studies is gratefully acknowledged. We are extremely grateful to Dr. M. Malone, HRI, Wellsbourne for kindly allowing us to use his micropressure probe. During this work, J.S. was the holder of a scholarship from CONACYT and CICY, Mexico.

References

1. Blackman PG and Davies WJ (1983) The effects of cytokinins and ABA on stomatal behaviour of maize and *Commelina*. J Exp Bot 34:1619–26
2. Brainerd KE and Fuchigami LH (1982) Acclimatization of asceptically cultured plants to low relative humidity. J Am Soc Hort Sci 106:515–18
3. Edwards M, Meidner H and Sheriff DW (1976) Direct measurements of turgor pressure potentials of guard cells. J Exp Bot 27:163–71
4. Hopkins D, Jackson A and Oates K (1991). The effect of aluminium coating on elemental standards in X-ray microanalysis. J Electron Microsc Techn 18:176–182
5. Husken D, Zimmermann U and Steudle E (1978) Pressure probe technique for measuring water relations of cells in higher plants. Plant Physiol 61:158–63
6. Incoll LD and Jewer PC (1987) Cytokinins and stomata. In: E Zeiger, GD Farquhar and IR Cowan, eds. Stomatal Function, pp 281–92. Stanford: Stanford University Press
7. Jones T and Mansfield TA (1970) Suppresion of stomatal opening in leaves treated with abscisic acid. J Exp Bot 21:714–19
8. MacRobbie EAC (1987) Ionic relations of Guard cells. In: E Zeiger, GD Farquhar and IR Cowan, eds. Stomatal Function, pp 125–62. Stanford: Stanford University Press
9. Murashige T and Skoog F (1962) A revised medium for rapid growth and bioassays with tobacco tissue cultures. Physiol Plant 15:473–95

10. Ogunkanmi AB, Tucker DJ and Mansfield TA (1973) An improved bio-assay for abscisic acid and other antitranspirants. New Phytol 72:277–82

11. Palevitz BA (1981) The structure and development of stomatal cells. In: PG Jarvis and TA Mansfield, eds. Stomatal Physiology, pp 1–23. Cambridge: Cambridge University Press

12. Preece JE and Sutter EG (1991) Acclimatization of micropropagated plants to the greenhouse and field. In: PC Debergh and RH Zimmerman, eds. Micropropagation, pp 71–93. Dordrecht: Kluwer Acad Publ

13. Quarrie S, Whitford PN. Appleford NE, Wang TL, Cook IE, Henson IE and Loveys BR (1988) A monoclonal antibody to (S)-abscisic acid: its characterization and use in a radioimmunoassay for measuring abscisic acid in crude extracts of cereals and lupins leaves. Planta 173:330–39

14. Santamaria JM, Davies WJ and Atkinson CJ (1993) Stomata from micropropagated Delphinium plants responded to ABA, CO_2, water potential and light but fail to close fully. J Exp Bot 44:99–107

15. Sharpe P, Wu H and Spence R (1987) Stomatal mechanics. In: E. Zeiger GD Farqhuar and IR Cowan, eds. Stomatal Function 1987, pp 91–114. Stanford: Stanford University Press

16. Tal M and Imber D (1970) Abnormal stomatal behaviour and hormonal imbalance in flacca, a wilty mutant of tomato. II. Auxin- and abscisic acid-like activity. Plant Physiol 46:373–76

17. Taylor IB (1991) Genetics of ABA synthesis. In: WJ Davies and HG Jones, eds. Abscisic Acid; Physiology and Biochemistry, pp 23–35. Oxford: Bios, Scientific Publishers

18. Turner NC (1974) Stomatal responses to light and water under field conditions. In: R Bieleski, ed. Mechanisms of Regulation of Plant Growth, Bulletin 12. 1974, pp 423–32. Royal Society of New Zealand

19. Wardle K and Short KC (1983) Stomatal response of *in vitro* plantlets. Responses in epidermal strips of Chrysanthemum to environmental factors and growth regulators. Biochem Physiol Pflanzen 178:619–24

20. Wright STC (1977) The relationship between leaf water potential and the levels of ABA and ethylene in excised leaves. Planta 134:183–9

21. Zeevart JA and Creelman RA (1988) Metabolism and physiology of abscisic acid. Ann Rev Plant Physiol Plant Mol Biol 39:439–73

Components of the gaseous environment and their effects on plant growth and development *in vitro*

J.M.C. BUDDENDORF-JOOSTEN[1,2] and E.J. WOLTERING[1]

[1] *Agrotechnological Research Institute (ATO-DLO), P.O. Box 17, 6700 AA Wageningen, The Netherlands*
[2] *Department of Horticulture, Agricultural University, Haagsteeg 3, 6708 PM Wageningen, The Netherlands*

1. Introduction

Several aspects of plant biotechnology have led to an increasing interest in *in vitro* culture research of plants. Micropropagation is a popular and expanding area for commercial production because it allows rapid production of genetically identical plant material and it facilitates the production of disease-free plants. In order to raise plants from genetically altered cells, *in vitro* regeneration and micropropagation systems are a prerequisite. Anther culture is an important aid for plant breeding. Furthermore, biosynthesis of secondary metabolites from plant cell cultures is an important area of research.

In order to protect the aseptic culture from infection and to prevent desiccation of the plant and the nutrient medium, *in vitro* culture is carried out in closed vessels, which unintentionally restricts the exchange of gases between the vessel atmosphere and the outside air. Growth and development of plants or explants depend not only on the composition of the nutrient medium, but may also be affected by the composition of the gaseous atmosphere [e.g. 5, 36]. However, aspects of the gaseous atmosphere have received relatively little attention. Several components, produced by either the plant material or other parts of the system, may accumulate in the vessel atmosphere. The most important factors that affect the accumulation of gases are the type and amount of plant material, the physical properties of the container and sealing, the components in the nutrient medium and aspects of the macroclimate (e.g. temperature, ventilation, light intensity).

In commercial practice many different types of vessels and sealings are used. Culture vessels are made of various materials such as glass, polypropylene, polyvinylglycine and polycarbonate with volumes ranging from 15 to 500 ml. Depending on the type of sealing, the gas exchange will be more or less restricted. Sealings can be made of several different materials, each having specific physical properties, like gas permeability and light transmittance. Sealing materials used *in vitro* are, among others, cotton plugs, cellulose

P.J. Lumsden, J.R. Nicholas and W.J. Davies (eds.), Physiology, Growth and Development of Plants in Culture, 165–190, 1994.
© *1994 Kluwer Academic Publishers.*

stoppers, transparent film, aluminum foil, screw caps made of e.g. poly-propylene or polyvinylglycine, or combinations of caps with foil or film. Due to the restriction of gas exchange by the sealing, the gaseous composition inside the vessel may significantly differ from the outside air. Gaseous components produced by the plant material, nutrient medium or container material may accumulate inside the vessel and components that are consumed or absorbed will be depleted. It has been observed that the type of vessel and particularly the type of sealing used, may have significant effects on appearance and development of the plantlets [5, 37, 62, 95]. A nutrient medium giving optimal results in one type of system (vessel and sealing) may fail to do so in another one. This indicates that the nutrient medium, the gas phase composition and the plants interact with each other to form favourable or less favourable conditions. It also suggests that manipulation or optimization of plant growth may be possible by controlling the gaseous composition in tissue culture vessels.

Several gaseous components are present in the vessel atmosphere. The *in vivo* atmosphere contains mainly nitrogen (78%), oxygen (21%) and carbon dioxide (0.035%). During photosynthesis plants use carbon dioxide and produce oxygen, during respiration however, plants produce carbon dioxide and use oxygen. Carbon dioxide, ethylene, and a number of other hydrocarbons have been reported to accumulate during culture [e.g. 16, 62, 79, 95]. With increasing carbon dioxide concentration oxygen depletion is observed [2, 37, 77]. However, there are also reports that show a decreased carbon dioxide concentration inside the vessels during *in vitro* culture [e.g. 23, 48].

A large number of reports on the effects of manipulation of the gaseous composition on plant quality have appeared in the last decade. High carbon dioxide levels (2%) were favourable in promoting embryo production in anther cultures from various species [39]. Other authors found positive effects on plant growth of carbon dioxide levels up to approximately 0.3% [e.g. 17, 47, 48, 50, 54]. Carbon dioxide enrichment (CDE) during *in vitro* culture is often used to promote net photosynthesis as a way to prepare the plants for transfer to soil and growth *ex vitro*. When plants are transferred, they have to adapt to *ex vitro* conditions, in particular to the lack of sucrose in the nutrient medium and reduced air humidity; this is called acclimatization. The effects of ethylene on plant growth may vary depending on the process studied. Generally, a certain amount of ethylene is required for initiation of organs, but ethylene inhibits the growth of these organs and induces senescence. An important part of the research on ethylene was done by using inhibitors of ethylene action or biosynthesis. About the presence and effects of other components which may accumulate in the vessel atmosphere relatively little is known.

Here the occurrence of different gaseous components in tissue culture vessels is discussed, with special emphasis on the effects of the different gases on growth and development of *in vitro* cultured (ex-)plants.

2. Presence and effects of different gaseous components

2.1 Carbon dioxide

Carbon dioxide concentrations inside the vessels alter due to respiration and photosynthesis of the plants. In the dark, CO_2 concentrations inside culture vessels increase due to respiration [e.g. 23, 37, 83]. During the light period the CO_2 concentrations may decrease depending on the photosynthetic activity of the plants.

2.1.1 Carbon dioxide concentrations in tissue culture vessels

In the dark, the CO_2 concentrations in vessels with different ornamental plant species increased with time and reached levels of 0.3 to 0.9% at the end of the dark period [23]. In culture vessels containing *Magnolia* plantlets a CO_2 concentration of 2 to 5% was reached at the end of the dark period [16].

For a long time it was thought that *in vitro* cultured plantlets do not have photosynthetic activity because of the ample supply of carbohydrates through the nutrient medium. However, in semi-closed vessels containing ornamental plants cultured on a sucrose-containing medium, illumination was accompanied by a significant decrease in CO_2 concentration [23]. In similar research, during the photoperiod at a photosynthetic photon flux (PPF) of 65 μmol m^{-2} s^{-1}, CO_2 concentrations in the vessels were about one third of the normal atmospheric CO_2 concentration [45]. Similar decreases in the CO_2 concentration during the photoperiod were reported by other authors [17, 46, 48, 83]. Carbon dioxide concentrations may therefore drop to levels that are generally considered to be limiting to plant growth (*in vivo*). However, it is not known whether these conditions affect plant growth *in vitro*. Blazková et al. [5] did not actually measure the CO_2 concentrations inside culture vessels but they studied growth in sealed vessels with and without addition of a CO_2 absorber (KOH). In both cases, plant growth was retarded compared to growth in vessels with a cellulose stopper; the latter allows diffusion of CO_2 into the vessel. From these results the authors argued that the retarded plant growth in tightly sealed vessels might be due to CO_2 deficiency.

Despite the decrease in CO_2 concentration that occurs during the day, an overall increase in CO_2 concentration over time is often observed. In different types of semi-closed containers with *Gerbera jamesonii* plantlets on rooting medium, CO_2 concentrations of 1.3 and 1.8% were found after four weeks; in sealed containers concentrations increased up to 13% [95]. After a culture period of 18 days of *Prunus* shoots, the CO_2 concentration in the (probably sealed) jars was over 20% [79]. In sealed flasks with shoot-forming *Pinus radiata*, the CO_2 concentration reached 20% within three weeks [53]. Jackson et al. [37] measured the CO_2 concentrations in culture vessels of *Ficus lyrata* with different types of sealing. At the end of the dark period, after 23 days of culture, concentrations of approximately 0.5, 3.4 and 8.5% CO_2 were found in loose, intermediate and tightly sealed vessels, respectively. The concentrations in the same vessels at the end of the light period were 0.1, 1 and 8.5%, respectively. In all these situations the CO_2 concentrations probably are not limiting for photosynthesis. These high CO_2 concentrations (> 1%) are generally toxic to plants *in vivo*.

2.1.2 Carbon dioxide enrichment (CDE)

The effects of elevated CO_2 concentrations on plant growth and development *in vitro* are summarized in Table 1. This table shows the applied CO_2 concentrations, light intensities and the application method of the different treatments. If allowed by the data, a ventilation rate was calculated.

There is no standard method for CDE in *in vitro* culture. In many cases the routinely used culture vessels are placed in a chamber with elevated CO_2 concentration. Diffusion of the CO_2 into the culture vessel results in an increased CO_2 concentration in the vessel atmosphere. In most of the reports where this method is used, the concentration inside the vessel is not the same as the concentration in the chamber. Information about the concentrations in the vessels is often lacking. For example, in culture vessels with plantlets placed in a room with 0.3% CO_2, the CO_2 concentration was between 0.2 and 0.1% at midday, under a light intensity of 80–250 μmol m^{-2} s^{-1}. When culture vessel ventilation is not high enough, the internal CO_2 concentration decreases during the photoperiod and may still be limiting for photosynthesis when the light intensity is high. A modification of this method uses a gas diffusible film or filter. This ensures more rapid diffusion of gases while preventing contamination of the culture. Besides more rapid diffusion of the applied CO_2 into the vessel, there is another important difference between vessels with and without a gas diffusible film or filter i.e. the normal accumulation of other volatiles, such as ethylene, is much less. This may explain differences observed in the effects of CDE in different vessels. Carbon dioxide applied to routinely used vessels may act, for instance, through inhibition of ethylene action leading to a beneficial effect on growth. Application of CDE in vessels with a film or filter may have no additional effect because no accumulation of ethylene is apparent.

Another method to increase the CO_2 concentration is by flushing gas mixtures through the *in vitro* culture system. This is called forced ventilation (Table 1). The composition of the gas mixture is exactly known and other components normally present in the culture vessel are removed. This means that, for example, ethylene and possibly other components as well, will be flushed out effectively. In case of forced ventilation, the air movement around the plants will be higher than in case of CDE through diffusion. This may have pronounced effects on net photosynthetic rate (NPR) and plant growth. The humidity of the air could also be affected when gas mixtures are flushed through the containers. The results of these three methods of CDE are difficult to compare. It is remarkable that in the cases where forced ventilation was used in CDE studies, often no effects of CDE on plant growth were observed compared to forced ventilation with 0.035% CO_2 (Table 1).

In culture tubes with asparagus and strawberry plantlets under normal culture conditions, the CO_2 concentration during the light period was well below the photosynthetic compensation point [54]. Restoration of the CO_2 level (to approximately 0.05%) through diffusion into the tubes, increased fresh weight of the plants. A similar effect of CDE on growth of these plant species was reported by Desjardins *et al.* [17] who used CDE and supplemental lighting. Growth of tobacco plantlets was promoted when CO_2 concentration was restored (and elevated up to 0.56%) at a light intensity of 40 μmol m^{-2} s^{-1} [84].

The concentrations of CO_2 that are used for enhancement of photosynthesis are usually below 0.3%. There are other reports, however, in which much higher CO_2 concentrations were applied. Woltering [95, 96] treated *Gerbera* and rose plantlets with up to 5% CO_2. In these experiments, plants of both genera were greener and the leaves showed less senescence when concentrations were over 1%. Micropropagation of cacao is promoted by a CO_2 concentration of 2% [21]. Also in anther and cell suspension cultures, relatively high CO_2 concentrations were found to be beneficial. Johansson and Eriksson [39] found an increased production of microspore-derived embryos when they incubated anthers of several *Anemone, Clematis* and *Papaver* species with 2% CO_2. The most favourable CO_2 concentration varies with the tested species and genera from 2 to 5% (the two highest concentrations in the tested range). Cell suspension cultures of soybean grow best when air with a CO_2 concentration between 0.4 and 5% is flushed through [31]. The effects of high CO_2 concentrations on embryo production in anthers could not be explained by changes of the pH in the culture medium [39]. The mechanism of the beneficial effects of high concentrations of CO_2 is unknown. It is suggested that this CO_2 might work in two ways: in promoting photosynthesis (CO_2 concentrations up to 0.15%) [e.g. 17, 54, 84] and as an inhibitor of ethylene action (concentrations above 0.15%) [22, 96].

Most studies using CDE have only been concerned with the effects *in vitro* but not with the effects *ex vitro*. There is a report on the effects of CDE in acclimatization on plants treated with CDE during *in vitro* culture [55]. It seems that the growth of the plantlets which were treated with high CO_2 concentrations *in vitro* is also enhanced in the acclimatization stage. Furthermore CDE usually enhances plant growth after transfer of the plants *ex vitro* (Table 2).

In spite of the beneficial effects of CDE described above, there are also authors who could not find any positive effect of CDE on plant growth. Cuello *et al.* [13, 14] and Walker *et al.* [93] investigated the effects of a controlled environment for micropropagation of *Buddleia, Chrysanthemum* and sugarcane. From their experiments they concluded that neither CDE nor supplemental lighting enhanced growth of the plants. Also growth of *Daucus carota* and *Catharanthus roseus* cell cultures were not influenced by concentrations of CO_2 up to 2% [87].

Table 1. Effects of carbon dioxide enrichment (CDE) in relation to other experimental parameters on plant growth and/or development of (ex-)plants *in vitro*

Species	CO_2 conc. (%)	Light intensity (μmol m^{-2} s^{-1})	Realization [1]
Cymbidium	0.04 - 0.3	19 - 125	Diff.
Nicotiana tabacum	0.045 - 0.09	70	Diff.
Asparagus officinalis *Fragaria x ananassa* *Rubus idaeus*	0.033 - 0.3	80 - 250	Diff.
Asparagus officinalis *Fragaria x ananassa* *Rubus idaeus*	0.06 - 0.3	80 - 250	Diff.
Cymbidium 'Reporsa'	0.1	51 - 230	Diff.
Cymbidium 'Reporsa'	0.035 - 0.3	35 - 226	Diff.
Dianthus caryophyllus	0.1 - 0.15	150	Diff.
Fragaria x ananassa	0.035 - 0.2	200	Diff.
Nicotiana tabacum	0.033 - 2.2	40	Diff.
Gerbera jamesonii *Rosa (hybrid)*	0.06 - 5	15	Diff.
Theobroma cacao	0.08 - 2	45 - 200	Diff.
Actinidia deliciosa	0.033 - 0.45	30 - 250	CO_2 injection
Buddleia alternifolia	0 - 0.12	100	F.V.
Chrysanthemum	0.035 - 0.12	75 - 125	F.V.
Rhododendron	0 - 0.1	39	F.V.
Saccharum officinale	0.035 - 0.14	40 - 300	F.V.
Solanum tuberosum	0.034 - 2	60	F.V.

[1] CDE can be realized by diffusion (Diff.), injection and forced ventilation (F.V.)

Table 1. (continued)

Realization [1]	Ventilation rate [2] (h^{-1})	Effect [3]	Influenced parameters	References.
Diff.		+	appearance (greenness)	30
Diff.		+	growth of all parts most for roots	66
Diff.		+	growth (fresh weight) CO_2-fixation	54, 55
Diff.		+	fresh weight CO_2-fixation stomatal density	17
Diff.	0.97	+	growth (dry weight) NPR [4]	50
Diff.	0.97	+	NPR	51
Diff.	1.4	+	growth (fresh + dry weight) NPR	48
Diff.	3.7	+	fresh + dry weight NPR	47
Diff.		+	growth rate dry weight leaf/area + thickness	84
Diff.	4 - 5	+	chlorofyll content inhibition senescence	96
Diff.		+	elongation	21
CO_2 injection		+	NPR	35
F.V.	9	0		13
F.V.	9	±	dry weight	14
F.V.	15,3	0		94
F.V.	6,3	0		93
F.V.	10	+	growth starch/sucrose ratio	12

[1] CDE can be realized by diffusion (Diff.), injection and forced ventilation (F.V.)
[2] Ventilation rate is not always mentioned
[3] The effects of CDE on the parameters mentioned in the next column are positive (+), small but positive (±) or absent (0)
[4] NPR = net photosynthetic rate

Table 2. Effects of carbon dioxide enrichment (CDE) during acclimatization of *ex vitro* plants

Species	CO_2 conc. (%)	Light intensity (μmol m^{-2} s^{-1})	Effect [1]
Asparagus officinalis *Fragaria x ananassa* *Rubus ideaus*	0.33 - 0.3	80 - 250	+
Pelargonium *Spathiphyllum floribundum* *Rosa rugosa*	0.04 - 0.16	50 - 60	+
Asparagus officinalis	0.033 - 0.15	ambient ambient + 80 SL [2]	+
Fragaria x ananassa	0.033 - 0.15	ambient ambient + 150 SL	+
Vitis (hybrid)	0.035 - 0.12	150	+

[1] The effects of CDE on the parameters in the next column are positive (+)

2.1.3 Photoautotrophic growth

In vitro cultured plants that are provided with carbohydrates through the nutrient medium, theoretically do not need CO_2 and light for the production of dry matter. However, there are a number of reports that showed better growth of *in vitro* cultured plants under photoautotrophic conditions than under conditions with sucrose in the nutrient medium.

Growth under photoautotrophic conditions can be stimulated by CDE in combination with high light intensities. The major part of the work in this respect was published by T. Kozai and colleagues. Because of the low CO_2 concentrations observed during the light period, Kozai and Iwanami [48] concluded that *in vitro* plantlets are apparently photosynthetically active. They argued that a carbon source in the nutrient medium is only necessary when CO_2 concentration and/or light intensity are too low to sustain photoautotrophic plant growth. For optimization of photosynthesis, together with CDE the light intensity is often increased up to 200 to 300 μmol m^{-2} s^{-1}. The light intensity at plant level is important when the effects of CDE are investigated, but it is not always measured in the same way in the different reports. In some reports it is estimated at plant level [e.g. 84], but there are also examples of light intensity measured at vessel periphery [12], above vessels [94] or at an empty shelf in the culture room [e.g. 49]. Kozai and co-workers [49, 52] postulated that the light intensity on the empty shelf is roughly two times that

Table 2. (continued)

Effect [1]	Influenced parameters	Remarks	References.
+	growth fresh + dry weight	CDE in vitro effects in acclimatization	55
+	CO$_2$-fixation		75 76
+	root + shoot dry weight shorter nursery period	effects on dry weight only with SL	18
+	NPR [3] root + shoot dry weight shorter nursery period		19
+	dry weight root growth leaf area root/shoot ratio		56

[1] The effects of CDE on the parameters in the next column are positive (+)
[2] SL = Supplemental lighting
[3] NPR = net photosynthetic rate

of light intensity at plant level in their culture system. When light intensity is measured outside the culture vessels it is important to consider the type of container and cap materials used. Each material has a different light transmittance, therefore the light intensity and spectrum at plant level cannot easily be predicted. Another aspect of increasing light intensity is that the temperature in the culture vessels may rise. In some reports this is taken into account, but in others it is not [54]. When temperature is changed due to altered light intensities, the air movement around the plants will also be affected and possibly the diffusion of components into and out of the vessels. These changes may affect plant growth and development and it is not possible to determine which of the observed effects is caused by changes in light, temperature, air movement or CO$_2$ concentration. Sometimes many variations, e.g. in light intensity, sucrose concentration of the nutrient medium, CO$_2$ concentrations are made at the same time and treatments with more than one factor changed are compared [30, 49].

Kozai and Iwanami supplied a higher CO$_2$ concentration to carnation plantlets [48] by increasing the CO$_2$ concentration outside the vessels (0.1 to 0.15%). The resulting increase in CO$_2$ concentration inside the vessel combined with a high photon flux (150 μmol m^{-2} s^{-1}) and a low sucrose concentration (1%) in the nutrient medium promoted plant growth and NPR. Measurements of sugar concentrations in the nutrient medium (sucrose, glucose, fructose) revealed that only 2 to 8% of the initial sugar content is absorbed after 30 days of culture of carnation plantlets under a PPF of 150 μmol m^{-2} s^{-1} [48]. In liquid medium with *Cymbidium* protocorm-like-bodies under similar conditions, almost all the sugar in the medium was consumed within 14 days of culture [30]. During the remaining culture period, the sucrose concentration is very low and CDE would be very effective. Under low PPF and CO_2 non-enriched conditions the fresh weight of potato plantlets increased with increasing sucrose concentration. When high PPF and CDE were applied, however, the sucrose concentration in the medium did not affect plant growth [49].

Photosynthesis may also be affected by the carbohydrate status of the medium; photosynthetic activity of *in vitro* grown *Clematis* plants is inhibited in sucrose enriched media, compared to plants grown *in vivo* [57]. In the first period after transfer to soil, *Vitis* plants show a delayed development which is ascribed to limited photosynthesis [56]. NPRs of *in vitro* plantlets are reported to be very low in comparison with NPRs of seedlings, as e.g. for strawberry [28]. However, net photosynthesis of *in vitro* cultured plants of *Asparagus* is similar to photosynthesis in seedlings grown in a greenhouse [20]. Also *Primula malacoides* plantlets *in vitro* have a photosynthetic ability comparable to that of plants *in vivo* [82]. Our own results (unpublished) showed that photosynthetic activity of *Gerbera* plantlets is little affected by the sucrose concentration in the nutrient medium.

However, photoautotrophy does not always lead to the best plant growth compared to growth on a medium containing sucrose. Some reports show a positive influence of sucrose on plant growth despite of CO_2 enriched conditions. For example, under CO_2 enriched conditions carnation grows best when 1% sucrose is added to the medium, compared to 0 or 2% [48]. Also in tobacco plantlets growth is optimal when CO_2 enriched conditions are combined with sucrose-containing (2%) medium [84]. Kozai *et al.* [51] reported on photosynthetic characteristics of *Cymbidium* plantlets *in vitro*. They showed that CDE (0.3%) at relatively high PPFs (100 and 225 μmol m^{-2}s^{-1}) promotes photosynthesis and, hence, growth of chlorophyllous shoots or plants *in vitro*. These *Cymbidium* plantlets were cultured on a nutrient medium containing 2% sucrose. Better growth of *in vitro* plants under photoautotrophic conditions, than under conditions with sucrose in the nutrient medium, might be explained by the higher uptake of ions under photoautotrophic conditions, as found in strawberry plantlets [47]. However, the role of sucrose in the nutrient medium is not clear yet. Despite high CO_2 concentrations and high light intensities, sucrose may still be beneficial for plant growth *in vitro*.

To improve plant growth and survival after transfer from *in vitro* culture to soil, stimulation of photosynthesis during *in vitro* culture may be useful. Another possibility is to stimulate photosynthesis during the acclimatization stage *ex vitro* (see Table 2). For this purpose CDE is often used in combination with supplemental lighting [19, 24, 46, 55, 56, 75]. These conditions are in most cases favourable for plant growth. Plants should be able to produce their own assimilates when sucrose is no longer available in the nutrient medium. Because of increased growth of the plantlets by CDE, the acclimatization period often can be shortened.

2.2 Ethylene

The effects of ethylene *in vitro* are diverse. For cell, callus and anther cultures, accumulation or addition of ethylene is often inhibitory. Addition of inhibitors of ethylene production or action is often stimulatory to growth and embryogenesis (for review see 4). Here we mainly focus on the concentrations that occur in culture vessels and the effects of ethylene on *in vitro* cultured (ex-)plants.

2.2.1 Ethylene concentrations in tissue culture vessels

Depending on the ethylene production of the plantlets, the ambient ethylene concentration in the air during transplanting and the type of sealing (ventilation rate) of the culture vessels, ethylene concentrations may be high or reach high values. After three weeks of culture of tomato, peach x almond hybrid, and sweet cherry, ethylene concentrations inside the vessels were between 0.4 and 0.5 μl l^{-1}, concentrations in vessels with grapevine were about 0.1 μl l^{-1} [41]. In vessels with miniroses concentrations up to 6 μl l^{-1} were found [15]. In vented tubes with *Brassica campestris* the ethylene concentration was 0.01 μl l^{-1}, whereas sealed tubes contained 0.30 μl l^{-1} after 20 days of culture [58, 59]. The vented treatment was obtained by placing a foam plug in a 25-mm aperture in the lid; in the sealed treatment the aperture was closed with a silicone stopper. In sealed vessels with carnation accumulation up to 0.7 μl l^{-1} occurred after 4 weeks, whereas unsealed vessels contained less than 0.1 μl l^{-1} [62]. In callus cultures of *Hevea brasiliensis* ethylene concentrations were over 5 μl l^{-1} under confined conditions in the dark after 5 weeks [3]. In flasks with shoot-forming *Pinus radiata* cotyledons 20 μl l^{-1} ethylene was found after three weeks of culture, whereas only 2 μl l^{-1} accumulated in similar flasks under non shoot-forming conditions [53]. In vessels with *Magnolia* plantlets the ethylene concentration gradually increased from 0.5 μl l^{-1} after one week to 2 to 3 μl l^{-1} after 9 weeks [16]. In vessels with *Gerbera* plantlets on rooting medium, accumulation of 0.02 to 1.3 μl l^{-1} ethylene was found, depending on the type of container and type of sealing used [95]. Ethylene concentrations inside culture vessels with *Prunus* shoots were between 5.6 and 8.4 μl l^{-1} immediately after transplanting *in vitro* as a result of fumes produced by the gas flame, applied for sterilization of tools in the laminar flow cabinet [77].

2.2.2 Ethylene production

Ethylene is produced by cultured plant cells, tissues, organs and complete plants. Differences in the rate of ethylene production *in vitro* were attributed to the type of tissue, the physiological state of the biological material and the nature and concentration of the added growth regulators to the nutrient medium [25]. In culture flasks with *Dahlia* leaf segments and callus, grown on medium with α-naphtaleneacetic acid (NAA) and kinetin, ethylene production was proportional to the amount of tissue and was stimulated by high NAA concentrations [26]. The amount of ethylene that is produced by lavandin explants depends on the culture stage and 6-benzylaminopurine (BA) concentration in the nutrient medium [68]. The ethylene production greatly varies with species [41] and may be influenced by light [60, 68]. Tobacco callus produced about 500 and 1000 nl ethylene per gram fresh weight per 24 h in the dark and light, respectively [33]. The ethylene production by rose explants was between 3.5 and 6.5 nl per explant per 24 h [25]. The ethylene production by bromeliads is about 4.8 nl per plant per 24 h [15], the production may vary between 1.7 and 20 nl per plant per 24 h, depending on the time in culture and the concentrations of BA and indole-3-acetic acid (IAA) in the nutrient medium [91].

   In addition to the plant material, other ethylene sources can be present in an *in vitro* culture system. Agar was identified as a significant source of ethylene [60, 63]. Different brands of agar released considerable amounts of ethylene when they were exposed to light [60]. The ethylene release from agar is concentration dependent: the higher the agar concentration, the higher the ethylene release [63]. Carbon dioxide cylinders used in CDE studies, may be contaminated with ethylene ranging from less than 0.1 up to 475 μl l^{-1} [65]. Another possible source of ethylene is rubber after autoclaving [38], often used in caps to facilitate gassampling. Introduction of ethylene in tissue culture vessels may also come from combustion gases produced by a gas burner used in combination with ethanol, to sterilize tools during transplanting [77].

2.2.3 Manipulation of ethylene production, ethylene action, and accumulation in the vessels

Ethylene accumulation in culture vessels is affected by the physical properties of the system (mainly the type of sealing) and the release by the plant material or other components in the system. Several methods are available to manipulate ethylene levels or production rates (Fig. 1).

   In plants ethylene is produced from S-adenosylmethionine (SAM) through the intermediate 1-aminocyclopropane-1-carboxylic acid (ACC). The enzyme, that converts SAM into ACC, is considered to be the rate limiting step in ethylene biosynthesis. The activity of this enzyme, ACC synthase, can effectively be blocked by treating tissues with aminoethoxyvinylglycine (AVG) or amino-oxyacetic acid (AOA), hereby manipulating ethylene production and, hence, accumulation in the vessel. Similarly, the enzyme converting ACC into ethylene, ACC oxidase, can be blocked by cobalt ions (e.g. $CoCl_2$).

   Addition of ACC to the medium often results in increased rates of ethylene production. Also auxins and cytokinins may, through their direct or indirect effects on ACC synthase activity, stimulate ethylene production. Carbon dioxide, present in the container atmosphere, may stimulate ethylene production through its effect on ACC oxidase activity. Once produced, the

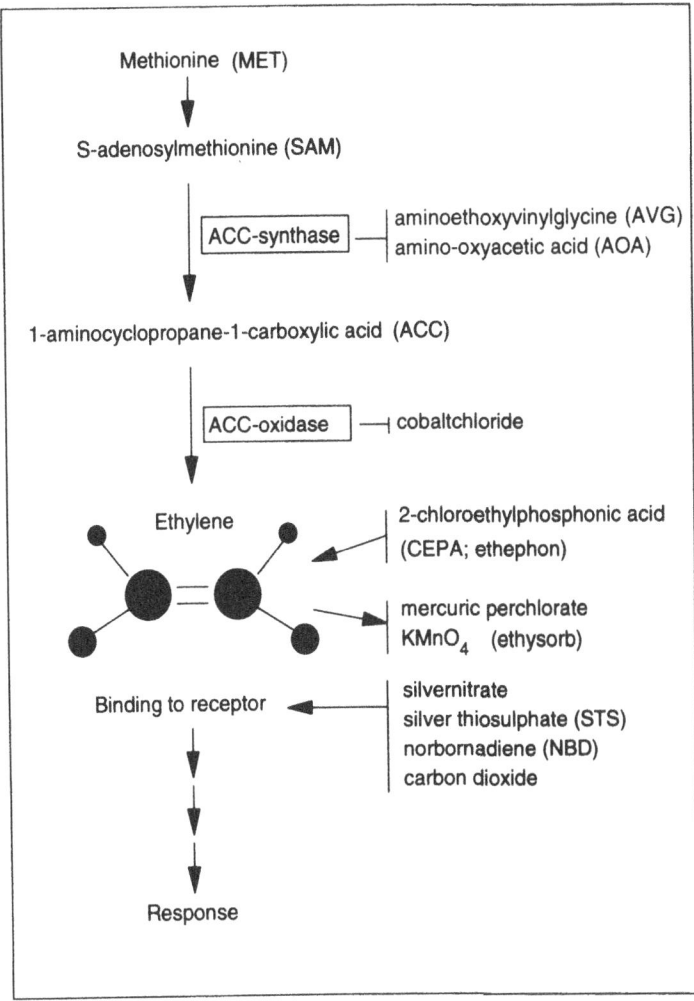

Fig. 1. Ethylene biosynthesis, action and inhibition [1, 69]

ethylene is thought to bind to a receptor protein which mediates the response. The binding of ethylene to the receptor can be blocked by 2,5-norbornadiene (NBD) and Ag^+ ions, either applied as $AgNO_3$ or as silver thiosulphate (STS). In this way the response is blocked without a direct effect on the production. Also increased CO_2 concentrations are known to suppress ethylene sensitivity. The precise mode of action is at present unknown. The ethylene produced may diffuse out of the tissue to the gaseous atmosphere where it can be chemically removed by using, for instance, a solution of potassium permanganate ($KMnO_4$), Ethysorb (activated aluminum oxide coated with $KMnO_4$) or a mercuric perchlorate solution. These chemicals cannot be dissolved in the nutrient medium and are therefore applied by placing e.g. a vial with the chemicals in the culture vessels. Finally, ethylene may be introduced into the vessels by using ethylene gas from pressurized bottles or by adding components that generate ethylene (e.g. ethrel or ethephon, both containing 2-chlorethylphosphonic acid (CEPA) or CEPA itself) to the nutrient medium (for references see 1, 69).

2.2.4 Effects of ethylene on plant growth and development in vitro

The positive and negative effects of ethylene on growth and development of *in vitro* cultured plants or explants are summarized in Tables 3 and 4. In most cases proliferation is promoted by ethylene (Table 4), but growth of the induced organs is inhibited by ethylene (Table 3). In *Brassica* species, which are generally recalcitrant *in vitro*, the effects of ethylene have been investigated by several authors. For example, Chi *et al.* [7] reported on enhanced shoot regeneration from seedling explants of several *Brassica* genotypes in the presence of the ethylene inhibitors AVG and $AgNO_3$ in the nutrient medium. These authors suggest that poor regeneration of cultured cells and tissues of *Brassica* may, at least partially, be attributed to ethylene produced by cultured plants. They also reported that the presence of a low amount of ethylene is required for root initiation. By using ethylene inhibitors and a number of recalcitrant *Brassica* genotypes, other authors showed also that ethylene was one of the causes of recalcitrance [8, 9, 58, 59, 73, 81].

Other negative effects of ethylene *in vitro* are the inhibition of plant growth and the enhancement of senescence. In sealed vessels with *Ficus lyrata*, ethylene accumulation was accompanied by a decrease in leaf area [37]. Also in sealed vessels with *Ficus lyrata*, ethylene was thought to increase callus formation at the expense of shoot proliferation, because the effects could be reversed by placing $KMnO_4$ in the vessel [40, 42]. Addition of STS to the culture medium of potato plantlets increased leaf size [64, 71], as well as shoot and root growth [71]. Tuberization of potato is promoted when $KMnO_4$ is added [32]. These results indicate that tuberization of potato is inhibited by ethylene. Growth of carnation explants is supposed to be inhibited by accumulated ethylene because growth enhancement occurred when $KMnO_4$ was added [62]. Shoot elongation of roses is promoted when mercuric perchlorate is present in the culture vessels, indicating that ethylene inhibits shoot growth [44]. Epinasty of a peach x almond hybrid was ascribed to accumulation of ethylene [42, 43]; adding $KMnO_4$ diminishes the epinasty. Ethylene treatment induces senescence of mini-roses [15] and causes yellowing of *Gerbera* plantlets [95]. Also in carnation, leaf yellowing is stimulated when ethylene is added to the vessel atmosphere [62].

Also positive effects of ethylene *in vitro* have been described. In bromeliads ethylene suppresses the apical dominance and thereby increases the number of axillary shoots that are formed [15]. Ethylene treatment increased the bud number per bulb-scale explant of *Lilium speciosum in vitro* [90] and enhanced the development of bulb primordia at the base of regenerated shoots of tulip [85]. Shoot-forming ability of lavandin is positively correlated with the rate of ethylene production in callus tissue and micropropagation explants [68]. Gaspar *et al.* [25] observed a peak in ethylene production which coincided with the initiation of lateral shoot outgrowth from basal axillary buds in cultured rose explants. Suppression of the ethylene production by addition of AVG or $CoCl_2$ did not change this process, indicating that there is no functional relationship between axillary budding and ethylene biosynthesis. Further investigations showed that pulse treatment of ethylene enhanced proliferation depending on concentration and time of application [44]. Ethylene removal with mercuric perchlorate resulted in a reduced shoot proliferation rate, but increased shoot elongation, thereby generating a higher number of rose shoots suitable for propagation [44].

Cachita *et al.* [6] reported of several positive effects of ethylene. They investigated shoot and root formation in the presence of CEPA and concluded that ethylene promotes organogenesis (roots and shoots) and growth of carnation meristems, *Cymbidium* protocorms, *Chrysanthemum* and potato minicuttings and *Gerbera* shoots.

Ethylene was shown to be an important factor in rooting, but the experimental results have been contradictory. In some studies ethylene stimulated rooting [6, 27, 61, 70] while in others ethylene had an inhibitory or no effect on rooting [11, 42, 71, 73]. The effects of ethylene on rooting probably depend on the concentration and on the physiological stage of the explants.

2.3 Oxygen

There are very few reports on the influence of oxygen concentration on plant growth *in vitro*. It is expected that a moderate decrease of oxygen concentration will not effect plant growth and development. Generally with increasing CO_2 concentration in the vessels, a comparable decrease in oxygen can be expected.

In Petri dishes with rice callus the oxygen concentration was 2 to 5 % after 24 days of culture [2]. In tightly sealed vessels with *Ficus* plantlets, oxygen concentrations of approximately 10 % were observed [37]. Several cultivars of *Prunus avium* (cultured at a PPF of 30 μmol m^{-2} s^{-1}) differed very much in the changes in O_2 concentration during *in vitro* culture. Two cultivars showed drastic O_2 consumption until the concentration reached a level of about 10 % after 30 days, whereas two other cultivars showed only a slight increase in O_2 concentration up to 24 % in the same period [78].

In *Torenia* stem segments adventitious buds were initiated after a short term treatment with 100 % nitrogen, similar treatment with air showed no effect [86]. An oxygen tension of 5 % or less stimulated plantlet production from anther cultures of tobacco [34]. In *Spiraea* the shoot number increased after treatment with 100 % nitrogen (no oxygen). The stimulation of shoot proliferation by hypoxia may be due to the effects of these conditions on ethylene biosynthesis according to Norton [67]. Plantlets of *Primula* and *Chrysanthemum* cultured in 10 and 1 % oxygen showed a higher NPR than those cultured in 21 % [82]. Hypoxia (of 0.5 to 2 % O_2) can also be applied to slow senescence of *in vitro*

Table 3. Presumed positive effects of ethylene on plant growth and/or development *in vitro*

Plant species	Treatment [1]	Effect [2]	Influenced parameters	References
Dianthus caryophyllus	CEPA	+	bud + shoot formation	6
Cymbidium	CEPA	+	organogenesis	
Chrysanthemum morifolium	CEPA	+	organogenesis	
Solanum tuberosum	CEPA	+	organogenesis	
Gerbera	CEPA	+	organogenesis + growth	
Brassica	AVG	-	root elongation	7
Achmea dactilyna	AVG	-	number of shoots	15
	ethyleen	+		
Corylus avellana	methionine	0	rooting	27
	ACC	0		
	CEPA	+		
	AVG	-		
GF677 (peach x almond hybrid)	sealing	+	proliferation	43
	KMnO$_4$			

Species	Treatment	Effect[2]	Parameter	
Rosa	ethylene AVG ACC $CoCl_2$	+ 0 0 0	proliferation	44
Pinus radiata	sealing $KMnO_4$	+ −	differentiation of buds	53
Helianthus annuus	AVG STS ACC ethephon	− − + +	root formation	61
Tulipa	ethylene ACC	+ +	bulb primordia	85

[1] For abbreviations of the treatments see text 2.2.3

[2] Effects on the parameters mentioned in the next column are stimulatory (+), inhibitory (−) or absent (0)

Table 4. Presumed negative effects of ethylene on plant growth and/or development *in vitro*

Plant species	Treatment [1]	Effect [2]	Influenced parameters	References
Brassica	AgNO$_3$ Ag$_2$SO$_4$ AVG AOA DNP ethephon	+ + + 0 0 -	shoot regeneration	7, 8, 9
Helianthus annuus	AgNO$_3$ CoCl$_2$ ethephon ethylene	+ + - -	shoot regeneration	10
Rosa	AVG ethylene	0 +	senescence	15
Solanum tuberosum	KMnO$_4$	+	tuberization	32
Ficus lyrata	sealing	-	leaf area	37
Ficus lyrata	sealing	-	shoot profileration	42
Peach x almond hybride	sealing KMnO$_4$	+ -	epinasty senescence	42,43
Rosa	AVG CoCl$_2$ ACC ethylene	+ + - -	shoot elongation	44

Brassica campestris	NBD sealing	+ -		58, 59
Dianthus caryophyllus	ethylene KMnO₄	-, + +, -	growth, yellowing growth, yellowing	62
Solanum tuberosum	STS	+	leaf size shoot + rootgrowth	64, 71
Cruciferae	AgNO₃ AVG	+ +	shoot regeneration	73
Gerbera jamesonii	sealing ethylene	+ +	yellowing	95

[1] For abbreviation of the treatments see text 2.2.3
[2] Effects on the parameters mentioned in the next column are stimulatory (+), inhibitory (−) or absent (0)

(peach) plantlets during storage in order to delay the subculture and to limit the risk of genetic variation [74].

Oxygen limiting conditions resulted in suppressed growth of *Catharanthus roseus* cell cultures. However, the growth yield (g cell mass generated per g glucose consumed) was not affected by oxygen deprivation. This means that under these conditions the cells did not convert significant amounts of sucrose to fermentative products such as ethanol and this may therefore not explain the observed repressed growth [87].

An advantage of decreased oxygen concentrations is that this inhibits the growth of fungi and bacteria in the culture vessel [82]. This facilitates working under sterile conditions. However, several processes in higher plants are inhibited when oxygen concentration is low. The effects of low oxygen concentration on plant growth *in vitro* can (partly) be explained by its effect on photorespiration. For *in vivo* plants, the loss of photosynthetically fixed carbon due to photorespiration can be up to 50 % in normal atmospheric O_2 concentration [98]. Photorespiration is repressed with decreasing O_2 concentration and is almost completely absent in 2 % O_2 [98]. *In vitro* C_3 plantlets have similar photosynthetic characteristics to that of C_3 plants grown *in vivo* [20,82]. This means that the net photosynthetic rate of C_3 plantlets *in vitro* should increase with decreasing oxygen concentration (as measured in 82).

2.4 Ethanol, acetaldehyde and other components

Besides CO_2, ethylene and oxygen, other components that may influence plant growth or development can be present in the gaseous environment. There are not many reports on these 'other' components, they seem to be regarded as of minor importance in normal *in vitro* culture.

When wild cherry cultures are multiplied in closed vessels, several volatile substances can be detected [79]. At the beginning of the culture period the shoots start to form ethylene and CO_2. Shoots begin to release ethanol and acetaldehyde after 6 and 14 days, respectively. After 30 days of culture of wild cherry, CO_2 reaches a concentration of 30 % [80]. Formation of ethanol and acetaldehyde increased when the CO_2 concentration exceeded 20 %. Accumulation of ethanol and acetaldehyde is followed by rapid tissue deterioration. At the time tissue deterioration occurred, ethane and acetone were detected inside the vessels. In control vessels (glass jars) without plants but with nutrient medium traces of ethylene, ethane, propylene, propane, methanol, acetaldehyde, ethanol and acetone were found during a three month incubation in a growth room under normal culture conditions [80].

Rice callus produces CO_2, ethylene and ethanol; one of the tested cultivars also produces small quantities of acetaldehyde and ethane [2]. Ethanol and acetaldehyde were detected in callus cultures grown in medium containing the auxin 2,4-dichlorophenoxyacetic acid (2,4-D). Independent of auxin concentration, embryogenesis in *Daucus carota* and *Phoenix dactylifera* showed an inverse relationship with the concentration of ethanol in the cultures.

Exogenous ethanol reversibly inhibited embryogenesis in wild carrot callus [89]. Callus of a non embryogenic carrot strain produced much more ethanol than an embryogenic strain [88], indicating that there is a negative correlation between ethanol concentration (or ethanol production) and embryogenesis.

Volatile components that are not produced by plant material, may be present inside the culture vessels, depending on the container material used and the way tools are sterilized during transplanting. In the cases where vessels or lids are made from plastic, components, which are added to the plastic to improve weakening (plasticizers), may be released into the container atmosphere and affect plant growth. The plasticizer di-butyl-phthalate (DBP), for instance, was found to be very toxic to plants [29, 92]. *In vitro*, many different types of vessels are used and problems may arise resulting in poor growth of plants as a consequence of phytotoxic components. In different container types with *Gerbera* plantlets, ethane, ethylene and propane were detected. The tested container types were made of glass, polypropylene (PP) or polyvinylchloride (PVC). In PP and PVC containers also propylene and butane were present [97]. In the PP containers an additional component, which was not present in the other vessels, was detected. This component was shown to be produced by the PP container material itself, irrespective of the presence of medium or plant material. This component was found to be phytotoxic to *Gerbera*, causing severe yellowing within two weeks of culture. The component has not yet been identified, although it is not the plasticizer DBP [97]; it is however removed by Ethysorb.

When hydrocarbons are partially oxidized or unevenly combusted and when ethanol is used for sterilization of dissecting instruments during transplanting, large quantities of biologically active components are produced [77]. This caused that culture vessels with *Prunus* shoots contained (besides ethylene and CO_2) acetaldehyde, ethanol, iso-butane, n-butane, propane and propylene immediately after transplanting.

3. Conclusions

Generally it can be stated that, in order to obtain optimal growth, the sealing of the vessels must allow sufficient ventilation to prevent significant built up of ethylene and depletion of CO_2. In addition, the container material used should be free of phytotoxic components that may diffuse into the container atmosphere.

A low concentration of ethylene *in vitro* seems to be necessary for organo-genesis. Higher ethylene concentrations may have negative effects on growth and development of plants and induce senescence.

Despite the numerous reports on the beneficial effects of CDE on plant growth and development *in vitro*, the mode of action is still not clear. It is pointed out that CDE enhances NPR of plantlets *in vitro*, but theoretically there is no direct need for photosynthetic activity because of the supply of carbohydrates through the nutrient medium. Another aspect is that high CO_2 concentrations (over 1%) have a beneficial effect on plant growth and development *in vitro*, whereas these concentrations have an inhibitory effect on photosynthesis and growth of *in vivo* plants [72]. The mode of action of these high CO_2 concentrations are also not clear yet. Possibly this CO_2 may act in inhibiting ethylene action.

To really unravel the relative contribution of the different gaseous components to plant growth and development and to investigate the physiological background, the experiments should be carried out under strictly controlled conditions. In such experiments it is important that appropriate controls are part of the treatments.

Acknowledgements

The authors are grateful to Professor Dr R.L.M. Pierik, Dr ir H.J. Scholten and Dr H.C.M.P. van der Valk for critically reading the manuscript. We also thank Braam Vitro BV, Cultiss Holland BV, Maurits & Mons VOF, PermX Multiplant BV, PhytoNova, Stichting Bedrijfslaboratorium voor Weefselkweek, Sovitro BV, Van Staaveren BV, Vigro Biotechnologie, Vitro Lab. Zuidgeest and Zaadunie BV for their financial support.

References

1. Abeles FB (1973) Ethylene in Plant Biology. Academic Press, New York, 302 pp.
2. Adkins SW, Shiraishi T and McComb JA (1990) Rice callus physiology – Identification of volatile emissions and their effects on culture growth. Physiol Plant 78:526–531
3. Auboiron E, Carron MP and Michaux-Ferriere N (1990) Influence of atmospheric gases, particularly ethylene, on somatic embryogenesis of *Hevea brasiliensis*. Plant Cell Tiss Org Cult 21:31–37
4. Biddington NL (1992) The influence of ethylene in plant tissue culture. Plant Growth Regul 11:173–187
5. Blazková A, Ullman J, Josefusova Z and Machackova I (1989) The influence of gaseous phase on growth of plants *in vitro* – the effect of different types of stoppers. Acta Hortic 251:209–214
6. Cachita CD, Achim F, Cristea V and Gergely K (1987) 2-chloroethylphosphonic acid and morphogenesis. Symposium Florizel 87 Arlon-Belgium. Plant micropropagation in horticultural industries. pp 234–237
7. Chi G-L, Barfield DG, Sim G-E and Pua E-C (1990) Effect of $AgNO_3$ and aminoethoxyvinylglycine on *in vitro* shoot and root organogenesis from seedling explants of recalcitrant *Brassica* genotypes. Plant Cell Rep 9:195–198
8. Chi G-L and Pua E-C (1989) Ethylene inhibitors enhanced de novo shoot regeneration from cotyledons of *Brassica campestris* ssp *chinensis* (Chinese cabbage) *in vitro*. Plant Physiol 96:178–183
9. Chi G-L, Pua E-C and Goh C-J (1991) Role of ethylene on de novo shoot regeneration from cotyledonary explants of *Brassica campestris* ssp *pekinensis* (Lour) Olsson *in vitro*. Plant Physiol 96:178–1839.
10. Chraibi KMB, Roustan J-P, Latché A and Fallot J (1991) Stimulation of shoot regeneration from cotyledons of *Helianthus annuus* by the ethylene inhibitors silver and cobalt. Plant Cell Rep 10:204–207

11. Coleman WK, Huxter TJ, Thorpe A (1980) Ethylene as an endogenous inhibitor of root regeneration in tomato leaf discs cultured *in vitro*. Physiol Plant 48:519–525

12. Cournac L, Dimon B, Carrier P, Lohou A and Chagvardieff P (1991) Growth and photosynthetic characteristics of *Solanum tuberosum* plantlets cultivated *in vitro* in different conditions of aeration sucrose supply and CO_2 enrichment. Plant Physiol 97:112–117

13. Cuello JL, Walker PN, Heuser CW and Heinemann PH (1991) Controlled *in vitro* environment for stage-II micropropagation of *Buddleia alternifolia* (Butterfly Bush). Transactions of the ASAE 34:1912–1918

14. Cuello JL, Walker PN and Heuser CW (1992) Controlled *in vitro* environment for stage-II micropropagation of *Chrysanthemum*. Transactions of the ASAE 35:1079–1083

15. De Proft MP, Van Den Broek G and Van Dijck R (1985) Implications of the container-atmosphere during micropropagation of plants. Med Fac Landbouww Rijksuniv Gent 50:129–132

16. De Proft MP, Maene LJ and Debergh PC (1985) Carbon dioxide and ethylene evolution in the culture atmosphere of *Magnolia* cultured *in vitro*. Physiol Plant 65:375–379

17. Desjardins Y, Laforge F, Lussier C and Gosselin A (1988) Effect of CO_2 enrichment and high photosynthetic photon flux on the development of autotrophy and growth of tissue-cultured strawberry, raspberry and asparagus plants. Acta Hort 230:45–53

18. Desjardins Y, Gosselin A and Lamarre M (1990) Growth of transplants and in vitro-cultured clones of asparagus in response to CO_2 enrichment and supplemental lighting. J Amer Soc Hort Sci 115:364–368

19. Desjardins Y, Gosselin A and Yelle S (1987) Acclimatization of ex vitro strawberry plantlets in CO_2-enriched environments and supplementary lighting. J Amer Soc Hort Sci 112:846–851

20. De Yue D, Desjardins Y, Lamarre M and Gosselin A (1992) Photosynthesis and transpiration of *in vitro* cultured asparagus plantlets. Sci Hort 49:9–16

21. Figueira A, Whipkey A and Janick J (1991) Increased CO_2 and light promote *in vitro* shoot growth and development of *Theobroma cacao*. J Amer Soc Hort Sci 116:585–589

22. Figueira A, Whipkey A and Janick J (1991) Elevated CO_2 facilitates micropropagation of *Theobroma cacao* L. International Cacao Conference, Kualalumpur, 25–28 sept 1991. App. B4

23. Fujiwara K, Kozai T and Watanabe I (1988) Development of a photoautotrophic tissue culture system for shoots and/or plantlets at rooting and acclimatization stages. Acta Hort 230:153–158

24. Fujiwara K, Kozai T and Watanabe I (1987) Fundamental studies on environments in plant tissue culture vessels. (3) Measurement of carbon dioxide gas concentration in closed vessels containing tissue cultured plantlets and estimates of net photosynthetic rates of the plantlets. J Agr Met 43:21–30

25. Gaspar T, Kevers C, Bouillenne H, Maziere Y and Barbe JP (1989) Ethylene production in relation to rose micropropagation through axillary budding. In: H Clijsters, M de Proft, R Marcell and M van Poucke, eds. Biochemical and Physiological Aspects of Ethylene Production in Lower and Higher Plants, pp 303–312. Dordrecht: Kluwer Academic Publishers

26. Gavinlertvatana P, Read PE, Wilkins HF and Heins R (1982) Ethylene levels in flask atmospheres of *Dahlia pinnata* Cav leaf segments and callus cultured *in vitro*. J Amer Soc Hort Sci 197:3–6

27. Gonzalez A, Rodriguez R and Sánchez Tamés R (1991) Ethylene and *in vitro* rooting of hazelnut (*Corylus avellana*) cotyledons. Physiol Plant 81:227–233

28. Grout BWW and Millam S (1985) Photosynthetic development of micropropagated strawberry plantlets following transplanting. Ann Bot 55:129–131

29. Hannay JW and Millar DJ (1986) Phytotoxicity of phthalate plasticisers. 1. Diagnosis and commercial implications. J Exp Bot 37:883–897

30. Honjo T, Futaya Y and Takakura T (1988) Effect of CO_2 enrichment on the growth of *Cymbidium* PLB in vitro. Acta Hort 230:185–188

31. Horn ME, Martin BA and Widholm JM (1992) Photoautotrophic growth of soybean cells in suspension culture. 2. Optimization of culture medium and conditions. Plant Cell Tiss Org Cult 30:85–91

32. Hussey G and Stacey NJ (1984) Factors affecting the formation of *in vitro* tubers of potato (*Solanum tuberosum* L). Ann Bot 53:565–578

33. Huxter TJ, Reid DM and Thorpe A (1979) Ethylene production by tobacco (*Nicotiana tabacum*) callus. Physiol Plant 46:374–380

34. Imamura J and Harada H (1981) Stimulation of tobacco pollen embryogenesis by anaerobic treatments. Z Pflanzenphysiol 103:259–263

35. Infante R, Magnanini E and Righetti B (1989) The role of light and CO_2 in optimising the conditions for shoot proliferation of *Actinidia deliciosa in vitro*. Physiol Plant 77:191–195

36. Jackson MB, Abbott AJ, Belcher AR and Hall KC (1987) Gas exchange in plant tissue cultures. Monograph British Plant Growth Regulator Group 16:57–71

37. Jackson MB, Abbott AJ, Belcher AR, Hall KC, Butler R and Cameron J (1991) Ventilation in plant tissue cultures and effects of poor aeration on ethylene and carbon dioxide accumulation oxygen depletion and explant development. Ann Bot 67:229–237

38. Jacobsen JV and McGlasson WB (1970) Ethylene production by autoclaved rubber injection caps used in biological systems. Plant Physiol 45:631

39. Johansson L and Eriksson T (1984) Effects of carbon dioxide in anther cultures. Physiol Plant 60:26–30

40. Jona R and Gribaudo I (1988) Environmental factors affecting *in vitro* propagation of *Ficus lyrata*. Acta Hort 226:59–64

41. Jona R and Gribaudo I (1990) Ethylene production in tissue culture of peach x almond hybrid tomato sweet cherry and grape. Acta Hort 228:445–449

42. Jona R, Gribaudo I and Vigliocco R (1984) Effects of naturally produced ethylene in tissue culture jars. In: Y Fuchs and E Chalutz, eds. Ethylene: Biochemical Physiological and Applied Aspects, pp 161–162. The Hague: Martinus Nijhoff/Dr W. Junk Publishers

43. Jona R, Gribaudo I and Vigliocco R (1988) Natural development of ethylene in air tight vessels of GF 677. Plant Micropropagation in Horticultural Industries, pp 61–66, Arlon, Belgium

44. Kevers C, Boyer N, Courduroux JC and Gaspar T (1992) The influence of ethylene on proliferation and growth of rose shoot cultures. Plant Cell Tiss Org Cult 28:175–181

45. Kozai T (1991) Micropropagation under photoautotrophic conditions. In: PC Debergh and RH Zimmerman, eds. Micropropagation: Technology and Application, pp 447–469. Dordrecht: Kluwer Academic Publishers

46. Kozai T, Hayashi M, Hirosawa Y, Kodama T and Watanabe I (1987) Environmental control for acclimatization of in vitro cultured plantlets. (1) Development of the acclimatization unit for accelerating the plantlet growth and the test cultivation. J Agr Met 42:349–358

47. Kozai T, Iwabuchi K, Watanabe K and Watanabe I (1991) Photoautotrophic and photomixotrophic growth of strawberry plantlets *in vitro* and changes in nutrient composition of the medium. Plant Cell Tiss Org Cult 25:107–115

48. Kozai T and Iwanami Y (1988) Effects of CO_2 enrichment and sucrose concentration under high photon fluxes on plantlet growth of carnation (*Dianthus caryophyllus* L) in tissue culture during the preparation stage. J Jap Soc Hort Sci 57:279–288

49. Kozai T, Koyama Y and Watanabe I (1988) Multiplication of potato plantlets *in vitro* with sugar free medium under high photosynthetic photon flux. Acta Hort 230:121–127

50. Kozai T, Oki H and Fujiwara K (1988) Effects of CO_2 enrichment and sucrose concentration under high photosynthetic photon fluxes on growth of tissue-cultured *Cymbidium* plantlets during the preparation stage, pp 135–141, Plant Micropropagation in Horticultural Industries, Arlon, Belgium

51. Kozai T, Oki H and Fujiwara K (1990) Photosynthetic characteristics of *Cymbidium* plantlets *in vitro*. Plant Cell Tiss Org Cult 22:205–211

52. Kozai T and Sekimoto K (1988) Effects of the number of air changes per hour of the stoppered vessel and the photosynthetic photon flux on the carbon dioxide concentration inside the vessel and the growth of strawberry plantlets *in vitro*. Environ Control Biol 26:21–29

53. Kumar PP, Reid DM and Thorpe TA (1987) The role of ethylene and carbon dioxide in differentiation of shoot buds in excised cotyledons of *Pinus radiata in vitro*. Physiol Plant 69:244–252

54. Laforge F, Desjardins Y, Graham MED and Gosselin A (1990) Miniature growth chambers for the study of environmental conditions *in vitro*. Can J Plant Sci 70:825–836

55. Laforge F, Lussier C, Desjardins Y and Gosselin A (1991) Effect of light intensity and CO_2 enrichment during *in vitro* rooting on subsequent growth of strawberry, raspberry and asparagus in acclimatization. Sci Hort 47:259–269

56. Lakso AN, Reisch BI, Mortensen J and Roberts MH (1986) Carbon dioxide enrichment for stimulation of growth of *in vitro*-propagated grapevines after transfer from culture. J Amer Soc Hort Sci 111:634–638

57. Lees RP, Evans EH and Nicholas JR (1991) Photosynthesis in *Clematis* 'The President' during growth *in vitro* and subsequent *in vivo* acclimatization. J Exp Bot 42:605–610

58. Lentini Z, Mussell H and Earle ED (1987) Ethylene effect on *in vitro* development of *Brassica campestris*. Plant Physiol 83:154

59. Lentini Z, Mussell H, Mutschler MA and Earle ED (1988) Ethylene generation and reversal of ethylene effects during development *in vitro* of rapid-cycling *Brassica campestris* L. Plant Sci 54:75–81

60. Leonhardt W and Kandeler R (1987) Ethylene accumulation in culture vessels – a reason for vitrification? Acta Hort 212:223–227

61. Liu JH and Reid DM (1992) Auxin and ethylene-stimulated adventitious rooting in relation to tissue sensitivity to auxin and ethylene production in sunflower hypocotyls. J Exp Bot 43:1191–1198

62. Melé E, Messeguer J and Camprubí P (1982) Effect of ethylene on carnation explants grown in sealed vessels. Proc 5th Intl Cong Plant Tissue & Cell Culture. pp 69–70

63. Mensuali-Sodi A, Panizza M and Tognoni F (1992) Quantification of ethylene losses in different container-seal systems and comparison of biotic and abiotic contributions to ethylene accumulation in cultured tissues. Physiol Plant 84:472–476

64. Möllers C, Zhang S and Wenzel G (1992) The influence of silver thiosulfate on potato protoplast cultures. Plant Breeding 108:12–18

65. Morison JIL and Gifford RM (1984) Ethylene contamination of CO_2 cylinders: Effects on plant growth in CO_2 enrichment studies. Plant Physiol 75:275–277

66. Mousseau M (1986) CO_2 enrichment *in vitro*: Effect on autotrophic and heterotrophic cultures of *Nicotiana tabacum* (var Samsun). Photosynth Res 8:187–191

67. Norton CR (1988) Metabolic and non-metabolic gas treatments to induce shoot proliferation in woody ornamental plants *in vitro*. Acta Hort 227:302–304

68. Panizza M, Mensuali-Sodi A and Tognoni F (1988) '*In vitro*' propagation of lavandin: ethylene production during plant development. Acta Hort 227:334–339

69. Pengelly WL and Su LY (1991) Ethylene and Plant Tissue Culture. In: AK Mattoo and JC Suttle, eds. The plant Hormone Ethylene, pp 259–278. New York: CRC Press

70. Perez-Bermudez D, Cornejo MJ and Segura J (1985) A morphogenetic role for ethylene in hypocotyl cultures of *Digitalis obscura* L. Plant Cell Rep 4:188–190

71. Perl A, Aviv D and Galun E (1988) Ethylene and in vitro culture of potato: suppression of ethylene generation vastly improves protoplast yield plating efficiency and transient expression of an alien gene. Plant Cell Rep 7:403–406

72. Porter MA and Grodzinski (1985) CO_2 enrichment in protected crops. Hort Rev 7:345–398

73. Pua EC, Chi GL and Barfield DG (1991) Effect of ethylene inhibitors on plant regeneration from seedling explants and somatic embryos of recalcitrant genotypes in *Cruciferae*. Abstracts VIIth International Congress on Plant Tissue and Cell Culture, Amsterdam, The Netherlands. pp 284

74. Regnard JL, Dorion N and Bigot C (1990) Hypoxic storage of *in vitro* micropropagated peach shoots cv 'Armking'. XXIIIrd International Horticultural Congress, Firenze, Italy. Abstract nr. 3112

75. Reuther G (1988) Comparative anatomical and physiological studies with ornamental plants under *in vitro* and greenhouse conditions. Acta Hort 226:91–98

76. Reuther G (1991) Stimulation of the photoautotrophy of *in vitro* plants. Acta Hortic 300:59–75

77. Righetti B (1990) Air pollutants from hydrocarbons and derivatives iin micropropagation

laboratories: toxicity symptoms on tissue culture of the cherry rootstock Colt (*Prunus avium x P. pseudocerasus*). Plant Cell Rep 9:374–377

78. Righetti B, Magnanini E and Facini O (1990) O $_2$, CO_2, C_2H_4 evolution during *in vitro Prunus avium* shoot cultures. XXIIIrd International Horticultural Congress, Firenze, Italy. Abstract nr. 3115

79. Righetti B, Magnanini E and Maccaferri M (1988) Ethylene and other volatile substances produced by *in vitro* cultured *Prunus avium*. Acta Hort 227:402–404

80. Righetti B, Magnanini E, Infante R and Predieri S (1990) Ethylene, ethanol, acetaldehyde and carbon dioxide released by *Prunus avium* shoot cultures. Physiol Plant 78:507–510

81. Sethi U, Basu A and Guha-Mukherjee S (1990) Control of cell proliferation and differentiation by modulators of ethylene biosynthesis and action in *Brassica* hypocotyl explants. Plant Sci 69:225–229

82. Shimada N, Tanaka F and Kozai T (1988) Effects of low O_2 concentration on net photosynthesis of C3 plantlets *in vitro*. Acta Hort 230:171–175

83. Solárová J (1989) Photosynthesis of plant regenerants: Diurnal variation in CO_2 concentration in cultivation vessels resulting from plantlets photosynthetic activity. Photosynthetica 23:100–107

84. Solárová J, Pospísilová J, Catsky J and Santrucek J (1989) Photosynthesis and growth of tobacco plantlets in dependence on carbon supply. Photosynthetica 23:629–637

85. Taeb AG and Alderson PG (1990) Shoot production and bulbing of tulip *in vitro* related to ethylene. HortSci 65:199–204

86. Tanimoto S and Harada H (1983) Promotive effects of anaerobic treatment on adventitious bud initiation in *Torenia* stem segments. Z Pflanzenphysiol 113:85–90

87. Tate JL and Payne GF (1991) Plant cell growth under different levels of oxygen and carbon dioxide. Plant Cell Rep 10:22–25

88. Thomas D des S and Murashige T (1979) Volatile emissions of plant tissue cultures. II. Effects of the auxin 2,4-D on production of volatiles in callus cultures. In Vitro 15:659–663

89. Tisserat B and Murashige T (1977) Effects of ethephon, ethylene and 2,4-dichlorophenoxya-cetic acid on asexual embryogenesis *in vitro*. Plant Physiol 60:437–439

90. Van Aartrijk J, Blom-Barnhoorn GJ and Bruinsma J (1985) Adventitious bud formation from bulb-scale explants of *Lilium speciosum* Thumb *in vitro*. Effects of aminoethoxyvinylglycine, 1-aminocyclopropane-1-carboxylic acid and ethylene. J Plant Physiol 117:401–410

91. Van Dijck R, De Proft MP and De Greef JA (1988) Role of ethylene and cytokinins in the initiation of lateral shoot growth in bromeliads. Plant Physiol 86:836–840

92. Virgin HI, Holst A and M örner J (1981) Effect of di-n-butylphthalate on the carotenoid synthesis in green plants. Physiol Plant 53:158–163

93. Walker PN, Harris JP and Gautz LD (1991) Optimal environment for sugarcane micropropagation. Transactions of the ASAE 34:2609–2614

94. Walker PN, Heuser CW and Heinemann PH (1989) Micropropagation: effects of ventilation and carbon dioxide level on *Rhododendron* 'PJM'. Transactions of the ASAE 32:348–352

95. Woltering EJ (1989) Effect of the gaseous composition on development of *Gerbera* plantlets grown *in vitro*. Acta Hort 261:377–383

96. Woltering EJ (1990) Beneficial effects of carbon dioxide on development of *Gerbera* and rose plantlets grown *in vitro*. Sci Hort 44:341–345

97. Woltering EJ (1990) Phytotoxic component in polypropylene tissue culture containers. Sci Hort 44:335–340

98. Zelitsch I (1975) Improving the efficiency of photosynthesis. The opportunity exists to increase crop productivity by regulating wasteful respiratory processes. Science 188:626–633

Measuring shortcomings in tissue culture aeration and their consequences for explant development

MICHAEL B. JACKSON, ANN R. BELCHER and PHILIP BRAIN
Department of Agricultural Sciences, University of Bristol, AFRC Institute of Arable Crops Research, Long Ashton Research Station, Bristol BS18 9AF, UK

1. Introduction

An outstanding characteristic of plant tissue culture and micropropagation methodology is the high level of precision with which cultural requirements are usually defined. A large and growing literature describes, in detail, ingredients of culture media needed to achieve particular developmental ends for a given species or cultivar [3]. However, despite authors' carefully assembled protocols, it can often be difficult to reproduce results from other laboratories. One of several factors contributing to the problem could be the absence of any clear definition of the ventilation characteristics of culture vessels. The gaseous atmosphere in culture vessels is susceptible to several influences, but is dominated by the consumption and production of gases by the tissue, by the volume of the enclosing vessel and by the extent to which the vessel is sealed. Gases of particular significance for plant growth include carbon dioxide, especially in relation to respiration and photosynthesis, and ethylene (ethene), which is physiologically active at sub-parts per million (v/v) concentrations. For plants growing in the open, with their roots in freely draining soil, impedances to gas exchange are too small to decrease inward diffusion rates sufficiently to slow respiration and photosynthesis, or to entrap sufficient ethylene to change developmental patterns. However, the need to protect tissue cultures from desiccation and contamination by micro-organisms requires at least partial sealing. This inevitably impedes the inward and outward movement of gases and thus may interfere with gaseous exchange necessary for normal growth and development. There is also the risk of entrapping potentially toxic volatiles emitted from any plastics used in the construction of tissue culture vessels or their closures.

We have previously reviewed the literature describing effects of enclosure on the gases present in culture vessels, and outlined a theoretical procedure for comparing the extent to which different methods of sealing interfere with culture vessel ventilation [4]. The method is based on estimating the delay before half a measured amount of an injected gas is lost from the vessel by diffusion. In a subsequent publication [5], we used this approach to assess different vessels

191

P.J. Lumsden, J.R. Nicholas and W.J. Davies (eds.), Physiology, Growth and Development of Plants in Culture, 191–203, 1994.

and closures containing *Ficus lyrata*, *Gerbera jamesonii* and *Solanum tuberosum* cultures. In the present paper we extend work with *S. tuberosum*, and include results using cultures of apple (*Malus domestica*).

2. Materials and methods

Apple shoots (*Malus domestica* Bork. cv. Bramley's Seedling) were established from meristems according to Abbott and Whitely [1] while cultures of potato (*Solanum tuberosum* L. cv. Red Craig's Royal) were made from single nodes of cultured shoots [5]. Both were grown on Murashige and Skoog basal medium at pH 5.2 set with 0.64% (wt/vol) Oxoid purified agar. No hormones were needed for potato, but for apple, 5 μM benzyladenine was included. Forty millilitres of medium were dispensed into 100 ml Erhlenmeyer flasks and autoclaved at 121 °C for 10 min. Apple shoots proliferated by axillary bud outgrowth and were sub-cultured every 6 weeks using 20 mm shoot tips. Potato explants grew as single leafy stems from the nodal bud present on each single-node explant. Six nodes were grown in each flask and sub-cultured every 4 weeks. Cultures were maintained at 20 °C (potato) or 25 °C (apple) under warm-white fluorescent tubes with a 16 h photoperiod of 160 μmol m^{-2} s^{-1} P.A.R.

Ethylene and carbon dioxide were measured by gas chromatography as described previously [5]. Briefly, ethylene was measured in 1-ml gas samples withdrawn from culture vessels in 2-ml glass hypodermic syringes and injected into an FID gas chromatograph fitted with a column containing activated alumina, and capable of detecting >0.01 μl l^{-1} ethylene after separation from ethane and methane. For carbon dioxide, a 1-ml sample of gas was injected into a gas chromatograph with a katherometer detector and two connected columns (Porapack Q and molecular sieve 2a) that separated carbon dioxide from oxygen and nitrogen. The system readily detected >0.01 % (v/v) carbon dioxide.

Measuring gases in the atmosphere of tissue culture vessels by gas chromatography requires volumes of gas to be removed. To minimise reductions in pressure within the vessel and possible contamination by outside air entering the vessel as pressures are balanced, culture vessels were fitted with two side arms. The first was sealed with a silicon rubber septum through which gas was withdrawn. A 20-ml plastic syringe containing sterile air was attached to the second side arm (Plate 1). The syringe was depressed as gas samples were withdrawn to maintain normal pressure. Some vessels were equipped with a third side arm (Plate 1) to which a 10-ml round-bottomed flask was fused and partly filled with an ethylene absorbing material based on alkaline potassium permanganate ('Ethysorb', Stayfresh Ltd., London, UK).

Contrasting levels of sealing the vessels were compared for their effects on ethylene and carbon dioxide concentrations, and on the growth of the cultures. The most complete sealing was achieved with 'Nescofilm' (Bando Chemicals, Tokyo, Japan) stretched tightly over the neck of the flask (SEAL A). Sealing, in

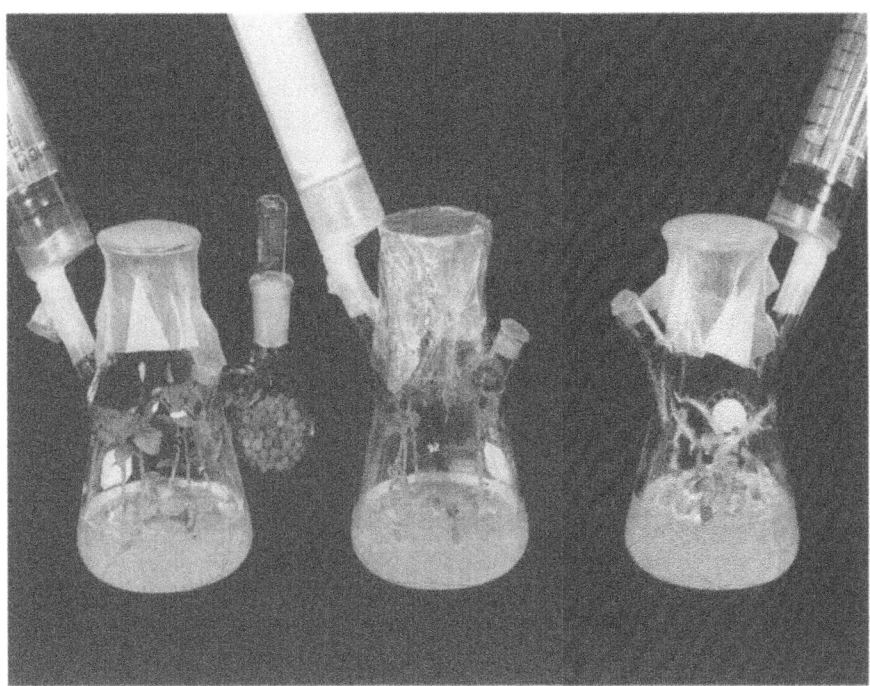

Plate 1. Modified 100-ml Erhlenmeyer flasks used to culture explants of potato and apple. Each vessel contains 40 ml of culture medium and has two side-arms, one sealed with a septum for removing gas samples for analysis, the second attached to a 20-ml syringe containing sterile air used to compensate for the pressure drop caused by sampling. The left-hand vessel has a third side-arm fused to a 10-ml round-bottomed flask containing the ethylene absorbent 'Ethysorb'. Also illustrated are two methods of sealing the vessels. The centre vessel is closed with aluminium foil covered with one layer of PVC cling film (SEAL C). The two remaining flasks are sealed with 'Nescofilm' stretched tightly over the rim (SEAL A).

decreasing order of completeness, was achieved by applying aluminium foil covered with PVC adhesive tape (SEAL B), or aluminium foil covered with 12.5 μm-thick PVC cling film (SEAL C), or cotton wool plugs (SEAL D).

3. Results and discussion

3.1 Effect of closure on half time (t_{50}) of gas retention

An attraction of using the half time for gas retention (t_{50}) is that it allows the level of ventilation to be compared in vessels sealed in different ways. It is easy to determine and provides bench-marks against which previously untested vessels can be compared. Values for t_{50} are readily obtained by tracking the decline in concentration, over time, of a tracer gas such as ethylene or carbon dioxide previously injected into the vessel. There is no need for initial concentrations to be exactly the same in each of the containers being compared. Since gas losses are predominantly by diffusion, decay kinetics are exponential and can, therefore, be transformed to a linear relationship by logging the data ($\log_e$). The slope of the linear line, which is itself a measure of ventilation, is obtained from $\log_e (C-C_A) = \log_e (C_o-C_A) - kt$, where C = the concentration of the gas at a given time in $\mu l\ l^{-1}$, A = the ambient concentration, C_o = the initial concentration in $\mu l\ l^{-1}$, k = the slope of the line and t = time elapsed. The half time can then be calculated from $t_{50} = \log_e(2)/k$. The term integrates several variables such as starting concentration, frequency of sampling, overall length of sampling period, and volumes of the culture vessel. This technique is more straightforward than estimating the number of gas exchanges per hour by the method of Kozai et al. [6]. Table 1 shows the effect of contrasting methods of sealing culture vessels on the t_{50} obtained after injecting sufficient gas to raise the initial internal concentrations to approximately 2% carbon dioxide or 5 μl l^{-1} ethylene. Values ranging from over 400 h down to less than 0.5 h were obtained (Table 1). These sealing methods have been used to test the impact of different t_{50}s on gas concentrations in tissue cultures and on plant growth. Closely comparable t_{50} values were obtained with either ethylene or carbon dioxide. The method also gave similar values at the two growing temperatures (20 °C or 25 °C) used for apple or potato explants (Table 1).

Table 1. Influence of different methods of sealing 100-ml glass culture vessels (modified 100-ml Erhlenmeyer flasks) on the time in hours required for half of injected ethylene or carbon dioxide to escape. This t_{50} value is a measure of the ventilation characteristics of the tissue culture vessels. The tests were run at 20 °C and at 25 °C

	Experiment 1 (20 °C)		Experiment 2 (25 °C)	
	CO_2	C_2H_4	CO_2	C_2H_4
t 50				
Sealing method				
A	429.6±40.2	439.6±49.4	312.2±43.9	435.4±56.4
B	8.7±0.8	9.2±0.8	9.4±1.5	8.4±0.7
C	5.1±0.2	5.7±0.5	4.6±0.3	4.6±2.5
D	0.3	0.3	0.3	0.3

Means and standard errors (n = 3)
Initial concentrations of ethylene and carbon dioxide were approximately 5 $\mu l\ l^{-1}$ and 2% (v/v), respectively.

3.2 Effects of sealing method, light and age of culture on accumulation of carbon dioxide

Carbon dioxide concentrations in apple cultures were measured weekly for 6 weeks at the end of the daily 8-h dark period and again towards the end of the 16-h photoperiod. Concentrations were always least at the end of the photoperiod, suggesting photosynthetic re-fixing of respiratory carbon dioxide entrapped within the vessels during the preceding 8 h of darkness (Fig. 1). Although the amount of carbon dioxide accumulating in the dark period increased in all flasks, as the cultures grew over the 6-week culture period, the effects were evident from the earliest days of the culture. Vessels with cotton wool plugs and thus with the smallest t_{50} (SEAL D) accumulated the least carbon dioxide. Levels of the gas in vessels with increasingly tight sealing and correspondingly larger t_{50} values (SEALS C-A) were considerably greater. The most carbon dioxide (2.8%) was found in the most tightly sealed vessels, and when tissue mass was largest at the end of the experiment. Within the 16 h light period, photosynthetic carbon dioxide consumption was sufficient to return concentrations to those of ambient air, or below, in all flasks, regardless of sealing method.

Time courses of the changes in carbon dioxide in light and dark were

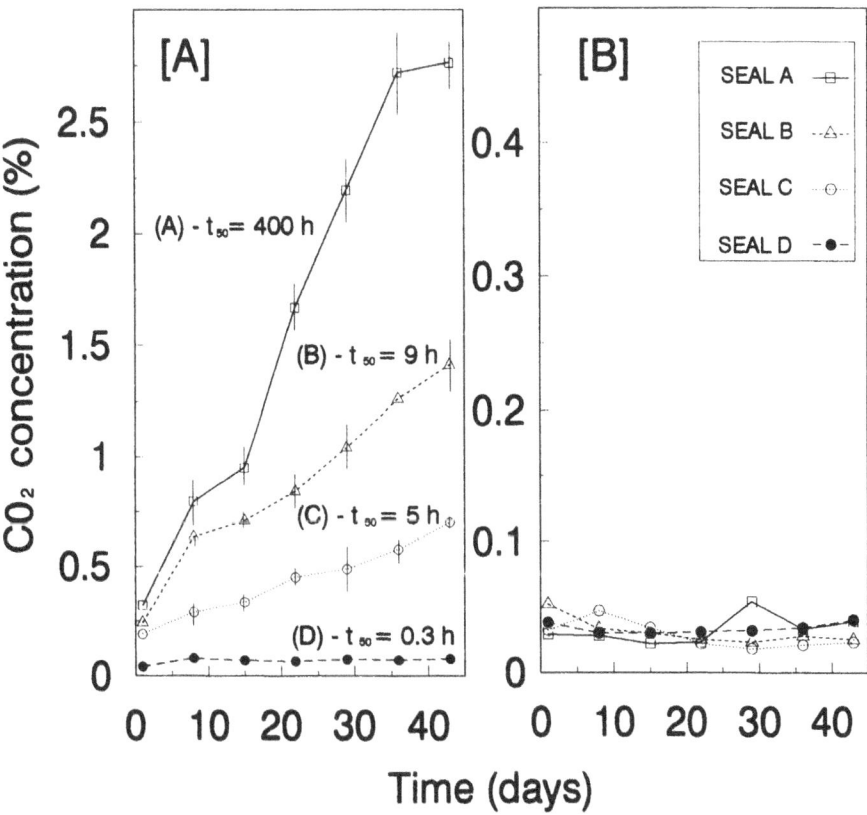

Fig. 1. Effect of different levels of sealing on concentrations of carbon dioxide in atmospheres of apple cultures measured at the end of the 8-h dark period (A) or at the end of the 16-h photoperiod (B) at weekly intervals for 6 weeks. Means of 10 replicates with x2 standard errors. Initial explant fresh weight was 0.235 g and final fresh weights of the explants were 3.0 g (A), 2.9 g (B), 2.8 g (C) and 1.6 g (D).

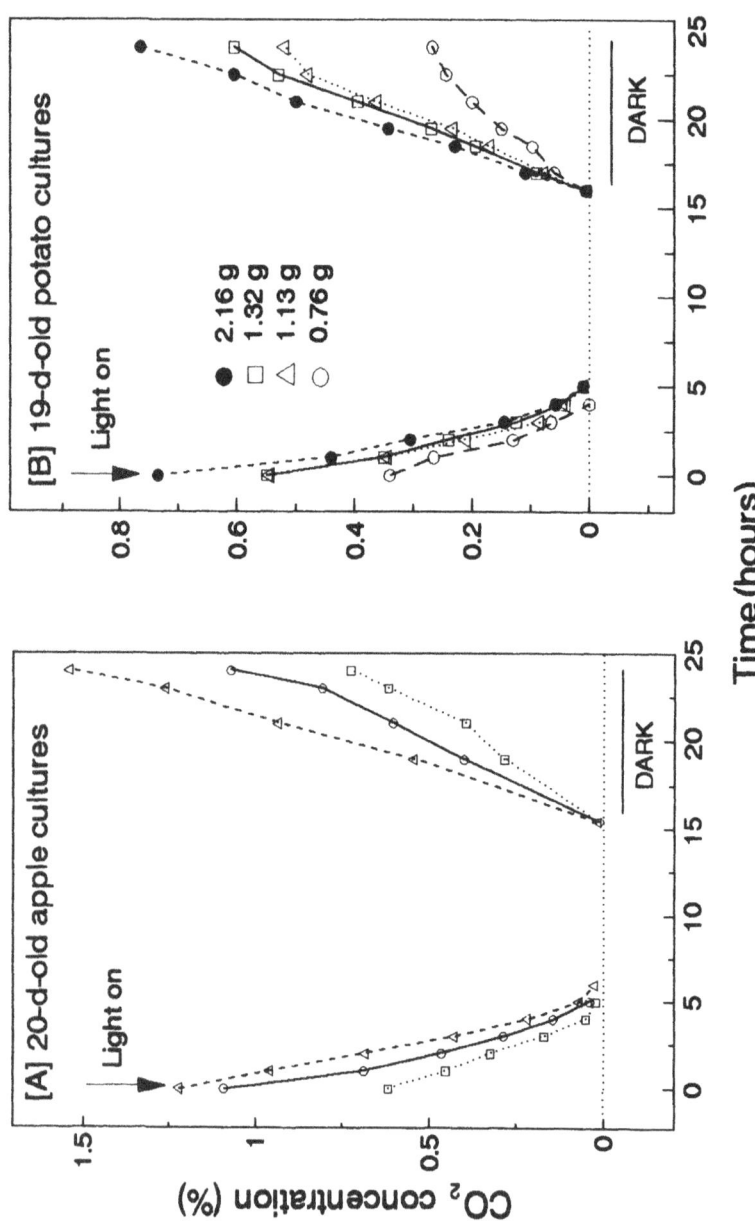

Fig. 2. Time courses of carbon dioxide depletion and enrichment of the atmospheres of cultures of (A) apple and (B) potato during a 16 h light period followed by 8 h darkness. Culture vessels were 100 ml in volume and sealed with Nescofilm (SEAL A) to create a t_{50} of approximately 400 h. Each line is from a separate culture flask. Fresh weights of the potato cultures are given in (B).

obtained by measuring the gas at hourly intervals for up to 8 h into the light or dark periods. The most tightly sealed flasks (SEAL A) were used. Similar patterns were found for both apple and potato cultures (Fig. 2). Under the low irradiance conditions of the experiment (160 μmol) approximately 6 h was required for most of the carbon dioxide to be consumed. In the remaining 10 h of the photoperiod carbon dioxide concentrations were small and often below ambient levels. In the dark, carbon dioxide accumulated steadily throughout the 8 h. Similar results from cultures grown on sucrose-containing media have been reported by Solarova [6]. The heaviest cultures produced the most carbon dioxide (Fig. 2B). Measurements of the increase in carbon dioxide can be used to give an estimate of the dark respiration rate, either by linear regression if the leakage rate is negligible, or using the method described in the Appendix where there is leakage. Similarly, measurements of the decreases in carbon dioxide can be used to give an estimate of the net rate of carbon fixation in a similar manner to the dark respiration rate. For both measurements, the approximately linear portion of the curve should be used, as it is unlikely that the net rate of carbon fixation is constant at low CO_2 levels, which would invalidate the model described in the Appendix. However, the rates can readily be adjusted for leakage from the container.

Thus the three main influences on carbon dioxide levels in tissue cultures are shown to be the age (mass) of tissue in the culture, light and dark, and the t_{50} of the culture vessel. It is also clear that all sealing methods, even with a t_{50} as low as 0.3 h impede gas exchange sufficiently to modify the atmosphere within culture vessels. In the most tightly sealed flasks, photosynthesis within the first few hours of each photoperiod is sufficiently vigorous to re-fix almost all respiratory carbon dioxide within 6 h at modest irradiance levels. The role of photosynthesis in the carbon economy of explants cultured on a sucrose-containing medium remains uncertain. However, there is considerable evidence that supplying cultures with additional carbon dioxide promotes growth in dry mass and improves success during subsequent growing-on of plantlets outside the culture vessel [7, 10, 2].

3.3 Effect of sealing method and age of culture on accumulation of ethylene

No marked differences were found between measurements of ethylene made during the light or dark periods. Results are therefore given for the end of the dark period. In apple cultures, concentrations increased over the 43 d of the culture (Fig. 3A), probably as a direct consequence of tissue-mass increment. As expected, the highest concentrations were present in the most tightly sealed flasks (up to 5 μl l^{-1}); flasks with a smaller t_{50} accumulated much less ethylene. Under SEAL B (foil + PVC tape), levels above 0.2 μl l^{-1} were common and this amount can be expected to be physiologically active. The ethylene present under SEAL C (foil + clingfilm) approached this level, while under SEAL D (cottonwool plug), less than 0.04 μl l^{-1} was measured. Results with potato shoot cultures were very similar to those for apple. Measurements taken weekly over 28 d from the most tightly sealed flasks containing potato plantlets are plotted against the fresh weight at the time of the measurements, to show the relationship between the accumulation of ethylene and the mass of tissue (Fig. 3B).

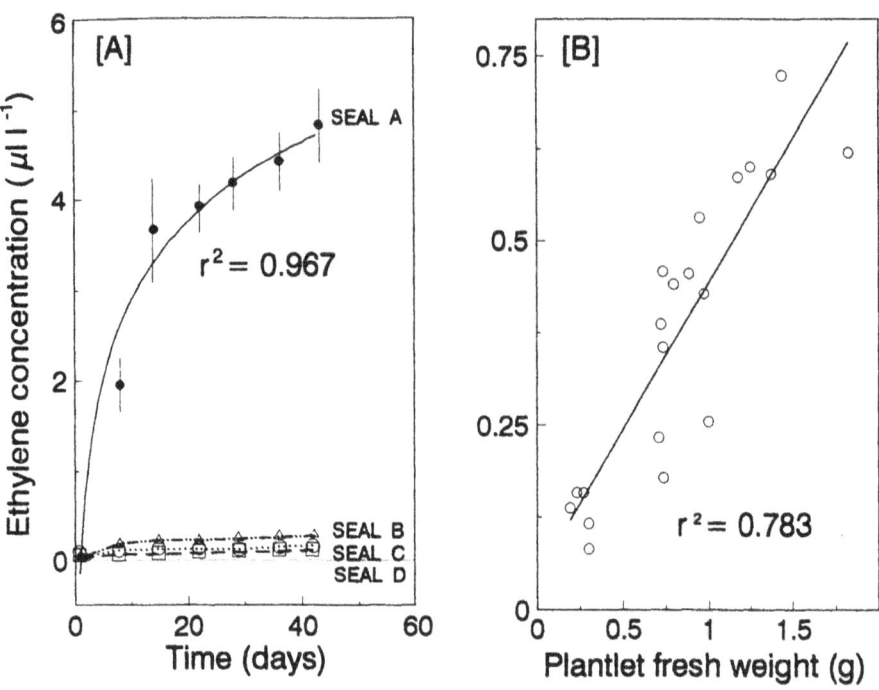

Fig. 3. (A) Time course of ethylene accumulation in cultures of apple sealed to create contrasting t_{50} values; means of 12 flasks with x2 standard errors. (B) Relationship between plantlet fresh weight and concentration of ethylene in culture flasks of potato sealed with Nescofilm (SEAL A, $t_{50} = 400$ h). Measurements were made after 1, 7, 14, 21 and 28 d.

3.4 Influence of t_{50} on growth of apple shoots

The proliferation of shoots was least in the most highly ventilated cultures i.e., those sealed with cotton wool plugs (SEAL D). Shoot production was also poor in the most tightly sealed flasks (SEAL A) and greatest under foil + cling film (SEAL C) (Table 2). Differences in the length of the tallest stem were less clear-cut except for a notable shortening under cotton wool plugs. The effect of SEAL A on the length of the longest shoot was surprisingly small in view of the large accumulations of ethylene in these flasks. There was little difference in the final fresh weight of the three most tightly sealed vessels with the t_{50} ranging from 400 h to 5.0 h. However, fresh weights were almost halved by the low t_{50} (0.3 h) achieved with the cotton wool plugs (SEAL D). Dry weights were not affected significantly by any of the sealing methods. The absence of clear differences in final dry weight suggests that the considerable amounts of photosynthetic fixation of carbon dioxide under the tighter seals made little direct impact on overall dry weight gain. The water content of the loosely sealed cultures (SEAL D) was about 10% less than those in the remaining treatments. This was linked to a halving of the water content of the agar culture medium (Table 2). Under more tightly sealed conditions, water loss by the medium was less than 1.5 %.

Table 2. Effect of different methods of sealing culture vessels on the growth and water content of apple explants after 43 d

	Flask seal				s.e.d. (33 d.f.)
	A	B	C	D	
Stems per explant	4.6	5.9	6.5	3.4	1.084/1.022
Height of tallest stem (mm)	45.7	50.5	38.4	16.7	0.391/0.368
Final fresh wt (g)	4.49	2.94	2.76	1.55	0.409/0.386
Final dry wt (g)	0.49	0.35	0.32	0.32	0.036/0.034
% Water content	89.0	87.1	88.3	79.2	0.0137/0.0129*
% Dehydration of medium	0.04	0.6	1.4	48.1	1.47

SEDs generated by ANOV are shown for comparison between SEAL A and any other sealing treatment (left), and for comparisons between treatments other than SEAL A. The t_{50}s for sealing methods A–D are as described in Table 1. SEDs marked * are of logged data ($\log_e$).

3.5 Influence of t_{50} on growth of potato cultures

For these experiments, sealing treatments were restricted to Foil + clingfilm (SEAL C), and to Nescofilm (SEAL A) with or without Ethysorb to remove accumulated ethylene. By the end of 28 d culture, all explants had formed a similar number of leaves, but there were large differences in their area (Table 3). Under SEAL A, potato shoots carried only vestigial leaves borne on swollen, diageotropic shoots in association with ethylene concentrations above 0.5 μl 1^{-1} ethylene. The inclusion of Ethysorb restored normal stem diameters and orientation. In the presence of Ethysorb, leaf areas exceeded those of explants

Table 3. Effect of sealing culture vessels with Nescofilm (SEAL A) or with Foil + clingfilm (SEAL C), and of the ethylene absorbent Ethysorb on the development of potato cultures after 28 d

	A Nescofilm	C Foil + clingfilm	Nescofilm + Ethysorb	s.e.d. (27 d.f.)
Number of leaves	7.7	7.1	6.4	0.332
Leaf area per plantlet (mm²)	1.73	24.1	43.6	0.1254*
Total fresh wt. (g)	1.23	1.63	1.28	0.137
% Fresh wt as leaves	13.9	25.5	37.8	—
Stem diameter (cm)	0.173	0.110	0.106	0.0231*

SEDs marked * are of logged data ($\log_e$).

grown under SEAL C, where ethylene accumulation was modest. The restoration was also more than complete when expressed in terms of the proportion of explant fresh weight that comprised leaf tissue. The inhibitory effect of ethylene on leaf expansion was much more marked than that reported previously for *Ficus lyrata* [5]. These results indicate that moderate sealing (SEAL C), creating a t_{50} of only 4–5 h, perturbed growth of this ethylene-sensitive species only slightly. However, in more tightly sealed flasks, ethylene accumulation had a substantial influence on shoot morphology, especially leaf expansion.

4. Conclusions

The half-time for the loss of a gas injected into empty culture vessels (t_{50}) is a convenient and useful measure of ventilation capacity. Carbon dioxide or ethylene, as tracer gases, give similar results. Sealing methods such as inserting a cotton wool plug into the neck of culture flasks that generate a low t_{50} (0.3 h) still impeded gas exchange sufficiently to favour small accumulations of carbon dioxide in the dark. In vessels with a larger t_{50}, marked fluctuations in carbon dioxide occur between light and dark periods. These result from the depletion, by photosynthesis, of sizable accumulations of carbon dioxide taking place in each preceding dark period. The inclusion of a light period of several hours each day thus serves to prevent excessive build-up of carbon dioxide in tightly sealed cultures. Carbon dioxide accumulates in proportion to the mass of tissue and thus increases during the growth of cultures as they enlarge over time. The slopes of carbon dioxide accumulation in the dark or depletion in the light coupled with the t_{50} value allow estimates of respiration and net photosynthesis. The extent to which ethylene accumulates in culture vessels also increases with t_{50}, and with the size and age of the cultures. Concentrations reach 0.2 μl l^{-1} at a t_{50} of 9 h, and attain over 5 μl l^{-1} in flasks with a t_{50} of approximately 400 h. Distortions to growth of potato plantlets resulting from a build-up of ethylene can be overcome by absorbing ethylene chemically. Thus, the principal damaging effects of tight sealing methods (t_{50} 400 h) on explant development, where a light period is provided, appear to be those of ethylene. This raises the possibility that some species could be grown successfully in completely sealed containers provided an ethylene absorbent is included and light is used to moderate carbon dioxide levels. The most appropriate sealing method may not be one that combines the lowest t_{50} compatible with maintenance of sterility since this may be insufficient to prevent dehydration of tissues and medium (e.g., t_{50} of 0.3 h), resulting in poor growth.

Acknowledgements

We thank Miss Ruth Butler for help with statistical analyses, Mrs Joan Llewellyn for typing and Stayfresh Ltd, London, UK for a gift of 'Ethysorb'. We also wish to acknowledge the support of Dr A. J. Abbott who died unexpectedly in 1991 and in whose laboratory our tissue cultures were grown.

Appendix

Predicting CO_2 accumulation or depletion in partially sealed culture vessels

1. Developing a model

Tissue cultures are grown in partially sealed flasks (see main text). When the flasks are illuminated there is a loss of CO_2 from within the flask due to photosynthesis, and when the flasks are in the dark there is a build-up of CO_2 due to respiration. However these flasks leak so that the measured CO_2 includes a component caused by leakage. A model is developed which allows the photosynthetic loss or respiration gain to be estimated where there is leakage.

The following assumptions are made in developing the model:
1. The instantaneous rate of loss of CO_2 by leakage is proportional to the difference between the amount of CO_2, C, in the flask, and that in the atmosphere, C_A. The proportionality constant is denoted by k.
2. The rate of CO_2 change caused by photosynthesis or respiration is constant for a given tissue culture and is independent of the CO_2 concentration in the flask. It is denoted by P (photosynthesis) or R (respiration).

Assumption 1 has already been implicitly used in the loss rate from a sealed flask, and in an empty flask leads to the equation quoted in the main text and allows the t_{50} to be obtained. This equation is written in a different form below:

$$(C-C_A) = (C_0-C_A) e^{-kt}$$

It should be recalled that $t_{50} = \ln(2)/k$, so that $k = \ln(2)/t_{50}$.

Assumption 2 is an approximation likely to hold at high CO_2 levels, but is obviously incorrect at lower levels of CO_2 as it implies that in the situation where there is no leakage the CO_2 concentration would become negative after a certain amount of time.

When these two assumptions are combined the rate of gain/loss of CO_2 from the flask, dC/dt is given below:

$$\frac{dC}{dt} = -k(C-C_A) - P$$

where there is photosynthetic loss,
and

$$\frac{dC}{dt} = -k(C-C_A) + R$$

where there is respiration occurring.

In the first equation the combined loss rate is that due to leakage and the amount being removed by photosynthesis. In the second, the loss rate is that due to leakage less the amount being added by respiration.

The equations can be solved and rewritten to give the CO_2 concentration, C, at time t:

$$C = C_o - (P + (C_o - C_A)k) \frac{(1 - e^{-kt})}{k} \quad \ldots\ldots\ldots\ldots (1)$$

where the flasks are illuminated.
Where the flasks are in the dark,

$$C = C_o + (R - (C_o - C_A)k) \frac{(1 - e^{-kt})}{k} \quad \ldots\ldots\ldots\ldots (2)$$

For both equations C_o is the concentration at time 0.

2. Using the model to predict changes in photosynthesis or respiration

The model can be used to predict plant photosynthesis or respiration in two cases, firstly when the leakage rate is known, and secondly when it is not. Both situations will be dealt with below. Under the assumption that CO_2 loss or gain caused by photosynthesis, or respiration, respectively occurs at a constant rate, plots of C versus time in the absence of leakage should be linear. If the plots show evidence of curvature then either there is leakage, or photosynthesis and respiration are affected by CO_2 concentration. The following results assume that the curvature, if present, is caused by leakage.

a) Leakage rate, k, known
In this case revised 'time' values, T, can be derived using the equation

$$T = \frac{(1 - e^{-kt})}{k}$$

for each time at which an observation is made. The two equations 1 and 2 are then linear in T. This can be confirmed by plotting the concentrations, C, versus T, which should linearise the previous non-linear plot. Linear regression can be used to estimate the intercept, C_0, and the slope, B. For the case where there is photosynthesis, the loss rate, P, is given by

$$P = -((C_o - C_A)k + B)$$

Where there is respiration, the respiration rate, R, is given by

$$R = (C_o - C_A)k + B$$

When k is very small, i.e., there is very little leakage, revised "times", T, are very close to the original times, t, and the correction is negligible.

b) Leakage Rate, k, Unknown
When k is unknown it can be estimated from the curvature of the plot. Equations 1 and 2 can be readily fitted by least squares using a statistical package such as Genstat [10]. Once k has been estimated the revised 'time', T, can be calculated as described in the previous section and the CO_2 concentration, C, plotted against it to check for linearity. The values of P and R can be estimated as before.

References

1. Abbott AJ and Whiteley E (1976) Culture of *Malus* tissues *in vitro*. 1. Multiplication of apple explants from isolated shoot apices. Sci Hort 4:183–189
2. Cournac L, Dimon B, Carrier P, Lohou A and Chagvardieff P (1991) Growth and photosynthetic characteristics of *Solanum tuberosum* plantlets cultivated *in vitro* in different conditions of aeration, sucrose supply, and CO_2 enrichment. Plant Physiol 97:112–117

3. George EF and Sherrington PD (1984) Plant Propagation by Tissue Culture. Basingstoke: Exergetics
4. Jackson MB, Abbott AJ, Belcher AR and Hall KC (1987) Gas exchange in plant tissue cultures. In: MB Jackson, SH Mantell and J Blake, eds. Advances in the Chemical Manipulation of Plant Tissue Cultures, British Plant Growth Regulator Group Monograph 16, pp 57–71. Bristol: British Plant Growth Regulator Group
5. Jackson MB, Abbott AJ, Belcher AR, Hall KC, Butler R and Cameron J (1991) Ventilation in plant tissue cultures and effects of poor aeration on ethylene and carbon dioxide accumulation, oxygen depletion and explant development. Ann Bot 67:229–237
6. Kozai T, Fujiwara K and Watanabe I (1986) Fundamental studies on environments in plant tissue culture vessels (2). Effects of stoppers and vessels on gas exchange rates between inside and outside of vessels closed with stoppers. J Agr Met 42:119–127 (in Japanese with English summary)
7. Kozai T and Iwanami Y (1988) Effects of CO_2 enrichment and sucrose concentration under high photon fluxes on plantlet growth of carnation (*Dianthus caryophyllus* L.) in tissue culture during the preparation stage. J Jap Soc Hort Sci 57:279–288
8. Payne RW, Lane PW, Ainsley AE, Bicknell KE, Digby PGN, Harding SA, Leech PK, Simpson HR, Todd A D, Verrier PJ, White RP, Gower JC, Tunnicliffe Wilson G & Paterson LJ (1987) Genstat 5 Reference Manual. Oxford: Oxford University Press
9. Solarova J (1989) Photosynthesis of plant regenerants. Diurnal variation in CO_2 concentration in cultivation vessels resulting from plantlets photosynthetic activity. Photosynthetica 13:100–107
10. Solarova J, Pospisilova J, Catsky J and Santrucek J (1989) Photosynthesis and growth of tobacco plantlets in dependence on carbon supply. Photosynthetica 23:629–637

The influence of the gas permeability of the vessel lid and growth-room light intensity on the characteristics of *Dianthus* microplants *in vitro* and *ex vitrum*

ALAN C. CASSELLS and TIMOTHY D. ROCHE
Department of Plant Science, University College, Cork, Ireland

Abbreviations: MVTR, Moisture Vapour Transmission Rate; LDPE, Low Density Polyethylene; PVC, Polyvinylchloride; RH, Relative humidity

1. Introduction

Published studies on the tissue culture of *Dianthus* include many references to the problem of vitrification [e.g. 16, 17, 18, 28, 29, 30] or hyperhydration [7]. Indeed, *Dianthus* is often chosen as a model for research on vitrification [e.g. see 13, 14, 20,]. These and other studies on vitrification have implicated both media factors and aspects of the *in vitro* gaseous atmosphere in this phenomenon. Keevers and coworkers [13] have implicated ethylene accumulation in the culture vessel as a contributory factor in vitrification and many of the strategies employed to reduce vitrification in cultures consciously, or subliminally, reduce ethylene accumulation or facilitate its escape [5, 10, 24 29].The water vapour pressure in the vessel has been implicated in poor establishment of micro plants [9, 22].

   Here, to evaluate the role the of relative humidity of the gaseous atmosphere in the vessel in relation to the quality of the biomass produced, the glass lid of a conventional culture vessel was replaced with membranes (films) of different gaseous permeabilities. Previous authors have reported the use of porous containers of unspecified permeability [25] or have used perforated lids with no specified selectivity [15]. Here, the membranes chosen had high oxygen, carbon dioxide and ethylene permeabilities, that is low barrier characteristics for these gases facilitating equilibrium with the ambient atmosphere. On the other hand, the membranes selected acted as barriers to moisture vapour loss to varying degrees (See Table 1, [2]). In the present work, the cultures were deliberately grown under low light intensities to enhance any stress effects.

2. Materials and methods

2.1 Plant material

Stock plants of *Dianthus* 'Mystere' were grown in soil under ambient environmental conditions. Cuttings were taken throughout the summer months

P.J. Lumsden, J.R. Nicholas and W.J. Davies (eds.), Physiology, Growth and Development of Plants in Culture, 204–214, 1994.
© 1994 *Kluwer Academic Publishers.*

and rooted in peat-based potting compost containing vermiculite and coarse sand (3:1:1; v:v:v). Rooted cuttings were potted up in 7 cm pots and subsequently planted outdoors.

2.2 Establishment and maintenance of tissue cultures

Rooted cuttings were grown in the glasshouse under stage O conditions [19]. Shoot tips were removed from the stock plants and surface sterilised by dipping in 80% (v/v) aq. ethanol for 30 sec then immersed in 10%(v/v) aq. hypochlorite solution (Domestos, Lever, U.K.) for 15 min followed by repeated washing in sterile distilled water. The apical meristem and first pair of leaf primordia were excised with the aid of a dissecting microscope and plated on meristem culture medium consisting of the basal medium of Murashige and Skoog [19] supplemented with 0.27 mM adenine sulphate and containing 1.32 μM kinetin, 0.65 μM gibberellic acid (GA_3), 8.7 mM sucrose and 8 g l $^{-1}$ agar at pH 5.8. The medium was sterilised by autoclaving at 105 kPa, 121 °C for 15 min. On bacterial indexing, primary nodal cultures derived from the latter were shown to be free of cultivable contaminants. These latter cultures were used to establish stock nodal cultures [3] as follows. Primary nodal cultures were established from proliferating meristem cultures by excising the nodes and plating the secondary nodes on half strength Murashige and Skoog [19] basal medium containing 2.6 μM gibberrelic acid (GA_3), 4.4 mM sucrose and 6 g l^{-1} agar. The medium was sterilised as above. Prior to cloning, the cultures were visually examined for the presence of contaminants and suspect cultures were rejected. Nodes from visibly clean cultures were subcultured on to fresh media, the basal node was retained for culture indexing as follows: the node was placed on bacteriological medium either nutrient glucose agar, yeast glucose calcium carbonate agar or peptone yeast agar [23]. The plates were incubated in the dark at 25 °C and examined at three and five day intervals. Cultures which were bacteria negative were used for clonal multiplication by repeated cycles of subculture on fresh media in 500 ml Kilner jars with five nodes per jar. Cultures were routinely culture-indexed on the above media to ensure freedom from contamination [3]. All cultures were incubated in a growthroom at 22 °C, 16h photoperiod at a mean light intensity of 32 μmol m^{-2}sec^{-1}, RH 40%.

2.3 Modified atmosphere container

To manipulate the atmosphere in the culture vessel, differentially gas permeable membranes (films; Table 1) were substituted for the glass lid of a standard 500 ml kilner jar. The control (glass) lid was tightly closed with a rubber O-ring in position. To ensure a gas-tight seal parafilm was applied to the outside of the lid. The films used were based on low density polyethylene and polyvinylchloride (PVC) packaging films. Moisture vapour transmission rates were determined by measurement of water loss from the containers lidded with the respective membranes. Oxygen transmission rates were supplied by the manufacturers.

Carbon dioxide and ethylene permeability was determined experimentally by the Davenport Apparatus (Davenport Ltd., London, U.K.). The membrane data is given in Table 1.

Table 1. Properties of membranes used to modify the gaseous atmosphere in the culture containers

| Membrane | Gaseous permeability (ml cm^{-2} 24h at 22°C; RH 40%) | | | |
	Water	Oxygen	Carbon Dioxide	Ethylene
LDPE	0.77	>1000	>1000	>1000
PVC	10.52	80.5	198.0	133.63
PVC	18.51	98.5	240.0	163.51

Each kilner jar containing 10 subapical nodes and lids of LDPE or PVC, as appropriate, was positioned in a randomised replicate block, 14 replicates of each, and incubated in a growth room (22 °C, 16 h day) with a forced air circulation system, under PAR light intensities of 32 or 1.7 mol m^{-2} sec^{-1}, respectively. The RH of the growthroom was 40%. After 42 days, five replicate jars from each of the treatments were selected at random for examination and quantification of the experimental parameters (see results). The remaining replicates were used for plantlet establishment trials . Computer-aided Image analysis was used to measure the growth parameters (DIAS Image Analysis System; Delta T Devices, Cambs., U.K.).

2.4 Determination of the dry weights

Dry weights were determined by infrared drying using a Mettler LJ 16 moisture analyser (Mettler- Toledo Ag, Greifensee, Switzerland).

2.5 Establishment of progeny plants

Plantlets for establishment were removed from the culture vessels and washed free of agar. They were then planted in propagators in a peat-based potting compost: fine gravel mix (3:1; v/v) containing BioP base fertilizer (Pan Britannica Ind., Herts, UK) at the manufacturer's recommended rate. The propagators were placed on a heated bench to ensure a minimum temperature of not less than 15 °C. The plantlets were sprayed twice daily with fine mist. The lid vents were opened after nine days and the lids removed after a further week when the propagators were removed from the heated bench.Plants were potted up in 7 cm diameter pots and grown on in the glasshouse.

3. Results

Under normal light mean node number per microshoot increased from 5.2 to 7.0 with increasing lid MVTR. In low light mean node number was reduced to 1.8 at low MVTR, increasing to 4.5 at the highest MVTR (Fig. 1). The incidence of multistemming i.e. reduced apical dominance, was the inverse of the latter, with 88% of the plantlets having multistems in cultures from low light/low MVTR (Fig. 2). Nodal separation was proportionate to node number in all treatments

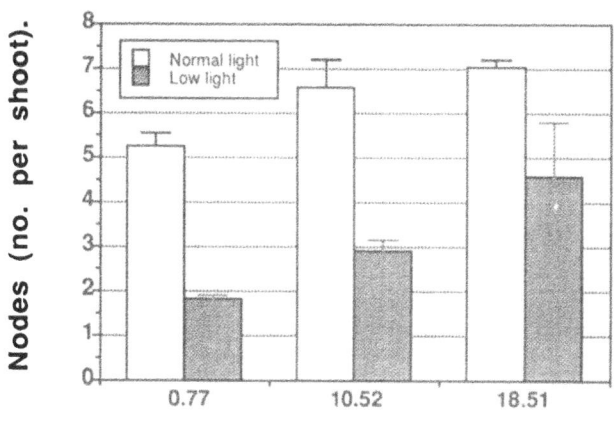

MVTR (mg/cm²/day @ 22°C).

Fig. 1. The effect of lids of different MVTR and light intensity on nodal increase.

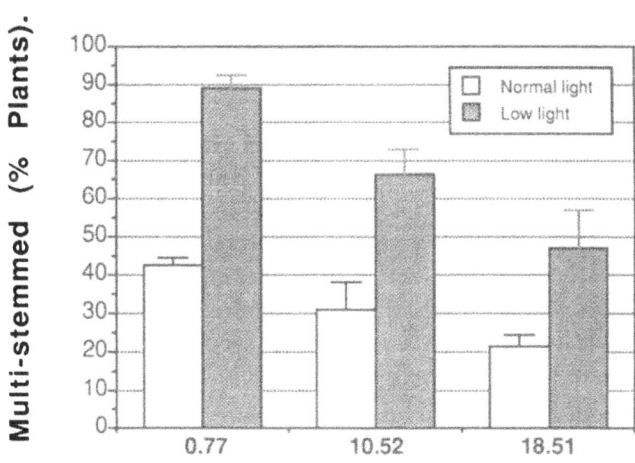

MVTR (mg/cm²/day @ 22°C).

Fig. 2. The incidence of multi-stemming in *Dianthus* microplants *in vitro*; the effect of MVTR of the lid and light intensity.

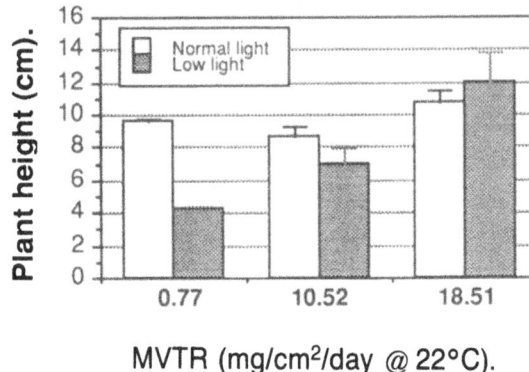

Fig. 3. The effect of lid MVTR and light intensity on *Dianthus* microplant height.

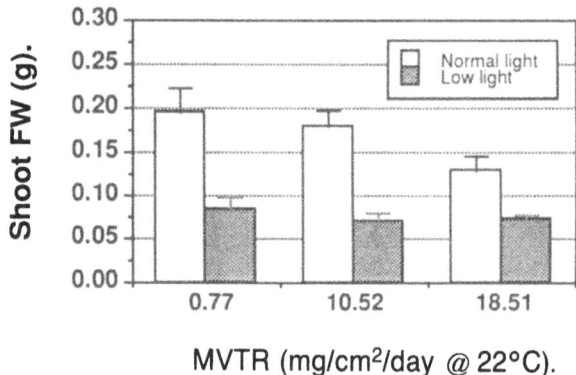

Fig. 4. The effect of lid MVTR and light intensity on *Dianthus* microshoot fresh weight.

(Fig. 3). In containers with lids with MVTRs in excess of 18.51 mg cm^{-2}day^{-1} nodal separation was poor and tissues were visibly desiccated (Cassells and Roche, unpublished).

Highest microshoot fresh weight was recorded in cultures in normal light under low MVTR lids. In low light grown cultures, fresh weight was lower and did not alter with increase in MVTR of the lid (Fig. 4). Microshoot dry weight paralleled the fresh weight data (Fig. 5). Microshoot dry matter content showed no significant differences in cultures from normal and low light at any lid MVTR value (Fig. 6).

The establishment of microplants *ex vitrum*, determined two weeks after planting out, was critically influenced by the light intensity under which the cultures were grown and, in the case of low light grown cultures, to the MVTR of the vessel lid. One hundred percent of the microplants from normal light cultures established *in vivo* whereas, only 20% of the microplants from the low light grown cultures with low MVTR lids survived. Percentage survival

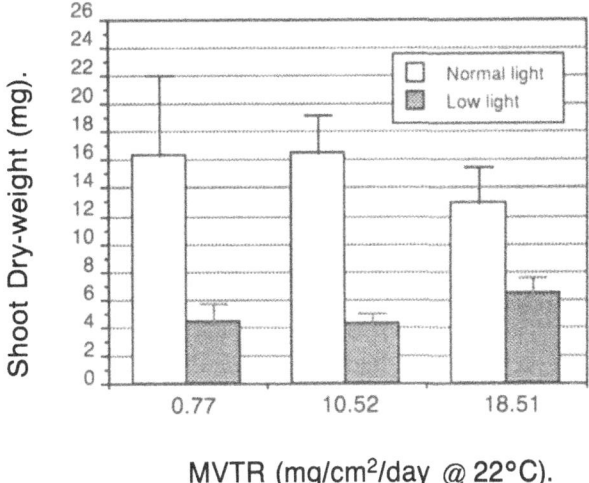

Fig. 5. The effect of lid MVTR and light intensity on *Dianthus* microshoot dry weight.

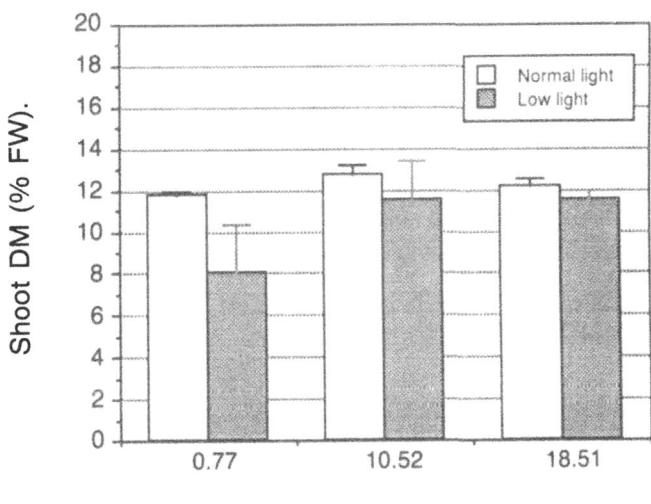

Fig. 6. The effect of lid MVTR and light intensity on the dry matter content of *Dianthus* microshoots.

increased in the latter to 80% in the case of microplants from vessels with lids with the highest MVTR (Fig. 7). There was a strong positive correlation between microshoot dry matter content and percentage establishment but not between microshoot fresh weight and establishment (Fig. 8).

Fresh weight determinations on *in vitro* microplants from the various

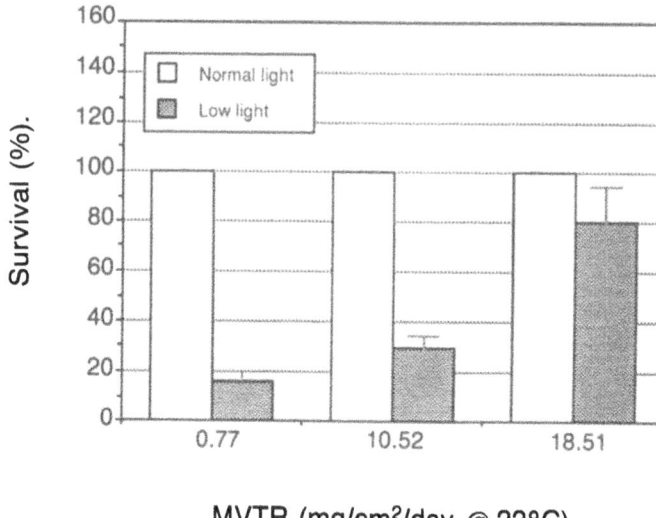

Fig. 7. The influence of lid MVTR and light intensity on the survival rates *ex vitrum* of *Dianthus* microplants

treatments and on surviving microplants two weeks after establishment, showed significant weight loss in plantlets from cultures which were grown in vessels with lids of low MVTR. This was especially so in the case of cultures grown in low light, low MVTR combination whereas, fresh weight increased significantly in progeny from both normal and low light treatments in vessels with the highest MVTR. lids. The plantlets from vessels with lids of intermediate MVTR, regardless of light conditions, showed intermediate behaviour (Fig. 9).

4. Discussion

The quality of biomass produced *in vitro*, arbitrarily expressed as the ability of such biomass to establish *ex vitrum*, is known to be influenced by the interaction between the specific genotype and the culture milieu. Important factors in the interaction are the composition of the culture medium, the agar concentration [1, 6] and the gaseous atmosphere in the container [28]. The latter is a reflection of the closure or lid and the porosity of the container [12]. Furthermore, the agar used [6] and the plastic of the container [11, 27] may release toxic materials. In the case of sensitive genotypes, vitrification and necrosis may result in the loss of cultures or reduced productivity. In less severe cases, subliminal vitrification may result in the need for costly weaning facilities [21].

Recent research suggests that vitrification may be reduced, or avoided, by allowing gaseous exchange with the ambient atmosphere [12]. While venting of cultures, long practised by micropropagators though largely empirical, may

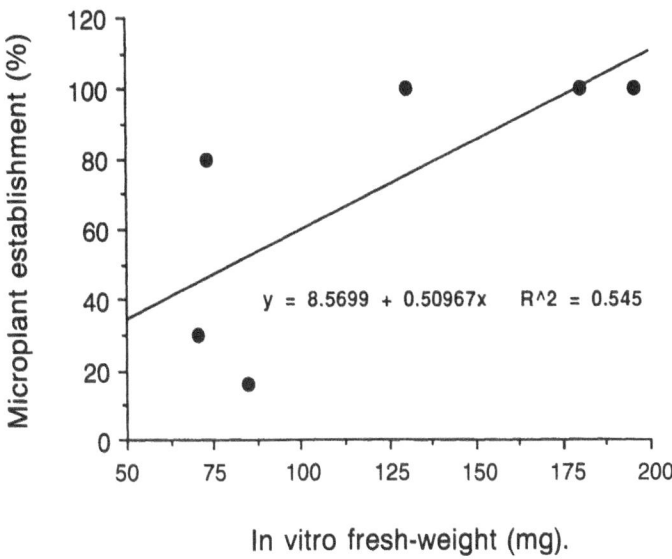

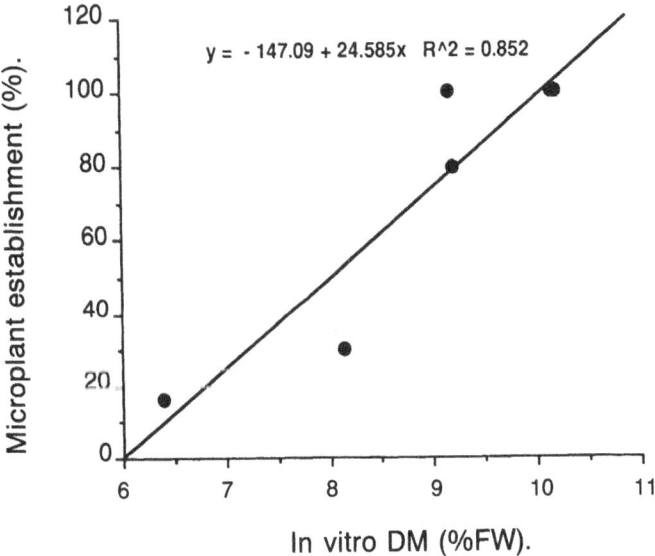

Fig. 8. The correlation between *Dianthus* microplant fresh weight and plantlet establishment (top) and microshoot dry matter content (bottom).

achieve this effect, unrestricted exchange with the atmosphere may result in contamination [3] and desiccation of the medium. Here, data is presented on the use of differentially permeable membranes which have the advantage of providing a barrier to the entry of contaminants and which facilitate oxygen and

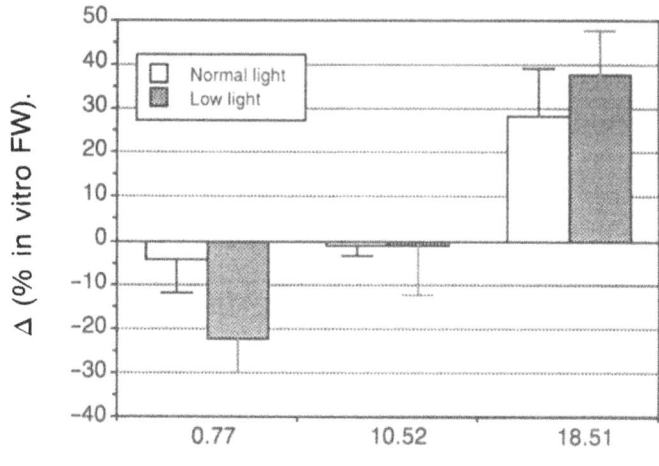

MVTR (mg/cm²/day @ 22°C).

Fig. 9. The influence of the lid MVTR and *in vitro* light intensity on weight changes in *Dianthus* microplants two weeks after establishment.

carbon dioxide exchange and ethylene escape while controlling moisture loss.

The data presented here represents part of an applied study to evaluate the potential of differentially permeable membranes to influence both the productivity and quality of *in vitro* biomass production in micropropagation. The results were obtained with membranes selected for their high permeability to carbon dioxide, oxygen and ethylene. The assumption that these membranes present no barrier to ethylene escape from the cultures is currently being confirmed by gas chromatograpic analysis of the vessel atmosphere. The use of ethylene binding agents provides circumstantial evidence that the latter assumption is valid. The membranes were chosen, specifically, to facilitate the study of changes in the water vapour component of the atmosphere on the performance of the cultures. The results obtained show that both productivity of the nodal cultures and quality of the microplants produced can be improved by increasing the MVTR of the vessel lid. The highest MVTR of 18.51 mg cm^{-2} day^{-1} was chosen on the basis of preliminary studies which indicated that water loss at higher lid MVTRs adversely affected both nodal separation , which increased labour time in subculture, and quality, as biomass became desiccated. Reduction in productivity in low light grown cultures was directly related to reduced apical dominance. The result is discussed further below. Dry matter content i.e. dry matter expressed as a percentage of fresh weight was not significantly different in any treatment and thus tissues raised in vessels with low MVTR lids were not hyperhydrated (see below). Such low MVTR membranes facilitated ethylene release. Percentage microplant establishment was directly correlated with reduction in fresh weight at weaning and presumably reflects poor cuticular development and stomatal malfunction in microplants raised in high humidity atmospheres as reported for rose cultures [9, 22]. As an aside,the present results call into question the use of hyperhydration as an alternative to vitrification [7]. Hyperhydration of the tissue was not associated with two aspects of the phenomenon previously described as vitrification, namely, reduced apical dominance and poorly developed cuticle and stomatal function As an aside the term vitromorphology is suggested as a generic term for this phenomenon.

A wide range of films of differing selectivities exists. Genetic diversity however, is also very wide. To optimise *in vitro* growth and establishment and post-establishment performance may ,as shown here, be possible with a single film. For other genotypes production may be optimised under one set of atmospheric conditions while weaning may require another achieved by a change of film at Stage III [2]. A further option is to use modified atmospheres [15] perhaps in association with ethylene binding agents [6], to achieve optimised conditions. The latter atmospheres, eg enriched in carbon dioxide, may be achieved by filling containers of high barrier materials with the gas mixture of choice. Alternatively, the atmosphere may be modified in the growth room and equilibrium achieved in the cultures by using high carbon dioxide permeability membranes as described here. The latter option could be implemented strategically during the production cycle.

Studies on the gaseous atmosphere in vessels with the above and other differentially permeable lids is in progress. These results should provide further insights into the influence of the components of the gaseous atmosphere in the tissue culture response including the contribution of volatiles such as ethanol, acetaldehyde etc. [26] to the physiological and morphological development of the *in vitro* plantlets with respect to the optimization of production and enhancement of establishment potential and subsequent growth performance.

Acknowledgements

The authors are grateful to Claire Walsh for skilled technical assistance. TDR acknowledges postgraduate awards from University College, Cork , Cork County Council and EOLAS.

References

1. Bornman CH and Velmann TC (1984) Effect of rigidity of gel medium on benzyladenine induced adventitious bud formation and vitrification in *Picea abies*. Physiol Plant 61:505–512
2. Cassells AC (1987) Plant container. European Patent no. 87 902578.1
3. Cassells AC (1991) Problems in tissue culture: culture contamination. In: PC Debergh and RH Zimmerman (eds). Micropropagation — Technology and Application, pp 31–44. Dordrecht: Kluwer
4. Cassells AC and Minas GJ (1983) Plant and in vitro factors influencing the micropropagation of Pelargonium cultivars by bud-tip culture. Sci Hort 21:53–65
5. Davis MJ, Baker R and Hanan JJ (1977) Clonal multiplication of carnation by micropropagation. J Amer Soc Hort Sci 102:48–53
6. Debergh PC (1983) Effects of agar brand and concentration on the tissue culture medium. Physiol Plant 59:270–276

7. Debergh P, Aitken-Christie J, Cohen D, Grout B, van Arnold S, Zimmerman R and Ziv M (1992) Reconsideration of the term 'vitrification' as used in micropropagation. Plant Cell Tiss Org Cult 30:135–140

8. Debergh PC and Maene LJ (1981) A scheme for commercial propagation of ornamental plants by tissue culture. Sci Hort 14:335–345

9. Ghashghaie J, Brenckmann F and Saugier B (1992) Water relations and growth of rose plants cultured *in vitro* under various relative humidities. Plant Cell Tiss Org Cult 30:51–57

10. Hakkaart FA and Versluijs JMA (1983) Some factors affecting glassiness in carnation meristem tip cultures. Neth J Pl Path 89:47–53

11. Hardwick R and Cole R (1986) Plastics that can kill plants. New Scientist 30 January:38–40

12. Jackson MB, Abbott AJ, Belcher AR, Hall KC, Butler R and Cameron J (1991) Ventilation in Plant Tissue Cultures and effects of poor aeration on ethylene and carbon dioxide accumulation, oxygen depletion and explant development. Ann. Bot. 67:229–237

13. Keevers C, Coumans M, Coumans-Gilles and Gaspar Th (1984) Physiological and biochemical events leading to vitrification of plants cultured *in vitro*. Physiol Plant 61:69–74

14. Keevers C. and Gaspar Th (1985) Vitrification of carnation *in vitro*; changes in ethylene production, ACC level and capacity to convert ACC to ethylene. Plant Cell Tiss Org Cult 4:215–223

15. Kozai T (1991) Micropropagation under photoautotrophic conditions. In: PC Debergh and RH Zimmerman (eds). Micropropagation — Technology and Application, pp 447–470. Dordrecht: Kluwer

16. Leshem B (1983) Growth of carnation meristems *in vitro*: anatomical structure of abnormal plantlets and the effect of agar concentration in the medium on their formation. Ann Bot 52:413–415

17. Leshem B (1983) The carnation succulent plant — a stable teratological growth. Ann Bot 52:873–876

18. Lehsem B and Sachs T (1985) Vitrified Dianthus — teratomata *in vitro* due to growth factor imbalance. Ann Bot 56:613–617

19. Murashige T and Skoog F (1962) A revised medium for rapid growth and bio-assays with tobacco tissue cultures. Physiol Plant 15:473–497

20. Paques M and Boxus Ph (1987) 'Vitrification' : review of literature. Acta Hortic. 212:155–166

21. Preece JE and Sutter EG (1991) Acclimatization of micropropagated plants to the greenhouse and field. In: PC Debergh and RH Zimmerman (eds). Micropropagation — Technology and Application, pp 77–94. Dordrecht: Kluwer

22. Sallanon H and Maziere Y (1992) Influence of growth room and vessel humidity on the in vitro development of rose plants. Plant Cell Tiss Org Cult 30:121–125

23. Schaad NW (1980) Laboratory guide for identification of plant pathogenic bacteria. St. Paul, Minnesota: American Phytopathological Society

24. Sutter E and Langhans RW (1979) Epicuticular wax formation in carnation plantlets regenerated from shoot tip culture. J Amer Soc Hort Sci 104:493–496

25. Tanaka M, Jinno K, Goi M and Higashiura T (1988) The use of disposable fluorocarbon polymer film culture vessel in micropropagation. Acta Hortic 230:73–80

26. Thomas D and Murashige T (1979) Volatile emissions of plant tissue cultures 1. Identification of major components. In Vitro 15:654–658

27. Woltering EJ (1990) Phytotoxic component in polypropylene tissue culture containers. Sci Hort 44:335–340

28. Ziv M (1991) Vitrification: morphological and physiological disorders of *in vitro* plants. In: PC Debergh and RH Zimmerman (eds). Micropropagation — Technology and Application, pp 45–70. Dordrecht: Kluwer

29. Ziv M, Meir G and Halevy AH (1983) Factors influencing the production of hardened glaucous carnation plantlets *in vitro*. Plant Cell Tiss and Org Cult 2:55–65

30. Ziv M, Schwartz A and Fleminger D (1987) Malfunctioning of stomata in vitreous leaves of carnation (*Dianthus caryophyllus*) plants propagated *in vitro*; implications for the hardening. Plant Science 52:127–134

Effect of light intensity and aeration during *in vitro* growth (stage III of micropropagation) of banana plants (*Musa* AAA cv. Petite Naine)

CUAUHTÉMOC NAVARRO*, CLAUDE TEISSON, FRANÇOIS CÔTE and JACKY GANRY

Institut de Recherches sur les Fruits et Agrumes / C.I.R.A.D. B.P. 5035 34032 Montpellier, Cedex, France
* *Present address: Centro de Investigación Científica de Yucatán, A.C., Apartado Postal 87 CORDEMEX, 97310 Mérida, Yucatán, México*

1. Introduction

In the commercial micropropagation of *Musa* spp., the growth phase (stage III), is generally carried out in a hormone-free medium in closed or semi-closed containers under a PPFD (photosynthetic photon flux density) of 30 to 50 μmol m^{-2} s^{-1} (i.e. less than 3% of full sunlight). It is very likely that *in vitro* conditioning of plants with poor vessel ventilation and low light intensity may negatively influence their growth rate when transferred to the acclimatization phase. The accumulation of C_2H_4 in air tight vessels produced *in vitro* by plant tissues has been reported to reduce leaf expansion [10, 5], to induce epinasty of leaves [6], shoot swelling [8] and explant vitrification [3]. Further, several authors have observed that micropropagated plants present reduced carbon fixation [2, 7] and low chlorophyll content [4] as compared to normal plants.

Hence, special attention should be given to these factors to determine their influence on the physiological status and growth of plants. The objective of the present work was to investigate how light intensity and ventilation in *in vitro* systems, modified plant development and could influence their subsequent *ex vitro* behavior.

2. Materials and methods

2.1 Plant material and culture conditions

Banana propagules (*Musa* AAA cv. Petite Naine) in the multiplication phase were transferred to the growth phase (stage III) in a Murashige and Skoog's [9] medium without growth regulators, supplemented with 200 mg l^{-1} KH$_2$PO$_4$, 40 g l^{-1} sucrose and 2 g l^{-1} gelrite and incubated under a 12h/12h (day/night) photoperiod, standard PPFD of 45 μmol m^{-2} s^{-1} and a temperature of 27±1 °C.

P.J. Lumsden, J.R. Nicholas and W.J. Davies (eds.), Physiology, Growth and Development of Plants in Culture, 215–219, 1994.

2.2 Effect of light intensity and ventilation

Plants were grown *in vitro* under two different PPFD's (measured with a Skye PAR radiation sensor): 45 μmol m^{-2} s^{-1} (standard conditions) and 380 μmol m^{-2} s^{-1} (equivalent to 30 and 240 μmol m^{-2} s^{-1} respectively, when determined inside the vessels), supplied by fluorescent lamps: TLD83, TL40W03 (PHILIPS, 40 W, white and blue cool light respectively) and GRO-LUX (SYLVANIA, 20 W). Under each light intensity level, the plants were cultivated either in air tight vessels (500 ml containers) firmly sealed with plastic caps surrounded with two layers of parafilm, or in ventilated vessels using MILLIPORE filters (0.2 μm, FG$_{50}$, 5 cm in diameter) attached horizontally to the plastic cap.

2.3 Analytical techniques

CO_2 concentration inside the culture vessels was determined by gas chromatography (GIRDEL chromatograph, Chromosorb 102 column) after withdrawing a 250 μl sample. C_2H_4 was measured by flame ionization gas chromatography (INTERSMAT I GC 120 DFL, porapak Q column) after withdrawing a 250 μl sample. Chlorophyll content was determined by triplicate by the method of Bruinsma [1]. Statistical analyses were performed with the programs STAT-ITCF 4 and MICROSTAT. The Newman and Keuls test was used to compare means at p = 0.05.

3. Results

Under high light intensity, the accumulated mass after 35 days by plants in the growth phase was significantly higher than under low light intensity (Table 1). This effect was enhanced by ventilation of the vessels. A significant increment in chlorophyll content was determined for plants cultured in ventilated vessels. The increment was more pronounced under high light intensity. In the ventilated vessels ethylene could diffuse through the filters and reached only 8% and 23% of the concentration found in sealed vessels at low and high light intensity, respectively.

Under low light intensity, the time course for CO_2-accumulation (Fig. 1) showed no change during the light period, whether the plants were cultivated in sealed or ventilated vessels. In air tight vessels, CO_2-concentration was approximately 300 times higher than in natural atmosphere, whereas in ventilated vessels it was 30-fold lower. At high light intensity, there was a dramatic decrease in CO_2-concentration during the first 4 to 6 h of the light period. After this period, CO_2-concentration remained constant. The plants *in vitro* exposed to a high PPFD and ventilation, showed the best *ex vitro* development (Table 1).

Table 1. Effects of *in vitro* light intensity and ventilation on some properties of plants after a 35 day period in stage III. Dry weight and %D.W. data are the mean ± S.D. of 15 replicates. Chlorophyll was analysed by triplicate. C_2H_4 was determined by gas chromatography in 10 vessels per treatment. *Ex-vitro* growth (dry mass and total leaf area) was measured in 30 plants per treatment

PPFD	30 μmolm^{-2}s^{-1}		240 μmolm^{-2}s^{-1}	
Container	Sealed	Ventilated	Sealed	Ventilated
Dry weight (mg plant^{-1})	130±40[c]	120±30[c]	220±60[b]	270±79[a]
% D.W. (DW/FW)x100	4.9±0.2[b]	5.2±0.2[b]	5.2±0.5[b]	5.7±0.3[a]
Chl a + b μg (g FW)$^{-1}$	108±55[c]	131±37[b]	168±46[b]	507±69[a]
C_2H_4 (μl l^{-1} flask^{-1})	2.8±0.5[a]	0.2±0.15[c]	2.7±0.95[a]	0.6±0.25[b]
Growth *ex-vitro* (Control=1)	1	1	1.05	1.20

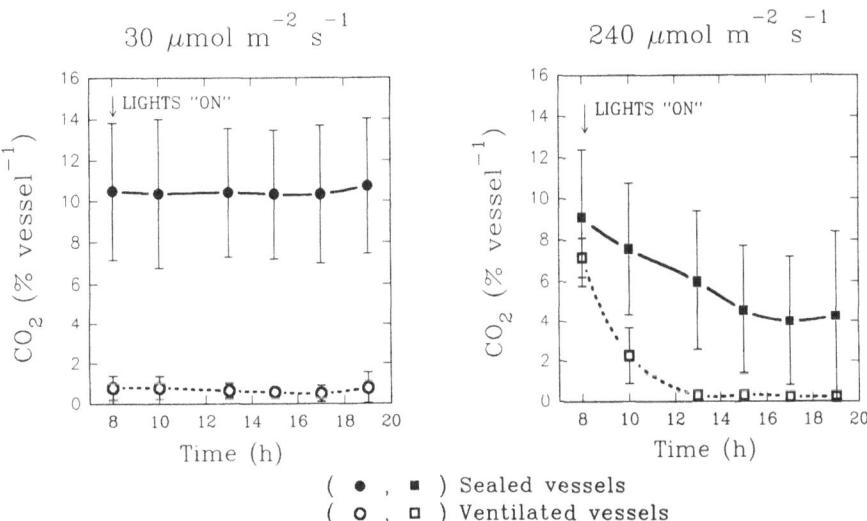

(● , ■) Sealed vessels
(○ , □) Ventilated vessels

Fig. 1. Time course of CO_2 in vessels (500 ml) containing banana plants (after 35 days in growth phase *in vitro*), cultured under two different light intensities (30 and 240 μmol m^{-2} s^{-1}) and with or without ventilation. Each vessel contained five plants. Data are the mean ± S.D. of 10 replicates per treatment.

4. Discussion

Inside the vessels, the gas concentrations were strongly dependent on the relation between plant mass and vessel volume. Frequently, extremely variable CO_2 levels were found. At low light intensity, stable CO_2 concentrations were observed. This suggests that photosynthetic activity can equal the respiratory activity. Presumably during growth phase, there is a gradual accumulation of CO_2 released by respiration in the night. In ventilated vessels, the situation is more complex due to the diffusion of CO_2 through the filters. At high light intensity, the partial (in air tight vessels) or complete (in ventilated vessels) CO_2-depletion observed at the end of illuminated period (Fig. 1) may indicate its photosynthetic fixation from dark respiration in the preceding night.

The first limiting factor of *in vitro* photosynthetic fixation of CO_2 is the low light intensity usually employed in culture incubation rooms. In contrast, when using a higher light intensity, respiratory CO_2 released during the night was rapidly fixed during the day and CO_2 availability became the limiting factor for photosynthesis. One might conclude that ventilation through the filters can improve CO_2 availability for plants inside the vessels but this quantity is negligible. The CO_2 diffusion rates trough MILLIPORE filters are less than 0.01 ml h^{-1} (and near zero for very small CO_2 concentrations). As a concentration of 7% CO_2 is present in the vessels at the beginning of the light period, this means that plants can consume *c.* 35 ml in 3 to 4 h.

Only when plants were grown under ventilated conditions did they increase significantly their dry mass *in vitro* and perform better *ex vitro* (Table 1). We assume that this is due to the lower C_2H_4 levels found in ventilated vessels as compared to air tight vessels. In ventilated conditions ethylene could diffuse through the filters. In standard conditions of *in vitro* culture, C_2H_4 accumulation in the vessels may cause negative effects to plants in the stage prior to acclimatization. Yellowing, epinasty and reduction of leaf area (data not shown) and chlorophyll content (Table 1) were associated to C_2H_4 accumulation in sealed vessels in our experiments. When millipore filters were used to promote the vessels' ventilation, these symptoms disappeared, at least partially.

This study shows how important are the *in vitro* growth conditions for micropropagated plants prior to acclimatization, for their performance *ex vitro*. Factors that have been neglected such as atmosphere and light inside the vessels, must be further investigated in order to provide a better *in vitro* environment to plants, which will facilitate their adaptation when transferred *ex vitro*.

Acknowledgements

The first author thanks the Mexican Government (CONACyT), for the scholarship awarded to do this work and Mr. Jean-Baptiste Dorval, manager of VITROPIC who kindly provided the vegetal material for this study.

References

1. Bruinsma J (1963) The quantitative analysis of chlorophylls a and b in plant extracts. Photochem. and Photobiol. (Chlor Metabol Sym), Vol 2 (pp 241–249). UK: Pergamon Press Ltd
2. Donnelly DI, Vidaver WE & Colbow K (1984) Fixation of $^{14}CO_2$ in tissue cultured raspberry prior to and after transfer to soil. Plant Cell Tiss Org Cult 3:313–317
3. Gaspar T (1988) Aspects physiologiques de l'organogénèse *in vitro*. In: JP Zryd, ed. Cultures de cellules, tissus et organes végétaux, pp 69–88. Suisse: Presses Polytechniques Romandes
4. Grout B and Aston H (1978) Transplanting of cauliflower plants regenerated from meristem culture. II. Carbon dioxide fixation and the development of photosynthetic ability. Hort Res 17:1–7
5. Jackson MB, Abbott AJ, Belcher AR, Hall KC, Butler R and Cameron J (1991) Ventilation in plant tissue cultures and effects of poor aeration on ethylene and carbon dioxide accumulation, oxigen depletion and explant development. Ann Bot 67:229–237
6. Jona TR, Gribaudo R and Vigliocco R (1987) Natural development of ethylene in air tight vessels of GF677. In: G Ducate, M Jacob and A Simeon (eds.). Plant Micropropagation in Horticultural Industries, pp 61–66. Belgium: Presses Universitaires
7. Kozai T & Iwanami Y (1988) Effects of CO_2 enrichment and sucrose concentration under high photon fluxes on plant growth of carnation (*Dianthus caryophyllus* L.) in tissue culture during the preparation stage. Jap Soc Hort Sci 57:279–288
8. Lentini Z, Mussell H, Mutschler MA and Earle ED (1988) Ethylene generation and reversal of ethylene effects during development *in vitro* of rapid cycling *Brassica campestris* L. Plant Sci 54:75–81
9. Murashige T and Skoog F (1962) A revised medium for rapid growth and bioassays with tobacco tissue cultures. Physiol Plant 15:473–497
10. Perl A, Aviv D and Galan E (1988) Ethylene and *in vitro* culture of potato: suppression of ethylene generation vastly improves protoplast yield, plating efficiency and transient expression of an alien gene. Plant Cell Rep 7:403–406

Ethylene and anther culture

N.L. BIDDINGTON and HELEN T. ROBINSON
Horticulture Research International, Wellesbourne, Warwick CV35 9EF, UK

Abbreviations: ABA, abscisic acid; ACC, 1-aminocyclopropane-1-carboxylic acid; AVG, aminoethoxyvinylglycine; 2,4-D, 2,4-dichlorophenoxyacetic acid; IAA, indole-3-acetic acid.

1. Introduction

It is becoming increasingly evident that endogenous ethylene can have large and diverse effects on growth and development in a wide range of plant tissue culture systems [2]. Anther culture in which haploid plants are regenerated from immature microspores is important because it allows the rapid production of homozygous lines for breeding and genetic studies [8]. However, it has proved unsuccessful with several food crops, and even with responsive species low and variable yields between and within genotypes may greatly limit its efficacy. Ethylene produced by the isolated anthers may accumulate in the culture vessels and its possible effects on anther culture need to be considered.

1.1 Ethylene effects in anther culture

The few reports to date suggest that, as with tissue culture generally [2], ethylene may promote or inhibit anther culture, depending to a large extent on the genotype studied. Both endogenous and exogenous ethylene appeared to enhance embryogenesis in anther culture of *Datura metel* [1] and *Solanum carolinense* [13]. In the former, embryo production was promoted by the ethylene-forming agent, ethephon and the ethylene precursor, methionine and inhibited by the ethylene biosynthesis inhibitor, $CoCl_2$, and the inhibitor of ethylene action, $AgNO_3$ [1]. With *S. carolinense*, differing responses to IAA and 2,4-D appeared to be related to different effects of the two auxins on ethylene production. IAA promoted both ethylene production and embryogenesis [13], whereas 2,4-D, which did not increase ethylene production, induced the formation of calli rather than embryos [14]. $CoCl_2$ partially inhibited IAA-induced embryogenesis, and ethephon and the ethylene precursor ACC promoted embryogenesis in the absence of IAA [13]. In *Solanum tuberosum* however $AgNO_3$ promoted and ethephon, at concentrations lower than those used with *S. carolinense*, inhibited anther culture [15].

A cultivar effect was shown with barley anther culture [6]; ACC or ethephon enhanced callus production in two cultivars whose anthers had the lowest

P.J. Lumsden, J.R. Nicholas and W.J. Davies (eds.), Physiology, Growth and Development of Plants in Culture, 220–226, 1994.

concentrations of ACC and produced ethylene the most slowly, whereas the ethylene biosynthesis inhibitor putrescine enhanced calli production in a third cultivar whose anthers contained the highest amounts of ACC and produced ethylene the most rapidly. The differences in ethylene production rates were measured at the very beginning of the culture period and it was suggested that the initial rate of ethylene production was important in optimising callus production [6].

The involvement of ethylene in tobacco anther culture is unclear. Although Horner *et al.* [10] found no effect on *Nicotiana* anther culture of ethylene removal from the culture vessel atmosphere, Dunwell [7] found removal either enhanced or retarded embryo induction, embryo survival and the number of plantlets produced, depending on the size of the vessel and the age of the anthers. Although $AgNO_3$ only increased embryo induction slightly, embryo yields without $AgNO_3$ were already high. This highlights the point that in highly responsive species or cultivars constraints imposed by endogenous factors such as ethylene may not present a problem. However, where the response is low, or, as with brassicas, variable [11], identifying such constraints is obviously important.

1.2 Ethylene and brussels sprouts anther culture

Brussels sprouts cultivars producing no or very few embryos in anther culture in the absence of $AgNO_3$ can be made highly productive by including $AgNO_3$ in the culture media [5, 12] (Table 1), a result which suggests ethylene blocks embryogenesis in this system. $AgNO_3$ has since been shown to improve anther culture of cauliflowers [9, 16], although the effect is not as great as with Brussels sprouts. In order to study the role of ethylene in anther culture we have used three cultivars of Brussels sprouts, one of which ('Hal') normally requires ($AgNO_3$) to induce successful anther culture, and two of which ('GA1xRDF2' and 'Gower') produce androgenic embryos without $AgNO_3$ [5, 12].

Table 1. The effects of $AgNO_3$ and AVG on embryo production[1] in anther culture of Brussels sprouts cv Hal

Concentration (μM)	$AgNO_3$	AVG
0	8	14
0.6	10	—
1.0	—	28
1.8	57	—
3.0	—	7
6.0	114	—
10.0	—	16
18.0	128	—

[1] Embryos per 100 anthers. Means of 7 experiments each consisting of 60 anthers per treatment. Five experiments showed a significant ($p < 0.05$) positive promotion with $AgNO_3$, one with AVG

2. Materials and methods

The methods used for producing donor plants, treating and culturing anthers, and measuring ethylene in culture petri dishes have been described previously [4]. For the latter, previously holed petri dishes were wrapped with a double layer of Nescofilm and culture atmosphere samples were removed with a 1.0 ml gas-tight syringe for ethylene measurement by gas chromatography. Dishes were rewrapped following each sampling. Where ethylene determinations were made after only 6 h of culture, dishes were immersed in water to allow measurable quantities of ethylene to accumulate, Nescofilm being relatively permeable to ethylene [4].

Embryo production is presented as the number of embryos per 100 anthers cultured. Experimental design and the statistical analysis of logarithmically transformed data have been described previously [4, 5]. Sixty anthers were used for each treatment in each experiment.

3. Results and discussion

The problems associated with the high level of variation both within and between Brussels sprouts anther culture experiments and the difficulty of combining results from individual experiments have already been discussed [3, 4, 5]. Where appropriate, the number of experiments done and those in which significant results were measured, are shown in the tables.

A period at an elevated temperature (e.g., 35 °C for 24 h) is usually essential for inducing embryogenesis in brassica anther culture [3, 11]. Embryos begin to emerge from Brussels sprouts anthers after about 20 days and the pattern of ethylene production during a 4-week period of anther culture is shown in Figure 1. Two peaks of ethylene production were measured, the first peak within the first 5 days of culture and the second peak from about 14 days onwards. Both peaks occurred whether or not the anthers were exposed to high temperature, although the 35 °C treatment reduced ethylene production during the first peak with 'Hal' and 'Gower' and increased it with 'GA1xRDF2' when compared to anthers given only 25 °C. The first peak was smaller in 'GA1 x RDF2', compared to 'Hal' and 'Gower'. The second ethylene peak (Fig. 1) tended to be larger in 'GA1xRDF2' and 'Gower' than in the normally unresponsive 'Hal'. However, the possibility that the second peak stimulates embryogenesis was dispelled by the finding that AVG reduced ethylene production to a low rate throughout the whole of the anther culture period with both 'GA1xRDF2' and 'Gower', without retarding embryogenesis [NL Biddington, unpublished].

Only anthers exposed to high temperature produce embryos. Thus if differences in embryogenic potential between the 3 cultivars is related to the rate of ethylene production, then it is ethylene production following the high temperature treatment which is obviously relevant. Differences in ethylene production between the 2 responsive cultivars and 'Hal' is most marked early

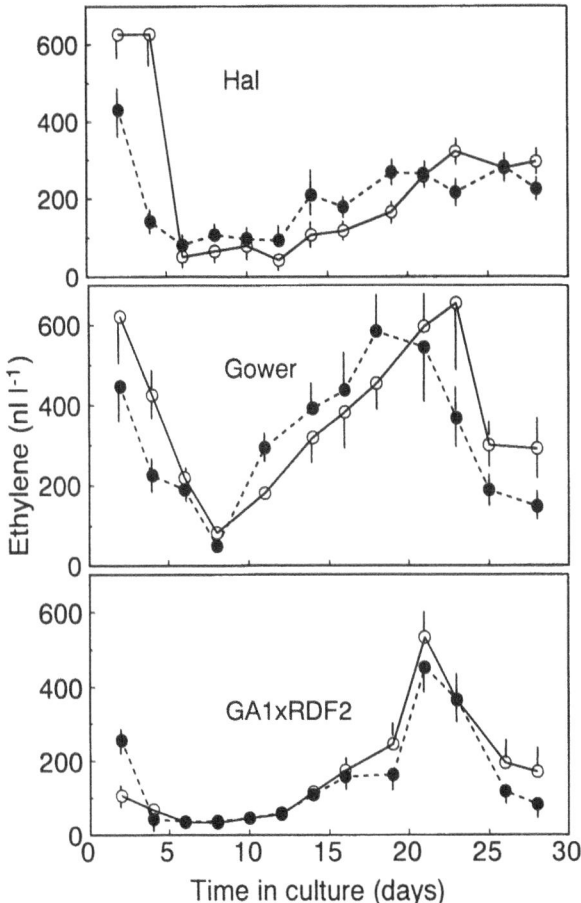

Fig. 1. Ethylene from anther cultures of Brussels sprouts cultivars Hal, Gower and GA1xRDF2 incubated continuously at 25 °C (o), or at 35 °C (●) for 24 h followed by 25 °C. Means and standard errors are from 10 replicate dishes. At the end of the 28 day period Hal had produced no embryos, Gower 52 embryos/ 100 anthers and GA1xRDF2 428 embryos/ 100 anthers.

on, e.g. at 6 h (Table 2) [4]. This coupled with the fact that $AgNO_3$ is effective only when given within the first 24 h (Table 3) suggests that it is a high rate of ethylene production early on in the anther culture period that is blocking embryo production in Hal. The inhibition of Brussels sprouts anther culture by abscisic acid appears to be due to its enhancement of ethylene biosynthesis (NL Biddington, unpublished), an effect that can be shown after the first 6 h of culture.

Unlike $AgNO_3$, the ethylene biosynthesis inhibitor AVG was generally ineffective in promoting embryo production in 'Hal' (Table 1), possibly because it only partially blocked ethylene production during the early stages of anther culture, particularly during the 35 °C treatment (Table 2).

High temperature reduces ethylene production early in the culture period

Table 2. Ethylene production[1] (nl l^{-1}) from anthers of Brussels sprouts, cvs Hal, Gower and GA1xRDF2 after 6 h culture at 25 °C or 35 °C. Hal anthers were also treated with AVG

Cultivar	Temperature (°C)	AVG concentration (μM)		
		0	1.0	3.0
Hal	25	78±12	42±6	19±6
	35	49±16	41±14	25±8
Gower	25	32±4		
	35	12±3		
GA1xRDF2	25	16±3		
	35	6±1		

[1] Means of 3 experiments ± standard errors

Table 3. The effects of time of application of $AgNO_3$ on embryo production in anther culture of Brussels sprouts, cv. Hal

	Embryos/100 anthers[1]
No $AgNO_3$	2
$AgNO_3$ at start of culture	103
$AgNO_3$ after 24 h of culture	9

[1] Means of 4 experiments

(Table 2) implying this may be the way in which it induces embryogenesis. However, neither $AgNO_3$ nor AVG promote embryo production in the absence of high temperature [4], suggesting that high temperature acts through factors other than ethylene biosynthesis. Alternatively, it could be argued that both $AgNO_3$ and AVG negate the ethylene effect far more slowly than high temperature, because of the time taken for the compounds to be taken up into the tissue, and hence they may act too late to prevent the ethylene inhibition of embryogenesis. However, it is not essential for high temperature to act immediately at the start of the culture period. We have shown that if the high temperature treatment is delayed for up to 8 h, the period when ethylene production is presumed to have a large inhibitory effect, embryo yields remain high, but if the delay is for 24 h then embryogenesis is almost completely blocked [3].

Nescofilm-sealed petri dishes are permeable to ethylene, 80% of an injected sample of ethylene being lost from 30 mm dishes within 4 h [4]. However, because the ethylene produced by the anthers exceeds leakage from the vessel, the gas accumulates to a high concentration inside the dish. It seems unlikely, however, that it is this general build up of ethylene that inhibits embryogenesis, but rather a localised response of the microspores to ethylene synthesised within the individual anther. Maintaining low ethylene concentrations in the petri dishes during the early stages of anther culture of 'Hal' by not sealing the plates did not induce embryogenesis (Table 4).

Table 4. The effects of not sealing the petri dishes during the first 24 h of anther culture of Brussels sprouts, cv. Hal, on ethylene concentrations in the dishes and embryo production

	Ethylene[1](nl l^{-1})	Embryos/100 anthers[2]
Seal at start of culture	866±140	0
Seal after 24 h of culture	43±23	2

[1,2] Means of 4 experiments ± standard errors

4. Conclusions

The capacity of ethylene to influence anther culture, either positively or negatively in different anther culture systems indicates an important role for ethylene in the control of microspore embryogenesis, and it is important that we learn more about this role. The fact that ethylene can be applied in the form of ACC or ethephon or its effects blocked by inhibitors of ethylene action or biosynthesis has already been used to good effect to improve anther culture yields of some species and may yet provide the means by which inroads may be made into some of the many species that are as yet recalcitrant.

Acknowledgements

We thank Dr R. A. Sutherland for help with the statistical analysis. This work was funded by the Agricultural and Food Research Council.

References

1. Babar SB and Gupta SC (1986) Putative role of ethylene in *Datura metel* microspore embryogenesis. Physiol Plant 68:141–144
2. Biddington NL (1992) The influence of ethylene in plant tissue culture (*review*). Plant Growth Reg 11:173–187
3. Biddington NL and Robinson HT (1990) Variations in response to high temperature treatments in anther culture of Brussels sprouts. Plant Cell Tiss Org Cult 22:48–54
4. Biddington NL and Robinson HT (1991). Ethylene production during anther culture of Brussels sprouts (*Brassica oleracea* var *gemmifera*) and its relationship with factors that affect embryo production. Plant Cell Tiss Org Cult 25:169–177
5. Biddington NL, Sutherland RA and Robinson HT (1988) Silver nitrate increases embryo production in anther culture of Brussels sprouts. Ann Bot 62:181–185
6. Cho UH and Kasha KJ (1989) Ethylene production and embryogenesis from barley anthers. Plant Cell Rep 8:415–428
7 Dunwell JM (1979) Anther culture in *Nicotiana tabacum*: The role of the culture vessel atmosphere in pollen embryo induction and growth. J Exp Bot 30:419–428

8. Dunwell JM (1986) Pollen, ovule and embryo culture as tools in plant breeding. In: LA Withers and PG Alderson, eds. Plant Tissue Culture and its Agricultural Applications, pp 377–404. London : Butterworths

9. Fuller MP, Turton S and Grout BWW (1990) Anther culture of winter-heading cauliflower. Acta Hort 280:329–332

10. Horner M, McComb JA, McComb AJ and Street HE (1977) Ethylene production and plantlet formation by *Nicotiana* anthers cultured in the presence and absence of charcoal. J Exp Bot 28:1365–1372

11. Ockendon DJ (1985) Anther culture in Brussels sprouts (*Brassica oleracea* var. *Gemmifera*). II. Effect of genotype on embryo yields. Ann Appl Biol 107:101–104

12. Ockendon DJ and McClenaghan R (1992) Effect of silver nitrate and 2,4-D on anther culture of Brussels sprouts (*Brassica oleracea* var. *Gemmifera*). Plant Cell Tiss Org Cult 32:41–46

13. Reynolds TL (1987) A possible role for ethylene during IAA-induced pollen embryogenesis in anther culture of *Solanum carolinense* L. Am J Bot 74:967–969

14. Reynolds TL (1989) Ethylene effects on pollen callus formation and organogenesis in anther culture of *Solanum carolinense* L. Plant Sci 61:131–136

15. Tianen T (1992) The role of ethylene and reducing agents on anther culture response of tetraploid potato (*Solanum tuberosum* L.). Plant Cell Rep 10:604–607

16. Yang Q, Chauvin JE and Hervé Y (1992) A study of factors affecting anther culture of cauliflower (*Brassica oleracea* var. *botrytis*). Plant Cell Tiss Org Cult 28:289–296

Leaf abscission in micropropagated sugar apple (*Annona squamosa* L.)

EURICO E.P. LEMOS and JENNET BLAKE

Unit for Advanced Propagation Systems, Department of Agriculture, Horticulture and The Environment, University of London (Wye College), Wye, Ashford, Kent TN25 5AH, UK

1. Introduction

Sugar apple (*Annona squamosa* L.), a tropical fruit tree, produces high fruit yields in drought-prone areas. As a drought-resistant plant, it has a physiological mechanism for leaf-shedding during dry seasons. Nodal explants have shown the same phenomenon as soon as they are placed in culture. Growth regulators are presumed to be involved in regulating abscission in leaves of many plants [3]. Ethylene is particulary important in the sequence of abscission phenomena [1]. Low levels of ethylene can promote abscission of fruits, leaves and buds in many plants [15]. Burg [5] has reported that ethylene is a hormonal regulator at $0.1–10.0\ \mu l\ l^{-1}$. In some tissue, however, concentrations as low as $0.01\ \mu l\ l^{-1}$ are effective [14]. Adams and Yang [2] have presented evidence that ACC is the immediate precursor of ethylene. It is known that silver ion is an antagonist for ethylene [4]. The present experiments were undertaken to determine the best inhibitor of the ethylene action in *Annona squamosa* explants in culture and to determine the ability of the tissues to convert ACC to ethylene.

2. Material and methods

2.1 Material and culture methods

Sugar apple (*Annona squamosa* L.) nodal cultures were initiated from eighteen-month-old plants grown in the glasshouse. Nodes were cultured on 10 ml of MS medium [12] with 2% sucrose, $0.5\ mg\ l^{-1}$ BAP, $0.5\ mg\ l^{-1}$ kinetin at pH 5.8 and gelled with 0.3% phytagel in 3x1 inch tubes. The cultures were maintained in a growth room at 25 °C with a 16 h photoperiod and light intensity of $65\ \mu M\ m^{-2}\ s^{-1}$. Unless specified, 15 replicates were used in each experiment.

227

P.J. Lumsden, J.R. Nicholas and W.J. Davies (eds.), Physiology, Growth and Development of Plants in Culture, 227–233, 1994.

2.2 Ethylene inhibitors and absorbents

Cobalt chloride (5.0 mg l^{-1}) or silver nitrate (5.0 mg l^{-1}) as inhibitors, and activated charcoal (3.0 g l^{-1}) or mercuric perchlorate (0.1 mg l^{-1}) as absorbents of ethylene, were added to the basic medium. In a later experiment, silver nitrate, at a concentration of 0.0, 2.5, 5.0 or 10.0 mg l^{-1}, was also added to the basic medium prior to autoclaving in tubes capped with polypropylene film or cling film. Explants were cultured for 4 weeks, when records were taken of leaf abscission, bud opening and dead shoots. Data on abscission are presented as a percentage of all leaves above 3 mm in length.

2.3 ACC treatments

ACC (5.0 mg l^{-1}), filter-sterilized in combinations with silver nitrate (0.0 or 5.0 mg l^{-1}), was added to the basic medium or used for dipping treatments of the explants before they were placed in the basic medium. The tubes were opened on alternate days in a sterile flow cabinet and allowed to exchange gas for one hour.

2.4 Ethylene measurement

Evolution of ethylene from ACC conversion was determined by gas chromatography (Vega 6000 GC, Carlo Erba Strumentazione) each day up to the sixth day and then alternate days up to the twelfth day. Three samples of 1.0 cm^3 air were withdrawn with a gas-tight syringe from each 3x1-inch tube capped with a 'suba seal' closure. Five replicates were used in each treatment. A stainless steel column (80 cm x 1.5 mm) filled with Poropack (80–100 mesh) was used. Column, injector and flame ionozation detector temperatures were 80, 150 and 150 °C, respectively. N$_2$ was used as a carrier gas (50 cm^3 min^{-1}).

2.5 Ethylene injection

Samples of 0.5 cm^3 of ethylene gas (ethylene 1% in nitrogen) were injected by gas-tight syringe into culture tubes closed with 'suba seal' caps. Concentration inside the culture tubes was 265 μl l^{-1}. Explants were grown in basic medium without or with 5.0 mg l^{-1} of silver nitrate. After 4, 8 and 12 days in culture, the percentage of leaf abscission in each treatment was recorded. Tubes were maintained closed during the whole experiment.

3. Results and discussion

3.1 Effect of inhibitors and absorbents

After 4 weeks in culture, silver nitrate proved to be the most effective treatment applied to control leaf abscission (Fig. 1). The silver ion has been reported as a

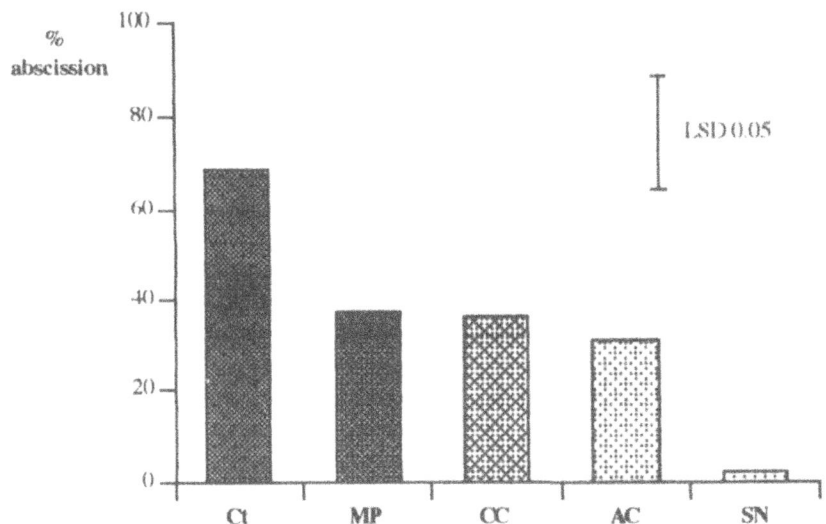

Fig. 1. Effect of inhibitors and absorbents of ethylene. % abscission of *Annona squamosa* leaves after 4 weeks in culture. 15 replicates per treatment. Ct-control; MP-mercuric perchlorate; CC-cobalt chloride; AC-activated charcoal; SN-silver nitrate.

strong inhibitor of ethylene action and as very effective in reducing leaf abscission in cotton [4]. Sisler and Yang [15] proposed that silver ions interfere with the ethylene receptor complex by removing an essential ligand of the binding receptor, resulting either in a biologically inactive complex or in a receptor that loses its capability to bind ethylene. Cobalt ion, which has been reported to interfere in ethylene biosinthesis by inhibiting ethylene production from its precursor ACC [10, 16], was not as effective as cobalt chloride at the concentration used (5 mg l^{-1}) in this experiment. Although reported to trap ethylene produced in culture [1, 11], neither mercuric perchlorate nor activated charcoal were able to counteract leaf drop. After 4 weeks in culture, a mean of 3 buds per explant had opened in the controls, but at least one of them had died (Fig. 2). With silver nitrate, an average of 1.5 buds opened per explant but no shoots died. It seems that the differences in the number of opened buds among treatments was due to the death of the tips, promoted by the effect of ethylene, thus inducing the development of axillary buds, rather than to a positive stimulation of bud production by ethylene.

3.2 Effect of silver nitrate vs closures

When added to the basic medium at 2.5, 5.0 and 10.0 mg l^{-1}, silver nitrate was effective in reducing leaf abscission at relatively high levels, 5.0 mg l^{-1} having a more persistent and less phytotoxic effect than 10.0 mg l^{-1} (Fig. 3). This

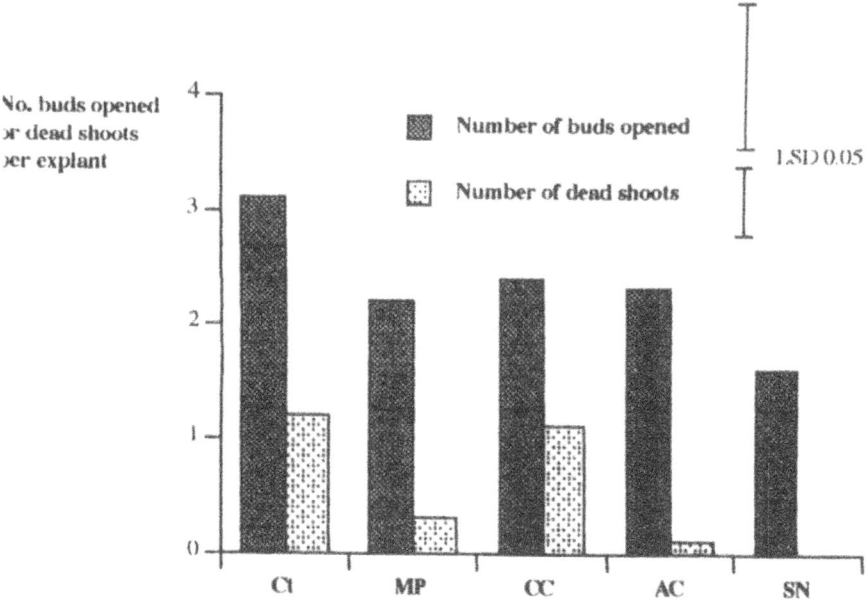

Fig. 2. Effect of inhibitors or absorbents of ethylene on nodal explants of *Annona squamosa* after 4 weeks in culture. 15 replicates per treatment. Ct-control; MP-mercuric perchlorate; CC-cobalt chloride; AC-activated charcoal; SN-silver nitrate.

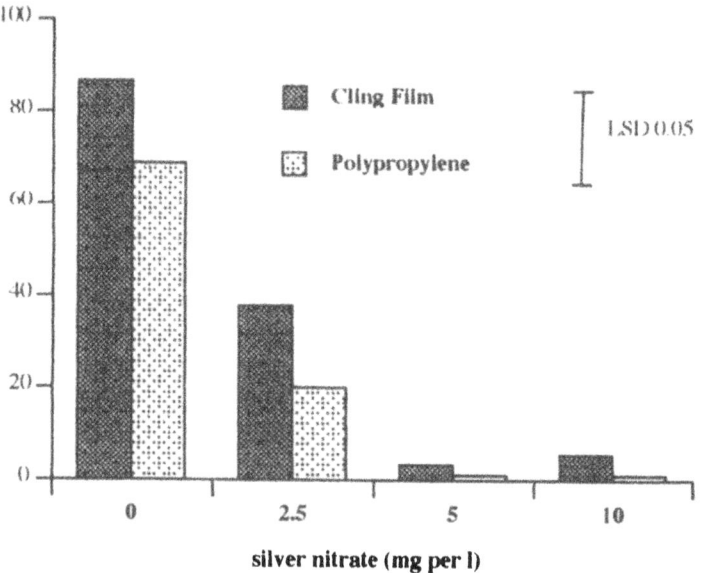

Fig. 3. Effect of silver nitrate on leaf abscission of *Annona squamosa* explants. Culture tubes capped with polypropylene film or cling film. % abscission after 4 weeks in culture. 15 replicates per treatment.

confirms Beyer's [4] report that the most outstanding anti-ethylene properties of silver ion are its persistence, specificity, and its lack of phytotoxity at effective concentrations.

The results indicated that vessel closure with cling film stimulated leaf abscission whether or not associated with silver nitrate. Authors have reported that ethylene has been shown to accumulate within closed containers to physiologically active concentrations [5, 7, 8]. Explants cultured under polypropylene film closure were healthier and showed less abscission than those placed in tubes covered with cling film. We have experienced that cling film is as impermeable to ethylene as 'suba seal' rubber caps (EEP Lemos, unpublished).

3.3 Effect of ACC and silver nitrate on evolution of ethylene

When ACC or silver nitrate was added to the medium ethylene production was enhanced compared to treatments where the explants were dipped in a filter-sterilized solution of one or other compound (Fig. 4). The dip treatments were not as effective in producing ethylene as those in which silver nitrate was added to the medium. ACC as the immediate precursor of ethylene was probably the source of most of the ethylene released. These relatively high concentrations of ethylene have only been produced when ACC was added or given as a dip. Even with high rates of ethylene being released between the 5th and the 10th day (16–20 μl l^{-1} day^{-1}), the silver nitrate applied helped to protect some of the leaves against abscission (Fig. 4). In the absence of silver nitrate, the ethylene production was lower (5.6 μl l^{-1} day^{-1}) because the explants were without leaves and could not convert ACC to ethylene effectively. Chi *et al.* [6], have

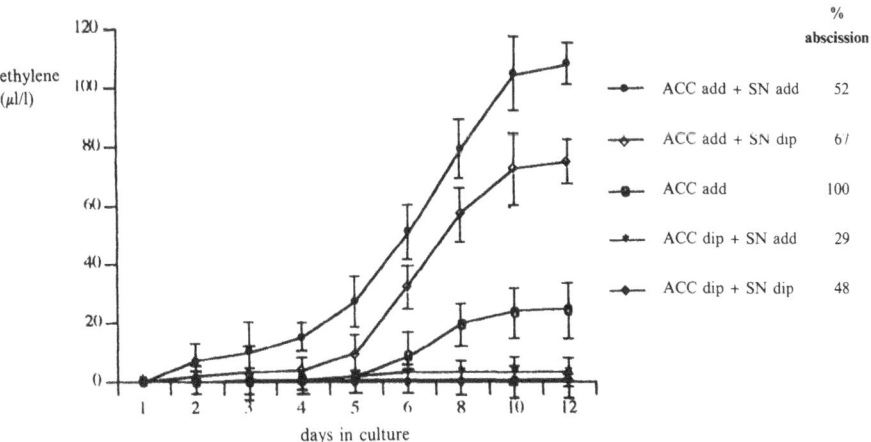

Fig. 4. Effect of ACC and Silver nitrate (SN) on evolution of ethylene and % abscission from nodal explants of *Annona squamosa* during 12 days in culture. Addition (add) of ACC (5 mg l^{-1}) or SN (5 mg l^{-1}). Explants dipped (dip) in ACC (5 mg l^{-1}) or SN (5 mg l^{-1}) solution before being placed in the medium. 5 replicates/treatment.

reported that silver nitrate enhanced ACC synthase activity, ACC accumulation and ethylene production. They claim that this is possibly due to an excess of silver ions in the medium, which may be phytotoxic. Stimulation of ethylene production is a common response of plants to stress [1]. Rostan *et al.* [13] have also reported that the addition of silver ions to the embryogenic medium led to a slight increase in ethylene production by the embryonic cells. Khalid *et al.* [9] reported a slight stimulation of ethylene production by high concentrations (25 μM) of silver nitrate in *Helianthus annuus* cotyledon cultures. Natural levels of ethylene produced by *Annona squamosa* nodal explants in culture are much lower (0.1 μl l^{-1} day^{-1}), but enough to produce physiological effects like leaf abscission.

3.4 Effect of ethylene injection

Explants cultured for 12 days in an atmosphere saturated with exogenous ethylene showed some similarities to those cultured with ACC. For instance, with silver nitrate they showed a better capacity to resist leaf abscission than those cultured without silver nitrate. Four days were enough for the latter explants to drop 100% of their leaves and after 12 days they had died whilst those with silver nitrate had lost only 50% of their leaves (Table 1).

Table 1. Effect of ethylene (265 μl l^{-1}) and silver nitrate on leaf abscission of *Annona squamosa* explants after 4, 8 and 12 days in culture. Culture tubes closed with 'suba seal' caps. 15 replicates per treatment

Silver nitrate (mg l$^-$)	% abscission		
	4	8 days	12
0.0	100	100	100
5.0	10	30	50

This study demonstrates that silver ions (a known inhibitor of ethylene action) can be used as an effective inhibitor of leaf abscission in the establishment of *Annona squamosa* in culture. The use of the appropriate combination of silver nitrate and type of closure of the vessel is required to allow further progress in micropropagating *Annona squamosa*.

Acknowledgements

EEP Lemos is grateful to the Brazilian 'Conselho Nacional de Desenvolvimento Científico e Tecnológico-CNPq' which supported this work.

References

1. Abeles FB (1973) Ethylene in Plant Biology. New York, Academic Press
2. Adams DO and Yang SF (1979) Ethylene biosynthesis: identification of 1-aminocyclopropane-1-carboxylic acid as an intermediate in the conversion of methionine to ethylene. Proc Nat Acad Sci USA 76:170–174
3. Addicott FT and Wiatr SM (1976) Hormonal controls of abscission: biochemical and ultrastructural aspects. In: PE Pilet, ed. Plant Growth Substances 1976, pp 248–257. Berlin: Springer-Verlag
4. Beyer EM (1976) A potent inhibitor of ethylene action in plants. Plant Physiol 58:268–271
5. Burg SP (1968) Ethylene, plant senescence and abscission. Plant Physiol 43:1503–1511
6. Chi G, Pua E and Goh C (1991) Role of ethylene on *de novo* shoot regeneration from cotyledonary explants of *Brassica campestris* ssp. *pekinensis* (Lour) Olsson *in vitro*. Plant Physiol 96:178–183
7. Jackson MB (1985) Ethylene and responses of plants to soil watterlogging and to submergence. Ann Rev Plant Physiol 36:145–174
8. Jackson MB, Abbott AJ, Belcher AR, Hall KC, Butler R and Cameron J (1991) Ventilation in plant tissue cultures and effects of poor aeration on ethylene and carbon dioxide accumulation, oxygen depletion and explant development. Ann Bot 67:229–237
9. Khalid M, Chraibi B, Latche A, Roustan JP and Fallot J (1991) Stimulation of shoot regeneration from cotyledons of *Helianthus annuus* by the ethylene inhibitors, silver and cobalt. Plant Cell Rep 10:204–207
10. Lau OL and Yang SF (1976) Stimulation of ethylene production in the mung bean hypocotyls by cupric ion, calcium ion and kinetin. Plant Physiol 57:88–92
11. Leonhardt W and Kandeler R (1987) Ethylene accumulation in culture vessels — a reason for vitrification? Acta Hort 212:223–229
12. Murashige T and Skoog F (1962) A revised medium for rapid growth and bioassays with tobacco tissues culture. Physiol Plant 15:473–479
13. Roustan JP, Latche A and Fallot J (1990) Control of carrot somatic embryogenesis by $AgNO_3$, an inhibitor of ethylene action: effect on arginine decarboxylase activity. Plant Sci 67:89–95
14. Saltveit ME and Yang SF (1987) Ethylene. In: L Rivier and A Crozier, eds. Principles and Practice of Plant Hormone Analysis 1987, pp 367–401. London: Academic Press
15. Sisler EC and Yang SF (1984) Ethylene the gaseous plant hormone. Bioscience 34:234–238
16. Yu YB, Adams DO and Yang SF (1979) Regulation of auxin-induced ethylene biosynthesis in mung bean hypocotyls. Role of 1-aminocyclo-propane-1-carboxylic acid. Plant Physiol 63:589–590

Surfactant stimulation of growth in cultured plant cells, tissues and organs

K.C. LOWE, M.R. DAVEY, L. LAOUAR, A. KHATUN,
R.C.S. RIBEIRO, J.B. POWER and B.J. MULLIGAN
Department of Life Science, University of Nottingham, University Park, Nottingham NG7 2RD, UK

1. Introduction

The culture of plant cells, tissues and organs under defined laboratory conditions has both pure and applied applications. A unique feature of plant cells is that they exhibit totipotency. This enables intact, fertile plants to be regenerated from tissues which are given the appropriate stimuli from exogenously supplied growth regulators, such as auxins and cytokinins, in the culture medium. The ability to regenerate plants is fundamental to the multiplication of elite individuals by micropropagation. Additionally, the genetic engineering of plants, through exposure of somaclonal variation, somatic hybridization by protoplast fusion and transformation involving *Agrobacterium*-mediated gene delivery or direct DNA uptake, also neccessitates reproducible plant regeneration.

A further feature of plant cells, tissues and organs is their ability to produce important secondary products which have applications in the agrochemical, food and pharmaceutical industries. Consequently, considerable attention is being focussed on the large scale culture of plant material and the improvement of media components and physical conditions to maximize secondary product synthesis. Thus, any such modifications which promote increased tissue growth, plant regeneration and new product synthesis, are of considerable interest and potential economic importance.

This paper discusses a novel approach to the improvement of growth of cultured plant protoplasts, cells, tissues and organs; specifically, the supplementation of culture media with non-ionic, co-polymer surfactants. The use of such compounds as cell-protecting agents in animal cell cultures has been known for over 30 years. However, in the case of plant cells, this area has been relatively poorly studied. This is surprising, given the biotechnological importance of plant cultures. It is only relatively recently that experiments have been carried out to redress the balance. The experimental work described focusses on studies using protoplasts, cells, tissues and organs from three species which have been selected because of their importance, either economically, or as models for studying the molecular basis of plant development. The results

234

P.J. Lumsden, J.R. Nicholas and W.J. Davies (eds.), Physiology, Growth and Development of Plants in Culture, 234–244, 1994.
© 1994 *Kluwer Academic Publishers.*

indicate considerable beneficial effects of using co-polymer surfactants in plant culture systems and this approach could be scaled-up with relative ease, to industrial production standards.

2. Co-polymer surfactants

2.1 General properties

Block co-polymers of ethylene oxide and propylene oxide were introduced commercially in the 1950s and are known generally as poloxamers. They are a group of non-ionic surfactants, consisting of a central hydrophobic poly-(oxypropylene) block sandwiched between two hydrophilic poly(oxyethylene) blocks. One such compound, Poloxamer 188, which is marketed by the BASF Corporation (Wyandotte, U.S.A.) as Pluronic F-68, contains 20% poly-(oxypropylene) and has an average molecular weight of approximately 8350 (Table 1). Other Pluronics, containing up to 90% poly(oxypropylene), are available and have a wide range of surfactant functions and physical properties [1]. They are used commercially in various applications as, for example, de-foaming agents, foam enhancers, emulsifiers and detergents [31].

Table 1. Some physico-chemical properties of commercial grade Pluronic co-polymer surfactants*

Compound	Average molecular weight	Melting point (°C)	[poly(oxypropylene)] content (% by weight)	HLB range**	Physical form at 20 °C
Pluronic L-35	1900	7	50	18–23	Liquid
Pluronic F-38	4700	48	20	18–23	Solid
Pluronic F-68	8350	50	20	>24	Solid

* Data from BASF Corporation, Wyandotte, U.S.A. [reference 1].

** The HLB range refers to the 'hydrophilic-lipophilic balance' and is an empirical measure of the emulsifying power of the surfactant [reference 27]. In general, compounds with low HLB values are hydrophobic, while those with higher HLB values are water soluble.

2.2 Biomedical applications

The surface-active properties of the Pluronics have been widely exploited in medicine; for example, Pluronic F-68 has been used in low concentrations to reduce blood viscosity and platelet adhesiveness during cardiopulmonary bypass [8]. More recently, the surfactant properties of Pluronic F-68 have been effectively employed in the emulsification of inert, gas-dissolving perfluorochemicals to produce so-called 'blood substitutes' [20, 21]. An important finding from related work was that Pluronic F-68 reduced red blood cell binding to vascular surfaces [33], suggesting additional clinical applications in the management of blood diseases, such as sickle cell anaemia or malaria.

Such indications that Pluronic F-68 may be rheologically active have led to the development by the Burroughs-Wellcome Corporation of RheothRx®, an injectable 15% (w/v) Pluronic F-68 formulation for human use. Preliminary studies have shown that RheothRx® can reduce blood viscosity, inhibit platelet aggregation, and improve the filterability of sickled human erythrocytes [4].

2.3 Applications in cell cultures

The earliest observations that Pluronic compounds could be used to protect animal cells against fluid-mechanical damage in agitated and aerated bio-reactors were made in the 1960s [11, 30, 34]. Pluronic F-68 is now commonly used as a medium additive in cultures of fragile animal cells grown in static, agitated and/or aerated animal cell cultures. There has been, however, much speculation about the mechanism(s) by which the Pluronics can protect cells and promote growth. In static cultures of chick embryonic fibroblasts or hamster melanoma cells, Pluronic F-68 had a concentration-dependent stimulatory effect on cell growth [3]. Furthermore, 2-deoxyglucose uptake and amino acid incorporation into chick fibroblasts was enhanced in the presence of Pluronic F-68 [5]. This result was consistent with the earlier studies of Mizrahi [23, 24], which suggested that the Pluronics could promote increased nutrient uptake into cultured human lymphocytes and lymphoblastoid cells. One explanation for these effects is that Pluronic F-68 interacts directly with the cytoplasmic membrane, producing alterations in permeability. This is supported by related studies using yeast in which Pluronic F-68 altered uptake of fluorescein diacetate [12, 14], and markedly increased the sensitivity to sub-lethal doses of antibiotics [19]. Furthermore, preliminary patch-clamp studies have shown that Pluronic F-68 can produce transient alterations in the permeability of artificial lipid bilayers [14]. While caution is needed in extrapolating from artificial bilayer systems to the intact yeast membrane, or from microbial to animal cell membranes, current evidence suggests that Pluronic F-68 may induce the formation of transmembrane pores, probably by affecting both lipid-lipid and lipid-protein interactions [14].

In studies using cultured insect cells, Murhammer and Goochee also proposed [26, 27] that the Pluronics interacted with the cytoplasmic membrane and conferred increased resistance to shear forces. The protective effect of the Pluronics correlated well with the hydrophilic-lipophilic balance (HLB; Table 1) of the individual compounds; polyols with the highest HLB balance were the best protectants [27]. Related work by Goldblum et al. [7] reported that Pluronic F-68 adsorbed onto the cytoplasmic membranes of cultured insect cells and thus protected them from damage. By comparison with mammalian cells, it would appear, however, that any such effects of Pluronic F-68 in strengthening cytoplasmic membranes only occurs when cells have been exposed to this compound for prolonged periods (c. 16 days) in culture [36].

In addition to any direct effects on the cells themselves, studies using agitated/aerated cultures have suggested that the Pluronics may offer protection against fluid-mechanical damage. It is generally believed that damage to suspended cells in such cultures is caused primarily by interactions with bubbles. Handa-Corrigan and colleagues proposed [9] that the Pluronics act to stabilize the foams which form on the surface of bubble-column bioreactors, thereby protecting cells from bubble disengagement events. A similar physical protective effect was proposed in the earlier paper of Kilburn and Webb [11].

It is thus reasonable to conclude that the cell-protecting and growth-stimulating effects of the Pluronics in animal cell cultures probably involves a combination of cellular *and* physical effects, but their relative contributions will depend upon the culture conditions (i.e. static versus agitated/aerated) and the cell type under study. A more detailed discussion of this topic can be found in the recent review by Papoutsakis [28].

3. Co-polymer surfactants in plant cell, tissue and organ cultures

3.1 Solanum dulcamara

S. dulcamara L. ('bittersweet', 'woody nightshade') is a medicinal plant which is a major source of economically-important steroidal alkaloids, such as solasodine [6] and, being a member of the *Solanaceae*, responds well in culture. Leaf callus-derived cells of *S. dulcamara* cultured in suspension grow readily in the presence of up to 1.0% (w/v) Pluronic F-68, with no apparent morphological changes [13] (Table 2). This initial work showed that *Solanum* cells were tolerant

Table 2. The effects of Pluronic F-68 on cell, tissue and organ cultures of different plant species

Species	Culture system	Conc. range (w/v)	Observations	References
A. thaliana	Root explant-derived cell suspensions	0.001–0.1	Enhanced growth	29
	Root explant-derived cell suspensions	0.001	Enhanced shoot production	29
C. capsularis	Excised seedling cotyledons with petioles	0.001–0.5	Enhanced shoot production	10
	A. rhizogenes-transformed roots	0.001–1.0	Enhanced growth and callus production; inhibition of necrosis	10
S. dulcamara	Cell suspension protoplasts	0.1	Enhanced plating efficiency	17, 18
	Leaf callus-derived cell suspensions	0.01–1.0	Normal growth	13
	Leaf callus	0.1	Enhanced growth	18
	A. rhizogenes-transformed roots	0.001–0.05	Enhanced growth	16, 18

of Pluronic F-68 to at least one order of magnitude greater than either cultured vertebrate or insect cells [3, 26, 27]. Protoplasts isolated from *Solanum* cell suspensions also grew readily in the presence of similar concentrations of Pluronic F-68 (AT King, MR Davey, BJ Mulligan, KC Lowe, unpublished).

The growth of *Solanum* roots transformed by *Agrobacterium rhizogenes* was stimulated by the addition of 0.001–0.05% (w/v) of freshly-prepared, commercial grade Pluronic F-68 to culture medium lacking growth regulators, with maximum increases in fresh and dry weights at 0.01% [16, 18] (Table 2). In contrast, higher concentrations (0.5–1.0% w/v) of freshly-prepared Pluronic inhibited root growth. A Pluronic F-68 fraction, prepared by passage through silica-Amberlite resin [2], similarly retarded root growth even at concentrations that were stimulatory with the commercial preparation. Furthermore, commercial grade Pluronic F-68 solutions that had been stored at 4°C or 22°C for 5 days ('aged') also inhibited root growth [18]. One explanation for these results is that peroxides, which can be formed in stored Pluronic F-68 solutions [22], had inhibitory effects on the growth of transformed roots.

Recently, a system for the culture of plant protoplasts on porous polypropylene membrane rafts has been described [15]. This technology has been exploited commercially by the production of partially hydrophilic membranes coated with an undisclosed surfactant, this culture system being claimed to be superior to agar-solidified media in terms of callus growth and both root and shoot proliferation [32]. Additionally, this arrangement facilitates the recovery of secondary metabolites following their release into the culture medium. We therefore compared the growth of *Solanum* transformed roots on commercial rafts, on rafts which had been washed with ethanol, and on ethanol-washed rafts soaked in Pluronic F-68 solution. Roots grew faster on Pluronic F-68-treated rafts compared with growth on the commercial rafts; such growth enhancement was comparable to that seen in medium supplemented with 0.01% (w/v) of freshly-prepared, commercial grade Pluronic F-68 [18] (Table 2).

Growth of *Solanum* callus was also stimulated by the addition of freshly-prepared, commercial grade Pluronic F-68 to agar-solidified culture medium, with maximum increases at 0.1% (w/v)(Table 2). In contrast, 1.0% (w/v) Pluronic F-68 was inhibitory to callus growth. Both root and callus soluble carbohydrates and proteins were increased by exposure to freshly-prepared, commercial grade Pluronic F-68. Similarly, the specific activities of malate dehydrogenase (MDH) and acid phosphatase (APase) were increased in Pluronic F-68-treated *Solanum* callus and transformed roots [18].

The mean plating efficiency (15 days after plating) of *Solanum* protoplasts cultured at densities of 0.1–2.0 x 10^5 cm^{-3} on agarose-solidified medium was increased up to 26% by 0.1% (w/v) Pluronic F-68, while 1.0% was inhibitory [17, 18] (Table 2). Further studies are in progress to assess the effects of Pluronics on the regeneration of intact, fertile plants from protoplast-derived tissues. A further objective of future work is to determine whether culture of *Solanum* cells and transformed roots in the presence of Pluronics will enhance the synthesis of secondary products.

3.2 Corchorus capsularis

Jute *(Corchorus* spp.) is a fibre-yielding plant which is of economic importance in many Third World countries. Cultures of *Corchorus* cells, tissues and organs provide the opportunity for the genetic engineering of this crop as an adjunct to conventional breeding. Thus, any improvements in the growth of such cultures could be of considerable commercial benefit. Therefore, the effects have been studied of Pluronic F-68 on the growth and organogenesis of cotyledons excised from seedlings, cotyledonary explants and *Agrobacterium*-transformed roots from two varieties of *C. capsularis* L. In addition, cotyledonary explants have been incubated with Pluronic F-38.

Significantly, supplementation of agar-solidified culture medium with 0.001– 0.5% (w/v) of commercial grade or purified Pluronic F-68 produced a marked stimulation of shoot production from the petioles of excised cotyledons of *C. capsularis* vars. D154 and C134 [10] (Table 2). In the case of D154 cotyledons grown in the presence of 0.5% (w/v) purified Pluronic F-68, 64.2 ± 0.4 shoots per explant (n = 3) were produced, whereas the corresponding figure with commercial grade Pluronic at the same concentration was 39.4 ± 5.4 shoots per explant (P <0.05) [10].

Shoot production from *C. capsularis* var. C134 cotyledons was also stimulated by culture with increasing concentrations of Pluronic F-68 [10] (Table 2). This effect was even more marked than in the case of D154 cotyledons, because of the failure of C134 cotyledons to produce shoots in the absence of Pluronic F-68. Again, the purified Pluronic fraction had the greatest effect [10]. Supplementation of culture medium with Pluronic F-68 concentrations above 0.5% (w/v) did not promote any further increase in the number of shoots per cotyledon in both varieties of *C. capsularis* [10]. Preliminary experiments have also shown that shoot formation from *C. capsularis* var. D154 cotyledons with attached petioles was also stimulated approximately 2-fold by supplementation of culture medium with 0.001% (w/v) Pluronic F-38 (A Khatun, MR Davey, JB Power, KC Lowe, unpublished). In addition, 0.001–0.1% (w/v) commercial grade Pluronic F-38 stimulated callus production from cotyledon explants of *C. capsularis* var. D154. In these experiments, excised cotyledons were cut longitudinally and transversely to give 4 individual explants [10]. This effect of Pluronic F-38 contrasted with earlier findings using Pluronic F-68 at the same concentrations, where callus growth from cotyledon explants did not differ significantly from control [10]. However, the extent of such callus growth depended upon the nature of auxins and cytokinins present in the culture medium.

There were no apparent morphological differences between plants grown from D154 cotyledons on untreated medium and those derived from the petioles of cotyledons exposed to either commercial grade or purified Pluronic F-68. Similarly, plants of C134 derived from cotyledons cultured in the presence of Pluronic were phenotypically normal.

As in the case of *S. dulcamara*, exposure of *C. capsularis* var. D154 roots transformed by *A. rhizogenes* to commercial grade or purified Pluronic F-68 also resulted in a stimulation of growth, with a maximum effect at 0.1% (w/v) in both cases [10] (Table 2). For example, the mean fresh weight of 60.3±2.8 g per flask (n = 3) with 0.1% (w/v) of commercial grade Pluronic recorded after 56 days of growth in MS-based liquid medium without growth regulators, was significantly greater (P <0.05) than the control (29.8±5.4 g).

An interesting finding from these experiments was that transformed roots of *C. capsularis* cultured in the presence of commercial grade or purified Pluronic F-68 synthesized chlorophyll and could be maintained without sub-culture for up to 70 days. In contrast, control cultures without Pluronic required sub-culture every 7 days, after which they underwent rapid necrosis [10]. A further observation was that *C. capsularis* var. D154 transformed roots cultured in the absence of Pluronic F-68 did not produce callus over a 56-day study period. However, addition of 0.001–1.0% (w/v) of either commercial grade Pluronic or its purified fraction to the culture medium resulted in callus growth, with a maximal effect at 0.001% (w/v) in both cases. Roots in their second passage in the presence of Pluronic F-68 (transferred to fresh medium with the same concentration of Pluronic at day 21), which were close to the surface of the medium, produced callus which was white, smooth and nodular in appearance. The production of nodular callus from transformed roots may provide a pathway for the regeneration, via somatic embryogenesis, of transgenic Jute plants.

3.3 Arabidopsis thaliana

A. thaliana L. ('thale cress') is a small cruciferous weed currently being exploited extensively in plant molecular biology [35]. The relatively simple organization of the *Arabidopsis* nuclear genome makes this species highly attractive for gene cloning studies that would be more difficult in many other plants. Thus, while *Arabidopsis* is of no economic value, it is nevertheless an ideal model for studies of the molecular basis of plant development. We have therefore examined the effects of Pluronic F-68 on the growth of *A. thaliana* cells and tissues cultured in both liquid and agar-solidified media, and on their subsequent regeneration into plants.

Growth of root explant-derived suspension cultures in Murashige and Skoog (MS)-based liquid medium [25] containing 2,4-dichlorophenoxyacetic acid, as measured by dry cell weight after 11 days, was reproducibly stimulated by the addition of 0.001–0.1% (w/v) of either commercial grade or purified Pluronic F-68 to culture medium [29] (Table 2). Even at the lowest concentrations of commercial and purified Pluronic tested, the mean dry cell weights were 0.27±0.01g and 0.26±0.02g respectively (n = 3) compared to 0.19±0.03g for untreated controls (P < 0.05).

Shoot formation from *Arabidopsis* cell aggregates after 4 weeks of growth on agar-solidified MS-based medium containing thidiazuron (TDZ) or 2, iso-pentenyl adenine (2,iP) as growth regulators, was stimulated by both commercial grade and purified Pluronic F-68 at 0.001% (w/v) [29]. In contrast, shoot regeneration in the presence of 0.01% Pluronic was similar to control and was inhibited by 0.1% Pluronic. Shoots induced on MS-based medium supplemented with Pluronic F-68 and either TDZ or 2,iP produced morphologically normal flowers.

The extent to which culture of *Arabidopsis* cells and tissues with Pluronics can influence the frequency of *Agrobacterium*-mediated transformation is currently under further study. Additionally, future work will determine whether the observed growth-stimulating effects of Pluronics involve alteration in gene expression. *Arabidopsis* is an excellent model for such work.

4. Possible mechanism of action of pluronics in plant culture systems

We have postulated that Pluronic F-68 induces the formation of short-lived, transmembrane 'pores' to explain the increased uptake of fluorescein diacetate into yeast in the presence of this surfactant [14]. Such membrane pores would also account for the increased uptake of 2-deoxyglucose and amino acids in Pluronic F-68-treated vertebrate cells [5]. These observations are consistent with the suggestions of Mizrahi [23] who proposed that, by lowering the surface tension in cultures of human cells, the Pluronics facilitated uptake of nutrients. It is therefore possible that the Pluronics similarly stimulated nutrient uptake into cultured plant cells, tissues and organs, thereby enhancing growth. Indeed, the present observations, using plant material cultured on semi-solid media supplemented with Pluronics, suggest that they have a direct rather than an indirect effect on cell, tissue and organ growth. While we are naturally cautious about extrapolating from studies with fungal and animal cells, it is nevertheless likely that similar membrane-associated surfactant effects occur in plant cultures. We are aware, however, that in suspension cultures of plant cells, the opportunity exists for Pluronics to protect cells indirectly, by stabilizing surface foams and reducing bubble-induced damage.

There are other explanations for the growth-enhancing effects of Pluronic F-68 in plant culture systems. The observed increases in MDH and APase activities in *S. dulcamara* transformed roots and callus cultured in the presence of Pluronic F-68 is evidence for an underlying biochemical basis for the response to this compound. Because MDH is involved in the oxidation of carbohydrates via the tricarboxylic acid cycle, increases in its activity promoted by Pluronic could enhance substrate utilization and hence, stimulate growth. Moreover, APase increases cellular phosphate utilization and therefore, stimulation by Pluronic would also contribute to increased metabolism and growth.

There are clear variations in the responsiveness of different plant cells, tissues and organs to increasing concentrations of Pluronic F-68 and its purified fraction. For example, culture of *C. capsularis* cotyledons with attached petioles in the presence of either commercial grade or purified Pluronic F-68 up to 0.5%

(w/v) increased shoot production, with no further stimulation at higher concentrations [10]. In contrast, maximum growth of *S. dulcamara* transformed roots occurred with 0.01% (w/v) commercial grade Pluronic F-68 [18], and that of leaf-derived callus with 0.1% [17]. Furthermore, cell suspensions of *S. dulcamara* grew normally in the presence of Pluronic F-68 up to 1.0% (w/v), although growth is inhibited at higher concentrations [13]. Similarly, the growth of root explant-derived cell suspensions of *A. thaliana* was also stimulated by the addition of 0.001–0.1% (w/v) of either commercial grade or purified Pluronic F-68 to culture medium [29]. Such variation in the responsiveness to Pluronic F-68 is broadly consistent with studies using cultured animal cells. In this regard, Pluronic F-68 markedly improves the shear robustness of insect cells, but has no comparable effect on mammalian hybridoma cells [7].

5. Conclusions

The work summarized in this paper demonstrates clear beneficial effects of using Pluronics in plant culture systems. One advantage of adding these compound to such cultures is that they are relatively cheap and their growth-stimulating effects occur at low concentrations. However, the composition of commercial grade preparations of Pluronic F-68 and other co-polymers can vary from batch to batch, depending on the supplier. This should be borne in mind when using these compounds as media supplements. To date, experimental work has focussed primarily on the growth-stimulating effects of one poloxamer, Pluronic F-68, in plant culture systems. Future studies should examine the responses to other Pluronics and determine the underlying mechanism(s) of any growth-stimulating effects, together with the optimum conditions for their use. While we are optimistic that co-polymer surfactants will become used routinely in plant cultures, further work should also determine whether Pluronics can positively affect plant cell transformation and gene expression, particularly those genes involved in morphogenesis.

Acknowledgements

The original work discussed in this paper was supported by grants from the Allied Lyons Research Fund, University of Nottingham. A.K. was supported by funds from the British Technical Assistance administered through the British Council, while on leave from the Bangladesh Jute Research Institute, Dhaka, Bangladesh; R.C.S.R. was also supported by the British Council. L.L. was supported by a scholarship from the Algerian Government. We are grateful to the BASF Corporation, Wyandotte, U.S.A., for their generous gifts of Pluronic F-68 and Pluronic F-38.

References

1. BASF Corporation (1989) Pluronic & Tetronic Surfactants. Wyandotte: BASF Speciality Products
2. Bentley PK, Davis SS, Johnson OL, Lowe KC and Washington C (1989) Purification of Pluronic F-68 for perfluorochemical emulsification. J Pharm Pharmac 41:661–663
3. Bentley PK, Gates RMC, Lowe KC, de Pomerai DI and Walker JAL (1989) *In vitro* cellular responses to a non-ionic surfactant, Pluronic F-68. Biotechnol Lett 11:111–114
4. Carter C, Fisher TC, Hamai H, Johnson CS, Meiselman HJ, Nash GB and Stuart J (1992) Haemorheological effects of a nonionic copolymer surfactant (Poloxamer 188). Clin Hemorheol 12:109–120
5. Cawrse N, de Pomerai DI and Lowe KC (1991) Effects of Pluronic F-68 on 2-deoxyglucose uptake and amino acid incorporation into chick embryonic fibroblasts *in vitro*. Biomed Sci 2:180–182
6. Fujita Y and Tabata M (1987) Secondary metabolites from plant cells — pharmaceutical applications and progress in commercial production. In: CE Green et al., eds. Plant Tissue and Cell Culture, pp 169–185. New York: Liss
7. Goldblum S, Bae Y-K, Hink WF and Chalmers J (1990) Protective effect of methylcellulose and other polymers on insect cells subjected to laminar shear stress. Biotechnol Prog 6:383–390
8. Grover FL, Heron WH, Newman MM and Paton BC (1969) Effects of a non-ionic surface active agent on blood viscosity and platelet adhesiveness. Circulation 39 (Suppl):I–249
9. Handa-Corrigan A, Emery AN and Spier RE (1989) Effect of gas-liquid interfaces on the growth of suspended mammalian cells:mechanisms of cell damage by bubbles. Enzyme Microb Technol 1:230–235
10. Khatun A, Laouar L, Davey MR, Power JB, Mulligan BJ and Lowe KC (1993) Effects of Pluronic F-68 on shoot regeneration from cultured jute cotyledons and on growth of transformed roots. Plant Cell Tiss Org Cult (in press)
11. Kilburn DG and Webb FC (1968) The cultivation of animal cells at controlled dissolved oxygen partial pressure. Biotechnol Bioeng 10:801–814
12. King AT, Lowe KC and Mulligan BJ (1988) Microbial cell responses to a non-ionic surfactant. II. Effects as assessed by fluorescein diacetate uptake. Biotechnol Lett 10:873–878
13. King AT, Davey MR, Mulligan BJ and Lowe KC (1990) Effects of Pluronic F-68 on plant cells in suspension culture. Biotechnol Lett 2:29–32
14. King AT, Davey MR, Mellor IR, Mulligan BJ and Lowe KC (1991) Surfactant effects on yeast cells. Enzyme Microb Technol 13:48–153
15. Kong Y and Chin CK (1988) Culture of asparagus protoplasts on porous polypropylene membrane. Plant Cell Rep 7:67–69
16. Kumar V, Laouar L, Davey MR, Mulligan BJ and Lowe KC (1990) Effects of Pluronic F-68 on growth of transformed roots of *Solanum dulcamara*. Biotechnol Lett 12:937–940
17. Kumar V, Laouar L, Davey MR, Mulligan BJ and Lowe KC (1991) Effects of Pluronic F-68 on callus growth and protoplast plating efficiency of *Solanum dulcamara*. Plant Cell Rep 10:52–54
18. Kumar V, Laouar L, Davey MR, Mulligan BJ and Lowe KC (1992) Pluronic F-68 stimulates growth of *Solanum dulcamara* in culture. J Exp Bot 43:487–493
19. Laouar L, Mulligan BJ and Lowe KC (1992) Surfactant effects on yeast growth and sensitivity to antibiotics. Soc Exp Biol, Lancaster, Abstract A3.22
20. Lowe KC (1991) Synthetic oxygen transport fluids based on perfluorochemicals: applications in medicine and biology. Vox Sang 60:129–140
21. Lowe KC (1992) Perfluorochemical blood substitutes: circulatory and biomedical applications. Clin Hemorheol 12:141–156
22. McCoy LE, Becker CA, Goodin TH and Barnhart MI (1984) Endothelial responses to perfluorochemical emulsion. Scan Electron Micros 6:311–319
23. Mizrahi A (1975) Pluronic polyols in human lymphocyte cell line cultures. J Clin Microbiol 2:11–13

24. Mizrahi A (1984) Oxygen in human lymphoblastoid cell line cultures and effect of polymers in agitated and aerated cultures. Dev Biol Stand 55:93–102

25. Murashige T and Skoog F (1962) A revised medium for rapid growth and bioassays with tobacco tissue cultures. Physiol Plant 5:473–497

26. Murhammer DW and Goochee CF (1988) Scaleup of insect cell cultures: protective effects of Pluronic F-68. Bio/Technol 6:1411–1418

27. Murhammer DW and Goochee CF (1990) Structural features of nonionic polyglycol polymer molecules responsible for the protective effect in sparged animal cell bioreactors. Biotechnol Prog 6:142–148

28. Papoutsakis ET (1991) Media additives for protecting freely suspended animal cells against agitation and aeration damage. Trends in Biotechnol 9:427–437

29. Ribeiro RCS, Laouar L, Kumar V, Davey MR, Power JB, Mulligan BJ and Lowe KC (1992) Effects of Pluronic F-68 on the growth of *Arabidopsis thaliana* in culture. Soc Exp Biol, Lancaster, Abstract A3.20

30. Runyan WS and Geyer RP (1963) Growth of L cell suspension in Warburg apparatus. Proc Soc Exp Biol Med 112:1027–1030

31. Schmolka IR (1977) A review of block co-polymer surfactants. J Am Oil Chem Soc 54:110–116

32. Sigma Plant Cell Culture (1990) Membrane rafts. Phytasource :1

33. Smith CM, Hebbel RP, Tukey DP, Clawson CC, White JG and Vercellotti GM (1987) Pluronic F-68 reduces the endothelial adherence and improves the rheology of liganded sickle erythrocytes. Blood 69:1631–1636

34. Swim HE and Parker RF (1960) Effect of Pluronic F-68 on growth of fibroblasts in suspension on rotary shaker. Proc Soc Exp Biol Med 103:252–254

35. Wilson, ZA, Dawson J, Russell J and Mulligan BJ (1991) *Arabidopsis thaliana*. Biologist 8:163–169

36. Zhang Z, Al-Rubeai M and Thomas CR (1993) The effect of Pluronic F-68 on the mechanical properties of mammalian cells. Enzyme Microb Technol 14:980–983

Micropropagation of *Narcissus*

B.M.R. HARVEY[1,2], C. SELBY[1,2], T.W. FRASER[2] and Y.N. CHOW[1]

[1] *Department of Agricultural Botany, The Queen's University of Belfast, Newforge Lane, Belfast BT9 5PX, Northern Ireland*
[2] *Department of Agriculture for Northern Ireland, Newforge Lane, Belfast BT9 5PX, Northern Ireland*

1. Introduction

Natural vegetative propagation of *Narcissus* is so slow [5, 6] that building up a commercial stock of a new cultivar takes fifteen to twenty-five years [9]. Micropropagation of *Narcissus* is also slow, e.g. Squires and Langton [7] obtained multiplication rates ranging from 1.44 to 2.30 for five cultivars over a series of 4–6 wk culture passages.

A further problem in micropropagation of *Narcissus* is that a large amorphous mass of achlorophyllous tissue develops at the base of shoot clump cultures and new leaves arise from it [1]. If this tissue is callus, true-to-type micropropagation cannot be guaranteed. Alternatively, the tissue could be derived from the basal plate of the original bulb explant. The structure of the basal plate of bulbs was therefore compared to the basal achlorophyllous tissue of shoot clump cultures. As microscopy showed that these tissues were similar and also showed primordia close to the base of each leaf of shoot clump cultures, the regenerative ability of single leaf plus achlorophyllous tissue secondary explants was investigated. Successful regeneration was obtained for all five cultivars of *Narcissus* tested.

2. Materials and methods

Longitudinal tissue slices 2 mm thick were cut from recently sprouted bulbs and from shoot clump cultures of *Narcissus* cultivar Hawera, to obtain median sections of the developing leaves and basal tissue. The tissue slices were fixed, dehydrated, embedded in LR White resin and 1 μm sections were cut for light microscopy. The shoot clump cultures had been initiated three years previously from bulb explants by the method of Squires and Langton [7] and sub-cultured every five weeks on *Narcissus* micropropagation medium [1].

Single leaf secondary explants (30–50 mm lamina) with basal achloro-phyllous tissue (2–3 mm long) were excised from shoot clump cultures by making 'wedge' cuts at the base of a leaf. These secondary explants were

245

P.J. Lumsden, J.R. Nicholas and W.J. Davies (eds.), Physiology, Growth and Development of Plants in Culture, 245–248, 1994.

prepared either intact or with the apical third of the leaf removed and then inoculated either upright or inverted into micropropagation medium. Five cultivars (Carlton, Fortune, Ice Follies, St. Keverne and Yellow Sun) were used with eight replicates of each for each treatment.

3. Results

Median longitudinal sections of the basal plate of bulbs showed complex anatomy with numerous vascular bundles visible in cross section, tangential and longitudinal section (Plate 1a). In the upper part of the basal plate the vascular bundles were surrounded by small densely packed parenchymatous cells (Plate 1a). Longitudinal sections of the amorphous basal tissue of shoot clump cultures also showed complex development of vascular tissue and quite densely packed parenchyma (Plate 1b). The main difference between the tissues was that primordia were clearly visible near the base of leaves in shoot clump cultures (Plate 1c) but not in bulbs.

Table 1. The number of new leaves produced by single leaf cultures of *Narcissus*. Both types of single leaf secondary explant included basal achlorophyllous tissue. Data presented are for the first two culture passages (0–10 weeks) and are the means of cultivars Carlton, Fortune, Ice Follies, St. Keverne and Yellow Sun

| | Explant type | | |
Inoculation treatment	Leaf intact	Leaf cut	Mean
Inverted	0.56	0.23	0.40
Upright	3.10	2.05	2.98
Mean	1.83	1.54	

Significance of effects and interactions: explant type *NS*, inoculation treatment $P<0.001$ (SE 0.313), cultivar *NS*, all interactions *NS*.

When single leaf secondary explants with basal achlorophyllous tissue were cultured on *Narcissus* micropropagation medium, organogenesis always occurred from the morphological base, whether the explants had been inoculated upright or inverted. Inverted explants produced fewer leaves (Table 1) but cutting the explant leaf had no significant effect on organogenesis. There was no significant difference between cultivars in multiplication (leaf production) and all five cultivars showed the same response to the treatments.

4. Discussion

The complex vascularisation of the basal plate of bulbs is consistent with it being a compressed stem axis [3, 8]. The basal achlorophyllous tissue of shoot

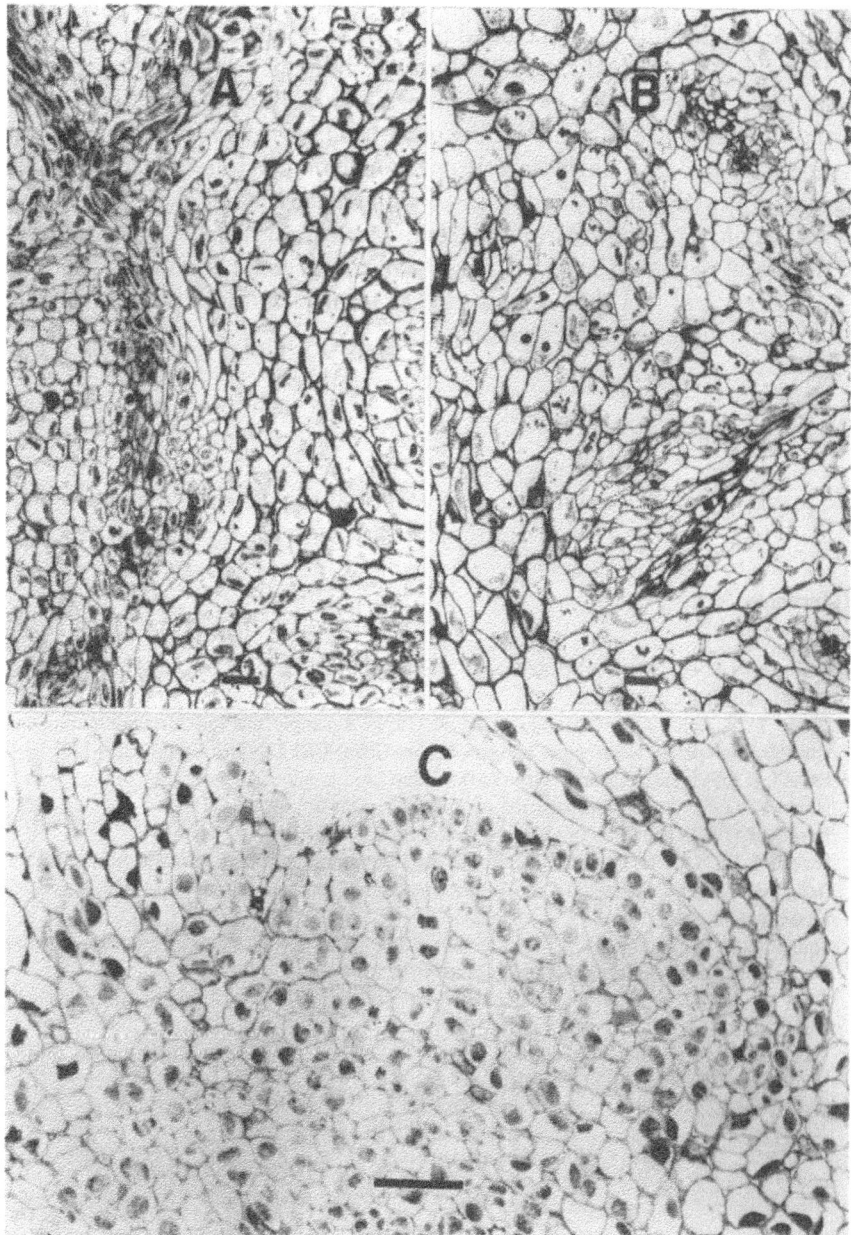

Plate 1. Longitudinal sections of (A) the basal plate of a sprouted bulb; (B) the basal achloro-phyllous tissue of a shoot clump culture and (C) the base of leaves and a primordium in a shoot clump culture of *Narcissus* cultivar Hawera. The scale bar indicates 50 μm.

clump cultures showed similar complexity. Overall, it did not resemble callus but appeared analogous to the basal plate tissue of bulbs.

Formation of primordia occurred by divisions in the surface cell layers of the basal plate tissue in shoot clump cultures. This differed from reports that primordia form on the abaxial surface of the leaf base in *Narcissus* bulb explants [4] and indicated that single leaves from shoot clump cultures could be used as secondary explants if they were excised with adjacent basal plate tissue.

As all cultivars tested showed the ability to regenerate from the basal tissue of single leaf secondary explants, single leaf culture could be widely applicable. Use of every leaf in a shoot clump culture for micropropagation of *Narcissus* could greatly improve multiplication.

Acknowledgements

We thank Mrs N Stevens for excellent technical assistance.

References

1. Chow Y N (1990) Micropropagation of Narcissus. PhD Thesis, The Queen's University of Belfast, Northern Ireland
2. Chow Y N, Selby C and Harvey B M R (1992) A simple method for maintaining high multiplication of *Narcissus* shoot cultures *in vitro*. Plant Cell Tiss Org Cult 30:227–230
3. Esau K (1965) Plant Anatomy. New York: Wiley
4. Hussey G (1982) *In vitro* propagation of *Narcissus*. Ann Bot 49:707–719
5. Rees A R (1972) The Growth of Bulbs. London: Academic Press
6. Seabrook J E A (1990) Narcissus (Daffodil). In: PV Ammirato, DA Evans, WR Sharp and YPS Bajaj, eds. Handbook of Plant Cell Culture Vol 5, Ornamental Species, pp 577–597. New York: McGraw Hill
7. Squires W M and Langton F A (1990) Potential and limitations of *Narcissus* micropropagation: an experimental evaluation. Acta Hortic 266:67–75
8. Stant M (1954) The shoot apex of some Monocotyledons II. Growth organisation. Ann Bot 18:441–447
9. Vreeburg P M J (1986) Chipping of Narcissus: a quicker way to obtain large numbers of small round bulbs. Acta Hort 177:579–584

Effect of explant stem length on potato (*Solanum tuberosum* L.) microtuber formation *in vitro*

F. PAPATHANASIOU, S. WATSON and B.M.R. HARVEY

Faculty of Agriculture and Food Science, Queen's University of Belfast, Newforge Lane, Belfast, BT9 5PX, Northern Ireland, UK

1. Introduction

The first report of induction of tuberisation *in vitro* was published by Barker as early as 1953 [1] but microtuber production from nodal explants of potato remains variable [2, 4, 6] even when care is taken to standardize the growth conditions of stock plantlets, age of stock plantlets, node position and the environment for microtuber production.

Variation in stem length of explants might affect tuberisation *in vitro* as various effects of stem length of cuttings on their subsequent growth have been reported. For potato stem cuttings Hepburn and Matthews [5] reported that a short basal internode was associated with tuber development. Cuttings with long basal internodes rooted prolifically but did not tuberise. Conversely sprouting and final tuber yields of sweet potato (*Ipomoea batatas*) cuttings increased with increasing cutting length [8].

The purpose of the present investigation was the assessment of effects of explant stem length on microtuber production. Microtubers formed either directly from the axillary bud (sessile tubers) or on the shoot which developed by extension of the axillary bud (aerial tubers).

2. Materials and methods

Plantlets of cultivar (cv) Arran Banner were grown *in vitro* as previously described [4] in individual 50 ml glass tubes, with 10 ml medium and one single-node stem segment per tube. The tubes were incubated in a controlled environment cabinet at 20 °C with a 16 h photoperiod (90–120 μmol m^{-2} s^{-1} PAR at bench height). When the plantlets had 5–7 nodes, after about 5 weeks, the first node was discarded and the second and third nodes were excised as secondary explants for experiments on tuberisation.

The three secondary explant stem lengths used were 4, 8, 16 mm above the node with 2 mm below the node (Expt. 1) or 4, 8, 16 mm below the node with 2 mm above the node (Expt. 2). These nodes were weighed aseptically,

P.J. Lumsden, J.R. Nicholas and W.J. Davies (eds.), Physiology, Growth and Development of Plants in Culture, 249–253, 1994.

inoculated in tuberisation medium [4] and incubated in darkness at 20 °C. There were 15 replicates for each treatment in experiment 1, and 40 in experiment 2. Tuberisation was assessed at weekly intervals. Microtuber, shoot and root fresh weights (FW) were determined after 6 weeks.

A total randomization design was used in all the experiments and the results were subjected to analysis of variance. Percentage tuberisation at week 6 was analysed by randomly grouping the replicate tubes into blocks of five; the data was analysed with and without arcsin transformation. Untransformed values are present as transformation did not affect the conclusions. Effects of explant fresh weight on growth parameters were examined by covariate analysis and calculation of correlation coefficients.

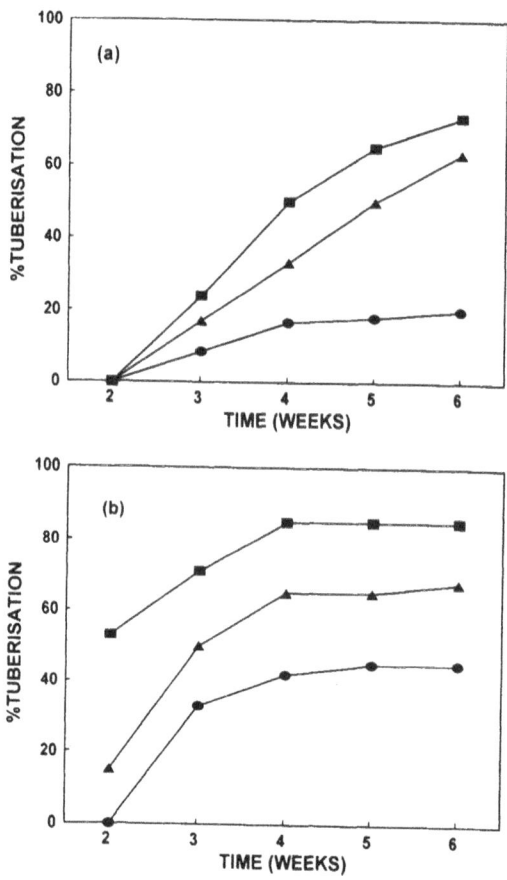

Fig. 1. Expts 1 and 2. The effect of explant stem length above the node (a) and below the node (b) on percentage tuberisation of potato cultivar Arran Banner (● 4 mm; ▲8 mm; ■16 mm).

3. Results

Varying the explant stem length above or below the node significantly affected the percentage of explants which tuberised (P < 0.001) and the tuber FW (P < 0.001) and root FW (P < 0.05) per explant but not the stem FW or stem height. Effects on percentage tuberisation (sessile and aerial tubers) are shown in Figure 1. Increasing explant stem length either above or below the node increased percentage tuberisation and the effect appeared stronger when length varied below the node. There were no significant effects on the proportion of sessile and aerial tubers produced (data not shown).

Effects on tuber FW per culture were expected because treatments affected percentage tuberisation, and were found whether stem length was varied above or below the node. However 'FW per tuber formed' was greatly increased by increasing stem length below the node (Table 1) and not significantly affected by increasing stem length above the node. Root formation and root FW were affected by explant stem length either above or below the node (Table 2). Covariate analysis in both the experiments showed no statistical significant relationship between explant FW and tuber FW.

Table 1. Effect of explant stem length above the node (Expt. 1) or below the node (Expt. 2) on fresh weight per tuber formed (mg) of potato cv Arran Banner

Explant stem length (mm)	Stem length varied	
	Above node	Below node
4	125[a]	90[x]
8	138[a]	149[y]
16	178[a]	222[z]

Superscripts within a column with no letter in common indicate values that differ significantly (P < 0.05)

Table 2. Effect of explant stem length above the node (Expt 1) or below the node (Expt 2) on root fresh weight of potato cv Arran Banner cultured in tuberisation medium

Explant stem length (mm)	Stem length varied	
	Above node	Below node
4	14.8	11.9
8	21.2	29.1
16	41.9	23.3

Standard errors and significances: Stem length above node 5.59 (P=0.004), Stem length below node 3.23 (P=0.033)

4. Discussion

When explant stem length below the node was varied, differences in percentage tuberisation were evident as early as 2 weeks after inoculation. Such early effects could be attributed to differences in content of nutrients and/or growth regulators in the explant or to differences in uptake of nutrients and/or growth regulators (BAP), assuming that explants with a longer stem below the node have a greater surface area for initial uptake from the culture medium and also may root more rapidly.

When explant stem length was varied above the node with only a short stem length below the node, tuberisation occurred more slowly than when stem length was varied below the node. This provides a further indication that initial uptake from the culture medium is important. Significant differences in percentage tuberisation occurred from four weeks onwards so that final effects on percentage tuberisation were similar whether stem length was varied above or below the node. However tuber FW (expressed per tuber formed rather than per explant) at 6 weeks after inoculation was significantly affected by explant stem length below the node but not by explant stem length above the node. As nutrients and/or growth regulator reserves would be similar whether stem length was varied above or below the node, this may indicate that stem reserves have a less important role in tuber development than uptake from the culture medium.

Uptake from the culture medium initially occurs through the explant stem surface and subsequently through the root system. Increasing explant stem length above or below the node increased root growth (FW at harvest) but tuber FW was increased only by increasing explant stem length below the node. This may provide additional evidence that initial uptake from the culture medium (by explants with a long stem below the node) has a very important role in tuber initiation and development.

Although stem length treatments which enhanced root FW did not necessarily result in increased tuber FW, a significant correlation was found between root FW and tuber FW at harvest (r=0.6, P< 0.001). Thus alternative treatments which accelerate and/or stimulate rooting might result in better tuber growth.

Auxin treatments enhance rooting *in vitro* and inclusion of auxin in culture medium has been reported to improve *in vitro* tuberisation of potato explants [3, 7]. As no direct regulatory role of auxin in tuber formation by potato has been demonstrated [9], stimulation of *in vitro* tuberisation by auxin could be explained by effects on rooting and uptake from the culture medium. A study of the interactive effects of explant stem length and auxin treatment is in progress and may lead to a method of improving microtuber production from plantlets with short internodes.

References

1. Barker WG (1953) A method for the *in vitro* culturing of potato tubers. Science 118:384–385
2. Chandra R, Dodds JH and Tovar P (1988) *In vitro* tuberisation in potato (*Solanum tuberosum* L.). International Plant Tissue Culture Association Newsletter:10–20
3. Harmey MA, Crowley MP and Clinch PEM (1966) The effect of growth regulators on tuberisation of cultured stem pieces of *Solanum tuberosum*. Eur Potato J 9:146–151
4. Harvey BMR, Crothers SH, Evans NE and Selby C (1991) The use of growth retardants to improve microtuber formation by potato (*Solanum tuberosum*). Plant Cell Tiss Org Cult 27:59–64
5. Hepburn HA and Matthews S (1986) Influence of proprietary rooting compounds and basal internode length on the rooting of potato stem cuttings. Potato Res 29:391–394
6. Hussey G and Stacey NJ (1984) Factors affecting the formation of *in vitro* tubers of potato (*Solanum tuberosum* L.). Ann Bot 53:565–578
7. Mangat BS, Kerson G and Wallace D (1984) The effect of 2, 4-D on tuberisation and starch content of potato tubers produced on stem segments cultured *in vitro*. Am Potato J 61:355–361
8. Sanchez VEE, Morales TA and Lopez ZYM (1982) The influence of stem-cutting length on tuber yield of sweet potato (*Ipomoea batatas*) clone CEMSA 74-228. Ciencia y Tecnica en la Agricultura, Viandas Tropicales 5:49–68
9. Vreugdenhil D and Struik PC (1989) An integrated view of the hormonal regulation of tuber formation in potato (*Solanum tuberosum*). Physiol Plant 75:525–531

Production of potato microtubers with and without growth regulators

K.M. NASIRUDDIN and JENNET BLAKE

Department of Agriculture, Horticulture & the Environment, Wye College (University of London), Wye, Ashford, Kent TN25 5AH

1. Introduction

Microtubers are of interest as propagules in seed potato production. With the increasing importance of their storability, phytosanitary advantages and ease of transportation, intensive attention is now being given to the production of microtubers. Culture environment and media including the use of growth regulators are the main factors exerting pronounced effects on tuberization. Varying concentrations of cytokinins, auxins and other regulatory substances have been used in culture media [4, 5, 11], while in the method developed at the International Potato Centre (CIP), the antigibberellin, chlorocholine chloride (CCC) together with BAP have been added [10]. A simple method for producing microtubers without growth regulators has been developed at Wye College [3] where all the twelve varieties tested produced microtubers in culture without difficulty [2]. The microtubers produced by the Wye College System have provided satisfactory results on field planting each year since 1984 [1]. The present research was undertaken to investigate the effects of growth regulators and daylength and to compare the Wye College and CIP systems.

2. Materials and methods

Nodal explants derived from VTSC quality potato tubers of cv Desiree were used. Fifteen explants were cultured in each 250 ml honey jar (closed with polypropylene film) containing 50 ml of Murashige and Skoog [7] medium (Flow Laboratories) supplemented with 37.6 mgl^{-1} Na-Fe-EDTA. Sucrose was added at 8% w/v. The pH of the medium was adjusted to 5.8 before addition of 0.6–0.7% w/v agar (Oxoid Technical) and autoclaving (121 °C at 100 kPa for 20 min). The plantlets were grown in a growth room at 19 °C with a 16h photoperiod illuminated with 1.83 m 'colour 84' fluorescent lamps (Philips) giving an average irradiance of 85 μmol m^{-2}s^{-1} at the tops of the vessels. After 4 weeks, half of the cultures (Wye treatment) were transferred to short (8h) days. The other half (CIP treatment) were transferred to dark conditions after the

P.J. Lumsden, J.R. Nicholas and W.J. Davies (eds.), Physiology, Growth and Development of Plants in Culture, 254–260, 1994.
© 1994 *Kluwer Academic Publishers.*

addition of 5 ml each of solutions of BAP (5 mgl^{-1}) and of CCC (500 mgl^{-1}) under aseptic conditions.

A Complete Randomized Design was followed and the treatment means were compared by LSD values. During the induction period growth parameters were measured. At harvest, data were taken on the number and weight of tubers which were also investigated anatomically. To examine the periderm, microtubers were fractured and immediately frozen in liquid nitrogen. The specimens were transferred to a Scanning Electron Microscope fitted with an Oxford Instruments/Hexland Cryotrans CT 1000 system and photographed at a temperature of −160 °C.

3. Results and discussion

There was no significant difference in the height of the primary shoots of microplants in the two treatments (Fig. 1). The number of primary roots per microplant in the Wye treatment was higher throughout the growing period, but the difference was only significant after the initial two weeks (Fig. 1). Although there were fewer secondary shoots in the CIP treatment, there were no significant differences in primary and secondary shoot development (Fig. 2). These results support the previous findings of inhibition of the root system in the presence of BAP [4]. The growth regulators also appeared to prevent the formation of shoots, stolons and microtubers inside the agar.

The CIP treatment produced a significantly higher number of microtubers initially but after the third week, the Wye treatment produced slightly larger

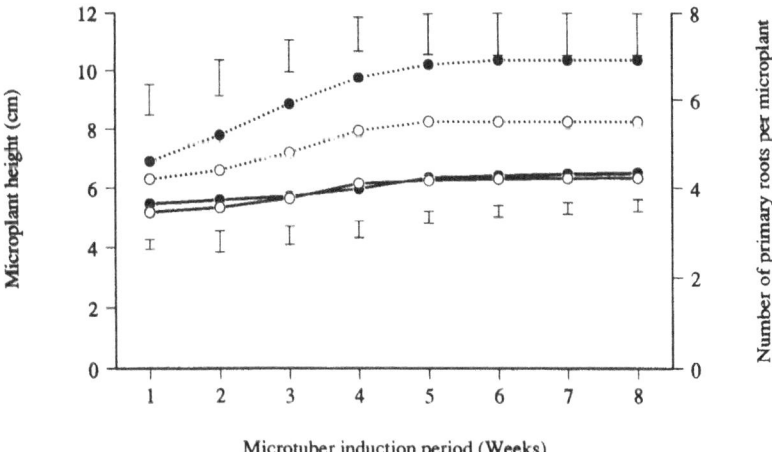

Fig. 1. Effect of growth regulators and photoperiods on height (__) and root number (...) of potato microplants. Vertical bars represent LSD$_{0.05}$ values. 15 explants per jar; 20 jars per treatment. WYE ●-●, CIP ○-○

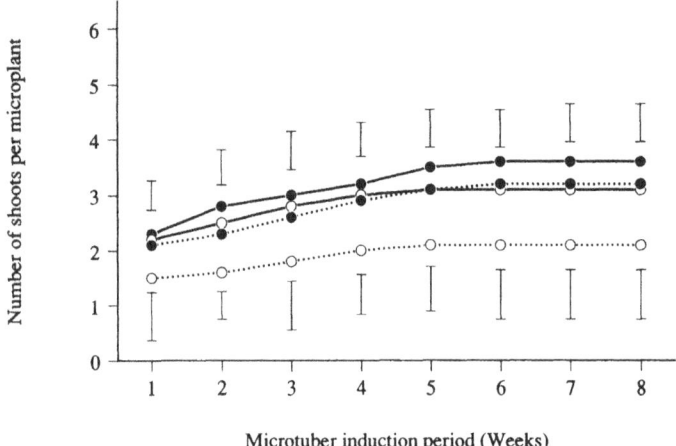

Fig. 2. Number of primary (__) and secondary (...) shoots of potato microplants with and without growth regulators. Vertical bars represent LSD$_{0.05}$ values. 15 explants per jar; 20 jars per treatment. WYE ●-○, CIP ○-○

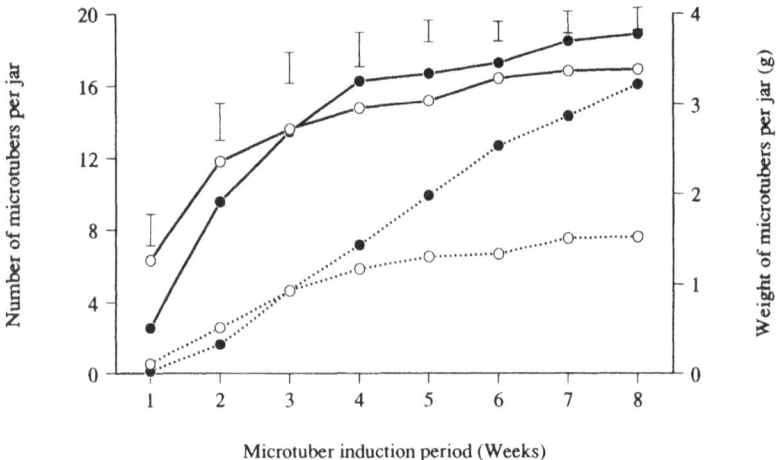

Fig. 3. Effect of growth regulators and daylength on number (__) and weight (...) of potato microtubers. Vertical bars represent LSD$_{0.05}$ values. 15 explants per jar; 20 jars per treatment. WYE ●-●, CIP ○-○

numbers (Fig. 3). The weight of microtubers in both the treatments showed a similar pattern, but the microtubers in the Wye treatment continued to grow and swell exponentially to give a much higher yield (Fig. 3). The present research supports the findings of the enhancement of tuberization with BAP [4, 6, 10], CCC [4, 9, 10] and total darkness [8]. It does not support the report describing BAP as a suppressant of tuberization [11]. The contradictory effect may be due to the subsequent reduction in growth of initiated microtubers and their ultimately smaller size at harvest as affected by growth regulators.

All the microtubers in the CIP treatment were found in basal, middle or terminal positions on the microplants, but some of the Wye microplants produced microtubers inside the agar (Table 1). Wye microplants produced more sessile microtubers but there was no significant difference between the treatments. The Wye microtubers sprouted more frequently than CIP microtubers (Table 1) and almost half of those were inside the agar.

Table 1. Position and types of microtubers produced with growth regulators (CIP treatment) and without growth regulators (Wye treatment). Mean values of 20 replicates

Parameter treatment	Number of microtubers per jar					No of sessile tubers	No of sprouted tubers
	basal	middle	upper	in agar	total		
WYE	3.90	8.25	2.70	4.00	18.40	4.55	2.55
CIP	4.45	9.65	2.85	0	16.80	3.80	0.15
LSD(0.05)	0.76	1.55	1.19	—	1.21	0.86	0.83

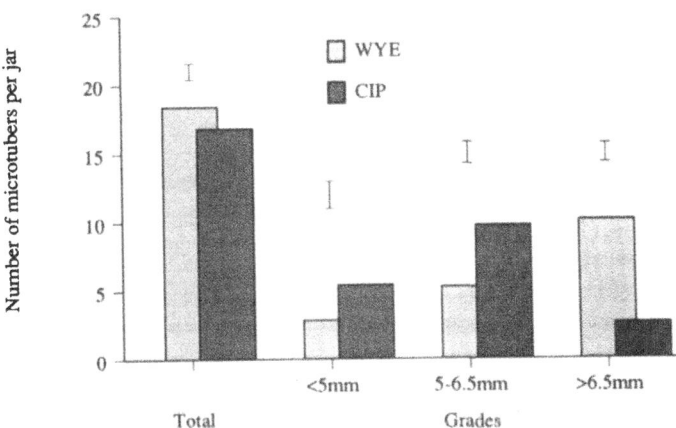

Fig. 4. Number of potato microtubers from WYE and CIP system in grades. Vertical bars represent $LSD_{0.05}$ values. 15 explants per jar; 20 jars per treatment.

Smaller and medium grade microtubers were more frequent in the CIP treatment but there were more large microtubers (> 6.5 mm) in the Wye treatment (Fig. 4). A similar pattern of variation was found in the fresh weight of microtubers (Fig. 5). The large increase in fresh weight in the Wye treatment may be attributed to their induction in short day (8h) conditions without addition of growth regulators which allowed accumulation of a higher content of dry matter than in the microtubers grown under the dark conditions of the CIP treatment. The microtubers grown with BAP and CCC in the dark (CIP) were almost spherical (Plate 1) and not of the characteristic shape for the

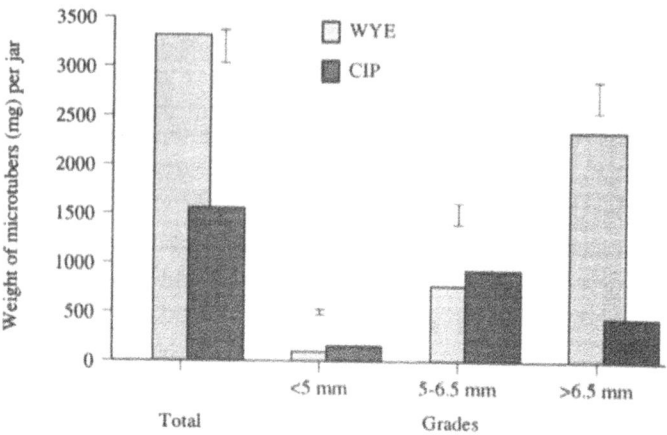

Fig. 5. Weight of microtubers from WYE and CIP system in grades. Vertical bars represent $LSD_{0.05}$ values. 15 explants per jar; 20 jars per treatment.

cultivar Desiree, whilst the microtubers produced in the Wye treatment had the characteristic elongated shape of tubers of this cultivar. The Wye microtubers were greenish in colour from the presence of chlorophyll produced in the light: in the absence of light the CIP microtubers remained white. It is not clear why the CIP microtubers should be spherical in shape but it is possible that the growth regulators inhibited elongation of the tubers, which is reflected in the lower number of eyes on these tubers (Table 2). It is possible to suggest that growth regulators might have suppressed bud formation as they did in the above-ground shoots.

Table 2. Characteristics of microtubers grown with growth regulators (CIP treatment) and without growth regulators (Wye treatment). Mean values of 20 replicates

Parameter Treatments	Lenticel		Cell no mm^{-2}	periderm thickness mm	periderm layers	No eyes/ tuber	Dry matter (%)
	number	size mm^2					
WYE	35.00	0.207	81.99	0.338	16.10	5.00	13.70
CIP	35.67	0.298	93.20	0.291	13.90	2.67	11.21
LSD(0.05)	6.95	0.039	10.39	0.018	1.37	2.33	1.98

Anatomical studies revealed that the number of lenticels did not vary significantly between the treatments but the size of lenticels in the CIP treatment microtubers was significantly greater than those in the Wye treatment (Table 2). In a cross-section of the microtubers, the CIP treatment had more cells per unit area indicating the smaller individual cell size than in the Wye microtubers. Microtubers in the Wye treatment had a thicker multi-layered periderm than those in the CIP treatment (Table 2). The periderm cells of the CIP treatment

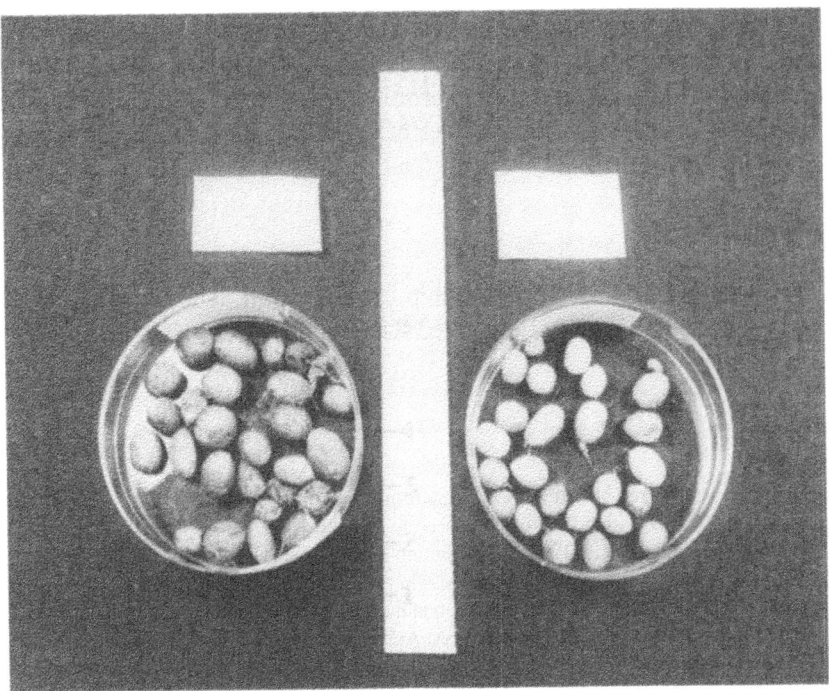

Plate 1. Harvested microtubers of Wye (left) and CIP (right) treatments.

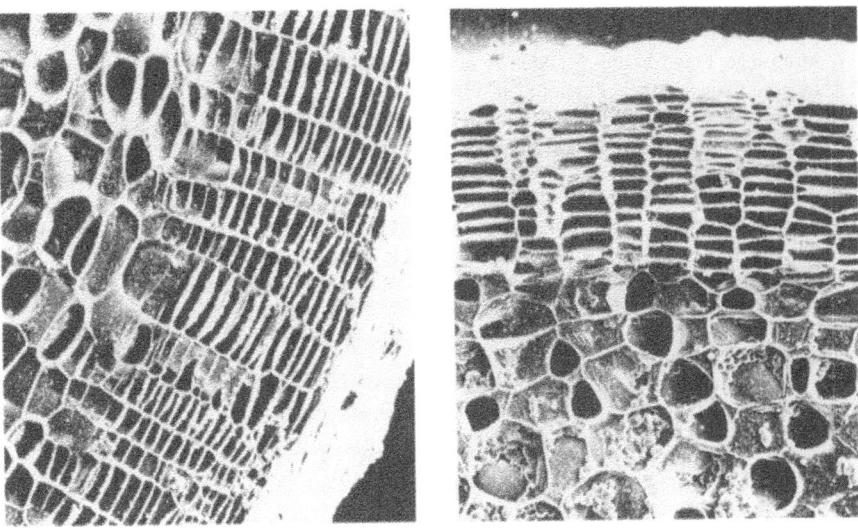

Plate 2. SEM micrographs of cross section of periderm of Wye (left) and CIP (right) microtubers.

were less well organized and aligned less regularly than the Wye treatment (Plate 2) and this, combined with the smaller cell size, may have been due to the rapid formation and development of microtubers and premature senescence of microplants without sufficient reserves being available for full development.

Acknowledgements

KMN acknowledges the studentship award from The Commonwealth Scholarship Commission in the United Kingdom.

References

1. Allard JM and Blake J (1990) The use of microtubers in the field. Abstract: 11th Conference of the European Association for Potato Research, Edinburgh 303–304
2. Allard JM and Blake J (1991) The use of potato microtubers in the field. Abstract: Conference of the Asian Potato Association, Indonesia (in press)
3. Garner N and Blake J (1989) The induction and development of potato microtubers *in vitro* on media free of growth regulating substances. Ann Bot 63:663–673
4. Hussey G and Stacey NJ (1984) Factors affecting the formation of *in vitro* tubers of potato (*Solanum tuberosum* L.). Ann Bot 53:565–578
5. Kim YC (1982) *In vitro* tuber formation from proliferated shoot of potato (*Solanum tuberosum* L.) as a method of aseptic maintenance. Ph.D. thesis, Jean Buk, National University, South Korea (Cited from Chandra R, Dodds JH and Tovar P (1988) IAPTC Newsletter 55:10–20)
6. Kostrica PB, Polreichova J and Domkarova (1985) The use of *in vitro* tuber formation for the maintenance of potato genetic resources. Genetika a Slechteni 21:269–278 (Cited from Chandra R, Dodds JH and Tovar P (1988) IAPTC Newsletter 55:10–20)
7. Murashige T and Skoog F (1962) A revised medium for rapid growth and bioassays with tobacco tissue cultures. Physiol Plant 15:473–497
8. Slimmon T, Machado VS and Coffin R (1989) The effect of light on *in vitro* microtuberization of potato cultivars. Am Potato J 66:843–848
9. Sladky Z and Bartosova L (1990) *In vitro* induction of axillary potato microtubers and improvement of their quality. Biol Plant 32:181–188
10. Tovar P, Estrada R, Schilde-Rentschler L and Dodds JH (1985) Induction of *in vitro* potato tubers. CIP Circular 13:1–4. International Potato Centre, Lima, Peru
11. Wang PJ and Hu CY (1982) *In vitro* mass tuberization and virus-free seed-potato production in Taiwan. Am Potato J 59:33–37

In vitro adventitious shoot production of *Beta vulgaris* and *Beta maritima*

SARAH MASH, JOHANNES VAN STADEN[1], TUDOR THOMAS and ZHONG ZHONGXIAN[2]

Broom's Barn Experimental Station, Higham, Bury St Edmunds, Suffolk IP28 6NP, UK
[1] *Department of Botany, University of Natal, Pietermaritzburg 3200, South Africa*
[2] *Horticulture Research Institute, Shangai Academy of Agricultural Sciences, Shanghai, PR China*

1. Introduction

Although micropropagation of sugar beet from leaf petioles, shoot buds, flower buds and other flower parts has been relatively successful, genotypic differences in response may be observed [6, 8]. Vitrification may be a problem esecially during long periods of culture [7], and is sometimes attributed to an accumulation in the culture vessel and/or overproduction by the explants of ethylene and cytokinins. The application of inhibitors of ethylene synthesis and action can improve the regeneration capacity of some species and prevent vitrification [6]. Most reported work on *Beta spp* has concentrated on sugar beet (*Beta vulgaris*), but micropropagation of wild beet species is of equal importance since they are known to be potential sources of resistance to major pests and diseased of cultivated sugar beet [10].

The work reported here investigated the effects of hormone concentration in the culture medium and of cultivar on shoot regeneration using seedling petioles and leaf segments of important UK sugar-beet cultivars as explants. The effects of inhibitors of ethylene action ($AgNO_3$) and synthesis (AVG, AOA) on the regenerative capacity of petiole explants was also studied. A method of regenerating shoots of wild beet (*Beta maritima*) from infloresence pieces and attempts to regenerate from mature petiole segments and leaf pieces are also described.

2. Materials and methods

Cultures of *Beta vulgaris*, derived from seeds were grown on MS medium as described previously [11]. After 28 to 55 days seedlings with 4–9 leaves provided petioles (1.0 cm long) or intact leaves (1.5 – 2.0 cm long) as explants for micropropagation.

Immature inflorescences from *Beta maritima* were cultured as described previously. Petiole and leaf pieces from mature greenhouse-grown plants (75 days old) were also used as explants. These were sterilised immediately

261

P.J. Lumsden, J.R. Nicholas and W.J. Davies (eds.), Physiology, Growth and Development of Plants in Culture, 261–266, 1994.
© *1994 Kluwer Academic Publishers.*

following excision and placed on MS media containing 3% sucrose over a range of NAA (1.0 to 5.0 mg l^{-1}) and BA (0.25 to 2.0 mgl-^{1}) concentrations. All cultures were incubated at 25±2°C under white fluorescent tubes giving 16 h photoperiods.

3. Results

3.1 Experiments with sugar beet (Beta vulgaris)

Micropropagation was more successful on a medium supplemented with BA as compared with kinetin as the cytokinin source. Shoot regeneration from both petioles and leaves was influenced by BA concentration, the optimum being from 0.5 to 1.0 mg l^{-1} for the two cultivars tested, cv Imogen and Amethyst [see 11]. In most explants and at all BA concentrations, hard green or white friable callus developed at the base of the petioles; only the latter was capable of regenerating shoots. Adventitious shoots were produced mainly from the concave surface of the petiole but sometimes directly from the callus, without transfer to other media being necessary.

Regeneration was more prolific in cv Amethyst that in cv Imogen. Indeed, no shoots were obtained from leaf explants of cv Imogen, whereas over 50% of the cv Amethyst leaves produced adventitious shoots at the optimum BA concentrations. In a further experiment, of six commercial cultivars used, Amethyst was again the most productive but in this case leaf explants of cv Imogen produced some, though few, adventitious shoots. Figure 1 shows that there is considerable difference in the regenerative capacity of the cultivars examined. Plantlets were rooted on a half-strength MS medium containing 3% sucrose and 1.0 mg l^{-1} NAA, with 70% success. However, some variations in growth habit, including multiple shooting, double apices and modified leaf shape were observed after transplanting into compost.

In the absence of inhibitors of ethylene synthesis (AOA and AVG) and ethylene action (AgNO$_3$) leaf and petiole explants produced callus at concentrations of BA above 1.0 mg l^{-1}. However, only petiole explants initiated adventitious shoots, either in the absence of BA or at a concentration of 1.0 mg l^{-1} (Table 1). The addition of AgNO3 to the medium was ineffective but AVG at both 0.5 and 1.0 mg l^{-1} stimulated shoot initiation particularly in the presence of 1.0 mg l^{-1} BA. Although bud nodules were initiated in the presence of 2.0 mg l^{-1} AOA, these showed no further development, even after transfer to fresh medium containing 0.25 mg l^{-1} BA.

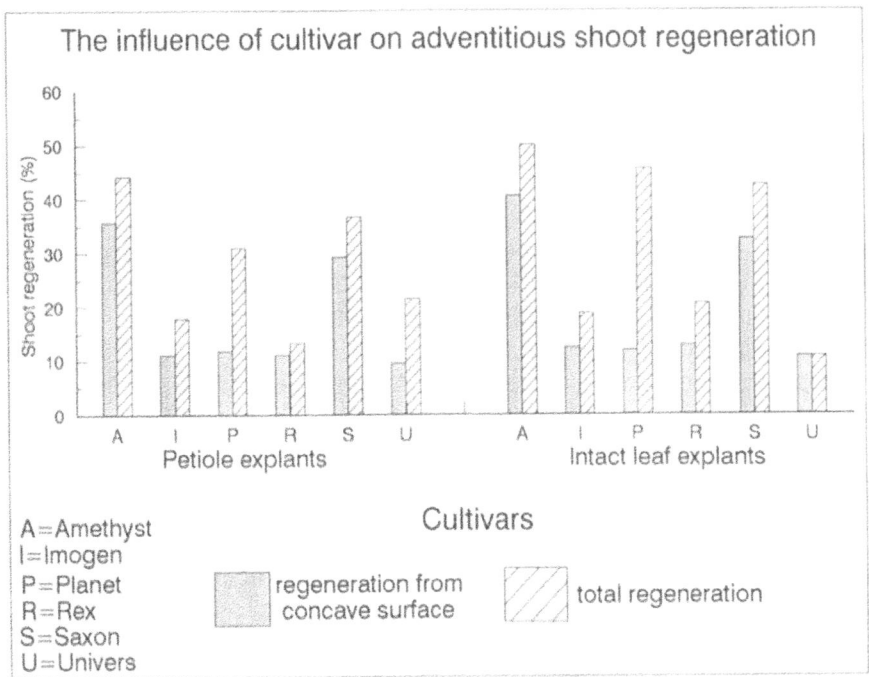

Fig. 1. The influence of cultivar on adventitious shoot regeneration in sugar beet.

Table 1. Effect of ethylene inhibitors on adventitious shoot and callus development on *Beta vulgaris* petiole explants after 36 days in culture

BA mg l^{-1}	Plant growth regulator concentrations							
	0		AgNO$_3$ 50 μM		AOA 2.0 mg l^{-1}		AVG 1.0 mg l^{-1}	
0	12	(0)	0	(0)	0	(0)	20	(0)
1.0	15	(75)	0	(40)	[a]100	(100)	60	(100)
5.0	0	(70)	20	(60)	[a]100	(100)	0	(60)

Values expressed as percentage explants forming shoots and callus (in parentheses).
[a] Bud nodules produced but not adventitious shoots.

3.2 *Experiments with wild beet* (Beta maritima)

Maximum adventitious shoot production from inflorescence tips and sub-apical segments occurred with 1.0 mg l^{-1} BA. Within a few days of culture, explants elongated and thickened rapidly, and after 2 weeks adventitious shoots emerged from the bases of the flower buds on the explants, i.e. from the receptacles. Combinations of BA with other hormones, including IAA, NAA and GA$_3$, gave no additional effects on regeneration. Vitrification was occasionally observed on media supplemented with BA levels above 0.5 mg l^{-1}.

Adventitious shoot induction was affected by the developmental stage of plants from which the explants were taken. Secondary, subapical inflorescences removed on 24 August, gave poor shoot regeneration and less shoots per explant, compared with inflorescences removed from plants on 23 July, i.e. 29% as compared with 82% explants formed adventitious shoots after 6 weeks of culture. Optimum rooting occured 2–3 weeks following transfer to half-strength MS medium with 3% sucrose and 1.0 mg l^{-1} NAA, all rooted plantlets surviving transplanting into compost, with no apparent morphological variation.

In experiments using mature petiole and leaf segments, there was little or no growth in the absence of NAA after 35 days of incubation, but both callus and roots developed with 1.0 to 5.0 mg l^{-1} NAA, and 0.25 to 2.0 mg l^{-1} BA. Most callus was produced with combinations of the higher concentrations, especially on petiole explants, whereas rooting was more prolific at the lower concentrations. However, transfer of callus from leaves or petioles to fresh MS media containing combinations of 0.1 and 2 mg^{-1} of BA and NAA did not result in adventitious shoot development. Many of the callus explants became black and subsequently died, although appearing extremely healthy when subcultured. Contamination losses were higher with leaf material than petiole segments.

4. Discussion

The optimum concentration of BA for adventitious shoot production from petioles and intact leaf explants of sugar beet is in the range 0.5 and 1.0 mg l^{-1}, as reported previously [3, 5, 9]. This optimum was consistent for all cultivars studied, but there were major varietal differences in regeneration capacity, which are both quantitative and qualitative. Shoots were mainly produced from the concave surface of petioles or from the callus formed on the base of the petioles, without transfer to new media. In contrast to a previous report [3], light did not seem to inhibit the production of callus. Variations in transplanted, clonal plants were similar to those reported previously for red beet [4] and sugar-beet [2].

A recent review [1] reported that ethylene from plant tissues accumulating in large quantities in culture vessels may influence growth and development, but that the effects may be prevented by incorporating inhibitors of both ethylene action and biosynthesis into the medium. In the present work, AVG stimulated the development of adventitious shoots, AOA induced adventitious bud nodules but AgNO$_3$ had little effect. The reason for these differences in action is not clear, though AVG is known to be a more potent inhibitor of ethylene synthesis than in AOA and so may exert a greater effect. However, such differences could also be related to varietal response. Further studies are required to assess the full benefits of incorporating ethylene-synthesis inhibitors into sugar-beet micropropagation systems.

The *Beta maritima* studies indicate that explants from inflorescence tips provide a source of healthy adventitious shoots *in vitro* but that concentrations of BA higher than 0.5 mg l^{-1} can cause leaf vitrification. This method can be used to multiply valuable *Beta* germplasm and elite hybrids to incorporate useful characteristics into new sugar-beet lines. *In vitro* shoot production was also attempted using petiole segments from mature, glasshous-grown plants and although callus and root development was prolific under optimal plant hormone regimes, shoot organogenesis did not occur. Investigations to identify a suitable medium for adventitious shoot production from mature petioles either directly of indirectly via callus production will continue.

Acknowledgements

Zhong Zhongxian and Johannes van Staden are grateful to the Horticultural Research Institute, Shanghai Academy of Agricultural Sciences and The Underwood Fund respectively for financial support during their sabbaticals at Broom's Barn Experimental Station.

References

1. Biddington NL (1992) The influence of ethylene in plant tissue culture. Plant Growth Regul 11: 173-187
2. Detrez C Sangwan RS and Sangwan-Norreel BS (1989) Phenotypic and karyotypic status of *Beta vulgaris* L. plants regenerated from direct organogenesis in petiole culture. Theor Appl Genet 77: 462-468
3. Freytag, AH, Anand SC, Rao-Arelli AP and Owens LD (1988) An improved medium for adventitious shoot formation and callus induction in *Beta vulgaris* L. *in vitro*. Plant Cell Rep 7: 30-34
4. Harms CT, Baktir I and Oertli JJ (1982) Conal propagation *in vitro* of red beet (*Beta vulgaris*) by multiple adventitious shoot formation. Plant Cell Tiss Org Culture 2: 93-102
5. Hussey G and Hepher A (1978) Clonal propagation of sugar-beet plants and the formation of polyploids by tissue culture. Ann Bot 42: 477-479
6. Jacq B, Telu T, Sangwan RS De Laat A and Sangwan-Norreel BS (1992) Plant regeneration from sugar beet (*Beta vulgaris L.*) hypocotyls cultures *in vitro* and flow cytometric nuclear DNA analysis of regenerants. Plant Cell Rep 11: 329-333
7. Keevers C, Coulmans-Gilles MF and Gasper T (1984) Physiological and biochemical events leading to vitrification of plants cultured *in vitro*. Physiol Plant 61: 69-74
8. Mezei S. Jelaska S and Kovacev L (1990) Vegetative propagation of sugar beet from floral rametes. J Sug Beet Res 27: 90-96
9. Saunder JW (1982) A flexible *in vitro* shoot culture propagation system for sugar beet that includes rapid floral induction of rametes. Crop Sci 22: 1102-1105
10. Thomas TH (1991) Sugar beet biotechnology. Br Sugar Beet Rev 59: 5-7

11. Zhong Zhongxian, Smith HG and Thomas TH (1992) *In vitro* culture of petioles and intact leaves of sugar beet (*Beta vulgaris*). Plant Growth Regul 12:59–66
12. Zhong Zhongxian, Smith HG and Thomas TH (1992) Micropropagation of wild beet (*Beta maritima*) from inflorescence pieces. Plant Growth Regul 12:53–57

Factors influencing the regeneration of *Solanum papita*

STEPHEN MILLAM, PETER DAVIE and WENDY CRAIG
Crop Genetics Department, Scottish Crop Research Institute, Invergowrie, Dundee DD2 5DA, UK

1. Introduction

Cell and tissue culture techniques provide a valuable link between cellular and molecular approaches to crop improvement and conventional plant breeding techniques. There have been numerous studies on the *in vitro* response of a range of Solanaceous species, notably *Solanum tuberosum* [3, 9, 10]. However, there exists a very wide range of wild potato germplasm amongst which there are many traits of agronomic interest. The series Longipedicella includes a number of species that have valuable characteristics for possible inclusion in a crop genetic improvement programme [1]. Among this series is the species *Solanum papita* (2n=48) which is a native of Northern Mexico where it is found in rocky coniferous forests. Among the most valuable traits of this species is a very high resistance to late blight (*Phytophthora infestans*). It has also been reported that this species shows resistance to some viruses and has some tolerance to heat and drought [2]. To enable the desirable traits of this species to be integrated efficiently into a gene transfer programme, the establishment, response and optimisation of *in vitro* regeneration techniques need to be evaluated.

2. Materials and methods

2.1 Plant material

The accessions of *S. papita* used in this study were obtained from the Commonwealth Potato Collection held at SCRI. Tubers were sprouted under glasshouse conditions. When the plants had grown to approx. 30 cm in height, they were harvested and the lateral buds excised.

2.2 Establishment of in vitro cultures

The excised buds were surface-sterilised using a four-step procedure. Initially, the buds were washed in tap water for ten minutes, followed by immersion in

P.J. Lumsden, J.R. Nicholas and W.J. Davies (eds.), Physiology, Growth and Development of Plants in Culture, 267–271, 1994.

70% ethanol for 20 s, agitation in 10% Domestos (Lever Bros) for 17 min, and finally, three washes in sterile distilled water. The buds were inoculated onto nodal medium consisting of half-strength M + S [7] medium (ICN Flow), supplemented with 87 μm sucrose, 4.6 μm kinetin and 5 μm gibberellic acid (GA$_3$)solidified with 8.0 g/l Sigma agar. The medium was adjusted to pH 5.8 prior to autoclaving at 104 kPa for 20 min and aseptically dispensed into 90 mm Sterilin Petri dishes. The cultures were maintained under conditions of 22 °C, 16 h photoperiod at 160–180 μmol m^{-2} sec^{-1}. The plantlets were routinely subcultured onto fresh nodal medium every four weeks for a number of growth cycles prior to use in regeneration experiments.

2.3 Explant source

Explants used for the regeneration experiments were excised from the *in vitro* cultures and comprised:
1. 15 mm leaf sections, bisected so that the cut edges were in close spatial contact with the medium;
2. 15 mm internodal stem sections, bisected longitudinally with the cut surface in contact with the medium.
All plates were sealed using Nescofilm.

2.4 Regeneration efficiencies of S. papita on a range of growth regulator regimes

The system utilised was based on a two-stage method where the explants were cultured on an initial medium (for growth regulator components see Table 1) for fourteen days prior to subculturing onto a second medium containing 9 μm BAP and 13 μm GA$_3$ [3] to promote shoot elongation.

2.5 Effect of auxin exposure time and carbohydrate source in the culture medium on shoot regeneration

The regeneration system was expanded to determine if fourteen days was the optimum exposure to auxin prior to transfer to the second non-auxin containing medium. Leaf explants were subcultured from the medium containing 9 μm BAP onto the second medium after 7, 11 and 14 days culture. As a control, explants were left on the initial medium for the entire period of this experiment. A further factor investigated in this experiment was to compare the effects of substituting the sucrose component of the culture medium with maltose. Results are presented in Table 2.

2.6 Effect of silver thiosulphate on shoot regeneration

The stimulatory effects of silver have been reported elsewhere for a range of plant species including *Solanum tuberosum* [4, 6]. In our experiments silver was

Table 1. The effect of the growth regulator component of the initial media on subsequent regeneration response in Solanum. Means of ten explants per plate, assessed after 28 days (14 days on initial medium and 14 days on secondary medium)

Growth regulator concentration μm		mean no. shoots		% explants with shoots		+/– roots	
		leaf	stem	leaf	stem	leaf	stem
ZR	NAA						
11.4	10.7	0	0	0	0	+	+
11.4	5.3	0.1	0	10	0	+	+
11.4	0	0	0.7	0	20	–	+
5.7	10.7	0	0	0	0	+	+
5.7	5.3	0	0	0	0	+	+
5.7	0	1.4	1.2	30	50	–	–
2.8	10.7	0	0	0	0	–	–
2.8	5.3	0	0	0	0	+	+
2.8	0	0	0	0	0	–	–
ZR	2,4-D						
11.4	4.5	0	0	0	0	–	+
11.4	2.2	0	0	0	0	–	–
5.7	4.5	0	0	0	0	–	–
5.7	2.2	0.7	2.3	30	80	–	+
2.8	4.5	0.1	0	10	0	–	+
2.8	2.2	0.9	1.6	30	50	–	–
BAP	NAA						
9	1.0	4.2	3.7	60	50	–	–

Table 2. The effect of exposure to auxin on subsequent regeneration response in Solanum and the effect of substituting sucrose with maltose in the culture medium. Means of five explants per plate assessed after a total of 28 days culture

days exposure	mean no. shoots		% explants with shoots		+/– roots	
	sucrose	maltose	sucrose	maltose	sucrose	maltose
0	0	0	0	0	0	0
7	2.3	3.0	20	33	+	+
11	0.9	0.3	20	10	++	+
14	0.7	0	20	0	++	+

aseptically added to the initial culture media in the form of silver thiosulphate at a concentration of 6 mM [4]. Silver was not present in the second medium. Precise shoot lengths were recorded for every shoot on every explant after 28 days. Results are presented in Table 3.

Table 3. The effect of silver thiosulphate on shoot regeneration numbers and shoot length in Solanum. Explants recorded after 28 days culture

No. explants	Total no. shoots	% Explants with shoots	Mean shoot length (mm)
control			
stem 75	145	53	19.6±7.5
leaf 50	200	100	23.2±5.5
silver			
stem 75	256	80	26.3±9.7
leaf 50	184	90	22.3±3.3

3. Results and discussion

There have been no previous reports of the regeneration of *Solanum papita* through tissue explants other than one report of the regeneration of anther-derived material [8]. It can be seen from the results shown in Table 1 that, despite an extensive screening of growth regulators, the control medium of 9 μm BAP and 1.1 μm NAA outperformed all other treatments in terms of shoot production. The two-stage method of regeneration was reported previously [9] and was subsequently modified [3, 10]. There was no evidence of a non-callus direct shoot regeneration response; however, in some instances only a minimal callus presence (on the cut edges of the leaf or stem explants) was recorded. Generally root formation was at the expense of shoot production. It is apparent from the results presented in Table 2 that a period of 7 days exposure to the initial medium is optimum for shoot production. Longer periods resulted in large amounts of root production on the sucrose medium. Calculated total figures comparing the overall effect of sucrose with maltose show that the mean number of shoots from the sucrose treatments was 1.4 per explant (12% of explants with shoots) compared with a figure of 0.78 per explant (10% of explants with shoots) from maltose. Mean fresh weight per explant comparison was 94.57 mg from sucrose compared with 81.21 mg from maltose. The effects of substituting maltose for sucrose in plant tissue culture medium have been reported previously. Using flax explants [5], an enhanced regeneration rate attributable to maltose was reported. In the experiments reported here there was no enhancement; in fact, the regeneration rates were lower using maltose than sucrose.

The effects of the addition of silver to the culture medium have been reported as being beneficial for regeneration of shoots from leaf material in a range of *Solanum tuberosum* cultivars [4] and in the regeneration of shoots from protoplast-derived calli of dihaploid potatoes [6]. In our experiments the effect was unclear. For stem tissue the silver treatment enhanced shoot production and from leaf tissue slightly repressed regeneration rates. There was a trend towards longer shoots from the silver treated stem sections but no statistically significant differences were observed.

In summary, we present findings that will contribute to the establishment of an efficient shoot regeneration system for *Solanum papita*. The protocols described would be particularly suitable for systems of genetic transformation using *Agrobacterium*-based vectors.

Acknowledgements

This work was supported by the Scottish Office Agriculture and Fisheries Department.

References

1. Adiwilaga KD and Brown CR (1988) Introgression of tetraploid Mexican wild species germplasm into cultivated potato by means of triplandroids. Am Pot Journal 65:467
2. Bilski JJ, Nelson DC and Conlon RL (1988) Response of six wild potato species to chloride and sulfate salinity. Am Pot Journal 65:605–612
3. Fish N and Jones MGK (1988) A comparison of tissue culture response between related tetraploid and dihaploid *S. tuberosum* types. Plant Cell Tiss Org Cult 15:201–210
4. Higgins ES, Hulme JS and Shields R (1992) Early events in transformation of potato by *Agrobacterium tumefaciens*. Plant Sci 82:109–118
5. Millam S, Davidson D and Powell W (1992) The use of flax as a model system for studies of organogenesis *in vitro*: the effect of different carbohydrates. Plant Cell Tiss Org Cult 28:163–166
6. Mollers C, Zhang S and Wenzel G (1992) The influence of silver thiosulphate on potato protoplast cultures. Plant Breeding 102:12–18
7. Murashige T and Skoog F (1962) A revised media for rapid growth and bioassay with tobacco tissue cultures. Physiol Plant 15:215–243
8. Powell W and Uhrig H (1987) Anther culture of *Solanum* genotypes. Plant Cell Tiss Org Cult 12:291–297
9. Webb KJ, Osifo EO and Henshaw GG (1983) Shoot regeneration from leaflet discs of six cultivars of potato (*Solanum tuberosum* subsp. *Tuberosum*). Plant Sci Lett 30:1–8
10. Wheeler VA, Evans NE, Foulger D, Webb KJ, Karp A, Franklin J and Bright SWJ (1985) Shoot formation from explant cultures of fourteen potato cultivars and studies of the cytology and morphology of regenerated plants. Ann Bot 55:309–320

Somatic embryogenesis in guava (*Psidium guajava* L.)

A. GHAFFOOR and P.G. ALDERSON
Department of Agriculture & Horticulture, University of Nottingham, Sutton Bonington Campus, Loughborough, Leics LE12 5RD, UK

1. Introduction

The guava (*Psidium guajava* L.) is the premier fruit among the myrtaceous fruit plants and has gained economic importance in the fruit industry of many countries. It is an evergreen, tropical tree which bears fruit twice in a year. Its fruits are notable for their content of vitamin C, which is several times greater than that of citrus or tomato fruits. The production of guava is constrained by the lack of an efficient technique for vegetative propagation and therefore the lack of clonal cultivars. There is considerable interest in the development of tissue culture techniques for improving the breeding of this fruit plant, but as yet there has been no published report of a suitable regeneration strategy for guava which could be used on genetically manipulated tissues or cells.

Somatic embryogenesis, induced directly on explant tissues or in callus cultures is a potential regeneration strategy, and has been reported in many fruit tree species [5, 10]. Several chemical and physical factors have been shown to influence the initiation and developmental pattern of somatic embryos *in vitro,* e.g. sucrose concentration, organic nitrogen, growth regulator levels and culture conditions [1]. Factors controlling the initiation of somatic embryogenesis have been reviewed [12].

This paper reports the regeneration of guava by somatic embryogenesis and discusses the physiological basis of the influence of growth regulators, sucrose concentration and activated charcoal on this process.

2. Materials and methods

2.1 Preparation and culture of zygotic embryos

Seeds of *Psidium guajava* L. cv. Safeda (supplied by Gomal University, Pakistan) were surface sterilised in a 10% bleach solution (Domestos, Lever Bros.) for 30 min and rinsed with three changes of sterile purified water. They were kept in sterile water for two days to soften the seed coat and then re-

P.J. Lumsden, J.R. Nicholas and W.J. Davies (eds.), Physiology, Growth and Development of Plants in Culture, 272–277, 1994.
© 1994 *Kluwer Academic Publishers.*

sterilised. Zygotic embryos (Plate 1) were excised aseptically and cultured on a range of induction media which contained half strength MS salts [7] supplemented with 30 gl^{-1} sucrose and solidified with 8 gl^{-1} agar. Cultures were incubated at 28±1 °C in a 16 h photoperiod with light intensity of 150 μEm^{-2}s^{-1} provided by warm white fluorescent tubes.

For direct embryogenesis, basal medium was supplemented with NAA (1–5 mgl^{-1}) or picloram (0.5–5 mgl^{-1} in combination with 0.1 mgl^{-1}) BAP, with and without activated charcoal (AC) (3 gl^{-1}). The induction medium containing NAA (2 mgl^{-1}) or 2,4-D (1 mgl^{-1}) + BAP (0.1 mgl^{-1}) was supplemented with 20, 30, 40, 60, 90 or 120 gl^{-1} sucrose to compare the effect of sucrose concentration on somatic embryo production. Excised zygotic embryos were also cultured on basal medium supplemented with 2,4-D (1 mgl^{-1}) + BAP (0.1 mgl^{-1}) in order to induce indirect embryogenesis from callus. Ten replicates of 6 embryos were used for each treatment.

2.2 Development and assessment of somatic embryos

After 4 to 8 weeks from initiation, the cultures which had produced callus or somatic embryos directly were transferred to regeneration media containing BAP (0.5–2.0 mgl^{-1}) + NAA (0.1 mgl^{-1}) or without growth regulators for further growth and development of embryos into plantlets. Cultures were transferred to the same fresh media at intervals of 4 weeks. Data were recorded on embryogenesis frequency, i.e. the percent of explants producing somatic embryos, and on the number of somatic embryos produced per responsive explant.

Plantlets which had developed their root system *in vitro* were transplanted to compost (Levington M2, Fisons Ltd) and placed in a polythene enclosure in a glasshouse at 92% RH and 28±1 °C for 2 weeks for acclimation and establishment.

3. Results

3.1 Direct embryogenesis on zygotic embryos

Somatic embryos were induced directly from mature zygotic embryos without an intervening callus phase within 8 weeks of culture on medium containing NAA (2.0 mgl^{-1}) + BAP (0.1 mgl^{-1}) (Table 1). Each explant produced from 2 to 30 embryos which were variable in size and shape (Plate 2).

3.2 Effect of sucrose on somatic embryogenesis

The number of somatic embryos regenerated increased with increasing sucrose concentration, up to 60 gl^{-1} (Table 2). At 120 gl^{-1} sucrose, no embryos were produced.

Plates 1–5. (1) Mature seed of *Psidium guajava* L. cut longitudinally to show zygotic embryo (e). bar = 0.5 mm; (2) Somatic embryos at different developmental stages, originating from zygotic embryos after 8 weeks in culture. bar = 2.5 mm; (3) High frequency somatic embryogenesis in callus formed after 4 to 8 weeks on ½ MS medium supplemented with 2,4-D (1.0 mgl^{-1}) + BAP (0.1 mgl^{-1}) + AC (3 gl^{-1}). Note somatic embryos at various stages of development. bar = 2mm; (4) Primary embryos with secondary embryos (arrows) arising from the hypocotyl after culture on ½ MS medium containing BAP (1.0 mgl^{-1}) + NAA (0.1 mgl^{-1}). bar = 3mm; (5) Plants established from somatic embryos. bar = 20mm.

Table 1. Effect of auxin type and concentration on direct or indirect somatic embryogenesis on zygotic embryos of *Psidium guajava* L. cv. Safeda. Embryos cultured for 8 weeks at 28 °C on ½ MS medium containing 0.1mgl^{-1} BAP + different auxins. Number of zygotic embryos cultured per treatment = 60

Auxin concentration (mgl^{-1})		% of zygotic embryos producing somatic embryos	Mean number of somatic embryos per responsive zygotic embryo	Embryogenic callus
2,4-D	0.5	8	8	+
	1.0	20	15	+
	5.0	10	10	+
Picloram	0.5	0	0	—
	1.0	5	6	—
	5.0	3	8	—
NAA	1.0	17	12	—
	2.0	30	25	—
	5.0	13	18	—

Table 2. Effect of sucrose concentration on % of zygotic embryos of *Psidium guajava* L. c.v. Safeda producing somatic embryos when cultured on MS medium containing 0.1 mgl^{-1}2,4-D or 2 mgl^{-1}NAA. Number of zygotic embryos cultured per treatment = 60.

Sucrose concentration (gl^{-1})	2,4-D medium	NAA medium
20	10	13
30	22	32
40	23	33
60	28	38
90	13	17
120	0	0

3.3 Indirect embryogenesis via callus

Somatic embryos formed indirectly via callus derived from zygotic embryos in response to different auxins (Table 1). Yellow-green friable callus was formed from mature zygotic embryos after 8 weeks of culture in the presence of 2,4-D (1.0 mgl^{-1}) + BAP (0.1 mgl^{-1}). Subsequently, embryogenic callus produced several somatic embryos on the same fresh medium, some of which germinated to produce white, torpedo-like structures (Plate 3) and other developmental stages.

3.4 Secondary embryogenesis and development of plantlets

Primary somatic embryos when recultured on ½ MS medium with BAP (1.0 mgl^{-1}) + NAA (0.1 mgl^{-1}) gave rise to secondary somatic embryos within 4 to

8 weeks either on early stages of the germinated embryos or on hypocotyl tissue (Plate 4). Successive cycles of embryogenesis were also induced on regeneration medium (1.0 mgl^{-1} BAP + 0.1 mgl^{-1} NAA). In the absence of growth regulators, somatic embryos developed into plantlets, of which more than 80% established successfully in compost as primary, secondary and tertiary regenerants (Plate 5).

4. Discussion

Somatic embryogenesis in guava from zygotic embryos occurs directly or indirectly via a callus phase. It has been inferred that direct somatic embryogenesis systems have pre-embryogenic determined cells which are already committed or completely rejuvenated for embryo development and need only to be released into cell division and expression of embryogenesis by growth regulators or other sets of conditions [8, 12]. In the case of indirect embryogenesis, where callus proliferation is a pre-requisite to embryo development, differentiated cells dedifferentiate and then are re-determined as embryogenic cells after cell division, hence the term 'induced embryogenic determined cells'.

The results from this study support the concept of a young embryo as a group of pre-embryogenic determined cells [1, 8, 12] which are normally co-ordinated to act as a polarised group in response to an internal physiological gradient. When somatic embryogenesis occurs, single cells or small cell groups at a certain point on this gradient are presumed to escape from overall group control to express their own pre-embryogenic determined state independently. It would seem that, in guava, release of cells, regeneration and somatic embryogenesis are dependent on the state of the cells involved. Certain cells within explants, termed pre-determined embryogenic cells, possess the metabolic state which enables them to respond to a series of physiological signals (such as wounding of the zygotic embryo in conjunction with auxins like 2,4-D or NAA) by the sequence of cell division and developmental events.

Our studies show that the auxins, 2,4-D and NAA and activated charcoal in the medium were essential for the induction of embryogenesis in guava. Sharpe *et al.* [8] proposed that systems with pre-embryogenic determined cells either have no auxin requirements or require auxin only for the onset of mitosis. It has been suggested that auxins may remove any existing developmental constraints from cells or that they provide the signals which evoke the embryogenic response in suitably competent cells [4].

Cycles of recurrent embryogenesis observed directly from different parts of somatic embryos of guava, such as the hypocotyl, are common among culture protocols using 2,4-D as the auxin [8]. For guava such a regeneration strategy may be of use for genetic transformation and mass propagation.

Sucrose may play a role in somatic embryogenesis as an osmoticum that can stimulate and regulate morphogenesis and as a carbon source [6]. Induction of somatic embryogenesis and the early stages of somatic embryo development often require moderate to high concentrations of sucrose [1], also evidenced by the requirement for 30 to 60 gl^{-1} sucrose to maximise somatic embryo production in guava. Thorpe [9] believes that osmotic adjustment very early in the culture increases mitochondrial activity for ATP production.

The requirement for activated charcoal in the culture medium for the induction of somatic embryogenesis in guava may be attributed to the absorption of toxic metabolites from the culture environment [11]. Alternatively, the charcoal may release some substance(s) that stimulate(s) embryogenesis [1].

A regeneration strategy for guava by embryogenesis should assist future *in vitro* programmes aimed at genetic improvement. The mode of action of exogenous growth regulators is still unclear, although it is generally assumed to have a genetic basis which may be elucidated by future studies of the expression of active DNA synthesis and rapid cell division in relation to somatic embryogenesis. The genetic stability of guava plantlets obtained from different generations of somatic embryos is currently being investigated.

References

1. Ammirato PV (1983) Embryogenesis. In: DA Evans, WR Sharpe, PV Ammirato and Y Yamada, eds. Handbook of Plant Cell Culture. Vol. I. pp 82–123. New York: MacMillan
2. Ammirato PV (1987) Organisational events during somatic embryogenesis. In: Plant Tissue and Cell Culture 1987 (Green CE, Somers DA, Hackett WP and Biersboer DD eds.), Plant Biology 3, pp 57–81. New York: Alan R Liss Inc
3. Cruz GS, Canhoto JM and Abreu MAV (1990) Somatic embryogenesis and plant regeneration from zygotic embryos of *Feijoa sellowiana* Berg. Plant Sci 66:263–270
4. Henshaw GG, O'Hara JF and Webb KJC (1982) Morphogenetic studies in plant tissue cultures. In: MM Yeoman and DES Truman, eds. Differentiation *in vitro*, pp 231–251. Cambridge: Cambridge University Press
5. Litz RE (1985) Somatic embryogenesis in tropical fruit trees. In: RR Henke, KW Hughes, MP Constantin and A Hollaender, eds. Tissue Culture in Forestry and Agriculture, pp 179–193. New York: Plenum Press
6. Litz RE (1986) Effect of osmotic stress on somatic embryogenesis in *Carica* suspension cultures. J Amer Soc Hort Sci 111:969–972
7. Murashige T and Skoog F (1962) A revised medium for rapid growth and bioassays with tobacco tissue cultures. Physiol Plant 15:473–497
8. Sharpe WR, Sondahl MR, Caldan LS and Maraffa SB (1980) The physiology of *in vitro* asexual embryogenesis. Hortic Rev 2:268–310
9. Thorpe TA (1982) Callus organization and de novo formation of shoots, roots and embryos *in vitro*. In: DT Tomes, BE Ellis, PM Harvey, KJ Kasha and RL Peterson, eds. Application of Plant Cell and Tissue Culture to Agriculture and Industry, pp 115–138. Ontario: University of Guelph
10. Tulecke W and McGranahan G (1985) Somatic embryogenesis and plant regeneration from cotyledon of walnut *Juglans regia* L. Plant Science Letters 40:57–63
11. Wang PJ and Huang PC (1976) Beneficial effects of activated charcoal on plant tissue and organ cultures. In Vitro 12:260–262
12. Williams EG and Maheswaran G (1986) Somatic embryogenesis — factors influencing co-ordinated behaviour of cells as an embryogenic group. Ann Bot 57:443–462

Micropropagation of pea (*Pisum sativum* L.) – *in vitro* system and its practical applications

MIROSLAV GRIGA and JAROMÍR STEJSKAL

OSEVA — Research Institute of Technical Crops and Legumes, Sumperk, CZ-787 01 Czech Republic

1. Introduction

Since 1974 when the first successful results were published on plant regeneration from pea meristems [8] this system has been studied for different purposes [2, 6, 7, 9, 12, 13, 14]. However, there is relatively little information on the genetic stability of seed progenies of pea meristem-regenerants and on plants regenerated from long-term micropropagated shoot-tip culture. Based on nearly 10 years work with pea micropropagation in our laboratory we present here an overview of our previous work supplemented with some new (unpublished) data and discuss the possibilities of practical application of this system.

2. Materials and methods

2.1 In vitro *system*

Shoot apical meristems (0.3 to 0.5 mm), shoot apices and axillary buds (1 to 2 mm) from aseptically germinated seedlings were used as initial explants for establishment of micropropagation culture [4, 5, 6]. MSB basal medium ([11] — mineral salts; [1] — vitamins) was supplemented with cytokinin and/or auxin for multiple shoot induction and root formation. Rooted shoots were transferred into non-sterile conditions (hydroponics, soil) in a greenhouse and grown to maturity. The effect of different initial explants and genotype on multiplication capacity was studied.

2.2 *Evaluation of seed progenies (T1 to T4 generation) of meristem-derived plants*

The seeds harvested from T0-regenerants in 1985 were used as a starting material for field evaluation of seed progenies in 1986–1988 (yield and morphological traits) as well as for electrophoretic analyses of storage proteins and some isozyme systems [3, 14].

P.J. Lumsden, J.R. Nicholas and W.J. Davies (eds.), Physiology, Growth and Development of Plants in Culture, 278–283, 1994.

2.3 Evaluation of regenerants from 9 year micropropagation culture

The plants regenerated from long-term micropropagated shoot-culture were transferred to non-sterile conditions and the morphological and biochemical evaluation was carried out.

3. Results and discussion

3.1 In vitro *system*

The optimized protocol of micropropagation consists of: (1) induction of multiple shoot formation (Plate 1) from meristems and buds on MSB-medium with 20 μM BAP and 0.1 μM NAA, (2) long-term micropropagation on the same medium (4–5 week subculture intervals) via repetitive proliferation of axillary buds and *de novo* regeneration of buds from swollen explants, (3) root induction (Plate 2) on medium with half strength concentration of MS-salts, B5 vitamins, 1 μM NAA or IBA and 4% sucrose [10], (4) transfer of complete plantlets to non-sterile conditions with 2–3 week acclimatization period (high relative humidity > 90%). The system optimized for model variety Bohatýr is applicable routinely for a broad spectrum of *Pisum sativum* and *P. arvense* genotypes [3] as well as for some wild *Pisum* species (*P. elatius, P. fulvum, P. jomardii, P. abyssinicum, P. syriacum* — unpublished data). There are differences in multiplication capacity between various genotypes [3], also initial explants of various location on the plant (apical buds, axillary buds of cotyledons, primary scales and true leaves) have different capacity to form multiple shoots, which increases in the direction: shoot apex < axillary buds of leaves < primary scales < cotyledonary buds. During long-term culture (more than 9 years in cv. Bohatýr) there is no decrease in proliferation capacity of multiple shoots. However, shoots isolated from long-term culture are more difficult to root as compared to shoots isolated from initial cultures.

Cytological investigation of root tips of primary regenerants (TO) and root tips of germinated seeds (T1) harvested from these regenerants revealed diploid number of chromosomes 2n=14 [6].

3.2 Evaluation of meristem-derived seed progenies of cv. Bohatýr

During the field evaluation of three seed generations (T1 to T3, 1986–1988) we did not observe any significant differences in flowering time, length of vegetation period or sensitivity to diseases (virus and fungus infections recorded) of regenerant progenies as compared to control material (elite seeds of cv. Bohatýr). We also observed no morphological alterations in vegetative or generative organs. The T1-generation plants exhibited a more robust habit (Plate 3) and non-significantly higher average values of some traits (plant height, height of the first pod-bearing node, number of pods and seeds, weight

Plates 1–5. (1) Multiple shoot formation from pea axillary buds on MSB-medium with 20 μM BAP and 0.1 μM NAA. Upper row: explants cultured on agar medium in Erlenmayer flasks covered with aluminium foil; lower row: explants cultured on agar medium in Erlenmayer flasks covered with cotton wool; (2) Root formation on isolated pea shoots on half-strength MS medium with 1μM NAA and 4% sucrose; (3) T1-generation plants (seed progeny of primary meristem regenerants, cv. Bohatýr) evaluated in field conditions at Šumperk location, 1986. More robust habit of tissue-culture treated material (middle) as compared to controls (left and right); (4) Virus elimination (PSbMV) via meristem culture combined with thermotherapy. Left: infected plant; middle and right: healthy plants after virus elimination; (5) Morphological alterations of leaves detected in pea plants cv. Bohatýr regenerated from long-term cultures (9 years) micropropagated on MSB-medium with 20 μM BAP and 0.1 μM NAA. A, B, C, D, E – morphologically altered variants, F – control, standard leaf morphology.

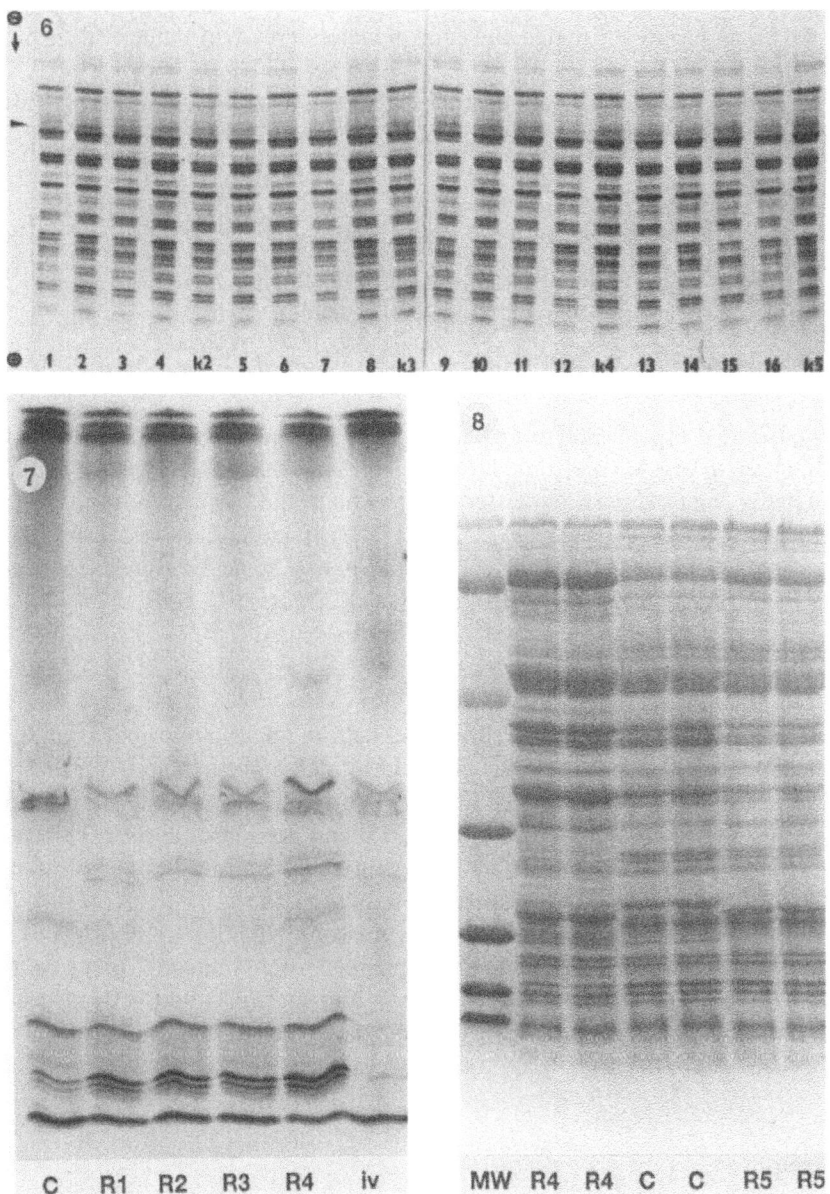

Plates 6–8. (6) SDS-PAGE analysis of 16 mericlones and 4 control plants (cv. Bohatýr) of T3-generation (i.e. T4-seeds). Arrowhead (left) indicates the position of the altered proteins of the samples 2, 7, 8 and 13; (7) IEF analysis of leaf isoperoxidases in long-term (9 years) micropropagated pea shoot culture. C — control plant, R1-R4 — pea regenerants transplanted into soil, iv — *in vitro* leaves; (8) SDS-PAGE analysis of seed storage proteins from direct regenerants after 9 years micropropagation. Abbreviations as above, MW-molecular weight markers.

of seeds and straw, weight of 1000 seeds). However, in following generations (T2, T3) the tissue culture — treated and control material reached similar values of 9 evaluated characters [3].

Isoelectric focusing analyses of leaf peroxidase, aspartate aminotransferase and malate dehydrogenase from T3-plants as well as SDS-PAGE analysis of storage proteins of T4-seeds (Plate 6) harvested from T3-plants showed minimum variation [14] confirming expected genetic stability of material passed through several micropropagation cycles (6 months of culture) followed by seed propagation. Based on these data we concluded that the system described above can be considered as a means of clonal propagation.

3.3 Evaluation of regenerants from 9 year old micropropagated culture

Rooted shoots from 9 year old shoot culture micropropagated on the MSB-medium with 20 μM BAP and 0.1 μM NAA developed into plants which exhibited a high proportion of sterile individuals and various morphological alterations of leaves (Plate 5). Analysis of leaf peroxidase, esterase, acid phosphatase and seed storage proteins also showed some differences between regenerants and control plants (from elite seeds) (Plates 6, 7, 8).

3.4 Applications of micropropagation system

The system of meristem and/or shoot bud culture and micropropagation has been successfully used for production of healthy plants from material infected by *Pisum* seed-borne mosaic virus (PSbMV, Plate 4) [3] and for multiplication of plants originated from somatic embryos [15] and organogenic calli. This last mentioned application enabled us to obtain sufficient somaclones for field testing of somaclonal variation in pea (unpublished data).

Recently we have attempted to use the micropropagation approach for (1) obtaining transformed non-chimaeric plants from chimaeric tissue after *Agrobacterium* — mediated transformation and multiplication of transgenic individuals as well as for (2) establishment of an *in vitro* pea selection system with the use of culture filtrates of phytopathogenic fungi (*Fusarium* sp.).

References

1. Gamborg OL, Miller RA and Ojima K (1968) Nutrient requirements of suspension cultures of soybean root cells. Exp Cell Res 50:151–158
2. Gould KS, Cutter EG, Young JPW and Charlton WA (1987) Positional differences in size, morphology and *in vitro* performance of pea axillary buds. Can J Bot 65:406–411
3. Griga M (1990) The study of regeneration systems *in vitro* in pea, faba bean and soybean. PhD Thesis 1990, Brno: Masaryk University
4. Griga M, Tejklová E and Novák FJ (1984) Hormonal regulation of growth of pea (*Pisum sativum* L.) shoot apices in *in vitro* culture. Rostl. Výr. 30:523–530
5. Griga M, Tejklová E and Novák FJ (1984) *In vitro* propagation of pea by axillary and

adventitious bud technique. In: FJ Novák, L Havel and J Doležel, eds. Proc Intern Symp Plant Tissue Cell Cult Application to Crop Improvement, pp 507–508, Olomouc

6. Griga M, Tejklová E, Novák FJ and Kubaláková M (1986) *In vitro* clonal propagation of *Pisum sativum* L. Plant Cell Tiss Org Cult 6:95–104

7. Kartha KK and Gamborg OL (1978) Meristem culture techniques in the production of disease-free plants and freeze-preservation of germplasm of tropical tuber crops and grain legumes. In: H Maraite and JA Meyer, eds. Diseases of Tropical Food Crops, pp 267–283. Louvain, Université Catholique de Louvain

8. Kartha KK, Gamborg OL and Constabel F (1974) Regeneration of pea (*Pisum sativum* L.) plants from shoot apical meristems. Z Pflanzenphysiol 72:172–176

9. Kartha KK, Leung HL and Gamborg OL (1979) Freeze-preservation of pea meristems in liquid nitrogen and subsequent plant regeneration. Plant Sci Lett 15:7–15

10. Kubaláková M, Tejklová E and Griga M (1988) Some factors affecting root formation on *in vitro* regenerated pea shoots. Biol Plant 30:179–184

11. Murashige T and Skoog F (1962) A revised medium for rapid growth and bioassays with tobacco tissue cultures. Physiol Plant 15:472–497

12. Novák FJ, Lucretti S, Donini B, Afza R and Hermelin T (1984b) Flower and pod development in shoot tip and axillary bud culture of pea (*Pisum sativum* L.) In: FJ Novák, L Havel and J Doležel, eds. Proc Intern Symp Plant Tissue Cell Cult. Application to Crop Improvement, pp 133–134, Olomouc

13. Novák FJ, Lucretti S, Hermelin T, Donini B, Afza R, Daskalov S, Kubaláková M and Griga M (1984a) Gamma ray irradiation effects on multiple shoot cultures of pea (*Pisum sativum* L.). In: FJ Novák, L Havel and J Doležel, eds. Proc Intern Symp Plant Tissue Cell Cult Application to Crop Improvement, pp 453–454, Olomouc

14. Stejskal J and Griga M (1989) Total protein and some isoenzyme spectra of meristem-tip culture derived plant progenies of pea (*Pisum sativum* L.). In: E Ohmann, E Grimm, S Loffler, eds. Abstracts IV Symp Young Scientists in Physiology and Biochemistry of Plants, pp 56, Holzhau, GDR

15. Stejskal J and Griga M (1992) Somatic embryogenesis and plant regeneration in *Pisum sativum* L. Biol Plant 34:15–22

Uptake of 2,4-D in coconut (*Cocos nucifera*) explants

C. OROPEZA[1] and H.F. TAYLOR[2]

[1] *Centro de Investigación Científica de Yucatán, Apdo. Postal 87, Cordemex 97310, Yucatán, México*
[2] *Department of Horticulture, University of London (Wye College), Wye, Kent TN25 5AH, UK*

1. Introduction

Regeneration of coconut (*Cocos nucifera*) *in vitro* has proved to be very difficult, despite the efforts of several research groups [1, 7, 8]. Recent progress has sprung from basic studies on the histological development of the cultures [1, 8], suggesting that further basic research on tissue culture of coconut is required to achieve greater progress in planlet regeneration. The protocol for the regeneration of coconut from inflorescence explants requires, according to Blake [1], an initial preculture of the explants in liquid medium which includes 2,4-D (a key growth regulator for coconut regeneration). It is not known though whether 2,4-D is taken up by the explants during preculture or what the mechanism(s) of uptake of this growth regulator by coconut tissue cultured *in vitro* are. The present report deals with uptake of ^{14}C-2,4-D in inflorescence explants during preculture.

2. Materials and methods

2.1 Materials

Coconut inflorescences were harvested from dwarf coconut palms growing near Centro de Investigación Científica de Yucatán, México and sent immediately to Wye. Delivery took about five days. Explants were obtained from the rachillae by cutting 3–5 mm segments under aseptic conditions. The developmental stage of the inflorescences used was -6, which was numbered by counting back from the last opened inflorescence. The radiolabelled synthetic auxin 2-[^{14}C]-2',4'-dichlorophenoxyacetic acid (specific activity 55 μmol/μCi) was obtained from Amersham International (UK).

P.J. Lumsden, J.R. Nicholas and W.J. Davies (eds.), Physiology, Growth and Development of Plants in Culture, 284–288, 1994.

2.2 In vitro *culture and assessment of 2,4-D uptake*

The media formulation and culture conditions were those of Branton and Blake [2] except that charcoal was not added and 2,4-D concentration was 1.5 M. All operations were carried out under sterile conditions. Tissue was extracted and radioactivity assessed according to Ebert and Taylor [3]. Samples were frozen prior to homogenization in 80% acetone and extracted overnight. Aliquots both of the acetone extracts and of the used medium were then added with scintillant and radiocounted.

2.3 Chromatography

Samples were applied to silica-gel chromatograph plates and developed in xylene/acetic acid (50:8, v/v). Radioactivity on the chromatoplates was scanned and measured using a Dunnschicht-Scanner II (LB 2722–2 Berthold).

3. Results and discussion

Radiolabelled 2,4-D was found to be taken up by coconut inflorescence explants. Recovery of radioactivity from the explants increased with time (Fig. 1) and total

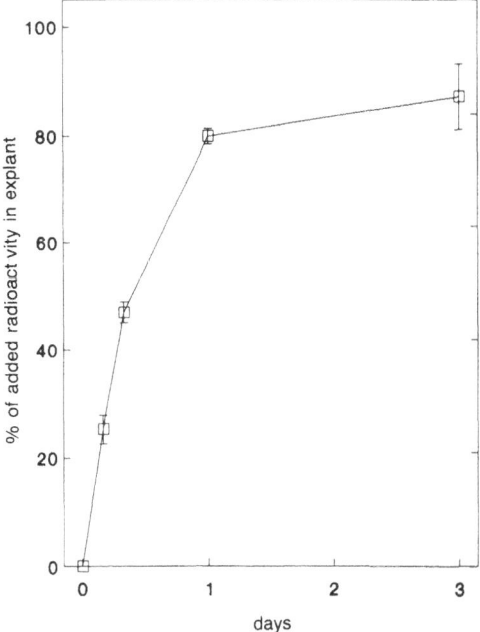

Fig. 1. Timecourse of the recovery of radioactivity from coconut inflorescence explants cultured in liquid medium containing ^{14}C-2,4-D. Bars denote standard deviations larger than the symbols (n=5)

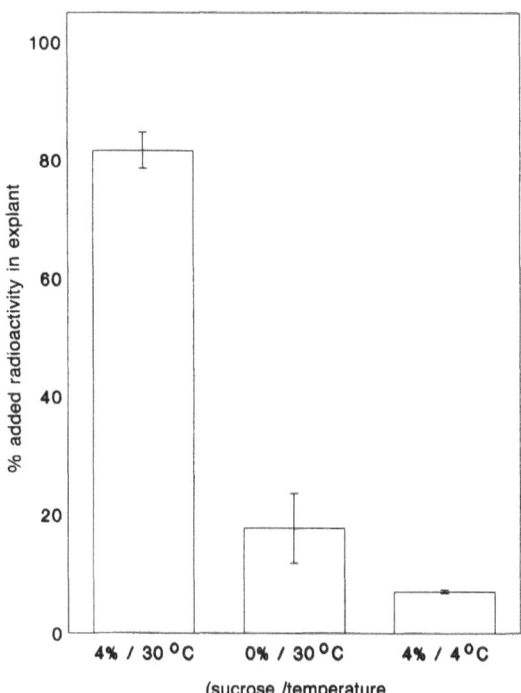

Fig. 2. Effect of removal of sucrose from the medium and low temperature on the recovery of radioactivity from coconut inflorescence explants cultured in liquid medium containing ^{14}C-2,4-D. Bars denote standard deviations larger than the symbols (n=5).

recovery of radioactivity approached 100% at all times studied. Most of the radioactivity was taken up by the tissue within 24 h. At this time the volume of the explant was only about one tenth of that of the external medium, so, uptake of 2,4-D occurred against a concentration gradient. Thus, uptake of ^{14}C-2,4-D by coconut inflorescence explants cannot be explained by simple diffusion. Alternatively, ^{14}C-2,4-D may be taken up by facilitated diffusion. Auxins, including indoleacetic acid and 2,4-D, have been shown to be taken up by different tissues by facilitated diffusion when the external pH is lower than the pH of the cytosol [4,6]. Uptake of auxins has also been shown to be dependent on energy and it has been suggested that some of this energy is required to maintain the pH gradient [5]. When sucrose was not included in the medium or the temperature was reduced to 4 °C, ^{14}C-2,4-D uptake fell several fold (Fig. 2), supporting a requirement for energy for uptake in coconut. Uptake of ^{14}C-2,4-D by coconut explants was four times greater at pH 3.5 than at pH 6.5 (Fig. 3A). In addition, the pH of the medium was found to decrease rapidly in the presence of explants (Fig. 3B). In contrast, when sucrose was not included in the medium, the pH decreased slightly after 24 h and returned to the intial value after 3 d (Fig. 3B). Therefore, uptake of ^{14}C-2,4-D by coconut explants (in preculture conditions)

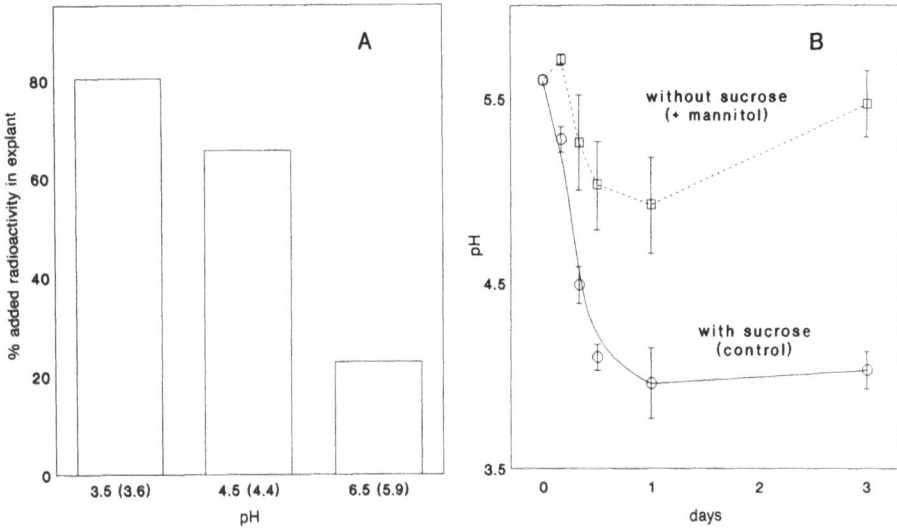

Fig. 3. (A) Recovery of radioactivity from coconut inflorescence explants cultured in liquid media containing ^{14}C-2,4-D and buffered at different pH values; (B) Timecourse of the changes of the medium pH during preculture with and without (4%) sucrose. Mannitol was added to keep osmotic potential constant. Bars denote standard deviations larger than the symbols (n=5).

would appear to be driven by a pH gradient and that an energy source like sucrose is required to maintain this gradient.

The results reported here are useful in understanding the fate of 2,4-D in coconut tissues cultured *in vitro*, but presently cannot be used directly to improve coconut regeneration. However, they point out the importance of pH for 2,4-D uptake by coconut explants. Further understanding of how 2,4-D uptake might affect regeneration, will require its study on coconut callus and embryogenic callus which are subsequently derived from the initial explant.

Acknowledgements

We wish to thank R. Cardeña, J.L. Chan and C. Talavera from CICY in Mérida for their help in providing the plant material which was used in this study. Also C. Ladley and C. Kemp from from Wye College for their tachnical assistance. Partial support for this research and a fellowship for CO was provided by a grant from the Commission of the European Communities.

References

1. Blake J (1990) Coconut (*Cocos nucifera* L.): Micropropagation. In: YPS Bajaj, ed. Biotechnology in Agriculture and Forestry, Vol. 10 Legumes and Oilseed Crops I, pp 538–554. Berlin: Springer-Verlag
2. Branton RL and Blake J (1983) Development of organized structures in callus derived from explants of *Cocos nucifera* L. Ann Bot 52:673–678
3. Ebert A and Taylor HF (1990) Assessment of the changes of 2,4-dichlorophenoxyacetic acid concentrations in plant tissue culture media in the presence of activated charcoal. Plant Cell Tiss Org Cult 20:165–172
4. Minocha SC and Nissen P (1985) Uptake of 2,4-dichlorophenoxyacetic acid and indoleacetic acid in tuber slices of Jerusalem artichoke and potato. J Plant Physiol 120:351–362
5. Raven JA (1975) Transport of indoleacetic acid in plant cells in relation to pH and electrical potential gradients, and its significance for polar indoleacetic transport. New Phytol 74:163–172
6. Rubery PH and Sheldrake AR (1973) Effect of pH and surface charge on cell uptake of auxin. Nature 244:285–288
7. Shirke SV, Kendurkar SV, Apte AV, Phadke CH, Nadgauda RS and Mascarenhas AF (1993) Regeneration in coconut tissue culture. In: Proceedings of the International Symposium of Coconut Research and Development-II. Oxford and IBH Publishing, India (In press)
8. Verdeil JL, Buffard-Morel J, Rival A, Grosdemange R, Huet C and Pannetier C (1993) Coconut clones through somatic embryogenesis. Proceedings of the International Symposium of Coconut Research and Development-II. Oxford and IBH Publishing, India (In press)

Peroxidase activity and endogenous free auxin during adventitious root formation

TH. GASPAR, CL. KEVERS, J.FR. HAUSMAN and V. RIPETTI
Hormonologie fondamentale et appliquée, Institut de Botanique B22, Université de Liège – Sart Tilman, B–4000 Liège, Belgium

1. Introduction

That endogenous free auxin has a central role in the process of adventitious root formation is generally accepted [27, 28, 35, 46, 49, 61]. The concept however has been developed more from the observation that exogenously applied auxins invariably induce the formation of a greater number of roots per cutting than other chemicals, than from established correlation between the auxin content or the auxin variation of cuttings and their rootability. In the mind of many plant physiologists, the exogenously supplied auxins, whatever synthetic, simply grow the bulk of endogenous auxins and act in a similar manner. This concept is out of date, as clearly shown on the one hand by the relationships between exogenous and endogenous auxins in cell cultures [52, 64], and on the other hand by recent analyses of the fate of exogenously supplied auxins in cuttings [7, 68]. The cell signaling eventual role of exogenous auxins supplied to cuttings thus has seldom been distinguished from the role of the endogenous ones. Literature concerning the estimation of endogenous auxin levels at the time of cutting excision in relation with their rooting capacities or concerning the variation of the levels in the course of adventitious rooting was until recently very discrepant [15, 36]. First, there were relatively few studies using unequivocal physico-chemical techniques of auxin analyses. Second, it was not clearly stated what developmental stage of rooting was sampled (induction preceding cell reactivation and division, organization of the primordia, growth of the newly formed roots), nor was the physiological condition of the stock plant taken into consideration.

In the past few years, considerable evidence has been acquired to show that auxin is not needed during all the stages of adventitious root formation and a number of results of auxin analyses using reliable physico-chemical techniques have been published. These recent data now allow a general picture of how endogenous free auxin varies in the course of root formation.

The variation of peroxidase activity in the course of adventitious root formation has been reviewed very recently [37]. Here we give a summary in order to show how it is correlated with the changes in endogenous auxin levels.

289

P.J. Lumsden, J.R. Nicholas and W.J. Davies (eds.), Physiology, Growth and Development of Plants in Culture, 289–298, 1994.
© 1994 *Kluwer Academic Publishers.*

2. Changes of peroxidase activity in the course of rooting

Several papers have reported a regulatory role of the so-called auxin-oxidase system for the auxin content [72], i.e. an inverse relationship between the endogenous IAA level of a plant tissue or organ and its IAA oxidase activity. Several groups then investigated the changes of IAA-oxidase activity in relation to rooting [57, 81, and references in 9 and 61]. The activity of the so-called IAA-oxidase system was later on attributed to one or several peroxidases [34]. Because the measurement of peroxidase activity was relatively more simple than that of IAA-oxidase activity, relationships between different aspects of rooting and changes of peroxidase activity and of isoperoxidases were investigated.

Analyses of changes of total peroxidase activity and in the isoperoxidase spectrum in the course of rooting by shoot cuttings raised *in vitro* from different *Prunus* [74] and *Asparagus* [80] species, and by epidermal layers of tobacco [40, 78] led to a general relationship [34] where it appeared that root formation occurred after the cutting has reached and passed a peak of maximum enzyme activity. The activity and/or number of acidic isoperoxidases increased continuously during the course of the process, which means that the peak of peroxidase activity was due to an inverse variation of the activity of the basic isoenzymes before and after the peak. Such changes of peroxidase activity in the course of root formation, most of them involving a passage through a peak, were measured in many other materials [5, 20, 25, 26, 42, 44, 51, 53, 58, 60]. They apparently corresponded to parallel changes of IAA-oxidase activity [3, 57, + refer in 9]. In some cases, peroxidase activity of the shoot cuttings dropped immediately after their transfer from a multiplication to an auxin-based rooting medium, which was followed by a rapid emergence of root primordia [74, 80]. It was shown later that the elevation and peaking of peroxidase activity preceding rooting might have taken place at the end of the foregoing multiplication /elongation cycle [30, 47].

There was an apparent coincidence between the peroxidase peak and the first visible signs of differentiation [39] which led us to hypothesize that the period of peroxidase elevation up to the peak corresponded to an inductive or preparatory phase of rooting, before any visible morphological or histological event [34]. The recognition and necessity of such an inductive phase for rooting were relatively new, although such a concept of induction had been investigated for flowering.

That the peroxidase peak corresponded to the first cell divisions i.e. that the peroxidase peak terminated the inductive period of rooting was questionned by Jarvis [50]. This author placed the end of the rooting induction phase (corresponding to the biochemical changes preceding cytological and histological events) before the peroxidase peak. Further studies by us (Fig. 1) confirmed that view [38, 62, 63]. Changes in specific peroxidase activity in *in vitro* cuttings of grapevine, on a non-hormone medium, showed a clear decrease with a minimum after 12 h, followed by a progressive increase and peak at 72 h, then a decrease. It was noted that some early cytological events such as nuclear swelling were visible at the time of peroxidase increase. Therefore the minimum of peroxidase activity coincided with the termination of the induction period. The period of high peroxidase activity corresponded to the early events of the initiation phase, itself preceding the expressive one [4]. Chemical or physical factors which enhance the peroxidase peak and its further decrease favor rooting [37].

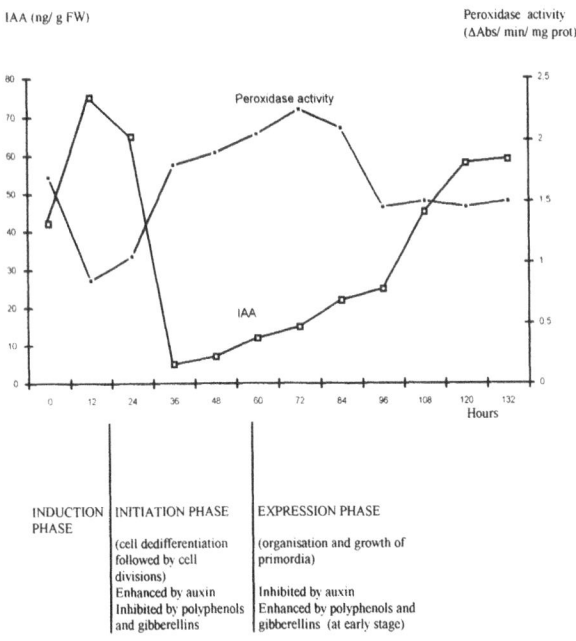

Fig. 1. Changes of peroxidase activity and of endogenous free auxin level along the rooting process by grapevine cuttings (redrawn from 38), which may be considered as also generally occurring along successive induction, initiation and expression phases of rooting in other types of cuttings (with varying durations of these phases).

Although the increase of peroxidase activity preceding rooting was generally noticed, some research workers were unable to recover the peak, that is the subsequent decrease of enzyme activity [29, 41, 71, 73]. The explanation of the discrepancies may be that in those studies, peroxidase activity was measured in purified extracts and not expressed per unit protein. When well evidenced, the (specific) peroxidase activity peak was measured in crude extracts where (poly)phenols apparently modulated the enzyme activity [6, 30, 38, 58]. We have further shown [1] that none of the cell fractions (extracellular, soluble, membrane, wall ionic, wall covalent) separated from a crude extract (of rhododendron shoots) exhibiting the peak showed the typical peroxidase variation. This indicated that the peroxidase peak did not only result in changes in enzyme activities but also in changes of their effectors. Results from Curir *et al.* [24] indirectly confirmed that view.

3. Changes of endogenous free auxin in the course of rooting

Pioneering research has suggested that the number of roots initiated per cutting may be a function of the amount of auxin-like substances in the regeneration zone [2, 48]. In a number of cases, endogenous auxin content has been reported to increase in the base of cuttings during the root inducing period [11, 59]. A relation has been shown between the endogenous acidic auxins and the rooting response to added root promoting substances [70], which suggested that optimum levels of auxin were required. A reduced level of auxin has been implicated in the failure of rooting in a number of species of plant cuttings [22, 75, 76]. However, in other cases auxin levels did not appear to be limiting [10, 45]. For example, Stoltz [77] found that in *Chrysanthemum* spp. endogenous auxin concentration did not positively correlate with rooting. In contrast, the number of lateral primordia arising from roots of *Pisum sativum* increased significantly after decapitation; however, a change in the level of IAA was not detected [17]. However, in some easy-to-root cuttings, the endogenous levels of auxin decreased appreciably over the course of rooting [16]. In another test, auxins in extract fractions that co-chromatographed with IAA were quantified by bioassays at 24 h intervals after pruning *Quercus rubra* seedlings [19]. Auxin activity sharply increased in the first 24 h and then decreased to pre-pruning levels. New laterals appeared 4–5 d after pruning. By means of gas chromatography, Brunner [18] identified IBA and IAA in hypocotyl cuttings of *Phaseolus vulgaris* L. In the control cuttings and all the auxin treated cuttings, a distinct increase in the contents of IAA and IBA became evident in the region of root regeneration in the first 24 h . With progressing regeneration, levels of the auxins decreased in all cases. Nakano *et al.* [66] also showed that auxin activity at the base of the cuttings diminished during propagation and was lower in well-rooted than in unrooted cuttings. They assumed that the auxin may have been metabolized before root differentiation. Seasonal changes in hormone levels have been determined in stem cuttings of the easy-to-root *Rhododendron ponticum* L. and the difficult-to-root *Rhododendron* 'Britannia'. The amount of IAA (and cytokinin) in the stem tissue of both groups was similar, with little seasonal variation in either [83].

Tréfois and Brunner [79] found a good positive correlation between endogenous auxin-like activity in *Prunus* spp. cuttings and percent of rooting of cuttings only in instances where auxin level was initially high at the time of auxin treatment. There was no effect of exogenous auxin treatment on rooting when the endogenous auxin level was low at the time cuttings were prepared. The rooting ability of easy and hard-to-root clones of *Sequoia sempervirens* was also correlated with their initial compounds of endogenous IAA [33]. Dunberg *et al.* [31] found that IAA content at the bases of IBA treated cuttings of *Pinus sylvestris* L. was three times higher than in untreated cuttings. They did not find differences in IAA metabolism or transport between the treated and untreated cuttings, and concluded that applied IBA was converted to IAA by the cuttings. The foregoing results were confirmed by Epstein and Lavee [32] in olive (*Olea europaea* L.) and grape (*Vitis vinifera* L.) cuttings through the use of radioactive IBA. Using a solid phase enzyme immunoassay, Weigel *et al.* [82] have found a large increase of auxin in the apical parts of *Chrysanthemum* spp. shoots after removing them as terminal stem cuttings from stock plants. Auxin level remained high until about the time when roots were formed. Synthesis of auxin probably occurred because the total amount increased, not only free auxin derived from bound forms. However, basipetal transport of IAA would have been arrested at the cut end of the cutting, which may have lead to an endogenous basal accumulation. In this system, prolonged high irradiance of stock plants (40 W. m^{-2}) delayed an IAA increase in the cuttings and concomitantly decreased the number of roots per cutting compared to controls (4.5 W.m^{-2}). However, root growth as determined by measuring root length or fresh weight were not affected in the latter test. A distinct relation was found between IAA content of stock plants at the time when cuttings were taken and the number of roots formed by the cuttings 20 d later. Similarly, rooting percentage correlated positively with IAA concentrations in *Cotinus coggygria* cuttings [12, 14]: IAA concentration in stem tissue of spring cuttings (which rooted well) was approximately 10 times the concentration in autumn cuttings (which rooted poorly). The concentration of IAA fell markedly after excision and before root appearance [12]. A transient increase in the concentration of IAA was also discovered in *Phaseolus aureus* hypocotyl cuttings, where root initiation occurs during the first 15 h [13]. Rhizogenesis by tobacco thin cell layers specifically (not floral or vegetative organogenesis) was also related to intracellular concentration of auxin [23].

All these results suggest that a transient rise in the level of IAA occurs before the first detectable event in root initiation, but that IAA falls prior to subsequent cell divisions, which finally might explain the apparently discrepant results. Using ELISA, Moncousin *et al.* [62, 63] and Gaspar [38] *et al.* reported changes in the levels of IAA in *in vitro* cuttings of grapevine, on a non-hormone medium. In the basal part of the shoot, they found a clear increase in IAA after 12 h, followed by a sharp decrease to a minimum reached after 36 h and maintained up to 72 h, then a very slow recovery. This early peak of IAA was not found in the apical part of the shoots, which confirmed the view that a peak of IAA must precede root formation. The authors noted that some early cytological events such as nuclear swelling were visibly at the time of the auxin decline but not before. Therefore the peak of auxin level coincided with the termination of the induction period. The period of low auxin level corresponded to the early events of the initiation phase. There was thus a nice inverse relationship between the changes in auxin level and the changes in specific peroxidase activity. Further support for the requirement of an early peak in IAA comes from the works of Maldiney *et al.* [56] with hypocotyl cuttings of *Craigella* and *Craigella* lateral suppressor tomatoes, of Label *et al.* [54] with *in vitro* explants of *Prunus avium* and of Noiton *et al.* [67] with apple micro-cuttings. Norcini and Heuser [68] and Berthon *et al.* [8] with *Sequoiadendron giganteum* cuttings confirmed that adventitious root initiation was accompanied by a fall in IAA corresponding with the rise of the peroxidase peak. In this analysis the timing of sampling dates was such that the former increase in IAA could easily have been missed. The presence of IAA in the root forming part of cuttings at concentrations above a certain level indeed has been shown to be a prerequisite for root formation to occur [69]. In other assays, it has been shown that factors which enhance the peak of auxin level and its further decrease also favour rooting [21, 65].

4. Conclusions

Root formation is no longer considered as a single physiological process, but has been broken down into successive interdependent phases. Auxin analyses using more reliable physico-chemical techniques as well as studies of peroxidase changes has allowed the involvement of endogenous free auxin in the process of adventitious root initiation to be more precisely defined. The scheme of Figure 1, drawn from results of Moncousin *et al.* [62], tentatively summarizes the changes of peroxidase activity and of endogenous free auxin level generally observed in the course of rooting by cuttings. Of course the timing changes from one cutting type to another one and in many studies, the early event might be missing due to the absence of earlier sampling.

The first stage of root formation is dedifferentiation and cell reactivation. The signal for this primary event might be the peak in the level of IAA in the rooting zone found by several authors. It should have been achieved through an inductive period where no cytological events are visible. External factors favorizing such an induction process are still not clear : an adequate auxin/ cytokinin ratio might be promotive [56, and reference in 27], and the temporary ethylene burst associated with cutting injury might be involved [38, 43, 55, 63]. In most cases the IAA peak was temporary and the level of IAA subsequently declined. This decline corresponds to the rise of peroxidase and IAA-oxidase activity observed by many authors. Early cell divisions and constitution of potentially morphogenetic fields (clusters of cells which show no polarity) follow cell reactivation during the rooting initiation phase. The completion of the second initiation phase might require external application of auxin and should be inhibited by external polyphenols and gibberellin. The following expression phase corresponding to individualization and growth of internal radical meristems requires an increased concentration of endogenous auxin. This requirement might be favoured by polyphenols and temporary application of gibberellin [39]. Repeated exogenous application of auxin at that stage will be inhibitory to root development. The fact that auxin is involved in the process of adventitious root formation is thus well established. There is also considerable evidence that endogenous (and exogenous) auxin is not needed during all the stages of this developmental phenomenon.

Acknowledgements

Work partly supported by the EEC ECLAIR project AGRE-CT91-0067 . JFH and VR gratefully acknowledge grants from the Luxembourg Ministry of Cultural Affairs and EEC respectively.

References

1. Aghmir A, Kevers C, Hausman JF and Gaspar T (1991) Peroxydase, compartimentation cellulaire et enracinement in vitro de pousses de Rhododendron catawbiense Michaux cv album. Arch Intern Physiol Biochem 99: pp9
2. Bastin M (1966) Root initiation, auxin level and biosynthesis of phenolic compounds. Photochem Photobiol 5:423–429
3. Ben Efraim J, Gad AE, Cohen P, Reymond P and Pilet PE (1990) The effect of 4-chlororesorcinol on the endogenous levels of IAA, ABA and oxidative enzymes in cuttings. Plant Growth Regul 9:97–106
4. Berthon JY, Ben Tahar S, Gaspar T and Boyer N (1990) Rooting phases of shoots of *Sequoiadendron giganteum in vitro* and their requirements. Plant Physiol Biochem 28:631–638
5. Berthon JY, Boyer N and Gaspar T (1987) Sequential rooting media and rooting of *Sequoiadendron giganteum in vitro*. Peroxidase activity as a marker. Plant Cell Reports 6:341–344
6. Berthon JY, Boyer N and Gaspar T (1990) Phenols as regulators and markers of root formation by shoots of *Sequoiadendron giganteum* raised *in vitro*. Arch Intern Physiol Biochem 98: pp28
7. Berthon JY, Boyer N and Gaspar T (1991) Uptake, distribution and metabolism of 2,4-dichlorophenoxyacetic acid in shoots of juvenile and mature clones of *Sequoiadendron giganteum* in relation to rooting *in vitro*. Plant Physiol Biochem 29:355–362
8. Berthon JY, Maldiney R, Sotta B, Gaspar T and Boyer N (1989) Endogenous levels of plant hormones during the course of adventious rooting in cuttings of *Sequoiadendron giganteum* (Lindl.) *in vitro*. Biochem Physiol Pflanzen 184:405–412
9. Bhattacharya NC (1988) Enzymes activities during adventitous rooting. In: TD Davis, BE Haissig and N Sankhla, eds. Adventitious Root Formation in Cuttings. Vol 2, pp 88–101. Portland: Dioscorides Press
10. Biran I and Halevy AH (1973) Endogenous levels of growth regulators and their relationship to the rooting of Dalhia cuttings. Physiol Plant 28: 244–247
11. Blahova M (1969) Changes in the level of endogenous gibberellins and auxins prcceding the formation of adventitious roots on isolated epicotyls of pea plants. Flora 160:493–499
12. Blakesley D, Alderson PJ and Weston GD (1985) Rooting studies on Cotinus. In: Abst 12th Int Conf Plant Growth Sub, p886. Heidelberg
13. Blakesley D, Hall JF, Weston GD and Elliott MC (1985) Endogenous plant growth regulators and the rooting of *Phaseolus aureus* cuttings. In : Abst 12th Int Conf Plant Growth Sub, p87. Heidelberg
14. Blakesley D, Weston GD and Elliott MC (1991) Endogenous level of indole-3-acetic acid and abscisic acid during the rooting of *Cotinus coggygria* taken at different times of the year. Plant Growth Regul 10:1–12
15. Blakesley D, Weston GD and Hall JF (1991) The role of endogenous auxin in root initiation. Part I: Evidence from studies on auxin application and analysis of endogenous levels. Plant Growth Regul 10:341–353
16. Bose TK, Basu S and Basu RN (1973) Changes in rooting factors during the regeneration of roots on cuttings of easy- and difficult-to root cultivars of *Bougainvillea* and *Hibiscus*. Indian J Plant Physiol 16:127–139

17. Böttger M (1978) Levels of endogenous indole-3-acetic acid and abscisic acid during the course of the formation of lateral roots. Z Pflanzenphysiol 86:283–286

18. Brunner H (1978) Einfluss verschiedener Wuchsstoffe und Stoffwechselgifte auf wurzel-regenerierendes Gewebe von *Phaseolus vulgaris* L. Veranderungen des Wuchsstoffgehaltes Sowie des Peroxydase- und der IAA-oxydase- Aktivität. Z Pflanzenphysiol 8:13–23

19. Carlson WC and Larson MM (1977) Changes in auxin and cytokinin activity in roots of red oak, *Quercus rubra*, seedlings during lateral root formation. Physiol Plant 41:162–166

20. Chandra GR, Gregory LE and Worley JF (1971) Studies on the initiation of adventitious root on mung bean hypocotyl. Plant Cell Physiol 12:317–324

21. Collet GF (1988) Improvement to induce rooting of fruit trees in vitro. Acta Hort 227:318–323

22. Cooper WC (1935) Hormones in relation to root formation on stem cuttings. Plant Physiol 10:789–794

23. Cousson A, Tran Than Van K and Hanh Trinh T (1992) Changes with time in the H^+ concentration of the culture medium influence morphogenesis in tobacco thin cell layers. Physiol Plant 85:102–110

24. Curir P, Van Sumere C, Termini A, Barthe P, Marchesini A and Dalei M (1989) Flavonoids accumulation is correlated with adventitious root formation in *Eucalyptus gunii* Hook micropropagated through axillary bud stimulation. Plant Physiol 92:1148–1153

25. Dalet F and Cornu D (1986) Les peroxydases : marqueurs enzymatiques de l'enracinement *in vitro* du merisier (*Prunus avium* L.). Bull Groupe Polyphenols 13:213–215

26. Dalet F and Cornu D (1989) Lignification level and peroxidase activity during *in vitro* rooting of *Prunus avium* L. Can J Bot 67:2182–2186

27. Davis TD and Haissig BE (1990) Chemical control of adventitious root formation in cuttings. PGRSA Quaterly 18:1–17

28. Davis TD, Haissig BE and Sankhla N (1988) Adventitious Root Formation in Cuttings. Portland: Dioscorides Press

29. de Klerk G, ter Brugge J, Smulders R and Benschop M (1990) Basic peroxidases and rooting in microcuttings of *Malus*. Acta Hortic 280:29–36

30. Druart P, Kevers C, Boxus P and Gaspar T (1982) *In vitro* promotion of root formation by apple shoots through darkness effect on endogenous phenols and peroxidases. Z Pflanzenphysiol 108:429–436

31. Dunberg A, Hsihan S and Sandberg G (1981) Auxin dynamics and the rooting of cuttings of *Pinus sylvestris*. Plant Physiol Suppl 65:5

32. Epstein E and Lavee S (1984) Conversion of IBA by cuttings of grapevine (*Vitis vinifera*) and olive (*Olea europea*). Plant Cell Physiol 25:697–703

33. Fouret Y, Arnaud Y, Maldiney R, Sotta B and Miginiac E (1986) Relation entre rhizogenèse et teneur en auxine et acide abscissique chez trois clônes de *Sequoia sempervirens* (Endl.) issus d'arbres d'âge différent. Comptes Rendus Acad Sci Paris,Sér III 303:135–138

34. Gaspar T (1981) Rooting and flowering,two antagonistic phenomena from a hormonal point of view. In: B Jeffcoat, ed. Aspects and Prospects of Plant Growth Regul, Monograph 6 , pp39–49. England: British Plant Growth Regul Group

35. Gaspar T and Coumans M (1987) Root formation. In: J-M Bonga and D-J Durzan, eds. Cell and Tissue Culture In Forestry Vol 2, pp 202–217. Dordrecht: Martinus Nijhoff

36. Gaspar T and Hofinger M (1988) Auxin metabolism during adventitious rooting. In: TD Davis, BE Haissig and N Sankhla, eds. Adventitious Root Formation In Cuttings, pp117–131. Portland: Dioscorides Press

37. Gaspar T, Kevers C, Hausman JF, Berthon JY and Ripetti V (1992) Practical uses of peroxidase activity as a predictive marker of rooting performance of micropropagated shoots. Agronomie 12:757-765

38. Gaspar T, Moncousin C and Greppin H (1990) The place and role of exogenous and endogenous auxin in adventitious root formation. In: B Millet and H Greppin, eds. Intra-and Intercellular Communications in Plants pp125–139. Paris: INRA ed

39. Gaspar T, Smith D and Thorpe TA (1977) Arguments supplémentaires en faveur d'une

variation inverse du ni veau auxinique endogène au cours des deux premières phases de la rhizogenèse. CR Acad Sc Paris 285:327–330

40. Gaspar T, Thorpe TA and Tran Than Van M (1977) Changes in isoperoxidases during differentiation of cultured tobacco epidermal layers. Acta Hortic 78:61–73

41. Gebhardt K (1985) Self rooted sour cherries *in vitro*: auxin effect on rooting and isoperoxidases. Acta Hortic 169:341–349

42. Gebhardt K (1986) The possible effect of peroxidase/ IAA oxidase stimulation on rooting of sour cherries *in vitro*. In: H Greppin, C Penel and T Gaspar, eds. Molecular and Physiological Aspects of Plant Peroxidases pp387–393. Genève: Univ. de Genève

43. Gonzalez A, Rodriguez R and Tames RS (1991) Ethylene and *in vitro* rooting of hazelnut (*Corylus avellana*) cotyledons. Physiol Plant 81:227–233

44. Gonzalez A, Tames RS and Rodriguez R (1991) Ethylene in relation to protein, peroxidase and polyphenol oxidase activities during rooting in hazelnut cotyledons. Physiol Plant 83:611–620

45. Greenwood MS, Atkinson OR and Yawney HW (1976) Studies of hard- and easy- to-root ortets of sugar maples: difference not due to endogenous auxin content. Plant Propagator 22:3–6

46. Haissig BE, Davis TD and Riemenschneider DE (1992) Reasearching the controls of adventitious rooting. Physiol Plant 84:310–317

47. Hausman JF, Kevers C and Gaspar T (1990) Rooting *in vitro* while multiplying. Arch Intern Physiol Bioch 98: pp42

48. Hemberg T (1954) The relation between the occurrence of auxin and the rooting of hypocotyls in *Phaseolus vulgaris* L. Physiol Plant 7:323–331

49. Jackson MB (1986) New Root Formation In Plants and Cuttings. Dordrecht: Martinus Nijhoff

50. Jarvis BC (1986) Endogenous control of adventitious rooting in non woody cuttings. In: M-B Jackson. ed. New Root Formation in Plants and Cuttings, pp191–222. Dordrecht: Martinus Nijhoff

51. Kevers C and Gaspar T (1992) Micropropagation of *Kalmia latifolia*: acclimation and rooting performance dependent on the preceding steps. Relationship with peroxidase activity. Med Fac Landbouww Univ Gent 57/3b: 977–985

52. Kevers C, Coumans M, De Greef W, Hofinger M and Gaspar T (1981) Habituation in sugarbeet callus : auxin content, auxin protectors, peroxidase pattern and inhibitors. Physiol Plant 51:281–286

53. Kevers C, Hausman JF, Hagege D and Gaspar T (1991) Post-effect of thidiazuron on peroxidase activity and rooting of microcuttings of *Kalmia latifolia*. Saussurea 22:27–31

54. Label PH, Sotta B and Miginiac E (1989) Endogenous levels of abscisic acid and indole-3-acetic acid during *in vitro* rooting of wild cherry explants produced by micropropagation. Plant Growth Regul 8:325–333

55. Lin J, Mukherjee I and Reid DM (1990) Adventitious rooting in hypocotyl of sunflower (*Helianthus annuus*) seedlings. III. The role of ethylene. Physiol Plant 78:268–276

56. Maldiney R, Pelese F, Pilate B, Sotta B, Sossountzov L and Miginiac E (1986) Endogenous levels of abscisic acid, indole-3-acetic acid, zeatin-riboside during the course of adventitious root formation in cuttings of *Craigella* and *Craigella* lateral supressor tomatoes. Physiol Plant 68:426–430

57. Mato MC and Vieitez AN (1986) Changes in auxin protectors and IAA oxidases during the rooting of chesnut shoots *in vitro*. Physiol Plant 66:491–494

58. Mato MC, Rua ML and Ferro E (1988) Biochemical changes during root formation in *Vitis* cultured *in vitro*. Physiol Plant 72:84–88

59. Michniewiez M and Kriesel M (1970) Dynamics of auxins, gibberellin-like substances and growth inhibitors in the rooting process of black poplar cuttings (*Populus nigra* L.). Acta Soc Bot Pol 39:383–390

60. Molnar JM and Lacroix LJ (1972) Studies on the rooting of cuttings of *Hydrangea macrophylla*: enzyme changes. Can J Bot 50:315–322

61. Moncousin C (1991) Rooting of *in vitro* cuttings. In: YPS Bajaj, ed. Biotechnology in Agriculture and Forestry 17. High-Tech and Micropropagation, pp 231 261. Berlin: Springer Verlag

62. Moncousin C, Favre JM and Gaspar T (1988) Changes in peroxidase activity and endogenous IAA levels during adventitious root formation in vine cuttings. In: M Kutacek, RS Bandurski and J Krekule, eds. Physiology and Biochemistry of Auxins in Plants, pp 331–337. Praha: Academia

63. Moncousin C, Favre JM and Gaspar T (1989) Early changes in auxin and ethylene production in vine cuttings before adventitious rooting. Plant Cell Tiss Org Cult 19:235–242

64. Mousdale DMA, Fidgeon C and Wilson G (1985) Auxin content and growth parameters in auxin-dependant and auxin autotrophic plant cell and tissue culture. Biol Plant 27:257–264

65. Nagy M, Tari I and Buban T (1991) IAA distribution in the hypocotyls and primary leaves of *Phaseolus vulgaris* L. treated with paclobutrazol in relation to their rooting capacity. Biochem Physiol 187:447–451

66. Nakano NE, Nakagawa Y and Nakagawa S (1980) Studies on rooting of hardwood cuttings of grapevine cv Delaware. Jap Soc Hort Sci 48:385–394

67. Noiton D, Vines JH and Mullins MG (1992) Endogenous indole-3-acetic acid and abscisic acid in apple microcuttings in relation to adventitious root formation. Plant Growth Regul 11:63–67

68. Norcini JG and Heuser CW (1988) Changes in the level of ^{14}C indole-3-acetic acid and ^{14}C indoleacetylaspartic acid during root formation in mung bean cuttings. Plant Physiol 86:1236–1239

69. Nordström AC (1990) Auxin dynamics during adventitious root formation in pea (*Pisum sativum* L.) cuttings. Doct Thesis, Stockholm Univ

70. Odom RE and Carpenter WJ (1965) The relationship between endogenous indole auxins and the rooting of herbaceous cuttings. Proc Amer Soc Hort Sci 87:494–501

71. Patience PA and Alderson PG (1987) Development of system to study peroxidases during the rooting of lilac shoots produced *in vitro*. Acta Hort 212:267–272

72. Pilet PE and Gaspar T (1968) Le Catabolisme Auxinique. Monographie de Physiologie Végétale No.1. 148p. Paris: Masson

73. Pythoud F and Buchala A-J (1989) Peroxidase activity and adventitious rooting in cuttings of *Populus tremula* L. Plant Physiol Biochem 27:503–510

74. Quoirin M, Boxus P and Gaspar T (1974) Root initiation and isoperoxidases of stem tip cuttings from mature *Prunus* plants. Physiol Veg 12:165–174

75. Smith DR and Wareing PF (1972) Rooting of hardwood cuttings in relation to bud dormancy and the auxin content of the excised stems. New Phytol 71:63–80

76. Smith DR and Wareing PF (1972) The rooting of actively growing and dormant leafy cutting in relation to the endogenous hormone levels and photoperiod. New Phytol 71:483–500

77. Stolz PL (1968) Factors influencing root initiation in an easy- and a difficult-to-root *Chrysanthemum*. Proc Amer Soc Hort Sci 92:622–626

78. Thorpe TA, Tran Than Van M and Gaspar T (1978) Isoperoxidases in epidermal layers of tabacco and changes during organ formation *in vitro*. Physiol Plant 44:388–394

79. Trefois R and Brunner T (1982) Influence du contenu auxinique endogène sur la réponse au bouturage et sur l'effet nanifiant de quelques *Prunus*. Bot Kozlem 69:197–204

80. Van Hoof P and Gaspar T (1976) Peroxidase and isoperoxidase changes in relation to root initiation of *Asparagus* cultured *in vitro*. Sci Hort 4:27–31

81. Vasquez A and Mato MC (1991) Effect of hydroxybenzaldehydes on rooting and indole-3-acetic acid-oxidase activity in bean cuttings. Physiol Plant 83:597–600

82. Weigel U, Horn W and Hock J (1984) Endogenous auxin levels in stem cuttings of *Chrysanthemum morifolium* during adventitious rooting. Physiol Plant 61:422–428

83. Wu FT and Barnes MF (1981) The hormone levels in stem cuttings of difficult-to-root and easy-to-root rhododendrons. Biochem Physiol Pflanzen 176:13–22

Stimulation of rooting *in vitro*: effects of inhibitors of abscisic acid synthesis

B.M.R. HARVEY[1], C. SELBY[1] and G. BOWDEN[2]
[1]Department of Agricultural Botany, The Queen's University of Belfast, Newforge Lane, Belfast BT9 5PX, UK [2]School of Biology and Biochemistry, The Queen's University of Belfast

1. Introduction

The herbicides fluridone [1-methyl-3-phenyl-5-(α, α, α,-trifluoro-*m*-tolyl)-4-pyridone and norflurazon] [4-chloro-5-methylamino-2-(α, α, α –trifluoro-*m*-tolyl)-pyridazin-3(2*H*)-one] inhibit phytoene desaturase and thus block carotenoid biosynthesis [6]. Consequent photobeaching of chlorophyll accounts for phytotoxicity of the herbicides [1]. Plants treated with fluridone or norflurazon and grown in darkness remain healthy [1, 8] but have very low endogenous concentrations of abscisic acid [3], abscisic acid being synthesized from carotenoid precursors [7]. Dramatic root proliferation was observed when these inhibitors were used to investigate the role of abscisic acid in tuberisation of potato *in vitro*.

2. Materials and methods

Stock plantlets of potato cultivars (cv) Arran Banner, Cara, Desirée and Spunta were grown as previously described [4] and single node stem segments from the middle of the plant were excised for experimentation. These were inoculated into potato tuberisation medium [4] which contained 0.6×10^{-6}M benzylaminopurine (BAP). When required, abscisic acid and fluridone or norflurazon dissolved in dimethyl sulphoxide (DMSO) were added after autoclaving and DMSO was included at the same concentration in control media. One stem segment was inoculated into each vial and there were ten replicates of each treatment and cultivar in each experiment. The vials were incubated in darkness for 4–6 wk, the roots were then counted and the data subjected to analysis of variance.

3. Results

Analysis of variance showed a significant interaction between inhibitor, concentration and cultivar ($p<0.001$). Primary root number of cv Arran

P.J. Lumsden, J.R. Nicholas and W.J. Davies (eds.), Physiology, Growth and Development of Plants in Culture, 299–302, 1994.
© 1994 *Kluwer Academic Publishers.*

Banner was significantly increased by fluridone at $10^{-6}M$ and $10^{-5}M$ and cv Cara showed a significant increase with $10^{-5}M$ fluridone. Desirée and Spunta responded significantly only to norflurazon, at $10^{-6}M$ and $10^{-4}M$ respectively (Fig. 1).

Cv Arran Banner and fluridone were selected for further experiments. The first showed that inclusion of $10^{-7}M$ ABA in the culture medium was sufficient to prevent fluridone stimulation of root proliferation (Fig. 2). In contrast, the second showed that although the BAP in potato tuberisation medium caused significant inhibition of rooting (p<0.001), it did not interfere with fluridone

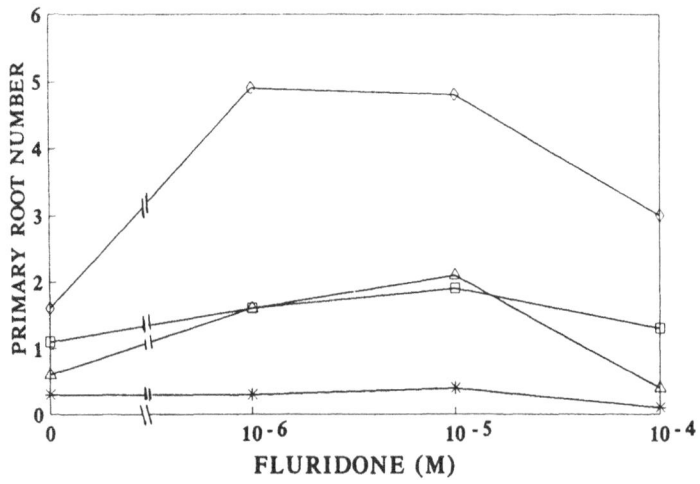

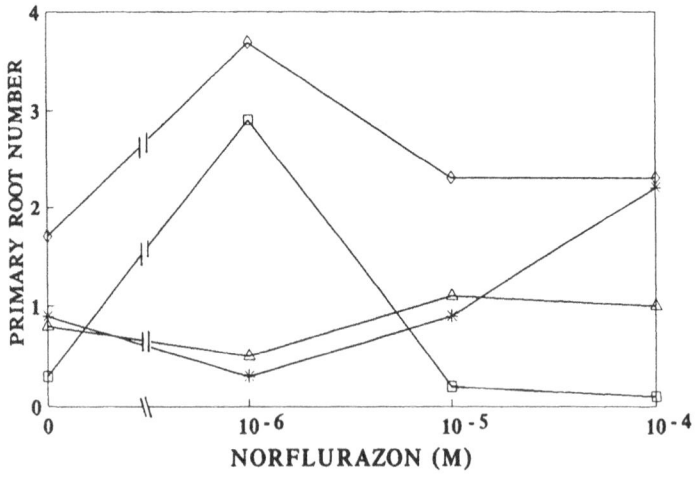

Fig. 1. The effects of fluridone and norflurazon on primary root number of nodal stem cultures of potato cultivars Arran Banner (◇), Cara (Δ), Desirée (*) and Spunta (□) incubated in darkness on tuberisation medium.

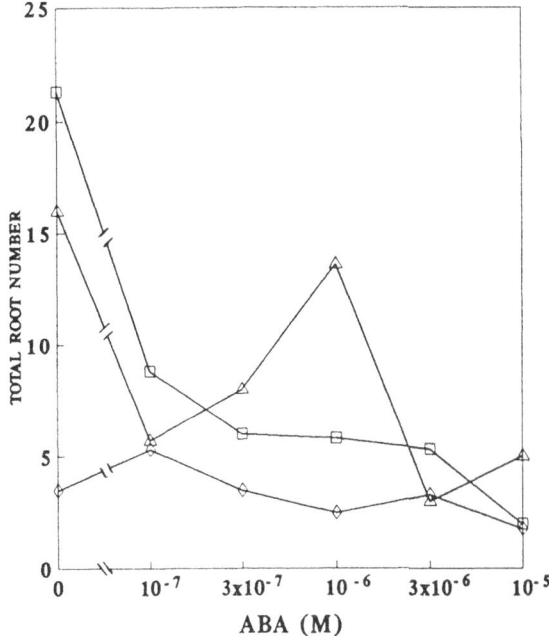

Fig. 2. The effects of exogenous ABA (log scale) and fluridone ($\diamond$,0; Δ,$10^{-6}M$; $\square$,$10^{-5}M$) on total root number of nodal stem cultures of cultivar Arran Banner incubated in darkness on tuberisation medium.

stimulation of rooting ($p<0.001$). There was no significant interaction between fluridone and BAP effects, indicating that fluridone was equally effective in the presence and absence of BAP (data not shown).

4. Discussion

The effective concentrations of fluridone and norflurazon differed greatly between the potato cultivars tested. As these cultivars have diverse genetic backgrounds, this could be due to differences in uptake, translocation or metabolism of the herbicides [2, 10]. However, all cultivars showed broadly similar responses i.e. a small but statistically significant increase in primary root numbers and a much larger increase in numbers of lateral roots.

Inclusion of $10^{-7}M$ ABA in the culture medium prevented fluridone stimulation of root proliferation, indicating that the fluridone effect was due to its activity as an inhibitor of ABA biosynthesis. This implies that ABA synthesis occurs when explants are cultured *in vitro*, perhaps as a stress response, and that ABA accumulates to levels sufficient to inhibit root formation. Reports of ABA inhibition of root production are infrequent but inhibition of rooting of stem cuttings by exogenous ABA has been noted in a few species [5, 9]. Both fluridone and norflurazon would prevent ABA synthesis in cuttings or explants, fluridone probably more effectively than norflurazon as it is a more potent inhibitor of phytoene desaturase than norflurazon [6].

The final experiment investigated the possibility that fluridone and norflurazon effects were artifacts of the presence of BAP in potato tuberisation medium. However, fluridone stimulation of rooting was obtained even in the absence of BAP. This indicates that fluridone was not simply overcoming BAP inhibition of rooting and is consistent with its action being due to inhibition of ABA biosynthesis. If ABA biosynthesis is a common stress response when plants are cultured *in vitro*, it may be possible to use ABA synthesis inhibitors to improve *in vitro* growth. However great care would be required as these inhibitors are herbicidal in light and so could only be used in darkness. This does not preclude brief post-inoculation treatments with fluridone to improve the *in vitro* rooting of recalcitrant species.

Acknowledgements

We thank Eli Lily and Company for the gift of fluridone, Sandoz AG for norflurazon and Mr C. Reavy and Ms S. Watson for assistance with experimentation and statistical analyses.

References

1. Bartels PG and Watson CW (1978) Inhibition of carotenoid synthesis by fluridone and norflurazon. Weed Sci 26:198–203
2. Berard DF, Rainey DP and Lin CC (1979) Absorption, translocation and metabolism of fluridone in selected crop species. Weed Sci 26:252–254
3. Gamble PE and Mullet JE (1986) Inhibition of carotenoid accumulation and abscisic acid synthesis in fluridone-treated dark-grown barley. Eur J Biochem 160:117–121
4. Harvey BMR, Crothers SH, Evans NE and Selby C (1991) The use of growth retardants to improve microtuber formation by potato (*Solanum tuberosum*). Plant Cell Tiss Org Cult 27:59–64
5. Heide OM (1968) Stimulation of adventitious bud formation in begonia leaves by abscisic acid. Nature 219:960–962
6. Mayer MP, Bartlett DL, Beyer P and Kleinig H (1989) The *in vitro* mode of action of bleaching herbicides on the desaturation of 15-cis-phytoene and cis- ζ –carotene in isolated daffodil chromoplasts. Pestic Biochem Physiol 34:111–117
7. Parry AD and Horgan R (1991) Carotenoids and abscisic acid (ABA) biosynthesis in higher plants. Physiol Plant 82:320–326
8. Sagar A and Briggs WR (1990) Effects of high light stress on carotenoid-deficient chloroplasts in *Pisum sativum*. Plant Physiol 94:1663–1670
9. Selby C, Kennedy SJ and Harvey BMR (1992) Adventitious root formation in hypocotyl cuttings of *Picea sitchensis* (Bong) Carr. – the influence of plant growth regulators. New Phytol 120:453–457
10. Strang RH and Rogers RL (1974) Behaviour and fate of two phenylyridazinone herbicides in cotton, corn and soybean. J Agric Food Chem 22:1119–1125

Rooting and acclimatization of chestnut by *in vitro* propagation

JOSÉ CARLOS GONÇALVES[1], SARA AMÂNCIO[2] and JOAO SANTOS PEREIRA[3]

[1] *Dep. de Biologia e Fisiologia Vegetal, Escola Superior Agrária, 6000 Castelo Branco, Portugal*
[2] *Dep. de Botânica e Eng. Biológica, I. Sup. Agronomia, Tapada da Ajuda, 1300 Lisboa, Portugal*
[3] *Dep. Eng. Florestal, Instituto Superior de Agronomia, Tapada da Ajuda, 1300 Lisboa, Portugal*

Abbreviations: IBA =indole-3-butyric acid; GD = Greshoff and Doy medium; MSm = Murashige and Skoog medium with half strength except nitrates at one quarter.

1. Introduction

Chestnut is a genus which is difficult to root by conventional means [12, 17, 18]. Under *in vitro* conditions it is necessary that the hormone IBA should be applied at a time when the rooting rate and the general health of the plantlet are so balanced as to maximize overall success after transfer it to soil [20]. Information on the transplanting and acclimatization is very scarce. Keys & Cech [7] report the difficulties found, Vieitez *et al.* [20] indicate only 35% of survival, and Mullins [9] refers to high mortality of juvenile material. This report describes the results obtained in a range of experiments on the rooting and acclimatization of chestnut by *in vitro* propagation.

2. Material and methods

The shoots used for these experiments were obtained from stock shoot multiplication cultures of adult hybrid chestnut clones 431 and M2 that were established and multiplied as described by Vieitez *et al.*, [19] and Gonçalves [5]. Their last subculture was in Murashige & Skoog [10] medium, with nitrates reduced to half strength and supplemented with 0.1 mgl^{-1} BAP. The shoots of 2 cm or more were decapitated at the beginning of the rooting treatment, and a warm drop of 30 mg l^{-1} BAP + agar was applied to the cut surface, ten days later, in an attempt to prevent necrosis and promote axillary growth [21]. Shoots were placed in 25 x 150 mm test tubes containing 15 ml of culture medium. We tested the mineral salts of Murashige & Skoog [10] with all macronutrients reduced to half strength, except nitrates which were reduced to one quarter, supplemented with the respective vitamins, 25 g l^{-1} sucrose and 6 g l^{-1} agar-agar, to be referred to as MSmod; and the mineral salts of Greshoff & Doy [6] with macronutrients at half strength, respective vitamins and sucrose and agar

P.J. Lumsden, J.R. Nicholas and W.J. Davies (eds.), Physiology, Growth and Development of Plants in Culture, 303–308, 1994.
© 1994 *Kluwer Academic Publishers.*

as above, to be referred to as 1/2GD. All media were adjusted to pH 5.5-5.6 before autoclaving.

Indole-3-butyric acid (IBA) was used as a growth regulator with concentrations and times of application dependent on the method used: IBA, at 1.5–5 mg l^{-1} in the culture medium during 7 days, or by dipping the base shoot in 1 g l^{-1} IBA for 30–120 s. In both cases, shoots were transferred to an auxin-free medium for root development.

All cultures were incubated in a growth chamber under day and night temperatures of 25 °C and 18 °C respectively, and a 16 h photoperiod of less than 30 μmol m^{-2} s^{-1} provided by cool white fluorescent tubes. Each rooting treatment was applied to 17 replicate cultures and all experiments were repeated twice; the results were analysed by two-way analysis of variance, and the averages compared by Duncan's multiple range test at 5%.

After 4–5 weeks, when the data were registered, the rooted plants were removed, rinsed gently to remove agar from the roots, and transplanted into 8 cm plastic pots containing a mix of peat:perlite (3:1, v:v) fertilized with osmocote. Plants were placed in a high humidity polyethylene tent with a fogging system, inside the greenhouse, during the first 8–9 weeks, with a gradually reducing RH. The plants were then transplanted to field pots containing peat:perlite:soil (2:1:2, v:v:v), and placed under automatic mist with a mist interval of 20 m (6 s duration).

3. Results

3.1 Application of IBA to the culture medium

The percentages of rooting were similar in both media (Table 1) but these values were significantly influenced by IBA concentration, with 5 mg l^{-1} giving the best results with 82.4% in both media. The number of roots formed was

Table 1. The effect of IBA concentration on two culture media on rooting percentage, root number, length of tallest root and average root length of micropropagated shoots of chestnut

IBA mgl^{-1}	Medium							
	1/2 GD				MS mod			
	Rooting (%)	Root number	Length of tallest root (mm)	Average root length (mm)	Rooting (%)	Root number	Length of tallest root (mm)	Average root length (mm)
1.5	5.9c	1.0c	20.5c	20.5b	8.9c	3.3a	21.0a	9.5a
3	50.0b	2.5b	49.3a	38.0a	32.4b	2.6a	13.9a	10.1a
5	82.4a	4.8a	36.2b	23.8b	82.4a	3.1a	19.4a	14.0a

Mean separation within columns by Duncan's multiple range test, 5% level. Percentages analysed after transforming them to arcsin values.

influenced both by the nutritive medium and by the concentration of IBA, with the combination 1/2GD + 5 mgl^{-1} IBA giving the best results, with 4.8 roots per rooted shoot (Table 1). 3 mgl^{-1} gave the longest roots and highest average root length for the 1/2GD medium.

Apparently, these rooted shoots are physiologically normal and have an also normal development, with a developing axillary shoot, stimulated by the drop with BAP, which plays the role of the decapitated shoot apex (Plate 1).

3.2 Application of IBA by dipping

This method produced less impressive results than those mentioned above. The percentage of rooting was influenced by the time of exposure to the solution, and the best result was achieved with 30 s for the clone 431, with 74% rooting, and 60 or 120 s for the clone M2, with 50 and 44% rooting respectively. There were no statistical differences betwen other variables in each of the two clones (Table 2).

Table 2. The effect of time of dipping on IBA concentrated solution on rooting percentage, root number, length of tallest root and average root length of two micropropagated clones of chestnut

Time (s)	Clone							
	M2				431			
	Rooting (%)	Root number	Length of tallest root (mm)	Average root length (mm)	Rooting (%)	Root number	Length of tallest root (mm)	Average root length (mm)
30	21.0b	2.1b	23.7a	17.0a	74.0a	2.2a	29.5a	26.5a
60	50.0a	3.3a	23.0a	19.2a	53.0ab	2.4a	28.3a	23.3a
120	44.0a	3.0ab	23.0a	18.2a	44.0b	2.7a	26.9a	23.2a

Mean separation within columns by Duncan's multiple range test, 5% level. Percentages analysed after transforming them to arcsin values.

3.3 Transplanting and acclimatization stage

Eight weeks after the *in vitro* regenerated plants were transferred to the greenhouse, the percentage survival was 60% and the morphological development of the plants seemed normal (Plate 2). Plants failing to survive did not show any sign of morpho-physiological activity, although they remained green. These plants did not form any new roots, and the roots that developed *in vitro* were dried and probably not functional. Another cause of death was the rotting of the base of the stem, probably due to fungal contamination.

Plate 1. Rooted chestnut shoots in 1/2GD + 5 mgl^{-1} IBA. Notice the axillary shoot development.

Plate 2. In vitro regenerated plant of chestnut after 8 weeks of acclimatization.

4. Discussion

The rooting stage remains a critical process in chestnut micropropagation, both because of problems with root induction and because the physiological status of the shoot during and after the rooting process. In fact, shoot-tip necrosis is a common disorder, not only in this genera [2, 3, 9], but also in others [1, 4, 13] which can seriously affect the *in vitro* culture. The methodology proposed by Vieitez *et al.* [21] to prevent these problems and used in these experiments, although it is a meticulous operation, permits the induction of axillary bud break, even though there is an inverse relation between the percentage of bud break and the elongation of the axillary shoot with the percentage of rooting. This could suggest incompatibility between the action of auxins and cytokinins in chestnut regeneration.

IBA plays an important role in the genesis of the root primordia that will affect the total number of differentiated roots, but has no effect on the morphological development of the root system, in which the nutritive medium seems more important. Both the media used have low levels of salts, which is an important aspect during the rhizogenesis process in many hardwoods [8, 11, 14, 19] but they differ in the ratio NH_4^+/NO_3^- and also in the form of nitrogen supplied (ammonium nitrate for MS and ammonium sulfate for GD). These aspects may modify the pattern of the adventitious root system [15]. Calcium can also be responsible for differences in root development, since it may play a role in the prevention of destruction of some substances that act as auxin protectors during rhizogenesis [16].

The survival rates obtained in these experiments are very encouraging. Media composition seems to be important in determining the physiological status of the *in vitro* regenerated plants, enabling plants to better resist the stress resulting from the transfer to greenhouse or field conditions.

Acknowledgements

Thanks are due to Dr Ana M. Vieitez and the Inst. Inv. Agrobiológicas de Galicia, Spain, for their valuable help and microculture stock of clone 431; to Dr M. Jesus Candeias for having read the English version of this paper.

References

1. Barghchi M, Alderson PG (1985) *In vitro* propagation of *Pistacia vera* L. and commercial cultivars Ohadi and Kallesghochi. J Hort Sci 60:423–430
2. Biondi S, Canciani L, De-Paoli G, Bagni N (1984) Studio della micropropagazione di cultivar da legno di *Castanea sativa* Mill.. In: G Savoia, ed. Propagazione *in vitro*: Ricerche su Alcune Specie Forestali, pp 108–114. Bologna: Azienda Regionale delle Foreste
3. Chevre AM (1985) Recherches sur la Multiplication Végétative *In Vitro* Chez le Châtaignier. These Doctoral (100 pp) Université de Bordeaux II

4. Druart P, Boxus P, Liard O, Delaite B (1981) La micropropagation du merisier a partir de la culture de méristême. In: M Boulay, ed. Proc Coll Int sur la Culture *in vitro* des Essences Forestières, pp 101–108. Fontainebleau: Ass Forêt-Cellulose

5. Gonçalves, JC (1990) Multiplicação *in vitro* de castanheiro (*Castanea* Miller) por rebentamento axilar. Proc 2nd Forestry National Conf (p 74) Porto

6. Greshoff PM, Doy CH (1972) Development and differentiation of haploid *Licopersicon esculentum* (tomato). Planta 107:161–170

7. Keys RN, Cech FC (1982) Propagation of American chestnut *in vitro*. In: Proc. USDA For Service Chestnut Research Cooperators Meeting, pp 106–110. Morgantown

8. Manzanera JA, Pardos JA (1990) Micropropagation of juvenile and adult *Quercus suber* L.. Plant Cell Tiss Org Cult 23:115–123

9. Mullins KV (1987) Micropropagation of chestnut (*Castanea sativa* Mill.). Acta Hort 212:525–530

10. Murashige T, Skoog F (1962) A revised medium for rapid bioassays with tobacco cultures. Physiol Plant 15:473–497

11. Quoirin M, Lepoivre PH, Boxus P (1977) Un premier bilan de 10 annés de recherches sur la culture de méristèmes et la multiplication *in vitro* de fruitiers ligneux. C R Rech, pp 93–117. Gembloux: Stn Cult Fruit Maraichers

12. Schad C, Solignat G, Grente J, Venot P (1952) Recherches sur le chataignier à la Station de Brive. Ann Amél Plantes 2:369–458

13. Sha L, McCown BH, Peterson LA (1985) Ocurrence and cause of shoot-tip necrosis in shoot cultures. J Am Soc Hort Sci 110:631–634

14. Skirvin RM, Chu MC Rukan H (1980a) Rooting studies with *Prunus* sp. *in vitro*. Hort Sci 15:83-
-415

15. Sriskandarajah S, Skirvin RM, Abu-Qaoud H (1990) The effect of some macronutrients on adventitious root development on scion apple cultivars *in vitro*. Plant Cell Tiss Org Cult 21:185–189

16. Stonier T (1971) The Role of Auxin Protectors in Autonomous Growth. In: Centre Nat Rech Sci (Ed) Les Cultures de Tissues de Plantes, pp 423–435. Paris

17. Urquijo P (1952) Hacia la solución del problema del castaño. Lombardero Press, La Coruña

18. Vieitez E (1952) Ensayos de reproducción vegetativa de híbridos de castaño: *Castanea sativa* x *C. crenata*. An Edafol Fisiol Veg 11:185–209

19. Vieitez AM, Ballester A, Vieitez ML, Vieitez E (1983) *In vitro* plantlet regeneration of mature chestnut. J Hort Sci 58:457–463

20. Vieitez AM, Vieitez ML, Vieitez E (1986) Chestnut (*Castanea* spp.). In: YPS Bajaj, ed. Biotechnology in Agriculture and Forestry Vol 1 Trees, pp 393–414. Berlin: Springer Verlag

21. Vieitez AM, Sánchez C, San-José C (1989) Prevention of shoot-tip necrosis in shoot cultures of chestnut and oak. Sci Hort 41:151–159

Acclimatization of micropropagated roses in multi-layer-cells: effect of different stage III conditions and CO_2-enrichment

J. DE RIEK and J. VAN HUYLENBROECK

Department of Plant Production, Laboratory of Horticulture, University Gent, Coupure 653, 9000 Gent, Belgium

1. Introduction

Recently, interest in the photosynthetic capacities of tissue cultured plants and shoots has been growing. It is often claimed that at the end of the *in vitro* period the shoots or plants should be as photoautotrophic as possible, so that acclimatization under greenhouse conditions occurs with maximum survival of plants. According to Kozaï [6] complete photoautotrophic micropropagation can be a solution. To obtain plants that are able to grow on a sugar free medium, the external CO_2-concentration and the light intensities in the culture rooms must be enhanced. Other results [1, 2] suggest that a sufficient carbon reserve in the form of a starch accumulation in the chloroplasts is more important; increased starch accumulation has been observed on media with a higher sucrose content. This carbon reserve can be used during the first days after transplantation to recover and to initiate growth.

In an attempt to understand carbon metabolism during micropropagation and acclimatization, the different culture stages for *Rosa multiflora* L. 'Montsé' were studied as a model system. The 'Double Layer Technique' was used; this involves the application of a liquid medium supplement to the culture, on top of the existing multiplication culture. Here 20 ml of a mineral solution [5] was added containing charcoal and sucrose. The purpose of this treatment is to induce shoot elongation and root induction but it seems also to function as an *in vitro* acclimatization stage. The aim of the experiments was to regulate in the last *in vitro* stage the carbon metabolism in such a way that a better behaviour during acclimatization could be obtained.

2. Materials and methods

Cultures of *Rosa multiflora* L. 'Montsé' were maintained under conditions as described by Cappelades [1]. To control the relative humidity inside the jars during Stage III, a bottom cooling system was used [4]. Other experimental details are given in the results.

309

P.J. Lumsden, J.R. Nicholas and W.J. Davies (eds.), Physiology, Growth and Development of Plants in Culture, 309–313, 1994.
© *1994 Kluwer Academic Publishers.*

3. Results

3.1 Experiments on stage III with ^{14}C-sucrose

To stimulate a carbohydrate accumulation at the end of the *in vitro* period, an extra amount of sucrose was added in the liquid Stage III supplement. Data for respiration, sucrose uptake and photosynthesis on Stage II (multiplication) [4] and Stage III (shoot elongation and root induction) cultures (Table 1) were obtained from experiments with radiolabeled ^{14}C-sucrose. Three different sucrose concentrations in the added liquid Stage III culture medium were tested (0%, 3% and 5% sucrose). Extra addition of sucrose in the liquid Stage III medium appeared to improve the growth rate during Stage III. This might be due to: 1. a higher sucrose incorporation; 2. an increase in stress-induced

Table 1. Overview of the different rates for biomass increase, sucrose uptake, sucrose incorporation, respiration and photosynthesis as determined from 3 Stage III experiments with ^{14}C-sucrose

STIII sucrose content	biomass increase	sucrose uptake	incorporation	respiration	photosynthesis
as absolute rates (mg C/jar/week)					
0%	114,098	123,383	85,610	37,772	28,487
3%	92,647	136,594	79,456	57,137	13,191
5%	144,16	158,574	116,491	42,082	27,669
relative to the biomass at the end of Stage II (%week)					
0%	35,840	38,757	26,892	11,865	8,948
3%	41,375	61,001	35,484	25,517	5,891
5%	47,708	52,478	38,551	13,927	9,157
relative to the biomass increase (%)					
0%	**100**	108,138	**75,032**	33,105	**24,967**
3%	**100**	147,434	**85,762**	61,672	**14,238**
5%	**100**	109,998	**80,807**	29,192	**19,193**
relatively to the sucrose uptake (%)					
0%	92,474	**100**	**69,385**	**30,613**	23,089
3%	67,827	**100**	**58,170**	**41,830**	9,657
5%	90,910	**100**	**73,462**	**26,538**	17,449

Note : These experiments were performed on 3 different Stage II cultures. After Stage II a average biomass content of 636.7 mg DM/jar was reached for the 0% sucrose Stage III treatment; 447.8 mg DM/jar for the 3% sucrose treatment and 604.3 mg DM/jar for the 5% sucrose treatment.

respiration, which can be supported by the extra available sucrose (e.g. the treatment with 3% sucrose). Table 1 shows that no clear decrease in the photosynthetic CO_2-fixation occurred in the presence of extra sucrose.

3.2 Acclimatization in multi-layer-cells at two CO_2-concentrations

An acclimatization experiment in multi-layer-cells (controlled light, temperature, relative humidity and CO_2-concentration conditions) was performed over 4 weeks on plants coming from the 3 different Stage III treatments. Growth rates, water retention capacities and concentrations of chlorophyll, protein, nucleic acid, sucrose and starch were examined during acclimatization with the plantlets transplanted to peat. Growth conditions in the multi-layer-cells were 16 h light (PAR $\pm$ 85 μmol m^{-2} s^{-1} at plant level) – 8 h dark. The temperature was 25 $\pm$ 1 °C. Two different CO_2-concentrations in the air were maintained: 400 and 1000 μl l^{-1}. During the first acclimatization period a high relative humidity was maintained : 95% R.H.. After rooting (day 17) occurred the relative humidity was lowered to 75%.

Growth. Three phases were observed in the growth rate (Fig. 1): 1. An adaptation period with some loss of biomass; 2. A period with slow shoot growth but a fast rooting; 3. A period with fast shoot growth. The growth was

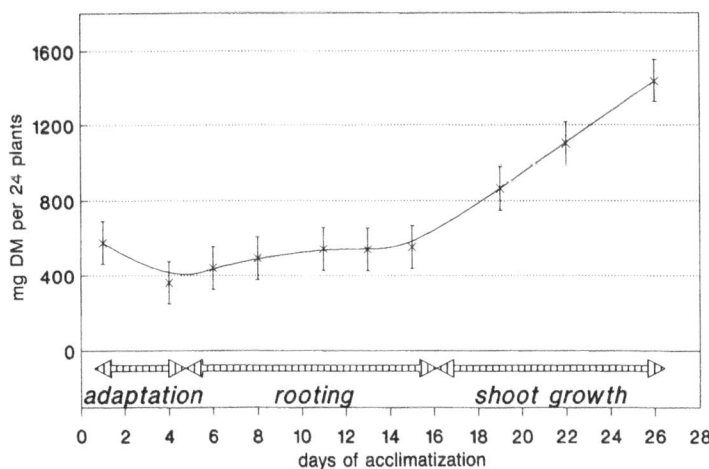

DM-increase of the shoots
(plantlets without the roots)

Fig. 1. Dry matter (DM)-increase of the shoots (plantlets without the roots) of *Rosa multiflora* over the acclimatization period as the average of the 6 treatments (error bars according to 90% LSD). Per treatment and per day a sample of 24 plantlets was taken. Three different growth phases are indicated. Losses of plantlets over acclimatization were neglectable.

faster for the treatments with a higher sucrose concentration in Stage III and for those plants grown under a higher CO_2-concentration during acclimatization (data not shown).

Water housekeeping. Water housekeeping was studied by following the waterloss of harvested plantlets exposed to a dry environment as a function of time; 24 plants were harvested per treatment, and were divided between 6 petridishes. The fresh weight was determined immediately. The water content during dessication was determined by weighing the petridishes at regular intervals when standing for 3 h in a room with 23 °C and 60% R.H. Afterwards the dry weight (24h at 105 °C) was measured. Water losses by transpiration (% waterloss per min) were calculated from these subsequent weighings [1].

During the adaptation a decrease in water content and in transpiration rate was observed. During rooting and shoot growth the water content and the transpiration rate increased again but slowly declined to the end of the acclimatization period.

The effect of the different sucrose treatments on water housekeeping was closely related to the relative humidity maintained in the cells. During the adaptation and rooting period (95% R.H.) the plants grown on Stage III media with a higher sucrose content showed a lower water content and a higher transpiration rate. During the shoot growth period (75% R.H.), those plants had a higher water content and a lower transpiration rate (significant at 90% LSD). CO_2-supplementation slighty lowered the water content, but the transpiration rate was significantly decreased.

Content of chlorophylls, carotenoids, nucleic acids, proteins, sucrose and starch. The content of the pigments gradually increased over the adaptation and rooting period. Before the start of shoot growth a decrease was observed. During shoot growth the pigment content increased again. CO_2-supplementation significantly increased the pigment contents. Nucleic acid content showed a peak in concentration on day 6 at the beginning of the rooting. This might be due to increased meristematic activity. Protein content was rather stable, but showed a peak at the end of the rooting. There were no significant differences between different sucrose and CO_2 treatments.

At the end of *in vitro* micropropagation the cultures grown with the higher sucrose treatments in Stage III showed a significantly higher content of total carbohydrates (90% LSD). Both starch and sugar (sucrose, glucose and fructose) concentrations were higher. During adaptation the total carbohydrate content first decreased because of starch decrease. Over the rest of the acclimatization period sucrose and starch gradually increased. Before shoot growth started, a small decrease in starch but a complete depletion of sucrose was observed. Finally both increased again. Plants grown with higher sucrose concentration in Stage III showed better translocation of sugars during acclimatization. This was indicated by a lower starch accumulation and a higher sucrose content at the end of the light period.

4. Discussion

Sucrose supplementation in Stage III stimulated growth during the last *in vitro* stage and did not block photosynthesis (Table 1), and improved acclimatization. Plants coming from media with a higher sucrose concentration showed:

- an increased growth
- better transpiration at high R.H.
- lower transpiration at low R.H.
- better translocation of the fixed carbon as sucrose

CO_2-supplementation during acclimatization resulted in:

- better water housekeeping
- higher pigment content
- higher carbohydrate content (sucrose and starch).

References

1. Cappelades MQ (1989) Histological and ecophysiological study of the changes occurring during the acclimatization of *in vitro* cultures, Dissertation, Gent State University, Belgium
2. Cappelades MQ, Lemeur R & Debergh PC (1990) Kinetics of chlorophyll fluorescence in micropropagated rose shootlets. Photosynthetica 24:190–193
3. Debergh PC & Read PE (1991) Micropropagation. In: PC Debergh and RH Zimmerman, eds. Micropropagation, Technology and Application. Dordrecht: Kluwer Academic Publishers
4. De Riek J, Van Cleemput O & Debergh PC (1991) Carbon metabolism of micropropagated *Rosa multiflora* L. In Vitro Cell Dev Biol 27:57–63
5. Knop W (1865) Quantitative Untersuchungen über die Ernahrungsprocesse der Planzen. Landwirtsch Vers Stn 7:93
6. Kozaï T (1991) Photoautotrophic micropropagation, Invited review. In Vitro Cell Dev Biol 27:47–51
7. Maene L & Debergh PC (1985) Liquid medium additions to established tissue cultures to improve elongation and rooting *in vivo*. Plant Cell Tiss Org Cult 5:23–33
8. Vanderschaeghe AM & Debergh PC (1987) Technical aspects of the relative humidity in tissue culture containers. In: G Ducate, M Jacobs & A Simeon, eds. Symposium Plant Micropropagation in Horticultural Industries, Preparation, hardening and acclimatization processes, Belgian Plant Tissue Culture Group, Florizel 87, 10–14 August 1987, Arlon, Belgium

Stage III techniques for improving water relations and autotrophy in micropropagated plants

A.V. ROBERTS, E.F. SMITH, I. HORAN, S. WALKER, D. MATTHEWS and J. MOTTLEY
Plant Biotechnology Research Unit, Department of Life Sciences, University of East London, Romford Road, London E15 4LZ, UK

1. Introduction

The precision with which the physical and chemical environment of plantlets cultured *in vitro* can be controlled has created unique opportunities for the manipulation of developmental processes such as adventitious regeneration, shoot multiplication and root initiation. Surprisingly, it has proved difficult to create environments that adequately prepare plantlets for the transition to life outside the culture vessel. It seems that, for this purpose, a wider range of conditions must be explored.

The pragmatic solution to problems of plant survival after transplantation to soil has been to acclimatize plants gradually in a humid atmosphere over a period of 4–16 weeks. However, acclimatization is expensive and mortality rates are often high. This has stimulated interest in the production of more robust plants by modification of Stage III, the final phase of *in vitro* culture [10]. The general requirement is for:
- improved water conservation by shoots, which requires
  - stomata that close when water-stressed
  - increased wax formation in the cuticle
- improved root systems to replace water lost by the shoots
- improved CO_2 fixation to support autotrophic growth after transplantation.

This paper considers the prospects and benefits of meeting these requirements.

2. Rooting in plugs

Gelling agents such as agar have commonly been used to provide plantlets with physical support at Stage III. However, as *in vitro* formed roots are thin and fragile, they are easily damaged when plantlets are removed from the gel and replanted in soil. This problem can be overcome by rooting plantlets at Stage III in plugs saturated with liquid culture medium (Plate 1). The plant and plug are then transferred together during transplantation and the roots remain

P.J. Lumsden, J.R. Nicholas and W.J. Davies (eds.), Physiology, Growth and Development of Plants in Culture, 314–322, 1994.

undisturbed in the protective matrix of the plug. For *in vitro* use, plugs must be clean, coherent and have high capillarity. Cellulose plugs known as Sorbarods, manufactured by Baumgartner Papiers, Lausanne, Switzerland, were found to meet these criteria and improve the ability of the root system to replace water lost by the shoots of chrysanthemum after transplantation [12].

Plate 1. Plantlet of chrysanthemum 'Pennine Reel' rooted in Sorbarod plug.

3. Culture media

Acclimatization problems are accentuated in plants that become vitrified in culture. These plants have reduced deposition of cellulose and lignin, which causes reduced cell wall pressure. This leads to hyperhydration and thus a glassy turgescence of leaves and stems. Causative factors include high concentrations of ammonium ions, calcium ions and cytokinins in the culture medium, and the production of ethylene by the plantlets [19]. These factors may also be detrimental to plantlets that do not show the extreme condition of vitrification and attention should be paid to that possibility.

Following reports of *in vivo* studies in which applications of growth retardants improved tolerance to water stress [4], the effect of eleven growth retardants on the wilting of micropropagated chrysanthemum after transfer to soil was investigated [16]. Six gave significant reductions in wilting when included at concentrations of 0.3 mg l⁻¹ in the Stage III culture medium and, at

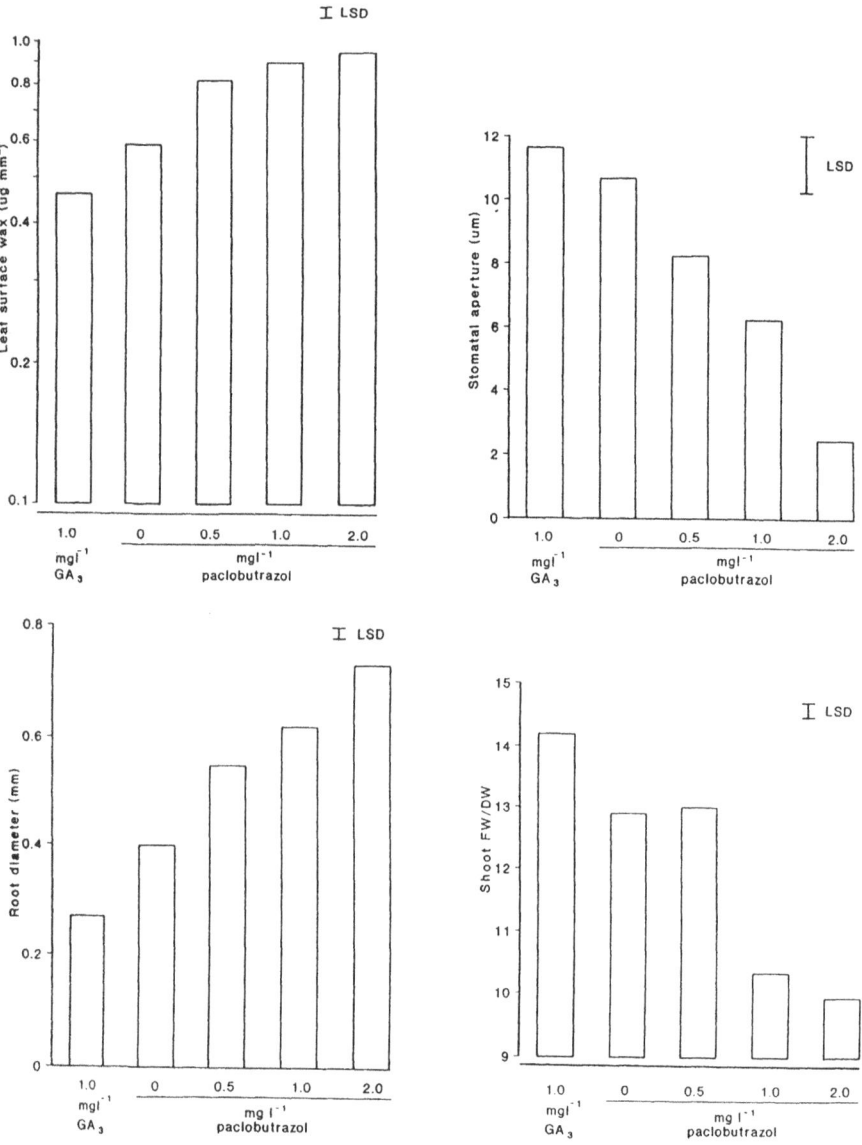

Fig. 1. The effect of GA_3 and paclobutrazol on surface wax of leaves (log transformed scale), stomatal aperture (fourth fully expanded leaf from apex, after exposure to 20% relative humidity for 18h at 23 °C), root diameter and shoot fresh weight/dry weight (FW/DW) of chrysanthemum 'Pennine Reel'. Least significant differences (LSD) are indicated. (After Smith *et al.* [13]).

concentrations of 3.0 mg l^{-1}, reduced wilting to negligible proportions over a period of 3 h at 27.5 °C and a relative humidity of 42%. One of these growth retardants, paclobutrazol, has been shown to induce increased deposition of epicuticular wax, improved stomatal closure in response to water stress and a thickening of roots [14] (Fig. 1). Similar responses to paclobutrazol were found in grapevine [13] and rose [Smith *et al.*, in press]. Paclobutrazol induced a reduction in the fresh weight/dry weight ratio in chrysanthemum (Fig. 1) which is consistent with evidence in other species that it reduces vitrification [18, 20]. The fact that gibberellic acid had the opposite effect of paclobutrazol on most of the characters studied in chrysanthemum (Fig. 1) implicated blockage of gibberellin synthesis as the causal mechanism. However, paclobutrazol and some other growth retardants are also known to block sterol biosynthesis [1]. Paclobutrazol consists in equal proportions of two enantiomers, the 2S, 3S form which blocks gibberellin synthesis and the 2R, 3R form which blocks sterol synthesis. When these two enantiomers were applied separately, only the 2S, 3S form significantly improved water relations in micropropagated chrysanthemum [Roberts and Matthews, in press]. The possible influence of growth retardants on abscisic acid synthesis has not been adequately tested, however.

4. Humidity

It has been known for many years [17] that high humidity in culture vessels is detrimental both to the development of surface wax and stomatal closure. Recently, commercially viable methods have been developed for hardening micropropagated plantlets by reducing the humidity of the atmosphere inside the culture vessel. One method is to chill the culture medium by the use of cooled shelves, so that water vapour condenses on the medium and the humidity in the warmer air above is reduced [9]. Another approach is to use culture vessels with closures that incorporate air filters [2, 8, 15]. The use of a culture vessel developed by Baumgartner Papiers, in which Tyvek inserts in the lid reduced the relative humidity from approximately 100% to 94% (Plate 2) led to an increased resistance to desiccation in chrysanthemum [15], grapevine [13] and rose [Smith *et al.*, in press]. This resistance to desiccation was associated with improved stomatal closure in each of these studies and an increased deposition of epicuticular wax in chrysanthemum and rose. In all three species, the combined use of Sorbarod plugs, paclobutrazol and reduced humidity at Stage III led to the production of plantlets that did not require acclimatization after transfer to soil.

There is little experimental evidence to assist in the interpretation of the observed responses to reduced humidity but it seems likely that:
- increased transpiration stimulates plant growth by enhancing acropetal transport of dissolved nutrients,
- evaporation from the cuticle draws wax precursors to the leaf surface,

Plate 2. Culture vessels (Baumgartner Papiers) containing 60 plantlets of rose 'Mountbatten' rooted in Sorbarods. The lid (opened) has five circular holes along each flank covered by a strip of Tyvek to permit gaseous exchange.

- developing guard cells are more completely closed while cellulose is being deposited and this influences their shape at maturity.

In connection with the latter postulate, it should be noted that, in vitreous carnation, stomatal closure is impaired by malformation of the guard cell wall, not by abnormal protoplast physiology [21].

5. Photosynthetic activity

For many years, plantlets have been cultured *in vitro* almost exclusively under artificial light of low irradiance, on media containing sucrose and in partially closed culture vessels, in which CO_2 concentrations fall to low levels during the photoperiod [7]. Photosynthesis is not sufficiently active to produce a net positive carbon balance under these conditions. In studies on cauliflower, this was attributed to low levels of chlorophyll and ribulose bisphosphate carboxylase activity [6]. If plantlets are exposed to a sufficiently high irradiance and CO_2 enrichment, mixotrophic growth occurs on sucrose-supplemented medium and autotrophic growth occurs on sucrose-free medium [7]. Surprisingly, under both mixotrophic and autotrophic conditions, growth of carnation is enhanced if media based on the complex MS [11] formulation are replaced by Enshi medium which is a simple formulation of salts used for hydroponic culture [7, 8].

In a recent investigation of autotrophic growth of roses at Stage III, the required irradiance for photosynthesis was obtained by placing culture vessels in an unshaded greenhouse in mid-summer [Walker *et al.*, in press]. Plantlets were cultured on Sorbarod plugs soaked in sucrose-free hydroponic culture medium in vessels with Tyvek inserts in the lid to allow gaseous exchange with the external environment. Under these conditions, faster shoot growth occurred over 4 weeks than in control plantlets grown on MS-based medium

Plate 3. Plantlets of rose 'Mountbatten' rooted in Sorbarods in vessels (Baumgartner Papiers, lids removed) four weeks after initiation from shoot tips. Growth on MS medium supplemented with 30 $g\,l^{-1}$ sucrose under cool-white fluorescent lights, irradiance $4\,W\,m^{-2}s^{-1}$, 16 h photoperiod at 23°C (top) is contrasted with growth on Hewitt's hydroponic salts without sucrose or temperature control in a greenhouse with mid-day irradiance of $16-189\,W\,m^{-2}\,s^{-1}$ (bottom). (After Walker *et al.*, in press).

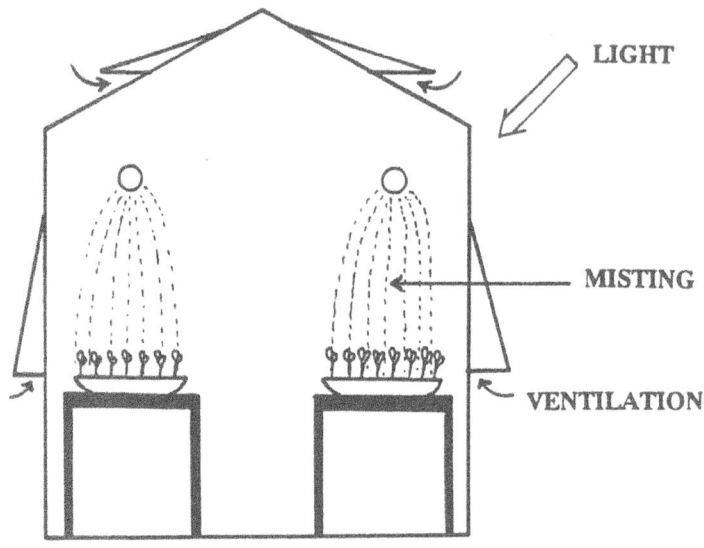

STAGE III CULTURES IN CONSERVATORY

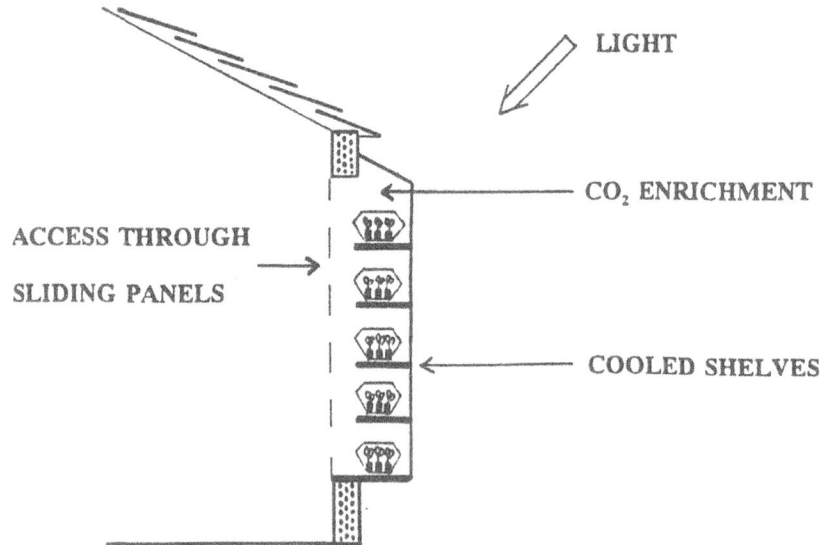

Fig. 2. Diagram contrasting the culture of seedlings under greenhouse conditions with the proposed culture of plantlets at Stage III of micropropagation in a conservatory. Seedlings are commonly cooled by ventilation and misting but cooled shelves (Maene and Debergh [9]) would minimize contamination of Stage III cultures. The volume of the Stage III conservatory has been minimized, to reduce cost of temperature control and CO_2 enrichment, by stacking shelves and eliminating corridors.

supplemented with 30 g l^{-1} sucrose in a 16 h photoperiod under growth room conditions (Plate 3). Furthermore, the root/shoot ratio was significantly higher in plantlets grown under greenhouse conditions, which indicates a better prognosis for replacement of water lost by shoots after transplantation to soil. No CO_2 enhancement or artificial light was used in the greenhouse environment and the superior growth of these autotrophic plantlets can clearly be achieved at lower cost than by conventional procedures. This establishes a base-line of cost-effectiveness against which the benefits of supplementary CO_2 and lighting, and improved greenhouse design can be assessed in future investigations.

The *in vitro* culture of plants in greenhouses has been previously suggested by De Fossard and Bourne [5]. Greenhouse culture of Stage III plantlets might appropriately be modelled on efficient practises currently used for greenhouse production of seedlings. Like *in vitro* plantlets, seedlings are of small size and are often grown on artificial substrates fed with mineral solutions. Various modifications of greenhouse design might be advantageous, as outlined in Figure 2.

6. Conclusions

At present, despite the advantages of year-round production of disease-free clones, commercial micropropagation is mainly restricted to horticultural crops and annual production, world-wide, has been estimated at only 258 million plants [3]. Expansion of micropropagation into the relatively huge market for agricultural and forestry species requires a reduction in production costs by 80–90% [7]. A radical revision of current practices would be needed to achieve this. The elimination of the need for acclimatization, the enhancement of the latent photosynthetic ability of plantlets, simplification of culture media and the use of low-cost greenhouse environments may contribute significantly to this goal.

References

1. Burden RS, Carter GA, Clark T, Cooke DT, Croker SJ, Deas AHB, Hedden P, James SS and Lenton JR (1987) Comparative activity of the enantiomers of triadimenol and paclobutrazol as inhibitors of fungal growth and plant sterol and gibberellin biosynthesis. Pestic Sci 21:253–267
2. Cassells A (1987) Plant package. US patent no 5, 054,234
3. Chu IYE (1992) Perspective of micropropagation industry. In: K Kurata and T Kozai, eds. Transplant Production Systems, pp 137–150. Dordrecht: Kluwer Academic Publishers
4. Davies TD, Steffens GL and Sankhla N (1988) Triazole plant growth regulators. Hort Rev 10:63–105
5. De Fossard RA and Bourne RA (1977) Reducing tissue culture costs for commercial propagation. Acta Hort 78:37–44
6. Grout BWW and Donkin ME (1987) Photosynthetic activity of cauliflower meristem cultures *in vitro* and at transplanting into soil. Acta Hort 212:323–327
7. Kozai T (1991) Autotrophic micropropagation. In YPS Bajaj, ed: Biotechnology in Agriculture and Forestry Vol. 17, pp 313–343. Berlin: Springer-Verlag

8. Kozai T, Kubota C and Watanabe I (1988) Effects of basal medium composition on the growth of carnation plantlets in auto- and mixo-trophic tissue culture. Acta Hort 230:159–166

9. Maene LJ and Debergh PC (1986) Optimization of Plant Micropropagation. Med Fac Landbouww Rijksuniv Gent 51/4:1479–1486

10. Murashige T (1974) Plant propagation through tissue cultures. Ann Rev Plant Physiol 25:135–166

11. Murashige T and Skoog F (1962) A revised medium for rapid growth and bioassays with tobacco tissue cultures. Physiol Plant 15:473–497

12. Roberts AV and Smith EF (1990) The preparation *in vitro* of chrysanthemum for transplantation to soil 1. Protection of roots by cellulose plugs. Plant Cell Tiss Org Cult 21:129–132

13. Smith EF, Gribaudo I, Roberts AV and Mottley J (1992) Paclobutrazol and reduced humidity improve resistance to wilting of micropropagated grapevine. Hort Sci 27:111–113

14. Smith EF, Roberts AV and Mottley J (1990) The preparation *in vitro* of chrysanthemum for transplantation to soil. 2. Improved resistance to desiccation conferred by paclobutrazol. Plant Cell Tiss Org Cult 21:133–140

15. Smith EF, Roberts AV and Mottley J (1990) The preparation *in vitro* of chrysanthemum for transplantation to soil. 3. Improved resistance to desiccation conferred by reduced humidity. Plant Cell Tiss Org Cult 21:141–145

16. Smith EF, Roberts AV, Mottley J and Denness S (1991) The preparation *in vitro* of chrysanthemum for transplantation to soil. 4. The effects of eleven growth retardants on wilting. Plant Cell Tiss Org Cult 27:111–113

17. Wardle K, Dobbs EB and Short KC (1983) *In vitro* acclimatization of aseptically cultured plantlets to humidity. J Amer Soc Hort Sci 108:386–389

18. Ziv M (1989) Enhanced shoot and cormlet proliferation in liquid cultured gladiolus buds by growth retardants. Plant Cell Tiss Org Cult 17:101–110

19. Ziv M (1991) Quality of micropropagated plants — vitrification. In Vitro Cell Dev Biol 27P:64–69

20. Ziv M and Ariel T (1991) Bud proliferation and plant regeneration in liquid cultured philodendron 'Burgundy' treated with ancymidol and paclobutrazol. J Plant Growth Regul 10:53–57

21. Ziv M, Schwarts A and Fleminger D (1987) Malfunctioning stomata in vitreous leaves of carnation (*Dianthus caryophyllus*) plants propagated *in vitro*; implications for hardening. Plant Sci 52:127–134

Physiological change and apparent rejuvenation of temperate fruit trees from micropropagation

O.P. JONES
Horticulture Research International, East Malling, Kent ME19 6BJ, UK

1. Introduction

Propagation *in vitro* (micropropagation) by shoot culture with proliferation of lateral shoots has been applied extensively to fruit tree rootstocks and scions [15]. Such propagation is a generally accepted method of rapid multiplication whilst maintaining the genetic integrity of a clone. Nevertheless, plants produced by micropropagation may differ physiologically from the mother plants from which the propagation was initiated. Such physiological change appears to be transient and with adult trees may involve some reversion to juvenile characters.

   This review is concerned with physiological change in temperate fruit trees during micropropagation and subsequently when the trees from this propagation are grown in the field in the long term. Such aspects of micropropagation have probably been studied more extensively with temperate fruit trees than with any other plants.

2. Juvenility

This is of major significance in relation to vegetative propagation of trees by both *in vitro* and conventional methods. The stage of growth of the seedling following germination is described as 'the juvenile phase'. This is the phase of very active growth when the tree does not initiate flowers and vegetative propagation is usually achieved readily. This phase continues for a number of years that varies according to species but eventually physiological change results in the tree attaining the capacity to produce flowers, fruits and seeds. The tree now enters 'the adult phase' which is characterised by decrease in vigour of growth and also decrease in capacity for vegetative propagation, features that usually become progressively more marked as the tree ages further. However, the transition from the juvenile phase appears to be reversible to some extent and it is common for some tissues of the adult tree to have the physiological characteristics of seedlings, such as high growth rates and high capacity for

P.J. Lumsden, J.R. Nicholas and W.J. Davies (eds.), Physiology, Growth and Development of Plants in Culture, 323–331, 1994.

adventitious root production. Such tissues, described as 'rejuvenated' may occur naturally, as with vigorous shoots (suckers) arising directly from the roots of adult trees, or may be induced by various treatments such as pruning, vegetative propagation, grafting onto seedlings or spraying with growth regulators [5]. Knowledge of the physiological and biochemical basis of juvenility and maturation remains very incomplete [8, 37].

3. Physiological change and apparent rejuvenation during shoot culture *in vitro*

An important feature of micropropagation with temperate fruit trees is that success is achieved with subjects which are difficult or impossible to propagate by conventional methods. The basis of this *in vitro* advantage appears to be a progressive increase in propagation potential with continued subculture of shoots to fresh culture medium at intervals of about a month. This increase in propagation potential could be regarded as a manifestation of a physiological process of rejuvenation in response to culture *in vitro*.

Effect of subculture has been studied extensively with shoots of scion and rootstock cultivars of apple which proved to be difficult to root during the initial stages of micropropagation. In the case of the scion cultivar Jonathan, nine subcultures were required before about 90% of the shoots could be rooted whilst 31 subcultures were required for the scion cultivar Golden Delicious to reach 79% rooted shoots [29]. In the case of the rootstock M.7, fewer subcultures were required than with cultivar Greensleeves to reach 70% rooted shoots [17]. Shoots of the rootstock Akerö required 8 subcultures before satisfactory rooting was achieved [36].

Effect of subculture on both shoot production and rooting has been studied with the important apple rootstock M.9 which is used extensively world-wide but nevertheless has the defect of being difficult to propagate by conventional methods. During micropropagation with this rootstock, there was a progressive increase in the rate of shoot proliferation with monthly subculture over 21 months. Moreover, this enhanced shoot production was accompanied by a progressive increase in rooting ability so that a maximum of about 70% rooted shoots was achieved in 10–11 months. An additional observation with M.9 was that shoot culture lines which originated from different shoot tip explants consistently exhibited different rates of shoot production and rooting ability irrespective of the changes with subculture [34]. This suggests that explants vary in their potential for propagation *in vitro*. Such differences with explant may be of physiological origin and could relate to the positions on the original mother plant of the buds used as sources of the initial explants as indicated with species of *Betula* and *Daphne* [22].

The basis of improved propagation with subculture remains obscure but there could be three possible explanations, viz: that culture *in vitro* leads to: a) elimination of viruses or other pathogens, b) genetic change, or c) tissue rejuvenation. Of these possibilities, rejuvenation seems to be the most likely since trees produced by shoot culture are generally true-to type [13]. Elimination of viruses or other pathogens seems unlikely since the beneficial effects of subculture were apparent with culture lines initiated from healthy virus-free trees. Moreover, improvement with subculture occurred gradually whereas virus or pathogen elimination would be expected to occur immediately upon taking the initial shoot tip explants into culture [18]. Further support for the view that rejuvenation occurs during shoot culture is provided by results of micropropagation with other trees such as *Sequoiadendron giganteum* [2, 24] and *Betula pendula* [3, 4]. Such trees, unlike apple, have morphological and biochemical markers of juvenility which were features of the micropropagated plants but were absent from the mother trees from which the propagation was initiated.

4. Performance *ex vitro* and apparent rejuvenation in the field

Plants from micropropagation exhibit juvenile characteristics when grown in the field. Aspects of such field performance in relation to vegetative growth, flowering and improved conventional propagation are now described.

4.1 Vegetative vigour, flowering and cropping

There are few reports of long-term field performance of plants from propagation *in vitro*. Possibly apple trees have been most studied. Composite apple trees were produced by grafting cultivar Greensleeves onto various rootstocks where both rootstock and scion were produced by micropropagation. These trees are being field-trialled in comparison with counterparts from conventional propagation throughout and all trees remain unpruned to avoid confounding effects of micropropagation and pruning.

Thus far the field-trial has continued for 7 years and there are no indications of mutation as a result of the *in vitro* propagation; micropropagated and conventional trees are indistinguishable in scion characteristics of bark, leaves, blossom and fruits and also rootstock characteristics of bark and graft union with scions. However, the micropropagated trees have grown much more vigorously and came into cropping a year later than the conventional counterparts. The delay in cropping might, however, have been a consequence of the small initial size of the micropropagated trees which were only about 50 cm in height at planting in comparison with about 1 m for the conventional trees. Growth analysis [12] indicated that relative growth rate as estimated by percent annual increase in girth of trunk was exceptionally large with the micropropagated trees for the first 4 years following planting with the result that these trees were larger than the conventional counterparts by this time. Thereafter, the micropropagated trees continued to exhibit greater absolute growth rates than the conventional trees so that by 7 y the micropropagated trees were about twice the size of the conventional trees. Nevertheless, after 4 y the relative growth rates of micropropagated and conventional trees were similar indicating that the enhanced vigour as a result of micropropagation was transient and the continued superior absolute growth of the micropropagated trees was merely a consequence of the size advantage at 4 years. Cropping of the trees remained generally in proportion to tree size, with micropropagated trees having about twice the crop and also larger fruits than the conventional trees at 7 years [16 and unpublished data].

Other studies have been carried out where only the scion or rootstock component of the grafted trees originated from *in vitro* propagation. Enhanced growth and delayed cropping were also a feature of two-year-old apple trees which were produced by grafting *in vitro* propagated scions of cultivar Cox's Orange Pippin and Bramley's Seedlings onto conventionally propagated MM.104 rootstocks. In this case the more vigorous growth represented an increase of about 30% over the conventional counterparts [33]. There have also been studies of trees with rootstocks of M.26 from micropropagation grafted with conventionally propagated scions. Here, rootstock performance with respect to tree vigour and cropping was similar to conventional counterparts except that the *in vitro* propagated trees produced more suckers [26].

The results of the investigations described above differ in the degree of vigour exhibited by the grafted trees. Such variation may merely reflect cultivar differences or that both rootstocks and scions or only one of these components originated from micropropagation. Alternatively there could have been different degrees of physiological change as a result of propagation *in vitro*.

Enhanced vigour and tardiness in flowering with some of the trees from micropropagation are characteristics of juvenile seedlings, therefore micropropagation appeared to lead to some degree of rejuvenation of adult trees. However, the delay in flowering with plants from micropropagation was not of the duration to be expected if the micropropagules had all the characteristics of juvenile seedlings. Thus the seedling from which the clone of cultivar Greensleeves originated did not initiate blossom until it was 4 years old and this was following grafting onto M.27 rootstocks which would be expected to hasten the onset of flowering by some years (Alston, personal communication 1992). This long period-to-flowering with the original seedling of Greensleeves contrasts with the micropropagules of Greensleeves which produced blossom in the second year following the propagation [16].

4.2 Improved conventional propagation

4.2.1 Shoot cuttings

A major advantage of micropropagation with fruit trees is the production of clones which have the juvenile characteristic of being exceptionally easy to propagate by conventional methods. There are several examples with temperate fruit trees of improved propagation by enhanced rooting of shoot cuttings from micropropagated plants. In the cases of the apple rootstocks M.9 and Ottawa 3 which are normally difficult to propagate, there was improved rooting of both summer and winter cuttings following micropropagation [35] and there were similar results with some rootstocks of *Pyrus* which are normally almost impossible to propagate [19]. There is evidence that such improved propagation may be sustained for long periods since with the plum rootstock cv. 'Pixy' (*Prunus insititia*) increased rooting from stockplants from micropropagation was still evident 9 years after establishment of the plants in the field [11].

4.2.2 Sucker and burrknot production
Some rootstocks from micropropagation have deleterious juvenile characters such as excessive production at the bases of the main stems of burrknots (nodules of callus tissue with root initials) and also express enhanced vigour through excessive production of suckers. Burrknots and suckers are undesirable with fruit tree rootstocks since they increase the cost of orchard management and susceptibility to disease, respectively [32]. Such deleterious characters are major problems with some important rootstocks such as M.9 for apple and this has undermined confidence in the use of apple rootstocks which come directly from micropropagation [15]. However, increased sucker and burrknot production appears to be transient and may be linked to the potential of micropropagated plants for improved conventional propagation. Clones of rootstock M.9 were produced from *in vitro* culture lines which had undergone different degrees of subculture in an attempt to influence the degree of subsequent rejuvenation of the micropropagated plants. These clones from micropropagation were evaluated as field hedge plants to provide shoot cuttings for conventional propagation. The hedge plants produced many suckers and burrknots and in general these deleterious features and also rooting of cuttings from the various clones increased with increase in duration of subculture of shoots *in vitro* during micropropagation [35]. Evidence for the transient nature of the burrknots and suckering with micropropagated rootstocks has been provided by studies with apple trees with Cox's Orange Pippin (Cox) scions which were raised with M.9 rootstocks which had been produced as a result of improved rooting of cuttings following micropropagation. After 3 years in the field these trees have no suckers and burrknots and are similar to counterparts with M.9 rootstocks from conventional propagation throughout. By contrast, counterparts with M.9 rootstocks which came directly from micropropagules have suckers and burrknots and also poor development of lateral shoots from the scions [14].

4.2.3 Stoolbeds
Improved conventional propagation by stoolbeds from micropropagated apple rootstocks has also been investigated. Management of stoolbeds involves burying the bases of shoots in soil to promote etiolation and rooting and then pruning the rooted shoots to soil level during the annual harvest [6]. Stoolbed shoots generally root more readily than cuttings from hedges but have the disadvantage of producing fewer shoots and being more expensive to maintain than hedges. Such disadvantages may be obviated, at least in part, through the

use of micropropagated plants. Evaluation over 5 years indicated that stoolbeds of M.27 rootstocks from micropropagation produced between 1.5 and 5.0 times as many shoots as the conventional counterparts and these shoots were better rooted [26]. Similar results were obtained over 8 years with M.9 rootstocks [35] and over 3 years with P.22 rootstocks [Czynczyk and Hodun, personal communication 1991]. The tendency of the micropropagated plants to have increased suckering is probably the basis, at least in part, of the increased productivity of stoolbeds. Moreover, burrknots remain confined to the main stem of the micropropagated plants used to establish stoolbeds and are therefore removed at the first pruning to soil level. Use of stoolbeds would therefore appear to be an effective method of exploiting the improved potential of some micropropagated plants for conventional propagation while eliminating deleterious effects of burrknots. In addition to application to fruit production, this increased production from stoolbeds would be particularly advantageous if applied to forest tree coppices for wood chip production [27].

5. Conclusions and future prospects

The studies with fruit trees provide substantial evidence that micropropagation may result in physiological change and some reversion to juvenile characteristics such as enhanced vigour and improved conventional propagation. Physiological change leading to rejuvenation *in vitro* as manifest by improved shoot growth and rooting with subculture would seem to be the likely basis of the juvenile characteristics of the plants in the field.

The persistence for at least up to 8 years of improved conventional propagation with shoot cuttings or stoolbeds following micropropagation contrasts with the transient increase in relative growth rates for only 4 years with unpruned micropropagated trees of cv Greensleeves. However, repeated severe pruning is routine during the management of hedges and stoolbeds for propagation. Such pruning is generally accepted as the basis of the lack of ageing with hedges or stoolbeds of fruit tree rootstock clones which were selected decades or even centuries ago [10, 31]. Repeated severe pruning is therefore likely to be the basis of the sustained improved conventional propagation following micropropagation.

Apparent rejuvenation leading to increased vigour with adult trees could be of value in the nursery with fruit trees although it might be disadvantageous for commercial orchards. The increased vigour, however, would be particularly advantageous if applied to forest trees. In this latter respect it is encouraging that growth of trees from propagation *in vitro* of élite selections of *Eucalyptus* was about 3 times faster and more uniform than that of trees from seed of the élite selections. Moreover, the trees from micropropagation flowered within 2.5 years in contrast to 4–5 years with seedling trees [7]. Furthermore, trees of *Betula pendula* which were micropropagated from a 20 year-old tree have been compared in the third year of field-trialling with a population of seedling trees from the same 20-year-old tree. Height and trunk girth measurement indicated that the micropropagated trees are growing at least as rapidly as the seedling trees and eighty three per cent of the micropropagated trees flowered in contrast to only 38% of the seedlings. This trial is part of the COST87 (EC) programme on micropropagation of woody plants and is repeated with similar results thus far at seven sites throughout Europe with micropropagated and seedling trees from the same 20-year-old mother tree [20].

In the cases of some fruit tree rootstocks such as M.9 which were selected decades ago, physiological change which appears to be similar to that induced by micropropagation has been reported with conventionally propagated clones from a number of countries. These M.9 clones have been classified according to differences in potential for conventional propagation and such differences appear to be, in some cases at least, the consequence of ontogenetic variation; the differences were sustained in severely pruned stoolbeds but disappeared when plants were grown as trees for some years [31]. Rooting of cuttings of selections of some M.9 clones which have particularly high capacities for vegetative propagation which may be the result of changed degree of rejuvenation has been compared with some micropropagated M.9 clones. Results indicated that the micropropagated M.9 had the highest propagation potential of all the clones tested [28]. Thus micropropagation may be a particularly effective and rapid method of achieving desirable physiological change of a type which is also achieved, although possibly to a lesser degree, during continued conventional propagation over decades.

Although fruit trees from micropropagation have some juvenile characters, onset of flowering does not appear to be retarded to the extent to be expected if the plants had all the characteristics of juvenile seedlings. Furthermore, there are important examples of accelerated flowering as a result of micropropagation with seedlings of some tree species as with loblolly pine (*Prunus taeda*) [23] and two species of bamboo (*Bambusa arundinaceae*, Willd and *Dendrocalanus brandisii*, Kurtz) [25]. The example with bamboo is remarkable in that flowers were produced *in vitro* after only 6 months of culture whereas the seedlings of the species concerned grow without flowering for about 30 years. This accelerated flowering with bamboo has made conventional breeding possible of one of the most important structural raw materials in the world. Recent investigation using *in vitro* culture of thin cell layer cultures from tobacco [30] and cotyledonary nodes of soybean [21] suggest that cytokinins play an important role in flower initiation with *in vitro* cultures and could be involved in accelerated flowering.

The physiological changes *in vitro* and *in vivo* that have been highlighted in this review suggest that an important future commercial application of micropropagation could be the production of rejuvenated clones of élite adult trees. Such clones may have enhanced vegetative vigour and improved potential for conventional propagation without the full extent of delay in flowering to be expected with seedling trees. Improved understanding of the physiological and biochemical basis of juvenility and maturation would therefore appear to merit some priority as a research topic which could lead to further exploitation of rejuvenation effects of micropropagation. Possibly progress with such understanding will be rapid in the near future in view of recent evidence for juvenile/mature phase dependent control of gene expression in woody plants such as *Hedera* helix [9] and *Prunus avium* [1].

References

1. Besford B and Hand P (1992) The molecular biology of juvenility in woody plants. Hort Res Intern, Ann Rep 1990–91. p 46
2. Bon M.-C. and Monteuuis O (1991) Rejuvenation of a 100-year-old *Sequoiadendron giganteum* through *in vitro* meristem culture. II. Biochemical arguments. Physiol Plant 81:116–120
3. Brand MH and Lineberger RD (1992a) *In vitro* rejuvenation of *Betula (Betulaceae)*: Morphological evaluation. Amer J Bot 76(b):618–625
4. Brand MH and Lineberger RD (1992b) *In vitro* rejuvenation of *Betula (Betulaceae)*: Biochemical evaluation. Amer J Bot 76(b):626–635
5. Franclet A (1979) Rejeunissement des arbres adultes en vue de leur propagation végétative. In: *Annales de Recherches Sylvicoles*, AFOCEL. Etudes Recherches 12: 6/79. *Micropropagation d'Arbres Forestiers*, 3–18
6. Garner RJ (1988) The Grafter's Handbook. Fifth Edition. Cassells Ltd, London
7. Gupta PK and Mascarenhas AF (1987) *Eucalpytus*. In: JM Bonga and DJ Durzan, eds. Cell and Tissue Culture in Forestry, pp 385–399. Dordrecht, The Netherlands: Martinus Nijhoff
8. Hackett WP (1985) Juvenility, maturation and rejuvenation of woody plants. Hort Rev 7:109–155
9. Hackett WP, Murray JR, Woo H-H, Stapfer RE, Geneve R (1990) Cellular, biochemical and molecular characteristics related to maturation and rejuvenation in woody species. In: R Rodriquez *et al.*, eds. Plant Ageing: Basic and Applied Approaches, pp 147–152. New York: Plenum Press
10. Hatton RG (1917) Paradise apple rootstocks. J Royal Hort Soc 42:361–399
11. Howard BH, Jones OP and Vasek J (1989) Long-term improvement in the rooting of plum cuttings following apparent rejuvenation. J Hort Sci 64:147–156
12. Hunt R (1981) Plant growth analysis: The Institute of Biology's Studies in Biology No 96. London: Edward Arnold
13. Hutchinson JF and Zimmerman RH (1987) Tissue culture of temperate fruit and nut trees. Hort Rev 9:273–349
14. Jones OP (1992) Field performance of micropropagated apple rootstocks. Hort Res Intern Ann Rep 1990–91 47–48
15. Jones OP (1993) Propagation *in vitro* of apple. In: R Ahuja, ed. Micropropagation of Woody Plants. Dordrecht, The Netherlands: Kluwer Academic Publishers (in press)
16. Jones OP and Hadlow WCC (1989) Juvenile-like character of apple trees produced by grafting scions on rootstocks produced by micropropagation. J Hort Sci 64:395–401
17. Jones OP. Marks TR and Waller BJ (1982) Propagation *in vitro*. Report East Malling Research Station for 1981. p 159
18. Jones OP and Vine SJ (1969) The culture of gooseberry shoot tips for elimination of virus. J Hort Sci 43:289–92
19. Jones OP and Webster CA (1989) Improved rooting from conventional cuttings taken from micropropagated plants of *Pyrus communis* rootstocks. J Hort Sci 64:429–434
20. Jones OP, Welander M, Ballester A and Meier-Dinkel A (1992) COST87 Micropropagation of *Betula* and *Quercus*. Initial reports of the woody plant working group. COST87 report. BRIDGE Commission of European Communities

21. Jullien F and Wyndaele R (1992) Precocious *in vitro* flowering of soybean cotyledonary nodes. J Plant Physiol 140:251–253
22. Marks TR and Meyer Pauline E (1992) Effect of explant location on early culture development *in vitro*. J Hort Sci 67:583–591
23. McKeand SE (1985) Expression of mature characteristics by tissue culture plantlets derived from embryos of Loblolly pine. J Amer Soc Hort Sci 110:619–623
24. Monteuuis O (1991) Rejuvenation of a 100-year-old *Sequoiadendron giganteum* through *in vitro* meristem culture. I. Organogenic and morphological arguments. Physiol Plant 81:111–115
25. Nadgauda RS, Parasharami VA and Mascarenhas AF (1990) Precocious flowering and seedling behaviour in tissue culture of bamboo. Nature 344:335–336
26. Navatel JC, Nio M, Vaysse P and Edin M (1988) Pommier. Comportement en marcottières et aux vergers des porte-greffes micropropagés de type *Malus*. Infos-Ctifl No 44:25–28
27. Potter CJ (1990) Coppice trees as energy sources. Report ETSUDIO 28: Crown Copyright UK
28. Sharma S (1989) The induction, development and strength of adventitious roots of *Malus* rootstocks M.9 in relation to tree establishment and anchorage. Thesis Wye College, University of London, UK
29. Sriskandarajah S, Mullins MG and Nair Y (1982) Induction of adventitious rooting *in vitro* in difficult to propagate cultivars of apple. Pl Sci Letters 24:1–9
30. Van den Ende G, Croes AF, Kemp Anke, Berendse GWM (1984) Development of flower buds in thin-cell layer cultures of floral stalk tissue from tobacco: role of hormones in different stages. Physiol Plant 61:114–118
31. Van Oosten HJ (1986) Intraclonal variation in M.9 rootstock and its effect on nursery and orchard performance. Acta Hort 160:67–78
32. Villeneuve F (1986) Le broissin dévelopment et moyens de lutte. Arbor Fruitière 33:44–48
33. Webster AD, Sparks TR and Belcher A (1986) The influence of micropropagation and chemical mutagens on the growth and precocity of Cox's Orange Pippin and Bramley's Seedling apples. Acta Hort 180:25–34
34. Webster CA and Jones OP (1989) Micropropagation of the apple rootstock M.9: effect of sustained sub-culture on apparent rejuvenation *in vitro*. J Hort Sci 64:421–428
35. Webster CA and Jones OP (1992) Performance of field hedge and stoolbed plants from micropropagated dwarfing apple rootstock clones with different degrees of apparent rejuvenation. J Hort Sci 67:521–528
36. Welander M (1985) *In vitro* root and shoot formation in the apple cultivar Akerö. Ann Bot 55:245–261
37. Zimmerman RH, Hackett WP and Pharis RP (1985) Hormonal aspects of phase change and precocious flowering. In: Pirson A, ed. Encyclopedia of Plant Physiology (NS) Vol II/IV. pp 79–115. Berlin and New York: Springer-Verlag

Propagation and physiological improvement of mature wild cherry (*Prunus avium* L.) and common ash (*Fraxinus excelsior* L.) by tissue culture

N. HAMMATT

Horticulture Research International, East Malling, West Malling, Kent ME19 6BJ, UK

1. Introduction

Continued rapid increase in world population is likely to increase demand for tree products, particularly through use of wood for fuel and construction. Such increased demand is rapidly depleting existing resources and necessitating research to provide global reafforestation with superior planting material. Most woodland and forest planting stock continues to derive from seedlings of variable phenotypic performance. However, significant increases in productivity are expected from clonal forestry following the development of techniques for the vegetative propagation of superior trees. Elite planting material for most species will derive in the near future from genotypes which have shown good long-term field performance, and in the longer term, from conventional breeding and modern methods of genetic manipulation with plant tissue culture [11].

As trees mature, they become more difficult to propagate vegetatively. Consequently trees are usually difficult to propagate clonally by the time it is apparent that they are superior. Despite these difficulties, significant advances have been made in the clonal propagation of mature fruit tree rootstocks through stoolbeds and severe pruning of hedges.

Micropropagation offers an alternative approach to clonally propagate mature fruit trees, and may also provide a means to improve subsequent conventional propagation by stoolbeds and hedges. Thus, both micropropagated shoots [2, 29, 31] and conventional cuttings taken from micropropagated stock plants [15, 20, 32] have improved potential for adventitious root formation when compared to conventional, mature counterparts. Improved conventional propagation after micropropagation has been reported for several other trees including *Ficus benjamina* [21] and *Rhododendron* [24]. Altered physiology of micropropagated, mature fruit trees has also been detected as enhanced vegetative vigour [e.g. 18, 30] and altered wood structure [23].

The means by which mature material becomes easier to root, and more vigorous during micropropagation have been discussed [18]. Although this may

P.J. Lumsden, J.R. Nicholas and W.J. Davies (eds.), Physiology, Growth and Development of Plants in Culture, 332–338, 1994.

have occurred as a result of pathogen elimination or somatic mutation, the weight of evidence points to tissue rejuvenation during micropropagation since vigour and enhanced rooting are characteristic of juvenile seedlings. This view is supported by observations with *Vitis* [25], *Sequoiadendron* [3] and *Betula* [4, 5], of other juvenile morphological and biochemical features among micro-propagated plants derived from mature trees.

If the improved propagation and enhanced vigour observed with fruit trees could also be achieved with forest species, micropropagation could have important implications for the development of clonal forestry. Furthermore, the increased sucker production with micropropagated apple stoolbeds [32] suggests the possible use of micropropagation to increase production with short-rotation forest tree coppices for energy production.

The aims of the research described in this paper were:
a) to achieve micropropagation of mature trees of some UK broadleaved tree species recommended for planting in farm woodlands, and
b) to study possible rejuvenation of the trees resulting from micropropagation.

2. Micropropagation of *Prunus avium* L. (wild cherry)

2.1 Introduction

Wild cherry was chosen because it is in the Rosaceae, species of which have been used most commonly to study physiological changes that occur during micropropagation. It is also an important timber tree and resistant to damage by squirrels. Furthermore, there are a number of possible sources of superior germplasm available for this species. Firstly, there are the cultivars F12/1 and Charger which have been used as rootstocks for sweet cherry, and which are vigorous and resistant to bacterial canker caused by *Pseudomonas syringeae* pv. *morsprunorum* [9], and available as virus-free clones. Secondly, there is material from a collection of phenotypically superior British wild cherry selected from existing woodlands. In the longer term, genetically improved material should become available from breeding programmes and modern *in vitro* techniques as already described [11].

P. avium selections for use as rootstocks have been propagated traditionally from shoot or root cuttings, but with only limited success [1, 10]. Shoots obtained from roots may have the additional advantage of being juvenile, but it is often difficult to identify from which tree a root originates. Following a study of propagation techniques for *P. avium*, Cornu and Chaix [7] concluded that micropropagation is the most effective means to rapidly produce self-rooted, clonal material from adult trees.

There have been several reports of protocols for the propagation *in vitro* of wild cherry [e.g. 7, 19, 28]. Shoot culture media employed previously for *P. avium* have differed with respect to pH, concentrations of gibberellic acid (GA_3) and benzyladenine (BA), agar strength and the inclusion of phloroglucinol (1,3,5-trihydroxybenzene; PG). The importance of each of these factors has been evaluated [12], and all were found to have significant effects upon the performance of shoot cultures of *P. avium*. Based on these results, the best formulation was the medium of Jones and Hopgood [19], but modified by reducing the pH to 5.0, halving the concentration of BA to 2.2 μM, and removing the GA_3. Using this medium, shoot cultures showed a progressive and significant increase in numbers of shoots per culture after each subculture period [12]. Enhanced vigour of the shoot cultures with time has been recorded previously with apple [31] and plum [2].

Ease of rooting at each subculture was also monitored in a rooting medium consisting of MS medium containing auxin (14.7 μM indole-3-butyric acid). In the case of the rootstock F12/1, both the percentage of shoots producing roots and the number of roots produced per rooted shoot significantly increased as the total time in culture increased [12].

The gradual acquisition of the juvenile characters of enhanced shoot vigour and improved rooting ability suggest that the tissues had become rejuvenated during micropropagation.

2.2 The use of phloroglucinol to promote adventitious rooting in micropropagated shoots of wild cherry

There are contradictory reports concerning the effectiveness of the phenolic compound PG to stimulate rooting *in vitro* with apple [16, 17, 34], cherry and plum [19]. Since shoots of *P. avium* were difficult to root in PG-free medium during the early life of the culture line, Hammatt and Grant [12] also inserted shoots into auxin-containing rooting medium supplemented with this compound (1 mM). PG increased the proportion of roots that rooted in the early stage of culture line establishment, but did not affect the numbers of roots. However, since the material acquired with time the ability to produce roots in medium without PG, any PG effect became undetectable. This finding that an effect of PG is detectable only during the first 700d of the culture line may account for previous observations with *P. avium* cv. F12/1 [19] and apple [34], that PG did not increase rooting in micropropagated shoots.

These results suggest that during the early establishment of a culture line of *P. avium*, phloroglucinol can substitute for an unknown factor acquired during successive subculturing, and which eventually leads to maximum responsiveness to auxin. Thus, understanding the process by which PG stimulates rooting may be useful in studies to investigate the mechanism by which juvenile characters are restored to adult tissues during micropropagation.

It has been suggested that PG may promote rooting either by influencing auxin metabolism or it may act as an antioxidant, maintaining tissues in a reduced state [16, 33]. During micropropagation, mature shoots of *P. avium* gradually acquired the ability to respond to exogenously supplied auxin by producing adventitious roots [12, 13]. However with PG alone, there was no corresponding change in the proportion of shoots that rooted with time [13]. Thus,

67-69% overall of shoots produced roots with PG alone, irrespective of their ability to respond to exogenously applied auxin. These results suggest that PG does not affect rooting by interacting with the metabolism of exogenous auxin.

Subsequent experiments also revealed that in the absence of exogenous auxin, PG can stimulate rooting in other genotypes of *Prunus* [13]. Furthermore, there was a positive correlation between PG concentration and both the proportion of shoots that rooted and the numbers of roots per rooted shoot. By contrast, in medium with auxin, only the numbers of roots per shoot were affected by PG concentration.

It has been suggested that during micropropagation, tissues become progressively more rejuvenated as subculturing continues. Therefore, given that non-rejuvenated shoots that did not respond to auxin, rooted better *in vitro* with PG alone, it is possible that the rooting of conventional cuttings from mature trees may also benefit from treatments with PG.

2.3 Performance of micropropagated and conventionally propagated wild cherry in soil

Despite several publications on the micropropagation of wild cherry, there are no reports on the performance of such trees in soil. Thus, a study has been initiated of the relative performance in pots, of *Prunus avium* propagated from seeds, conventional shoot cuttings or micropropagation. Seedlings were obtained in winter from a commercial nursery as bare rooted trees, such material being of the type usually used for planting new woodland. Summer softwood shoot cuttings were from hard-pruned field hedges of the rootstock F12/1. Micropropagated trees of the rootstock F12/1 were weaned in summer and overwintered in frost-free conditions (Hammatt, unpublished data).

Thus far, after two years of growth, the following conclusions have been drawn from this study. Firstly, the seedlings exhibited much reduced growth in comparison with clonal F12/1 from shoot cuttings or micropropagation. Furthermore, unlike micropropagated trees, more than 60% of trees from cuttings produced branches during their first year and these trees showed reduced elongation of the main stem. After two years, the micropropagated trees had grown taller than those from cuttings in spite of the possibility that shoots from hedges may be partially rejuvenated. These experiments have now been transferred to field conditions for further evaluation.

2.4 Conclusions on the micropropagation of mature wild cherry

Improvements in vigour and rooting potential *in vitro* with successive subculturing, combined with preliminary indications of enhanced growth of micropropagated plants in soil, provide evidence that rejuvenation may have occurred during micropropagation of *P. avium*. Thus, this technique could have important applications in the development of improved genotypes of this species for use in multiclonal plantations including farm woodlands.

3. Micropropagation of juvenile and mature common ash (*Fraxinus excelsior* L.)

In addition to wild cherry, common ash (Oleaceae) was included in order to extend our studies of rejuvenation *in vitro* beyond Rosaceae. However, unlike wild cherry, there are no readily available sources of superior UK timber ash.

Ash has seldom been clonally propagated with the exception of ornamental forms which traditionally have been grafted onto seedling rootstocks. Micropropagation of *F. americana* [27] and *F. excelsior* [6] has been described previously. As a preliminary to investigations with adult material of *F. excelsior*, we examined the effects of culture medium and BA concentration upon shoot culture development in common ash seedlings [14]. Shoot cultures on woody plant medium (WPM) [22] produced excessive callus, while those on MS medium [26] suffered from shoot tip necrosis. Excessive callus production on WPM has also been reported for *F. americana* [27]. Both of these phenomena were avoided by using DKW medium [8] supplemented with 22.2 μM BA, and on this medium, shoots of a selected clone (19) could be obtained from both shoot tip and nodal explants. Up to 80% of shoots of clone 19 produced roots in half-strength WPM with 4.9 μM IBA.

Progress with mature ash has been hampered by an inability to obtain axenic cultures due to contamination with bacteria. Furthermore, most of those explants that did not succumb to bacteria, turned brown and died. Nonetheless, two culture lines have been obtained from a 70 year-old tree of common ash. The methodology developed previously for ash seedlings has been used successfully to propagate these shoot culture lines when nodes were used as explants (Hammatt, unpublished data). Rejuvenation *in vitro* is being monitored by estimating rooting potential of the shoots at each subculture.

Acknowledgements

This farm woodland research was funded by the Ministry of Agriculture, Fisheries and Food and the Agricultural and Food Research Council.

References

1. Al Barazi Z and Schwabe WW (1985) Studies on possible internal factors involved in determining ease of rooting in cuttings of *Pistacia vera* and *Prunus avium*, cvs Colt and F12/1. J Hort Sci 60: 439-445
2. Baleriola-Lucas C and Mullins M (1984) Micropropagation of two French prune cultivars (*Prunus avium* L.). Agronomie 4: 473-477
3. Bon M-C and Monteuuis O (1991) Rejuvenation of a 100-year-old *Sequoiadendron giganteum* through *in vitro* meristem culture. II. Biochemical arguments. Physiol Plant 81: 116-120
4. Brand MH and Lineberger RD (1992) In vitro rejuvenation of *Betula* (Betulaceae): Morphological evaluation. Amer J Bot 79: 618-625

5. Brand MH and Lineberger RD (1992) In vitro rejuvenation of *Betula* (Betulaceae): Biochemical evaluation. Amer J Bot 79: 625-635

6. Chalupa V (1990) Micropropagation of hornbeam (*Carpinus betulus* L.) and ash (*Fraxinus excelsior* L.). Biol Plant 32: 332-338

7. Cornu D and Chaix C (1982) Multiplication par culture *in vitro* de merisiers adultes (*Prunus avium*): application à un large éventail de clones. Colloque International sur la culture *in vitro* des essences forestières. IUFRO S2.01.5: Fontainbleau 1981, pp 71-79. Nangis: AFOCEL

8. Driver JA and Kuniyuki AH (1984) *In vitro* propagation of Paradox walnut rootstock. Hort Sci 19: 507-509

9. Garrett C (1989) *In vitro* techniques for determining host resistance and susceptibility to bacterial canker. Ann Rep AFRC Inst Hort Res for 1988: 12

10. Ghani AKMO and Cahalan CM (1991) Propagation of *Prunus avium* from root cuttings. Forestry 64: 403-409

11. Hammatt N (1992) Progress in the biotechnology of trees. World J Microbiol Biotech 8: 369-377

12. Hammatt N and Grant N (1993) Apparent rejuvenation of mature wild cherry (*Prunus avium* L.) during micropropagation. J Plant Physiol 144: 341-346

13. Hammatt N (1994) Promotion by phloroglucinol of adventitious root formation in micropropagated wild cherry (*Prunus avium* L.). Plant Growth Reg (In Press)

14. Hammatt N and Ridout MS (1992) Micropropagation of common ash (*Fraxinus excelsior*). Plant Cell Tiss Org Cult 13: 67-74

15. Howard BH, Jones OP and Vasek J (1989) Long-term improvement in the rooting of plum cuttings following apparent rejuvenation. J Hort Sci 64: 147-156

16. James DJ and Thurbon IJ (1981) Phenolic compounds and other factors controlling rhizogenesis *in vitro* in the apple rootstocks M.9 and M.26. Z Pflanzenphysiol 105: 11-20

17. Jones OP (1983) *In vitro* propagation of tree crops. In Plant Biotechnology, Soc Exp Biol Seminar Series 18, Cambridge University Press: 139-159

18. Jones OP and Hadlow WCC (1989) Juvenile-like character of apple trees produced by grafting scions and rootstocks produced by micropropagation. J Hort Sci 64: 395-401

19. Jones OP and Hopgood ME (1979) The successful propagation *in vitro* of two rootstocks of *Prunus:* the plum rootstock Pixy *(P. insititia)* and the cherry rootstock F12/1 *(P. avium)*. J Hort Sci 54:63-66

20. Jones OP and Webster CA (1989) Improved rooting from conventional cuttings taken from micropropagated plants of *Pyrus communis* rootstocks. J Hort Sci 64: 429-434

21. Kristiansen K (1991) Post-propagation growth of cuttings from *in vitro* and *in vivo* propagated stock plants of *Ficus benjamina*. Sc Hort 46: 315-322

22. Lloyd G and McCown B (1980) Commercially-feasible micropropagation of mountain laurel, *Kalmia latifolia*, by use of shoot tip culture. Comb Proc Intl Plant Prop Soc 30: 421-427

23. Mackenzie KAD (1989) Cambial activity and wood structure of micropropagated vs conventionally propagated trees. Ann Rep AFRC Inst Hort Res for 1988: 32-33

24. Marks TR (1991) *Rhododendron* cuttings. I. Improved rooting following 'rejuvenation' *in vitro*. J Hort Sci 66: 103-111

25. Mullins MG, Nair Y and Sampet P (1979) Rejuvenation *in vitro*: Induction of juvenile characters in an adult clone of *Vitis vinifera* L. Ann Bot 44: 623-627

26. Murashige T and Skoog F (1962) A revised medium for the rapid growth and bioassays with tobacco tissue cultures. Physiol Plant 15: 473-497

27. Preece JE, Christ PH, Ensenberger L and Zhao J-L (1987) Micropropagation of ash *(Fraxinus)*. Comb Proc Intl Plant Prop Soc 37: 366-372

28. Ranjit M, Kester DE and Micke WC (1988) Micropropagation of cherry rootstocks: I. Response to culture. J Amer Soc Hort Sci 113: 146-149

29. Sriskandarajah S, Mullins, MG and Nair Y (1982) Induction of adventitious rooting *in vitro* with difficult to propagate cultivars of apple. Plant Sci Lett 24: 1-9

30. Webster AD, Sparks TR and Blecher, A (1986) The influence of micropropagation and chemical mutagens on the growth and precocity of Cox's Orange Pippin and Bramley's Seedling apple. Acta Hortic 180: 25-34

31. Webster CA and Jones OP (1989) Micropropagation of the apple rootstock M.9: effect of sustained subculture on apparent rejuvenation *in vitro*. J Hort Sci 64: 421-428
32. Webster CA and Jones OP (1992) Performance of field hedge and stoolbed plants of micropropagated dwarfing apple rootstock clones with different degrees of apparent rejuvenation. J Hort Sci 67: 521-528
33. Wilson PJ and Van Staden J (1990) Rhizocaline, rooting co-factors, and the concept of promoters and inhibitors of adventitious rooting – a review. Ann Bot 66: 479-490
34. Zimmerman RH and Broome O (1981) Phloroglucinol and *in vitro* rooting of apple cultivar cuttings. J Amer Soc Hort Sci 106: 648-652

Evidence for epigenetic inheritance of ontogenetic phase and tissue origin characteristics in the performance of plants derived through adventitious shoots

A.C. JAMES and S.H. MANTELL

Horticulture Section, Wye College, University of London, Wye, Ashford, Kent, TN23 3AH, UK

1. Introduction

Phase change and the stability of juvenile and adult phases in woody perennial plants has been regarded as a product of determination of apical meristems along two alternative pathways of development [17]. For example, from the morphological differences observed between juvenile and vegetative adult phase shoots of *Hedera helix* [15] it can be inferred that there are intrinsic differences in the apical meristems [5] since it is from these perennial zones of cell division that such morphological parameters as patterns of phyllotaxy [11] originate. Three hypotheses have been proposed to account for determination of phase differences in apices [17]:

1. the structure and properties of the apical meristem are controlled by influences arising from pre-existing differentiated tissue,
2. the stable characteristic properties of the apical meristem are controlled by the structure and organisation of the apex as a whole, and
3. there are intrinsic differences in the meristematic cells of juvenile and adult apices. There is positive evidence for all three hypotheses [6]. The present work aimed to assess the degree and kind of epigenetic change inherited in meristems derived *de novo* from various differentiated shoot tissues of both juvenile and adult phase phenotypes of the Australasian woody perennial *Solanum aviculare* Forst. To our knowledge the only report of epigenetic inheritance in regenerants is that found in *Saintpaulia* [2] plants regenerated from either epidermal or sub-epidermal tissue. There are no recorded instances of vegetative adult-phase meristems arising *de novo* (i.e. adventitiously) from adult tissue in Angiosperms [6].

There are several advantages for using *Solanum aviculare* as a model for phase change research. Like many Solanaceous plants, shoot cultures of *Solanum aviculare* grow well *in vitro* without the need for cytokinins [9] and, with the use of growth regulators, totipotency is possible in all tissues so that shoot meristems can be readily produced *de novo*. In addition there are marked and predicable heteroblastic leaf characters associated with the progression of the plant to the adult phase [1, 7, 8].

P.J. Lumsden, J.R. Nicholas and W.J. Davies (eds.), Physiology, Growth and Development of Plants in Culture, 339–345, 1994.
© *1994 Kluwer Academic Publishers.*

2. Materials and methods

2.1 Plant material

Clonal *S. aviculare* plants were obtained from selfed progeny derived from a stock of plants at the University of Bristol, UK Botanic Gardens, and were maintained at 20 °C under a 10-h photoperiod. Illumination was provided by banks of warm-white fluorescent tubes supplemented with 100W tungsten incandescent bulbs, giving an average irradiance of 240 μmol m^{-2} s^{-1}. Adult material was derived from a single clone of *Solanum aviculare*, which was propagated vegetatively using multi-node cuttings rooted under mist. Juvenile material was raised from seed of this clone grown under 16-h days. All plants were grown in a 9:1 peat/grit compost contained in 10 cm^2 plastic pots.

2.2 Tissue culture procedures

Explants were surface sterilised and prepared for culture using a standard method [10]. The medium used to induce adventitious shoots in *S. aviculare* material [9] consisted of full-strength Murashige and Skoog (MS) [13] media containing 2% sucrose, 0.8% Technical Grade 3 agar, 0.2 mg l^{-1} NAA and 2 mg l^{-1} BAP (SM Medium). For rooting of adventitious shoots, medium consisted of 1 / 2 x MS salts, 2% sucrose and 0.8% Technical Grade 3 agar without growth regulators (RM Medium). The pH was adjusted to 5.8 prior to autoclaving.

With the exception of shoot apices, explants were taken from the first fully developed internode below the shoot apex of each plant. Shoots were dissected so that explants from each shoot were allocated equally to 4 x 2 treatments. These consisted of 40 explants each of epidermal strips (derived from stem internodes), pith, vascular tissue and shoot apices (<1cm in length) from juvenile and adult phases. Epidermal strips were cut into rectangles of 1cm x 0.3cm each; pith into 3 mm^3 pieces, vascular tissues into explants of *c.* 4 x 2mm and shoot apices trimmed to 3–7 mm in length. After weighing, explants were implanted into SM medium. Cultures were incubated at a constant temperature of 19 °C, with 8 h photoperiods provided by banks of fluorescent lamps (Philips TDL colour 84 High Frequency 30 watts) giving an average irradiance of 83 μmol m^{-2}s^{-1}.

Shoots were rooted on RM Medium and weaned under fog for 2 weeks before tranfering to the constant environment room for the growth and development assessments.

2.3 Morphological assessment

Plants were assessed when flower primordia were visible and the two subtending axillary shoots had developed fully grown leaves. The following measurements were then taken:
1. Final plant height to the flowering node minus initial height at weaning.
2. The numbers of nodes to flower minus the numbers present before weaning.
3. The leaf lobing indexes of the 13 nodes below the flowering node (chosen as the best possible number in view of leaf senescence and abscission).
4. The length/breadth ratio of the terminal leaf lobes for 13 nodes below the flower primordia.

3. Results

3.1 Assessment of shooting response in the different explants cultured

No histological evidence was found for direct organogenesis from previously differentiated cells; instead shoots arose from meristematic centres in dedifferentiated tissue. In comparisons of juvenile and adult responses, a higher percentage of juvenile explants produced shoots. Juvenile explants also produced significantly more shoots per shoot-producing explant than adult explants (Table 1). Both juvenile and adult apices produced the highest morphogenic responses in their categories, with the vascular tissues showing similar trends.

Table 1. Comparisons of shoot morphogenesis in explants from different tissues in juvenile and adult phases, one month post induction on MS medium supplemented with cytokinin and auxin (SM Medium). Means sharing a common letter were not significantly different at P=0.05, following square root transformation (*t* test)

Explant tissue source		n	% of explants producing shoots	Mean no. shoots >2 mm per shoot producing explant	
Shoot tip	J	19	100	6.667 ± 0.707	A
	A	20	85	3.588 ± 0.594	B
Epidermal strip	J	19	94	3.722 ± 0.535	B
	A	21	24	2.000 ± 0.548	B
Pith	J	20	60	1.400 ± 0.221	C
	A	19	26	1.600 ± 0.400	C
Vascular stele	J	18	94	5.824 ± 0.837	AB
	A	20	55	3.182 ± 0.711	B

3.2 Assessments of epigenetic inheritance in regenerants ex vitro

Plants of juvenile origin, whether derived from pre-existing shoot apical meristems or *de novo* from juvenile tissues, required significantly (P=0.05) more nodes to flower than those of adult phase of origin (Table 2), they were also taller in general than adult origin plants. Plants of adventitious origin were both significantly taller and required significantly less nodes to flower than those derived from existing meristems (SED comparisons not shown).

Table 2. Nodes and height to the flowering node in plants derived adventiously (adv.) *in vitro* from different tissues, or from pre-existing shoot apices, in juvenile (J) and adult (A) phases of *Solanum aviculare.* * and ** denote significance at P=0.05 and 0.02 respectively (*t* test)

Adventitiously derived source		Total plant number	n	Leaf score	SED	Length/ breadth ratio	SED
Epidermal strips	J	12	152	8.76	0.078 ns	3.11	0.083***
	A	10	125	8.78		3.84	
Pith	J	11	139	8.26	0.075***	3.53	0.081***
	A	12	153	8.60		4.11	
Shoot tips	J	12	153	8.46	0.077***	3.54	0.082*
	A	10	127	8.82		3.72	
Vascular ring	J	10	128	8.21	0.079***	3.35	0.084***
	A	11	140	8.75		3.70	
Original shoot apices	J	10	123	8.14	0.101***	3.82	0.109***
	A	06	59	10.44		4.37	

3.3 Effects of ramet / explant origin on heteroblastic characters associated with phase change

Plants derived adventitiously from adult- phase tissue were rejuvenated in terms of leaf lobing score and the length / breadth ratio of the leaf terminal lobe (Table 3). There were also significant diffences between plants derived from the same phase but different tissue types (SED comparisons not shown). The behaviour of juvenile-origin plants, in terms of the l / b ratio, differed markedly according to the tissue of origin (Figs. 1 & 2). A low and stable ratio was found in plants derived from epidermal strips whilst those derived from vascular tissue exhibited a higher ratio with a linear increase upto the flowering node.

   Chromosome counts, using root tip material, indicated that the regeneranted plants studied above were diploid (2N=46) so polyploidy could not therefore have accounted for the differences described.

Table 3. Comparisons of overall leaf lobing scores and length/breadth ratios of the 13 nodes below the first flowering node in plants derived adventiously or from pre-existing meristems in juvenile (J) and adult (A) phases of *Solanum aviculare*. * and *** denote significance at P=0.05 and 0.001, respectively (*t* test).

Adventitiously derived source		n	Mean nodes to flower	SED	Mean hieght to flower (mm)	SED
Overall adventitious	J	46	22.5	0.47*	748.16	20.287 *ns*
	A	44	20.9		732.02	
Epidermal strips	J	12	22.3	1.26 *ns*	748.08	39.720 *ns*
	A	11	21.2		690.18	
Pith tissue	J	11	22.9	1.23 *ns*	721.91	39.720 *ns*
	A	12	21.3		756.67	
Vascular tissue	J	11	23.0	1.29 *	776.50	39.720 *ns*
	A	12	20.1		754.92	
Shoot tips (adv.)	J	12	21.5	1.33 *ns*	746.17	41.960 *ns*
	A	9	20.9		690.33	
Original shoot apices	J	12	28.8	1.51 **	462.08	47.601 *ns*
	A	6	24.8		409.17	

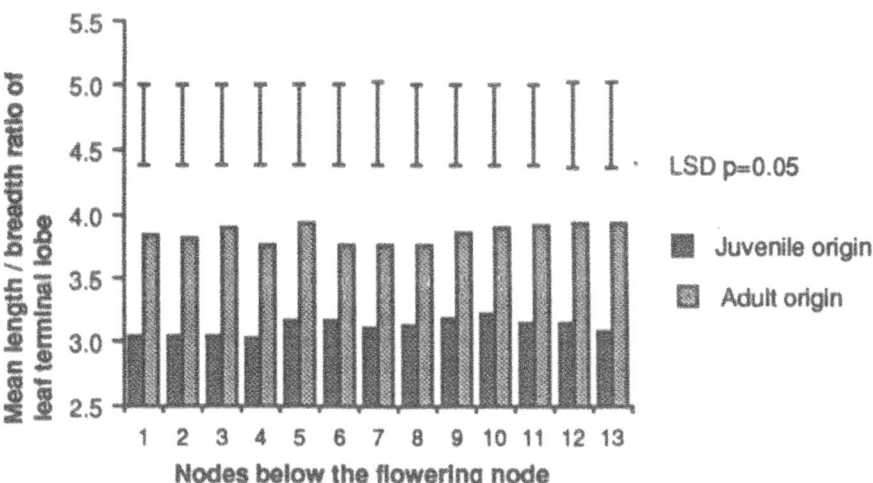

Fig. 1. Mean length / breadth ratios of the leaf terminal lobes in plants derived adventitiously from epidermal strips of juvenile and adult-phase *Solanum aviculare*. Each column represents the mean of *c.* 11 plants.

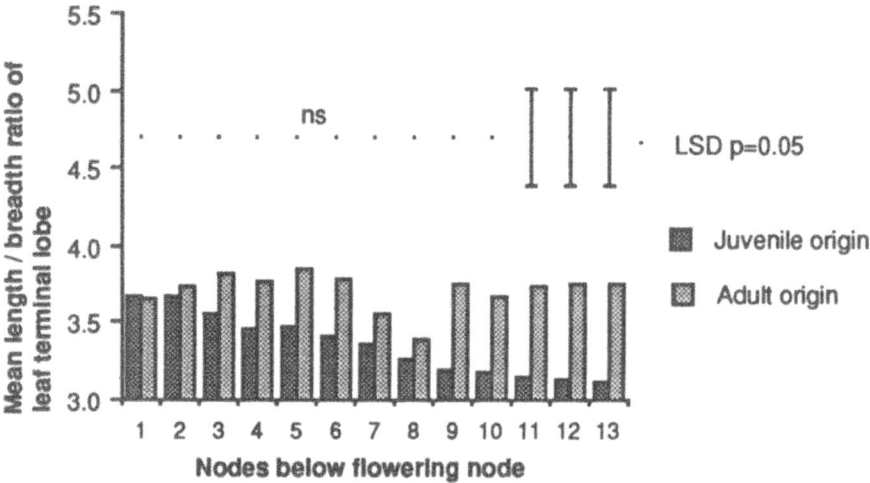

Fig. 2. Mean length / breadth ratios of the leaf terminal lobes in plants derived adventitiously from vascular tissues of juvenile and adult-phase *Solanum aviculare*. Each column represents the mean of *c.* 11 plants.

4. Discussion

Although this work does not provide direct evidence of intrinsic cellular differences between meristems of different phase, the inference is that if such differences exist they may be inherited through mitosis by subsequent differentiating tissue. The work of Raff [14] has clearly shown that tissue specific proteins are inherited in de-differentiated callus tissue, and Meins [12] has shown how epigenetic traits are lost during the re-setting of the developmental clock that accompanies regeneration from de-differentiated tissue. The present work demonstrates that some tissue – and phase – specific characters are inherited even after de-differentiation and subsequent shoot organogenesis. As phase-specific characters are superimposed on all regenerants, regardless of tissue origin, it is suggested that this infers intrinsic differences in the meristem initials. The epigenetic basis for the stability of these phase phenotypes at the cell level is unknown ; however evidence is available for both positive-feedback regulative mechanisms [12] and reversable forms of genetic change such as gene amplification [30]0 and transposition of genetic elements [4]. The hypothesis that the structure of the adult apex is a product of developmental history, maintained by positive feedback mechanisms, is implied by the rejuvenated nature of *de novo* meristems. The fact that in adventitiously derived plants, height to flower is of more importance than node number concurs with similar evidence found in *Ribes* [16] and *Hedera helix* [17]. The implication of these results is that the distance of the shoot apical meristem from the roots is likely to be a critical factor in phase change induction of plants derived from *de novo* meristems.

References

1. Baylis GTS (1963) A cytogenetical study of the *Solanum aviculare* species complex. Aust J Bot 11:168–177
2. Bilkey PC and Cocking EC (1981) Increased plant vigor by *in vitro* propagation of *Saintpaulia ionantha* Wendl. from sub-epidermal tissue. Hort Sci 16:643–644
3. Buiatti M (1977) DNA amplification and tissue cultures. In : J.Reinert and Y.P.S.Bajaj, eds. Plant Cell, Tissue, and Organ Culture, pp 338–372. Berlin: Springer-Verlag
4. Freeling M (1984) Plant transposable elements and insertion sequences. Ann Rev Plant Physiol 33:277–298
5. Hackett WP, Cordero RE and Srinivasan, C (1987) Apical meristem characteristics and activity in relation to juvenility in *Hedera*. In : J.G.Atherton, ed. Manipulation of Flowering, pp 93–99. London: Butterworths
6. Hackett WP (1983) Juvenility, maturation, and rejuvenation on woody plants. In: Hort Reviews 7:109–23
7. James AC (1993) PhD thesis, Wye College, University of London
8. James AC and Mantell SH (1991) Changes in phase phenotypes of the woody perennial *Solanum aviculare* Forst. following micropropagation. Revista Botanica Brasiliensis, (in press)
9. Kaddoura RL (1989) PhD thesis , Wye College, University of London
10. Mantell SH and Hugo SA (1989) Effects of photoperiod, mineral medium strength, inorganic ammonium, sucrose and cytokinin on root, shoot and microtuber development in shoot cultures of *Dioscrea alata* L. and *D. bulbifera* L. yams. Plant Cell Tiss Org Cult 16:23–37
11. Mc Donald EA (1973) PhD thesis, Wye College, University of London
12. Meins F (1987) Hormones and the molecular basis of determination in plants. In: Advances in the Chemical Manipulation of Plant Tissue Cultures, British Plant Growth Regulator Group, Monograph 16, pp 19–28
13. Murashige T and Skoog F (1962) A revised medium for rapid growth and bioassays with tobacco tissue cultures. Physiol Plant 2:473–497
14. Raff JW, Hutchinson JF, Knox RB and Clarke AE (1979) Cell recognition: Antigen determinants of plant organs and their cultured callus cells. Differentiation 12:179–186
15. Stein OL and Fosket EB (1969) Comparative developmental anatomy of shoots of juvenile and adult *Hedera helix*. Amer J Bot 36:346–331
16. Schwabe WW and Al-Doori AII (1973) Analysis of a juvenile-like condition affecting flowering in the black currant (*Ribes nigrum*) J Exp Bot 24:969–981
17. Wareing PF and Frydman VM (1976) General aspects of phase change with special reference to Hedera helix L. Acta Hort 56:57–68

Juvenility in micropropagated strawberries (*Fragaria ananassa* Duch)

THÉRÈSA JANE HUXLEY and P.M. CARTWRIGHT

Department of Agricultural Botany, University of Reading, P.O. Box 221, Reading, RG6 2AS, UK

1. Introduction

Strawberry plants are readily micropropagated from meristems [2]. However it is widely reported that micropropagated (MP) plants differ from their counterparts conventionally propagated (CP) from runners (stolons) in that they show more vigorous vegetative growth, especially stolon production, but delayed cropping [1, 3]. Also, MP plantlets of strawberry and other fruit crop species frequently display simpler leaf forms and other 'juvenile' characteristics and growth patterns [13] which, together with their initially smaller size suggests a closer resemblance to young seedlings than to CP plants. The possible association between such characters and juvenility in MP plants was therefore investigated and experiments conducted to distinguish those due to micropropagation *per se* from those which might be subject to manipulation by changing conditions in the *in vitro* or acclimatization stages.

2. Experimental methods

The main experiment was conducted in 1989–90 using the then newly released photoperiod-sensitive cultivar Malling Pandora. The micropropagation routine was similar to that developed at Horticulture Research International (HRI) East Malling [personal communication], which omits a rooting stage since *in vitro* grown strawberry plantlets become well-rooted during proliferation. On 14 August 1989 some 80 each of newly acclimatized MP and CP plants, all at the four leaf stage (i.e. MP plants had expanded four leaves since weaning from culture) were potted in Levington M2 potting compost and arranged on four trolleys in a glasshouse. All the plants received 8h natural daylight from 08.00 to 16.00h daily which was supplemented in January and February with HPMV lamps giving approximately 87 μmol m^{-2} s^{-1} PAR (400–700nm) at plant surface. From 16.00 to 08.00h daily they were in separate light proof cabinets, two with 16h darkness (the SD regime) and two with daylength extensions from 04.00h to 08.00h and 16.00h to 20.00h from tungsten lamps giving

346

P.J. Lumsden, J.R. Nicholas and W.J. Davies (eds.), Physiology, Growth and Development of Plants in Culture, 346–355, 1994.
© 1994 *Kluwer Academic Publishers.*

approximately 5 μmol m^{-2} s^{-1} PAR (the LD regime). The growth and development (leaf production, leaf area, fresh weight, branching) of the MP and CP plants in both SD and LD regimes were monitored by sampling at least six plants per treatment, three from each trolley, at intervals of 20–30 days until the final harvest at 198 days (28 Feb).

3. Results

3.1 Vegetative development

Although the CP and MP plants were initially at the same leaf stage the MP plants had considerably smaller leaf areas (-39%, Fig. 1) and fresh weights (-56%, Fig. 2) than their CP counterparts. However, the size difference was probably caused by the different environments in which they had grown (i.e. growth room cf. outdoors) rather than the method of propagation *per se.*

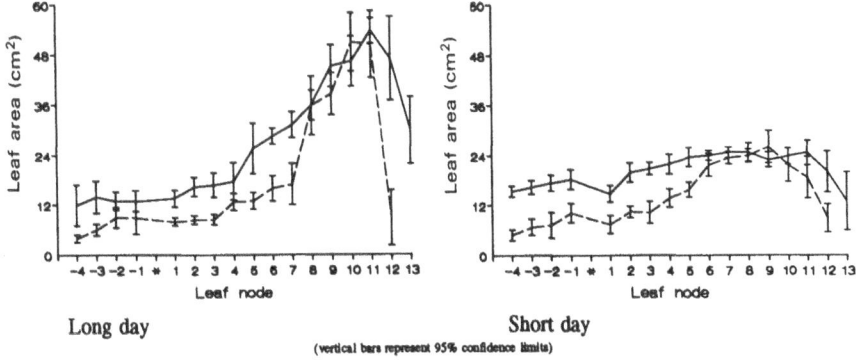

Long day   Short day

(vertical bars represent 95% confidence limits)

Fig. 1. Final leaf areas at successive nodes on conventionally propagated (CP) (———) and micropropagated (MP) (– – –) strawberry plants *cv.* Malling Pandora grown in long (16h) and short (8h) days in a glasshouse. * = day 0, 14 Aug '90. Leaves -4 to -1 present at the start of the experimental period in the glasshouse.

There were also clear differences in leaf form: the CP plants had only the trifoliolate leaves typical of the species while most of the MP plants bore monofoliolate, occasionally bifoliolate, leaves, resembling the juvenile leaves of young seedlings (Fig. 3). As new leaves expanded in the glasshouse the differences in their size and form were gradually lost since in the MP plants leaf size increased (Fig. 2) and the frequency of monofoliolate leaves decreased with advance in node number so that the size and growth habit of the MP and CP plants became increasingly similar.

There was an initial lag phase in the growth of the MP plants, indicated by the longer phyllochron during the first month (Table 1) and the absence of increase in weight over the first 26 days (Fig. 2). Subsequently their rates of leaf production were similar and their absolute growth rates exceeded those of the

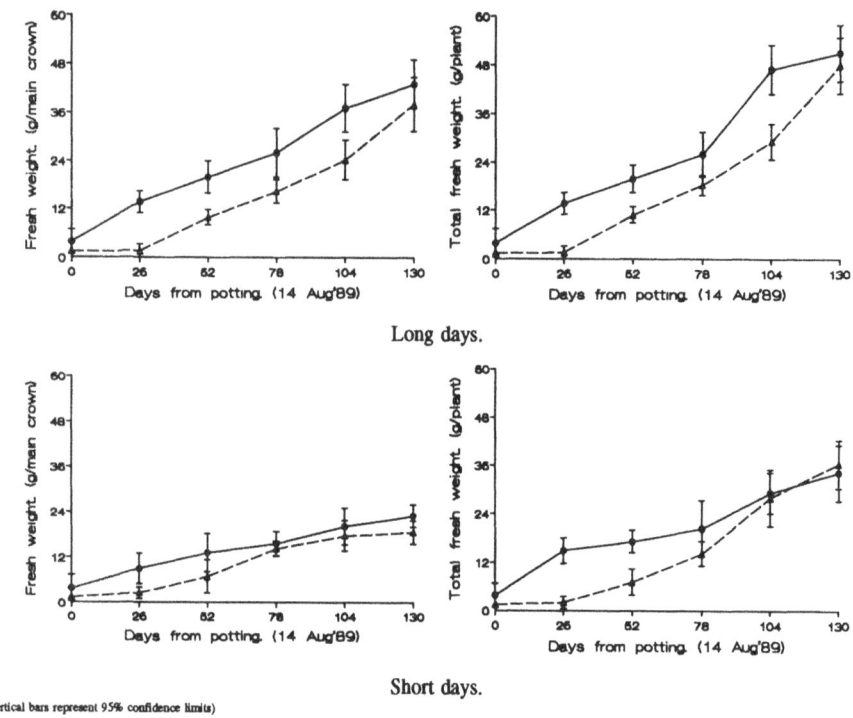

Long days.

Short days.

(Vertical bars represent 95% confidence limits)

Fig. 2. Fresh weight of main crown and whole plant of conventionally propagated (CP) (———) and micropropagated (MP) (— — —) strawberry *cv.* Malling Pandora grown in long (16h) and short (8h) days in a glasshouse.

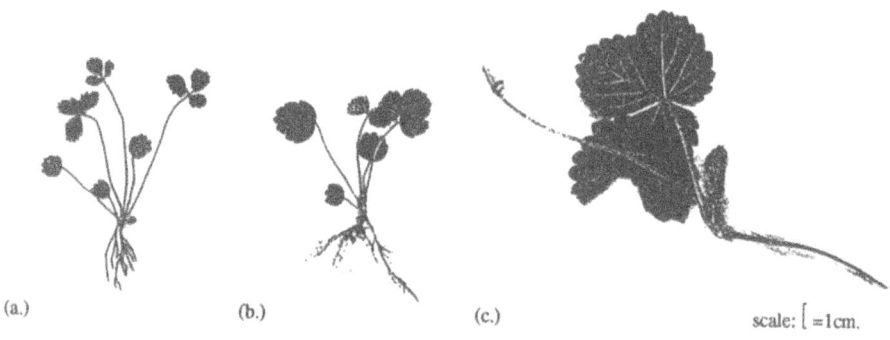

(a.)   (b.)   (c.)   scale: [=1cm.

Fig. 3. Photocopies of: (a) micropropagated plantlet (b) seedling and (c) conventionally runner-propagated strawberry *cv.* Malling Pandora.

CP plants in the same photoperiod regime. Also MP plants branched earlier than CP plants and so partitioned a larger fraction of their total biomass to branches (Fig. 4), to produce more branch crowns in the SD regime and to produce stolons in the LD regime (Fig. 5). Thus it can be argued that the MP plants were more vigorous since after an initial lag phase they had markedly higher relative growth rates and produced branches earlier and more frequently.

Table 1. Monthly leaf production per main crown of conventionally (CP) and micropropagated (MP) strawberry plants *cv.* Malling Pandora grown in the glasshouse

Month:	Aug-Sep'89	Oct'89	Nov'89	Dec'89	Jan'90	Feb'90
Days:	47	31	30	31	31	28
CP-Long day.	3.9	2.3	2.3	1.8	2.4	2.3
MP-Long day.	3.2	1.8	2.0	1.9	2.4	2.3
CP-Short day.	3.1	1.8	1.9	1.8	2.2	2.4
MP-Short day.	2.3	1.6	2.2	1.9	2.0	2.2

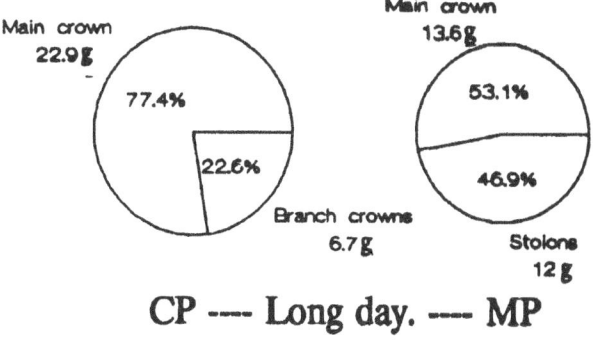

CP ---- Long day. ---- MP

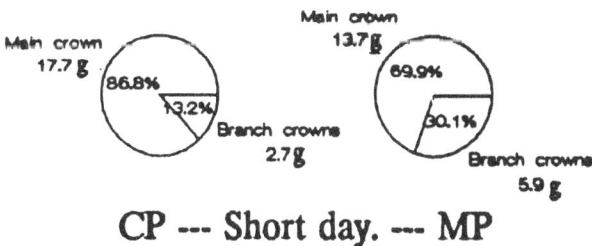

CP --- Short day. --- MP

Fig. 4. Partitioning of fresh weight per plant after 198 days, of conventionally propagated (CP) and micropropagated (MP) strawberry plants *cv.* Malling Pandora grown in either long (16h) or short (8h) days in a glasshouse.

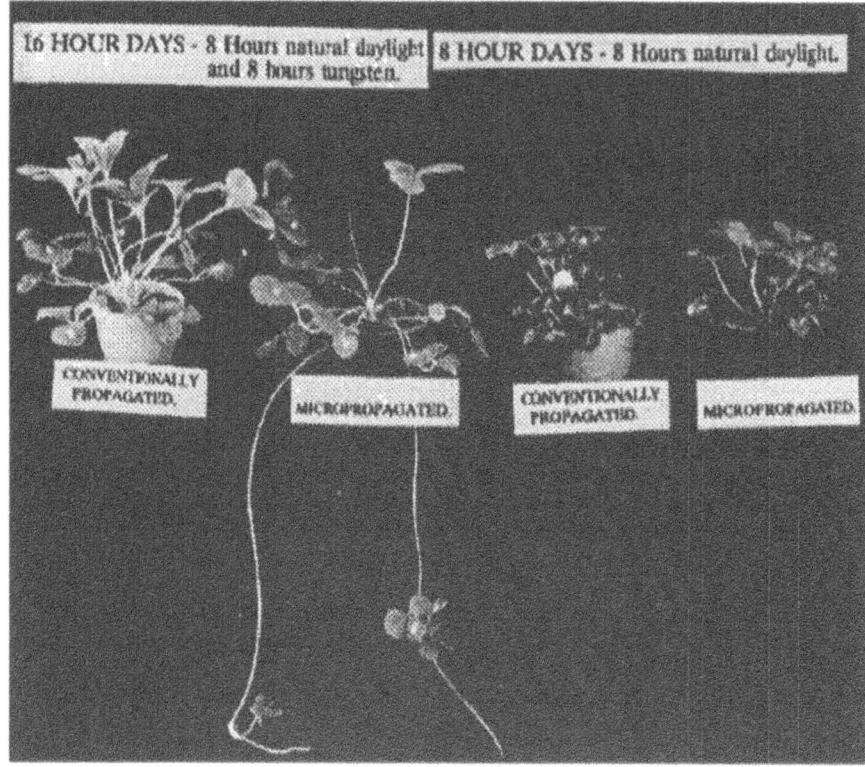

Fig. 5. Conventionally propagated and micropropagated strawberry plants *cv.* Malling Pandora after 196 days in either long (16h) or short (8h) days in a glasshouse.

3.2 Floral development

As well as recording conventional growth analysis data the microscopic apical development of the main crowns (i.e. main shoots) and branch crowns was scored [7]. As expected in a photoperiod-sensitive cultivar such as Malling Pandora, all the plants in the LD regime remained vegetative. Therefore only the observations on plants in the SD regime are presented, in Table 2.

The final leaf numbers indicate that floral initiation in the MP plants was delayed by four plastochrons suggesting a delay of at least four weeks in the attainment of competence to flower, which means that the MP plants must have remained juvenile for at least four weeks after weaning. This is consistent with the results from our studies with other photoperiod-sensitive cultivars (including Hapil and Cambridge Favourite), where it was found that SD treatments applied during the *in vitro* stage and during acclimatization were ineffective for inducing flowering.

Table 2. Apical development scores[a][7] of strawberry plants *cv.* Malling Pandora grown in the SD regime

(MC = main crown, BC = branch crown, AI = axillary inflorescence)			
Day (date):		Conventionally propagated	Micropropagated
53 (5 Oct)	MC:	V V V F0 F0 F2	V V V V V T1
	BC:	F	V
81 (2 Nov)	MC:	V F0 F0 F4 F5 F5	V F0 F2 F2 F3 F3
	BC:	F0 - F4	F0 - F5
	AI:	F0 - F2	
119 (10 Dec)	MC:	F6 F6 F6 F6 F6 F6	F6 F6 F6 F6 F6 F6
	BC:	F6	F6
	AI:	F6	F6
123 (14 Dec)	MC:	bracts visible	
	BC:	bracts expanded	
142 (2 Jan)	MC:	FB	
	BC:	FB	
198 (28 Feb)	MC:	FB FB FB FB FB FB	F6 F6 F6 F6 F6 F6
	BC:	FB	F6
	AI:	FB	F6
Final leaf number	MC:	10	14

[a] V = vegetative;
T = transitional.
F = floral: F0 = distinct 1° and 2° flower primordia; F1 = sepal primordia;
F2 = episepal primordia; F5 = whorls of primordia complete;
F6 = hairs on receptacle visible;
FB = primary flower blooming.

3.3 Duration of the juvenile phase

A further study was conducted to determine the duration of the juvenile phase i.e. the age or leaf stage at which MP plants of these cultivars can be induced to initiate flower primordia by short day treatment. Two photoperiod-insensitive cultivars i.e. Fern (day-neutral) and Rapella (everbearer) were also included in the study. The plants were micropropagated and the controls (T1) maintained in a LD regime (natural summer daylight) throughout. In T2 the plants were transferred to SD at weaning and remained in SD. In the remaining five treatments the plants were in LD but transferred to SD for 14-day periods, starting at 0, 7, 14, 21 and 28 days after weaning, in treatments T3, T4, T5, T6 and T7 respectively. Observations on these plants after 90 days are summarised in Table 3.

Table 3. Crown volume and apical development scores [7] of micropropagated, photoperiod-sensitive(S) strawberry cultivars, Hapil(H) and Malling Pandora(MP) and photoperiod-intensive(I) cultivars Fern(F) and Rapella(R) 90 days from weaning from culture

Treatment: Cultivar - Type:	Long day T1	Short day T2	T3	T4	T5	T6	T7
Crown volume(cm^3)^{a,b}							
H - S	289	**38**	282	314	312	316	315
MP - S	140	**41**	169	181	152	154	130
F - I	241	**36**	245	257	229	233	259
R - I	153	**98**	183	176	174	202	157
Apical development score^c							
H - S	V	**F0-F6**	V	V	V	V	F0-F6
MP - S	V	**F0-F6**	V	V	V	V	F0-F5
F - I	F0-F6	**F0-F6**	F6	F5-F6	F6	F5-F6	F6
R - I	F0-F6	**F0-F6**	F6	F5-F6	F6-F6	F6	F5-F6

[a] Values underlined are significantly different from T1 within cvs. at P = 0.05.
[b] Plant height (h) and spread taken as the mean of the north-south (N-S) and east-west (E-W) diameters (i.e. 2 x r) were used to calculate the crown volume of the shoot assuming an inverted conical shape.
[c] V = vegetative.
T = transitional.
F = floral: F0 = distinct 1° and 2° flower primordia; F1 = sepal primordia;
F2 = episepal primordia; F3 = petal and stamen primordia; F4 = peripheral pistil primordia;
F5 = whorls of primordia complete; F6 = hairs on receptacle visible.

As expected the two photoperiod-insensitive cultivars, Fern and Rapella, had initiated flower primordia in all the treatments but the photoperiod-sensitive cultivars had initiated only in T2 (continuous SD) and in T7 (SD from 28–42 days). Since all the plants in T6 (SD from 21–35 days) were still vegetative it appears that the SD were only inductive from about 35 days after weaning which is one week after the end of the acclimatization period. Thus it seems unlikely that changing the photoperiod regime *in vitro* could change the phenology of the MP plants. Although the alternative photoperiod regime i.e. SD during the *in vitro* and acclimatization stages did not influence time or leaf number to flower initiation in the MP plants, there was a marked effect on their vegetative growth habit similar to that shown in CP plants. Thus leaves produced in SD were smaller, darker and with shorter petioles, and they formed a more compact, flatter rosette with a smaller crown volume (compare T1 and T2 in Table 3). It is clear that these vegetative responses to photoperiod, shown in cultivars of all three phenological classes, are independent of the flowering responses shown by the photoperiod-sensitive cultivars. For example, Table 5 shows that the photoperiod-insensitive cultivars Fern and Rapella initiated flower primordia in all the photoperiod treatments (T1-T7) and the photoperiod-sensitive cultivars only in T2 and T7, but all showed similar responses in growth habit.

4. Discussion

It would appear that many of the features associated with micropropagation in strawberry plants, including the morphologically distinct leaf forms, enhanced shoot growth and branching observed in both this and other studies of strawberry [1, 3, 5] are primarily a consequence of micropropagation *per se* rather than features induced by particular *in vitro* conditions. Certainly such features are seen in studies of other members of the Rosaecae with trifoliolate leaves such as thornless blackberry *Rubus sp.* [12] and are also common in other fruit species including grape *Vitis sp.* [10], blueberry *Vaccinium sp.* [11], and lingonberry *Vaccinium vitis* [6]. They are already accepted as indicators of juvenility [15] and as in conventionally raised plants these juvenile characters become less apparent as the plants mature, so that the MP strawberry plants like other micropropagated species become less easily distinguished once weaned from culture. Indeed if juvenility is defined as incompetence to flower [16, 15] then this study confirms that MP strawberry plants, unlike CP plants but like seedlings [4], have a juvenile phase which appears to last for about four plastochrons, usually between four and five weeks, after weaning. During this phase newly weaned plants resemble seedlings in that they have smaller shoot apical domes and frequently display juvenile leaf morphology (Table 4, Fig. 3). Modifications to the micropropagation regime, including giving SD (instead of the standard LD) during the proliferation and acclimatization stages greatly decreased the size and vigour of the plantlets and prolonged the lag phase before growth resumed after potting. However it did not affect the leaf number to flower initiation, although the time taken was usually slightly greater. Thus the investigation showed that changing the photoperiod conditions during micropropagation did not affect the duration of the juvenile phase though it did modify plantlets' early growth, and influence the production of branch crowns and stolons by the weaned plants.

Table 4. Observations on micropropagated plantlets, seedlings and runner-propagated plants

Propagation method:	Micropropagation:	Conventional propagation:	
Plant source:	stolon meristem tip	seed	stolon (runner)
Character:			
Leaf size score(cm²)ᵃ	0.5 - 2.5 (small)	0.5 - 2.5 (small)	2.6 - 10 (medium)
Occurrence of juvenile leaves	Yes	Yes	No
Shoot apex diameter(mm)ᵇ	N-S:0.05 - 0.14 E-W:0.11 - 0.16	N-S: 0.03 - 0.16 E-W: 0.05 - 0.16	N-S: 0.15 - 0.20 E-W: 0.17 - 0.20
Minimum trifoliolate leaves pre-floweringᶜ	10	5 - 7ᵈ	3 - 4ᵉ

ᵃ Leaves were visually scored using a calibrated series of leaf outlines.
ᵇ Measured as north-south (N-S), east-west (E-W) diameter across the apical dome [7]. (seedling with less than 4 expanded leaves)
ᶜ Inductive photoperiod also required if a photoperiod-sensitive cultivar.
ᵈ [8]
ᵉ [9]

References

1. Boxus Ph (1989) Review on strawberry mass propagation. Acta Hort 265: pp 309–320
2. Boxus Ph, Damiano C and Brasseur E (1984) Strawberry. In: Ammirato, Evans, Sharp, and Yamada, eds. Handbook of Plant Cell Culture, pp 453–486. New York: MacMillan
3. Cameron JS and Hancock JF (1986) Enhanced vigor in vegetative progeny of micropropagated strawberry plants. HortSci 21:pp 1225–1226
4. Guttridge CG (1969) Fragaria. In: Evans, ed. The Induction of Flowering. Some Case Histories, pp 247–267. Melbourne, Australia: Macmillan
5. Hennerty MJ, Hunter SA and Fox MJ (1987) Field performance of tissue cultured strawberry plants. In: Boxus and Larvor, eds. *In vitro* Culture of Strawberry Plants, pp 41–46. Luxembourge: CEC Biological Series Rep. EUR 10871
6. Hosier MA, Flatebo G and Reade PE (1985) *In vitro* propagation of lingonberry. HortSci 20: pp 364–365
7. Huxley TJ (1992) The manipulation of flowering and vegetative growth in micropropagated strawberry plants, *Fragaria ananassa* Duch. PhD Thesis, University of Reading
8. Jonkers H (1958) Accelerated flowering of strawberry seedlings. Euphytica 7: p 41
9. Jonkers H (1965) On the flower formation, the dormancy and the early forcing of strawberries. Mededelingen van de Landbouwhogeschool Wageningen, Nederland 65: pp 1–59
10. Krul WR and Myerson J (1980) *In vitro* propagation of grape. In: Proceedings of the Conference on Nursery Production of Fruit Plants through Tissue Culture — Applications and Feasibility. USDA-SEA Agricultural Research Results ARR-NE-11. pp 35–43
11. Lyrene PM (1981) Juvenility and the production of fast-rooting cuttings from blueberry shoot cultures. J Amer Soc Hort Sci 112: pp 125–130
12. McPheeters K and Skirvin RM (1984) Field observation of micropropagated thornless 'Evergreen' blackberry. HortSci 19: p 544 (abs)
13. Swartz HJ and Lindstrom JT (1986) Small fruit and grape tissue culture from 1980–1985. Commercialisation of the technique. In: Zimmermann, Griesbach, Hammerschlag and Lawson,

eds. Tissue Culture as a Plant Production System for Horticultural Crops, pp 201–220. Dordrecht: Martinus Nijhoff

14. Swartz HJ, Lindstrom JT, Fiola JA and Zimmerman R H (1987) The use of tissue culture in the United States. In: Boxus and Larvor, eds. *In vitro* Culture of Strawberry Plants, pp 79–100. Luxembourge, CEC Biological Series Rep. EUR 10871

15. Wareing PF (1987) Juvenility and cell determination. In: Atherton, ed. Manipulation of Flowering, pp 83–92. Butterworths

16. Zimmerman RH, Hackett WP and Pharis RP (1985) Hormonal aspects of phase-change and precocious flowering. In: Pierson and Reid, eds. Encyclopedia of Plant Physiology (New Series) II, pp 79–115. Berlin and New York: Springer-Verlag

Flowering abundance of strawberry depending on the number of subcultures *in vitro*

Relationship with growth, rooting and peroxidase activity

A. JEMMALI[1], PH. BOXUS[1], CL. KEVERS[2] and TH. GASPAR[2]

[1] *Station des Cultures fruitières et maraîchères, CRA, 234, Chaussée de Charleroi, B – 5030 Gembloux, Belgium*
[2] *Université de Liège, Institut de Botanique B 22, Sart Tilman, B – 4000 Liège, Belgium*

1. Introduction

Some difficulties have arisen over the years with the use of micropropagated strawberry plants, or runner plants derived from them. These have included delayed fruiting and the size of individual fruits from such plants which tends to be smaller than from conventionnaly runner-propagated plants [13]. The smaller size of the fruits might come from a tendency of micropropagated strawberries to produce more flowers, particularly when originated from shoots raised after a high number of multiplication cycles [7]. The present work aimed to study the flowering abundance in relation with the number of subcultures on the multiplication medium and also correlate this flowering behaviour with other morphogenetic traits. Particular attention has been paid to rooting because flowering and rooting are known to be antagonistic developmental processes [3, 8, 9].

2. Materials and methods

Micropropagation of *Fragaria* x *ananassa* Duch–cv Gorella was initiated by *in vitro* introduction of meristem tips from field donor plants following Boxus [1]. Proliferation through axillary shooting continued in the presence of BAP (1 mg/l) , IBA (1 mg/l) and GA$_3$ (0.1 mg/l) by regular subcultures every three weeks. Shoots coming directly from the multiplication medium were transferred for 6 weeks on a rooting medium containing IBA (1 mg/l). The cultures were maintained under TL lamps providing 30 to 40 μmol m^{-2}s^{-1} 16 h per day, at a temperature of 24 ± 2 °C. Acclimatization and initial growth phase were performed in a controlled environment cabinet (light intensity : 180 μmol m^{-2}s^{-1} 16 h/day; temperature : 20 ± 1 °C). They were prolonged by a 4 weeks flowering induction period (8/16 day/night cycle) and a 6–8 weeks second growing phase (16/8 day/night cycle) successively under the same light and temperature conditions.

 Measurements were carried out on rooted vitroplantlets (total fresh weight

P.J. Lumsden, J.R. Nicholas and W.J. Davies (eds.), Physiology, Growth and Development of Plants in Culture, 356–362, 1994.

and root fresh weight per plantlet) and on adult plants (petiole length, leaf area, root dry weight). Flowering was examined before and at anthesis. A series of 15 individuals were sampled. The student T test was applied. Peroxidase activity per unit of protein was measured in crude extracts of shoots kept on the rooting medium for increased periods of time, according to techniques already used in collaboration [2, 12].

3. Results

3.1 Flowering

The flowering behaviour was compared between strawberry vitroplants after 15 or 45 subcultures. The floral distribution was drawn according to the phenologic microscopic stages as described by Jonkers [6]; thus before the emergence of the inflorescence, there was an evident advantage for the vitroplants raised after 45 subcultures (Fig. 1). Counting the number of flowers developing to anthesis on the main truss (Fig. 2) also showed a tendency to an hyperflowering habit in vitroplants from the 45th subculture.

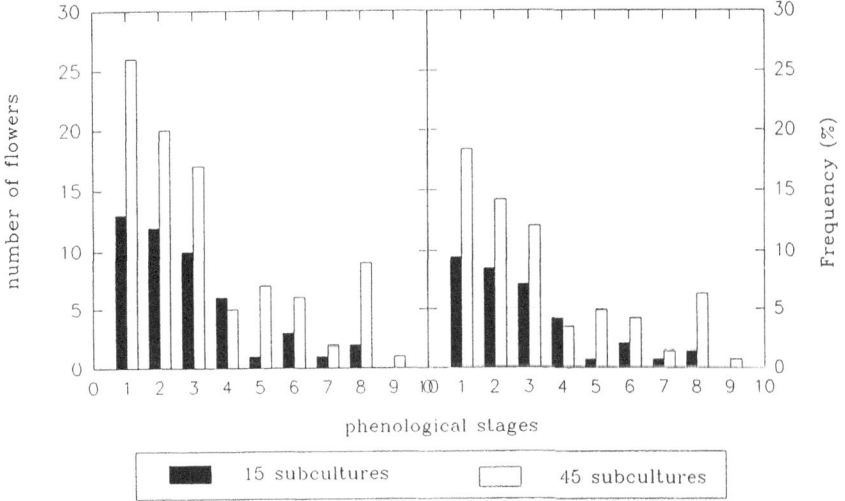

Fig. 1. Number of flowers, as shown by floral distribution according to the phenological stages described by Jonkers [6] , after 98 d of culture of strawberries issued from the 15th and the 45th multiplication cycles. 1= Round generative growing point; 2 = The first bract; 3 = Primordia of the calyx; 4 = Primordia of the corolla; 5 = Flowers of one truss in different stages of development; 6 = Primordia of the stamens; 7 = Primordia of the carpels but the receptacle is not complete; 8= Advanced stage of 7; 9 = Pubescent and light green flower buds.

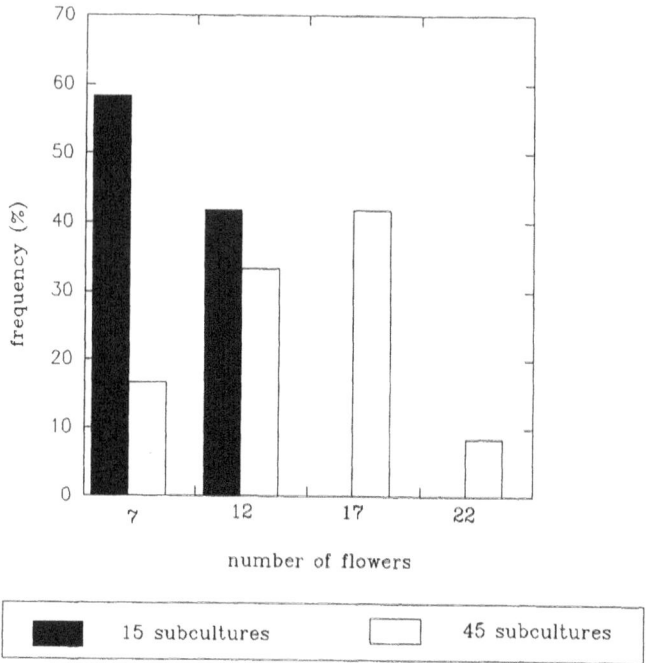

Fig. 2. Number of flowers developing to anthesis in the main truss of strawberries issued from the 15th and the 45th multiplication cycles.

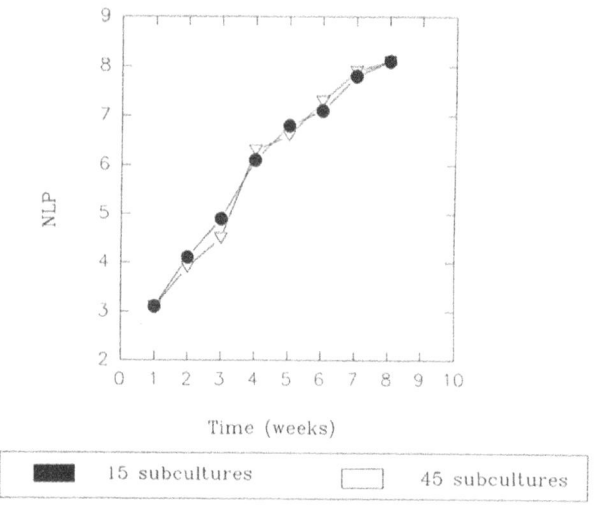

Fig. 3. Evolution of mean number of leaves in strawberries issued from the 15th and the 45th multiplication cycles.

3.2 Vegetative characteristics

Strawberry vitroplants in their initial growth phase at acclimatization appeared shorter with the increasing number of subcultures, but the number of subcultures did not influence the mean number of leaves per plant, until at least the 8th week (Fig. 3). There was however an influence of the number of subcultures on the leaf petiole length as shown in Figure 4. Fresh weights of the whole plantlets or of the root system were considerably lower in vitroplants from the higher number of subcultures (Table 1). The hyperflowering of adult plants raised from the higher number of subcultures (Table 2) was again associated with a reduced (dry) weight of the root system, and also with a reduced leaf area (Table 2).

Table 1. Incidence of the *in vitro* subculture number on the vitroplantlets vigour of micropropagated strawberry cv Gorella

Subculture number	Fresh weight/plantlet (mg)	Root fresh weight/plantlet (mg)
15	121.0 a	29.1 a
45	53.5 b	15.1 b
T-test	(P<.05)	(P<.05)
	(Means with distinct subscripts are significantly different)	

Table 2. Some vegetative and generative characteristics of adult strawberry plants as influenced by the number of subcultures

Subculture number	Leaf area (cm²/plant)	Root dry weight (mg/plant)	Number of flowers per plant at anthesis
15	600.0 a	123.0 a	9.3 b
45	445.0 b	94.0 b	13.7 a
T-test	(P<.05)	(P<.05)	(P<.05)
	(Means with distinct subscripts are significantly different)		

3.3 Rooting and associated changes of peroxidase activity

As shown by results in Figure 5, the higher number of subcultures accelerated root formation within the first days after transfer to rooting medium. Changes of peroxidase activity in crude extracts of the whole strawberry shoots undergoing rooting on the rooting medium showed the typical peak of activity as already observed in many materials undergoing rooting [3, 2, 12, 11]. The peak of activity occurred after the same number of days regardless of the number of previous subculture cycles (Fig. 6), but there was a slight overall increase in peroxidase activity in the plants from the higher number of subcultures.

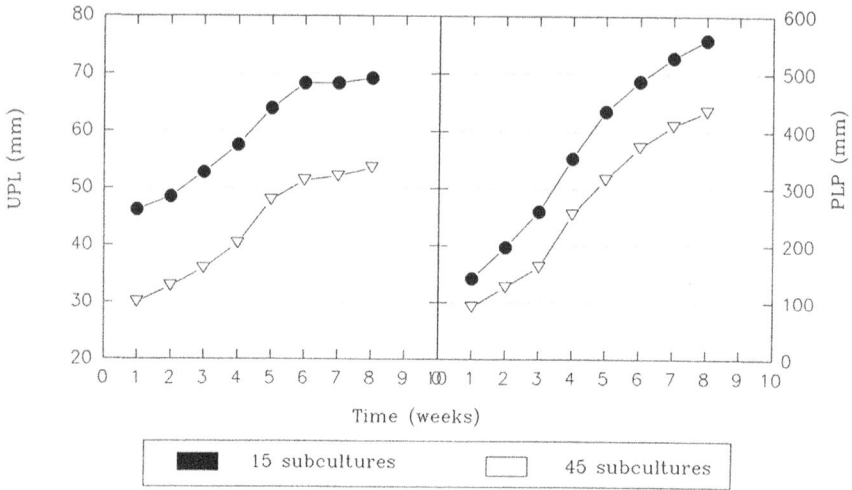

Fig. 4. Evolution of mean leaf petiole length of strawberries issued from the 15th and the 45th multiplication cycles. UPL : unitary petiole length. PLP : petiole length per plant.

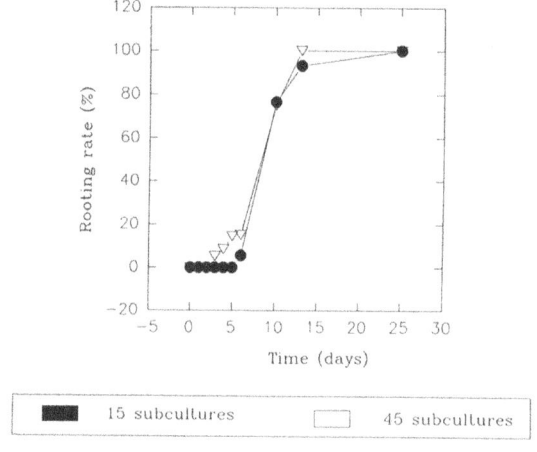

Fig. 5. Evolution of the rooting rate of strawberries issued from the 15th and the 45th multiplication cycles.

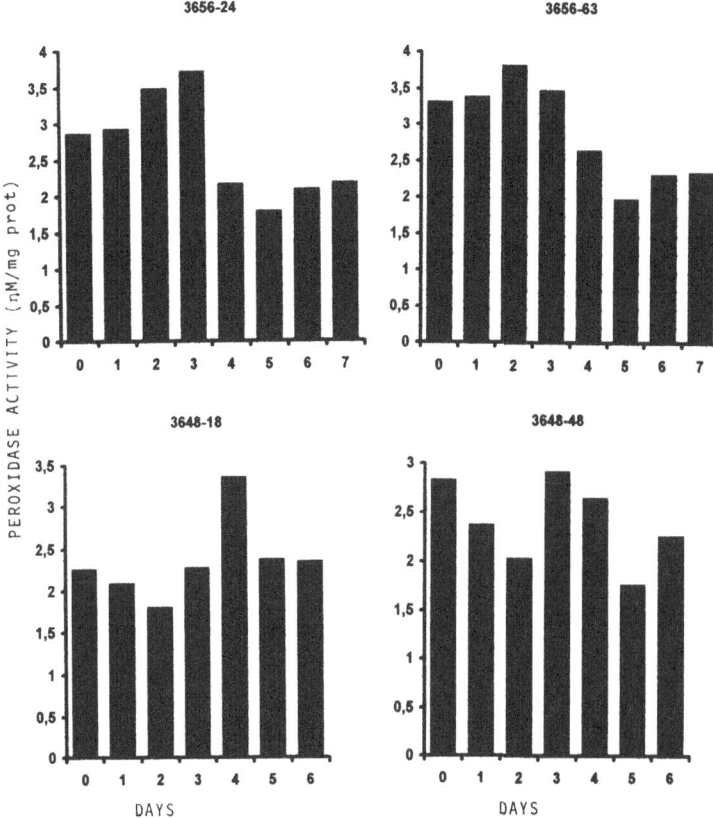

Fig. 6. Evolution of specific peroxidase activity in the course of rooting of two clones (3648 : 18 and 48 subcultures; 3656 : 24 and 63 subcultures) of strawberries in relation to the foregoing number of multiplication cycles.

4. Discussion and conclusion

The hyperflowering behaviour of strawberry vitroplants issued from a great number of subcultures [7] is here again confirmed. It is shown here that this flowering behaviour is associated with changes in morphological characteristics such as reduced global height due to reduced leaf petiole length (leaf area is also reduced but not the number of leaves formed) and reduced development of the root system. Such an antagonism in the development of the flowering and the root systems has already been observed in different types of plants, not necessarily issued from micropropagation [3, 8, 9]. Strawberry shoots issued from a higher number of subcultures however rooted more precociously, which does not seem favourable for the development of the root system. The transient rise in peroxidase activity of shoots undergoing rooting [3, 2, 12,11] is also confirmed. Interestingly, as shown for other materials [11], the precocity of the peroxidase peak is in correspondence with the precocity for rooting. Such a relationship between the number of subcultures, peroxidase changes and rooting has already been reported for globe artichoke [10, 11] – Peroxidase activity and related metabolisms, namely lignification and auxin catabolism [5] thus might be investigated further to explain the altered phenotypical changes but also as potential markers of the number of subcultures.

References

1. Boxus Ph (1974) The production of strawberry plants by *in vitro* micropropagation. J Hort Sci 49:209–210
2. Druart Ph, Kevers C, Boxus Ph and Gaspar Th (1982) *In vitro* promotion of root formation by apple shoots through darkness effect on endogenous phenols and peroxidases. Z Pflanzenphysiol 108:429–436
3. Gaspar Th, Kevers C, Hausman JF, Berthon JY and Ripetti V (1992) Practical uses of peroxidase activity as a predictive marker of rooting performance of micropropagated shoots. Agronomie: in press
4. Gaspar Th (1981) Rooting and flowering: two antagonistic phenomena from a hormonal point of view. In: B Jeffcoat, ed. Aspects and Prospects of Plant Growth Regul, Monograph 6, pp 39–49. Wantage: British Plant Growth Regul Group
5. Gaspar Th, Penel C, Hagège D and Greppin H (1991) Peroxidases in plant growth, differentiation, and development processes. In: J Lobarzewski, H Greppin, C Penel and Th Gaspar, eds. Biochemical, Molecular, and Physiological Aspects of Plant Peroxidases, 249–280. Genève: Univ de Genève
6. Jonkers H (1965) On the flower formation, the dormancy and the early forcing of strawberries. Meded Landbouwhogesch Wageningen 65–66
7. Kinet JM and Parmentier A (1989) The flowering behaviour of micropropagated strawberry plants cv Gorella: the influence of the number of subcultures on the multiplication medium. Acta Hort 265:327–334
8. Miginiac E (1978) Some aspects of regulation of flowering role of correlative factors in photoperiodic plants. Bot Mag (Tokyo) Special Issue 1:159–173
9. Miginiac E and Sotta B (1985) Organ correlations affecting flowering in relation to phytohormones. Biol Plant 27:373–381
10. Moncousin C (1982) Peroxidase as a tracer for rooting improvement and expressing rejuvenation during *in vitro* multiplication of *Cynara cardunculus scolymus* L.. In: A Fujiwara, ed. Plant Tissue Culture, pp 147–148. Proc 5th Intl Cong Pant Tissue and Cell Culture. Tokyo: Jap Assoc Plant Tissue Cult
11. Moncousin C and Ducreux G (1984) Activité peroxydasique et rhizogenèse dans le cas de *Cynara scolymus* L.: évolution au cours de repiquages successifs de boutures cultivées *in vitro*. Comparaison avec de jeunes plantes issues de graines. Agronomie 4:105–111
12. Quoirin M, Boxus Ph and Gaspar Th (1974) Root initiation and isoperoxidases of stem tip cuttings from mature *Prunus* plants. Physiol Vég 12:165–174
13. Zimmerman RH (1991) Micropropagation of temperate zone fruit and nut crops. In: PC Debergh and RH Zimmerman, eds. Micropropagation Technology and Application. pp 231–246. Dordrecht: Kluwer Academic Publishers

Dealing with microbial contaminants in plant tissue and cell culture: hazard analysis and critical control points

C. LEIFERT[1] and W.M. WAITES[2]

[1] Department of Plant & Soil Science, University of Aberdeen, Aberdeen AB9 2UE, UK
[2] Department of Applied Biochemistry and Food Science, University of Nottingham, Sutton Bonington, Loughborough, Leicester LE12 5RD, UK

1. Introduction

Microbial contamination is the most important reason for losses in both scientific and commercial micropropagation systems. In most laboratories losses averages between 3 and 15% of plants at every subculture [33]. Microbial contaminants can be introduced both with the explants used to initiate plant tissue cultures and at every stage of the tissue culture process in the laboratory [3, 4, 33]. However, it is often difficult to determine the exact source or sources of contamination from a visible assessment of the contaminated cultures. As in other industries which produce 'aseptic' or contamination-free products (such as the food or pharmaceutical industries), plant tissue culture laboratories require some form of 'microbiological production control', in order to monitor all possible sources of contamination continuously, with a view to detecting and preventing contamination at source.

We therefore applied the Hazard Analysis Critical Control Point (HACCP) system, which has been used to prevent or reduce microbial contamination in the food industry [49], to commercial plant tissue culture systems [33]. The establishment of HACCP involves 3 steps. Firstly, all potential microbiological hazards associated with individual stages in the production system are assessed and listed. Secondly, methods for early detection and prevention of hazards were introduced (if already available) or developed if no appropriate methods were available. Early detection involves the isolation and identification of indicator microorganisms which are known hazards of specific steps in the production process. Thirdly, methods for suppression or treatment of microbiological hazards are developed as a back-up, when early detection and prevention is not possible or has failed.

In this review we have therefore described how HACCP can be used in plant tissue culture and summarized the current knowledge about:
1. The microbiological hazards associated with the various stages of plant tissue culture;
2. The methods used to detect and prevent contamination at different stages and

P.J. Lumsden, J.R. Nicholas and W.J. Davies (eds.), Physiology, Growth and Development of Plants in Culture, 363–378, 1994.
© 1994 *Kluwer Academic Publishers.*

3. The methods available for the treatment of contamination in plant tissue culture.

2. Hazards caused by different contaminants in plant tissue and cell cultures

Mites, thrips, fungi, yeasts, bacteria and viruses are the most commonly described contaminants of plant tissue cultures [29, 33]. The hazards associated with the different organisms can be classified according to their effect on plants *in vitro* and the economic loss they can cause.

Nearly all fungal and yeast species are severe hazards *in vitro* because they grow well on plant tissue culture media and kill plants by reducing the pH and the production of toxic metabolites and increase the competition for nutrients [9, 14, 28, 29,].

Some bacterial contaminants are classified as severe hazards because they also grow on plant tissue culture media and kill *in vitro* plants [25, 26, 27, 29, 33]. This includes many bacterial species which naturally inhabit plant surfaces *in vivo* as saprophytes or pathogens, such as *Lactobacillus plantarum, Corynebacterium* spp., fluorescent Pseudomonads and various Enterobacteriaceae [12, 33].

Mites and thrips are considered severe hazards because they introduce fungi, yeasts and bacterial contaminants into plant tissue cultures. Mites and thrips are very small organisms (often <1 mm) and are able to enter most tissue culture vessels. Once growth rooms are infested, migrating stages of mites or thrips act as vectors and spread fungal and yeasts contaminants from contaminated to uncontaminated cultures [3, 23, 29].

A different type of severe hazard is caused by bacteria and viruses which stay 'latent' (not producing symptoms on the plant or visible growth on the medium) for long periods of time *in vitro* (Table 1). In many cases these 'latent' contaminants either reduce multiplication rates and, thereby the productivity of plant cultures, or become 'virulent' at a later stage (causing disease symptoms or plant death after long periods of time *in vitro* or after weaning of plants [7, 8, 26, 27]). When 'latent' contaminants are not detected, large numbers of infected plants are produced [7, 29]. 'Latent' persistence of bacteria and viruses in plant tissue cultures is found to occur as a result of a number of different problems.

Contaminants can be suppressed by plant resistance mechanisms which are active *in vitro*. In this context it is interesting to note that plant tissue and cell cultures are increasingly used to study plant resistance mechanisms and the production of anti-viral, anti-bacterial and anti-fungal plant metabolites [33]. Some resistance mechanisms have been shown to be more active *in vitro* and there is an increasing amount of evidence that this is due to (1) the juvenility of the cultured plant cells and tissues and/or (2) the growth regulators incorporated in the growth medium [33].

Some types of tissue cultures (especially callus cultures) lack the tissues

Table 1. Critical control points, indicator microorganisms and equipment and methods used for microbiological production control in plant tissue culture and to prevent contamination

Critical control points in plant tissue culture

Stage of plant tissue culture	Parameters which need to be monitored	Indicator micro-organisms	Equipment needed	References for methods
Stage 0	– symptoms of viroid, viral, bacterial and fungal pathogens – detection and indentification of viroids, viruses, bacteria and fungi		– hand lens – light microscope – phase contrast microscope – selective bacteriological media – seriological tests/ DNA probes – indicator plants	viruses: [18,36] bacteria: [35] fungi: [45]
Stage 1	– active chlorine content in hypochlorite solutions – activity of other biocides	Gram negative bacteria: *Pseudomonas fluorescens, Erwinia* spp., *Klebsiella* spp., *Serratia* spp., *Agrobacterium* spp.	– Whatman indicator paper – arsenite solution – titration column – depending on the compound	[19] [6,21,44,46]
Stage 2 Indexing (Critical control point 1)	– presence of viruses, bacteria and fungi on explants after surface disinfected		– indexing media – seriological tests – nuclei acid hybridisation tests	[29,33] [29,33] [33]

needed for the proliferation of certain microorganisms (for example, many of the fastidious plant pathogenic bacteria) which grow in specific plant tissues such as the phloem or xylem [15, 22, 40].

Many plant species grown *in vitro* reduce the pH of the medium from 5.6 to values between 3.8 and 4.5, which reduces or prevents the growth of bacteria such as *Agrobacterium tumefaciens, Bacillus* spp., *Pseudomonas maltophilia, Pseudomonas syringae, Staphylococcus saprophyticus*, and *Xanthomonas campestris* [7, 8, 25, 31].

Latency can also be caused by suppression of contaminants by the prophylactic incorporation of antibiotics, fungicides and other biocides into the plant growth media [29, 30, 32]. Virulence of 'latent' contaminants can be triggered *in vitro* by changes in incubation temperature, growth medium composition, and the removal of antibiotics from the growth medium. For example, a rise in incubation temperature due to breakdown of a growth room cooling system was associated with increased growth of *Lactobacillus plantarum* in *Hemerocallis* cultures with the result that the increased production of lactic acid by the *Lactobacillus* killed all infected plants within 14 days [33]. Transfer onto rooting media with lower sugar, mineral and growth regulator concentrations was also shown to induce growth and virulence of certain bacterial contaminants [7]. No visible contamination could be observed in *Hemerocallis* cultures infected with *Corynebacterium* growing on media containing combinations of antibiotics, but as soon as plants which remained infected during treatment were transferred onto antibiotic-free media they became visibly contaminated and died within 3 subcultures [30]. Micropropagation of 'latent' infected plants can therefore be totally unpredictable. It can result in larger economic losses than those that result from contaminants which kill plants instantly, because large numbers of infected plants are often produced before the contaminants become 'virulent'. This can cause sudden loss of whole crops from one subculture to the next [27, 33].

Contamination of plant tissue cultures with pathogenic viroids, viruses and bacteria, which stay 'latent' *in vitro*, is a severe hazard, because these contaminants can become 'virulent' *in vitro* or cause disease after weaning *in vivo* [33, 38, 41, 50, 51, 52]. For example, *A. tumefaciens* and *Xanthomonas* spp. and many common plant viruses (such as cucumber mosaic virus) can stay 'latent' throughout all *in vitro* stages, but become virulent once plants have been weaned. In some instances, disease symptoms only became apparent several months after weaning at a time when plants have already been dispatched to customers. This represents a hazard because it can result in the spread of plant diseases and the loss of customer confidence in commercial plant tissue culture [7].

Some 'latent' bacteria introduced in the laboratory (such as some *Bacillus*, *Micrococcus* as *Staphylococcus* spp. and certain *Pseudomonas* spp.) may not result in deterioration of tissue cultured plants. Most of these contaminants do not grow on plant tissues and are dependent on plant exudates for growth in the tissue culture medium [33, 25]. Plants are not usually killed by these types of bacteria and can be propagated in their presence. The degree to which these bacteria affect the multiplication and/or rooting rates in plants depends very much on the plant species. For species which have strong *in vitro* resistance and/or reduce the medium pH such as *Hemerocallis* spp. they were shown to cause little or no effect. However, for less 'resistant' plant species they have to be classified as intermediate to slight hazards, depending on the degree of inhibition they cause [25, 33].

3. Critical control points in pant tssue and cell culture

All plant tissue culture systems can be divided into different stages as described in Table 1 [10, 34]. Although contaminants can be introduced at every *in vitro* stage, different genera of fungi, yeasts and especially bacteria were found to be introduced at the different stages. Every stage in the micropropagation process was therefore found to be a Critical Control Point and had to be monitored constantly to avoid contamination. Microorganisms which were found to be exclusively or primarily introduced at specific stages can be used as 'indicator microorganisms' to pin-point the source of contamination (Table 1). Depending on the hazard associated with the different stages they are classified as either Critical Control Points I (insufficient control at this point could potentially prevent the production of individual or all crops in the laboratory) or Critical Control Points II (insufficient control at this point resulted in a proportion of plants being lost) (Table 1). These definitions for the Critical Control Points have been adopted to reflect the situation of plant tissue culture and are therefore different to those used in some other industries.

3.1 Stockplant treatment (stage 0)

Many of the severe microbiological hazards found in plant tissue cultures have originated from the explants taken from the stockplant (Table 2). Explants taken from (1) unprotected plants in the field, (2) plant parts infected with plant diseases and pests and (3) plant parts growing in or near the soil are often impossible or very difficult to disinfect [13, 14, 29, 33]. Overhead watering increases the populations of fungi, yeasts and bacteria on aerial plant surfaces and prevents or reduces the success of surface disinfection [16, 33]. Mites and thrips can be introduced into laboratory growth rooms with explants taken from infected plants [3, 9, 29, 33] and the use of stock plants infected with plant pathogenic bacteria and viruses can either prevent the establishment of tissue cultures or if the contaminants stay 'latent' result in the production of large numbers of infected plants [5, 7, 8]. New stock plants infected with diseases or pests can also be an infection source for other stock plants in the glasshouse. Due to the severe hazards that can be associated with the stock plant, this tissue culture stage is classified as a Critical Control Point 1.

To control hazards at this stage only stock plants grown under protected conditions (glasshouses, growth chambers) should be used. New stock plants from outside sources should be kept in quarantine in a separate quarantine glasshouse for at least 6 months. During quarantine they should be assessed and tested at regular intervals for plant diseases and plant pests (Table 1). Only disease and pest free stock plants should be transferred into the stock plant glasshouse. To reduce the number of microorganisms on the plant surfaces a minimal irrigation regime without overhead watering should be used and plants should be protected by strict chemical and/or biological disease and pest control [16, 29]. If diseased plants have to be used as explant sources they should be kept permanently in a separate quarantine glasshouse.

Table 2. Critical control points, indicator microorganisms and equipment and methods used for microbiological production control in plant tissue culture and to prevent contamination (continued)

Critical control points in plant tissue culture

Stage of plant tissue culture	Parameters which need to be monitored	Indicator micro-organisms	Equipment needed	References for methods
Stage 3 Media preparation	3.1. autoclaving (CCP1)	– *Bacillus* spp. (eg. *B. circulans*, *B. subtilis*, *B. pumilus*)	– physical methods (autoclave tape, temperature indicators) – biological methods (eg. spore strips)	[33]
	3.2. media pouring (CCP2)	– *Penicillium* spp. – *Candida* spp. – *Staphylococcus* spp.	– 'sterility' testing of media – testing of laboratory air and laminar flow (as described below)	[9] see below
Stage 4 Subculturing of plants in the laboratory	4.1. air 'sterility' (CCP2)	– *Rhodotorula* spp. (pink yeasts)	– air sampler and/or – particle counter	[17]
	4.2 laminar flow cabinets (CCP2)	– *Penicillium* spp.	– particle counter	[17]
	4.3 instrument sterilisation (CCP1 or CCP2)	– *Bacillus* spp.	– indexing media	[29,33]
	4.4 Aseptic technique of operators (CCP1 or CCP2)	– *Staphylococcus* spp. – *Streptococcus* spp. – *Candida albicans*	– recording of individual contamination rates	[33]
Stage 5 Storage of media and plants (CCP1)	as 4.1	as 4.1	as 4.1	as 4.1
Stage 6 Weaning of plants (CCP2)	as in stage 0	as in stage 0	as in stage 0	as in stage 0

3.2 Disinfection of explants and initiation of tissue cultures (stage 1)

As described above, most saprophytic and pathogenic mites, thrips, fungi, yeasts, bacteria, viruses and viroids inhabiting plants become severe hazards if they survive surface disinfection and are introduced into tissue culture [9, 11, 13, 14, 16, 26, 29, 33]. Since failure to eradicate contaminants at this stage can prevent successful plant tissue culture the disinfection of explants was also classified as a Critical Control Point 1 (Table 1). To avoid hazards at this stage, explants are usually disinfected with broad spectrum biocides such as hypochlorite, alcohols and mercuric chloride or chemicals with a more selective and restricted activity such as insecticides, fungicides and antibiotics (see [6, 16, 21, 33, 44, 46] for additional information about the methods and compounds used for disinfection). Most of the broad spectrum biocides are not systemic and/or kill explants if they were allowed to penetrate internal tissues of the explant. Successful disinfection using these compounds therefore relies on internal plant tissues being free of contaminating microorganisms. However, if internal plant tissues are infected with fungi, bacteria and viruses these are inevitably transferred into tissue culture.

Disinfection in most laboratories includes immersion into sodium or calcium hypochlorite or commercial bleach solutions, although the concentration and exposure time needed for successful disinfection varies greatly depending on the explant type and plant species [29]. It is important to remember that (1) hypochlorite solutions lose their activity rapidly at room temperature, (2) deterioration accelerates with increasing storage temperature and decreasing pH and (3) different commercial bleach preparations contain different amounts of 'active chlorine' [6, 29, 33]. To avoid variation of the 'active chlorine' content, it is therefore recommended that hypochlorite solutions are made freshly every time plant material is disinfected and/or the content of 'active chorine' is measured in commercial or stored bleach preparations prior to use [19]. Wetting agents are incorporated in many commercial bleach preparation and should be added to the hypochlorite solutions to aid penetration of the chemicals into small cavities on the plant surface [16].

Disinfection with mercuric chloride is especially useful where fungal contaminants are prevalent and on plant tissues which are highly sensitive to hypochlorite treatment [33]. However, because of its high toxicity and persistence, its use should be restricted to situations where all other chemicals have failed.

Disinfection with alcohol has often been combined with hypochlorite treatment because pre-treatment with alcohol was shown to allow better penetration of the hypochlorite [Leifert unpublished]. Alcohol has also been used to disinfect the explant surfaces on the mother plant before excision, to avoid surface microorganisms being smeared onto the cut surface when the explant is excised. This was found to be especially beneficial when explants were taken from plants infected with pathogens such as *Erwinia carotovora* [33].

Some of the antibiotics, fungicides and insecticides used for disinfection of explants are systemic and will kill or suppress the organisms within their activity range when taken into plant tissues. However to our knowledge, there are no reports of a significant increase in disinfection success by the use of antibacterial antibiotics. This could be due to the fact that most bacterial contaminants found on explants are Gram-negative species [29] which are less sensitive to most of the commonly used antibiotics [24, 39]. On the other hand, when fungicides such as benomyl were used together with mercuric chloride, significantly more explants could be disinfected than with mercuric chloride alone [37].

It was also possible to physically separate microorganisms from plant cells and protoplasts by sucrose gradient centrifugation [2] and this method could be combined with dilute hypochlorite and/or antibiotic treatment of cells before and after centrifugation. Disinfection of plant cells rather than tissues can therefore avoid the introduction of 'endogenous' extracellular contaminants which can survive in tissues of the explant which are not reached by surface disinfection. However, the relatively small number of plant species which can reliably be regenerated from cell suspension cultures restricts the widespread use of these methods.

Viruses and viroids and many bacterial pathogens which multiply in plant tissues or in plant cells are very difficult to eliminate by the disinfection methods described above. It has, however, long been known that very young actively dividing meristematic cells of infected plants are free of many viruses [38, 41, 50, 51, 52]. Excision of small (0.1–1 mm) meristem domes or domes with 1–5 leaf primordia has therefore been used since the late 1950's to eliminate viruses from plants. This is often combined (1) with heat treatment of stock plants before excision of explants or during the initial growth of explants *in vitro* or (2) incorporation of chemicals (such as virazole) in the growth medium to inhibit virus replication. Shoot tip culture can also eliminate viroids and certain bacterial diseases from plants. However, elimination of viroids increased when plants were incubated at low (5 °C) rather than the high temperatures (36–41 °C) used for virus elimination [43]. Elimination of fastidious plant pathogenic bacteria which only inhabit and grow in the phloem or xylem can be achieved by both meristem tip culture and long term culture of callus tissues which lack the organised tissues needed for their proliferation or by incorporation of antibiotics in the medium [22, 40, 48].

Like some fastidious bacteria, certain viruses do not persist when plants are grown in cell, callus or tissue culture. This is explained by activation of plant resistance mechanisms by the plant growth regulators in the tissue culture medium [42, 47, 48, 51, 53] and is one reason for the more vigorous growth habit of many micropropagated plants. However, the 'more active resistance mechanisms' are also probably the reason for the 'latent' persistence of viral and bacterial contaminants (see above). If viral and bacterial growth is only suppressed, rather than prevented, by such mechanisms, contaminant numbers might drop below the level needed for detection of contaminants by indexing or serological methods [33, 50, 51] and this could result in 'latent' persistence during the *in vitro* stages.

3.3 Indexing and sterility testing (stage 2)

It is important to stress that all the methods used for elimination of microbial contaminants, viruses and viroids only result in a proportion of plants becoming free of contaminants. It is therefore essential that plants are tested for the presence of 'latent' viroids, viruses or bacteria during and after the tissue culture stages (Table 1) to avoid the production of pathogen infected plants and/or the loss of large numbers of plants at a later stage of micropropagation. The 'indexing' or 'sterility testing' of plants for latent bacteria and viruses is therefore another Critical Control Point I.

Routine testing for 'latent' bacteria is usually done by placing plant material and growth medium into bacteriological 'sterility test' or 'indexing' media [29, 33]. However, these methods have several limitations:

1. Each bacteriological medium only detects a certain range of bacteria (for example *Lactobacillus plantarum* does not grow on nutrient broth and some *Xanthomonas* spp. do not grow on a wide range of liquid indexing media [7, 29];

2. Certain bacteria need to be inoculated in high numbers to allow growth and detection by indexing [Leifert unpublished];

3. Plant material from a range of plant species produce blackening compounds when transferred onto indexing media and make the assessment difficult or even impossible [33, Cooke & Leifert unpublished] and

4. Most mycoplasma-, spiroplasma-, and rickettsia-like plant inhabiting bacteria and, of course, all viroids and viruses do not grow on common bacteriological media [33]. Indexing alone can therefore only reduce the chance of using infected plant material.

Some fastidious plant pathogenic bacteria can be detected on serum containing indexing media and such media have been used as additional indexing media [29]. On the other hand, *Hyphomicrobium* spp. which are common contaminants of plant cell cultures do not grow on these complex media and can only be detected on mineral media with one or two carbon and energy sources such as methanol or methylene [33].

Inoculation of indicator plants, various serological tests and nucleic acid hybridization methods have therefore usually been used routinely for the detection of fastidious plant pathogenic bacteria, viruses and viroids in *in vivo* plants [18, 35, 36 and the references listed in Table 1]. These methods can also be used to detect the organisms on explants or established tissue cultures. Such tests are usually very sensitive, but only detect one species of virus or bacterium or sometimes very closely related species. Due to the high cost of such tests they were only used to check for organisms which are known to be a hazard to the particular plant species. For example many *Pelargonium* producing laboratories use *Xanthomonas pelargonium* specific ELISA-tests to detect 'latent' infection of tissue cultures with the organism [5, 33].

3.4 Media preparation (stage 3)

Since media become contaminated by (1) insufficient sterilisation and (2) during manipulations after autoclaving two separate Critical Control Points (CCP) are identified at the media preparation stage (Table 2).

3.4.1 Media sterilisation

Due to the heat sensitivity of many plant growth regulators, tissue culture media are usually autoclaved at the minimum temperature and time required (15 to 20 min at 120 °C) for sterilisation. Slight alterations to the setting of heating cycles, the loading (modern autoclaves adjust the heating cycle to the volume of containers loaded), the containers used for autoclaving and faulty autoclaves can result in insufficient heat treatment of media. Autoclaving media in the tissue culture containers can result in insufficient humidity inside of containers for sterilisation within a 15 to 20 min autoclave run (at the same temperature, sterilisation by dry heat takes much longer [29]). Insufficient heat treatment resulted exclusively in contamination with *Bacillus* spp. [33], because the heat resistant endospores produced by this bacterial genus can survive temperatures of 100 °C or more [33]. Since a single faulty autoclave run can be the reason for whole crops and/or a large proportion of the production becoming completely contaminated the media sterilisation is classified as a Critical Control Point 1. Contamination at this stage can be avoided by assuring that every autoclave run is carried out at the appropriate temperature. The sterilisation success of individual autoclave runs can be monitored by physical or the more accurate biological methods which are based on the survival of *Bacillus stearothermophilus* spores (Table 2).

3.4.2 Media manipulations

During cooling and manual pouring of media, contaminants can be introduced due to faulty aseptic technique and/or faulty laminar flow cabinets. Detection of bacteria introduced during media preparation is difficult because most bacteria do not grow on plant tissue culture media in the absence of plant material and do not form colonies which can be seen by visual assessment [11, 33]. However, media can be tested for sterility by adding extra nutrients in the form of nutrient broth or tryptone soya broth to the medium after pouring. This allowed microbial growth and enables visible detection of microorganisms as colonies in the medium [33]. Using this method we found that between 2 and 5% of media became contaminated during manual pouring after autoclaving. The contaminants found to be introduced during media pouring were mainly bacteria of the genus *Staphylococcus* and yeasts of the genus *Candida* (Table 2, [28, 33]). These organisms are shown to be typical inhabitants of human skin or the laboratory air [33]. Contamination at this control point results in a reduction of productivity, but usually does not prevent the tissue culture process altogether. It is therefore classified as a Critical Control Point 2.

Contamination during manual pouring of media can only be avoided by detailed training of employees in aseptic techniques together with the testing of poured media for the presence of contaminants. Mechanised and semi-mechanized media pouring systems usually result in much lower, contamination rates than manual pouring and are now used in many of the larger micro-propagation companies [34]. The equipment needed is, however, still relatively expensive.

3.5 Subculturing of plants (stage 4)

The subculturing of plants includes 4 separate Critical Control Points (Table 2). There is, however, every little information about the numbers and the types of organisms that are introduced during the subculturing of plant material in laminar flow cabinets [4, 9, 26, 28, 33]. The rate at which contaminants are introduced depends on the standard of aseptic technique and the equipment used. For example, the rate at which index negative plants (plants were tested on bacteriological media for the presence of microorganisms) became contaminated during repeated subculturing was different between different operators and higher if plants were subcultured towards the front of the laminar flow cabinet [33].

When contaminants were identified, similar species to those introduced during manual pouring (*Staphylococcus* spp. and *Candida albicans*) were found [33]. Again, these organisms are likely to have originated from the skin or respiration of operators since they are described as obligate inhabitants of humans and other animals [26, 28, 33]. Others contaminants introduced at this stage such as *Penicillium*, *Rhodotorula* (pink yeast) spp. and *Candida guilliermondii* were shown to have originated from the laboratory environment. For example, the outside surfaces of culture vessels are usually contaminated with similar fungal and yeast populations to those found in the growth room air (Leifert, unpublished). Danby and coworkers [9] showed a close correlation between the fungal genera isolated from the laboratory air and those found as contaminants in plant tissue cultures which had been *in vitro* for longer than 1 year.

Many of the bacteria introduced during subculturing do not cause visible symptoms on the plant but reduce growth rates of plants [27, 30]. The effect of many of the bacteria introduced during subculturing (e.g. *Bacillus*, *Staphylococcus* and *Pseudomonas* spp.) on tissue cultured plants depends on the plant species. Some plant species such as *Hemerocallis*, *Rosa*, *Coffea* (which even grew in the presence of fungal contaminants [9]) and *Rhododendron* suppressed or eliminated a wide range of introduced bacteria by reducing the medium pH to below 4 and can be micropropagated even if the standards of aseptic technique in the laboratory are low [7, 25, 31, 33]. On the other hand, plants such as *Delphinium* and *Cheiranthus* which show poor resistance *in vitro* and increase the pH to about 6.0 are found to be severely affected by most bacteria introduced in the laboratory [7, 25, 31]. These plant species can only be produced if the standards of aseptic technique in the laboratory are very high

[33]. Faulty filters in the laminar flow cabinets and insufficient sterilisation of instruments were also shown to introduce contamination in the laboratory.

Because contaminants were rarely found to be introduced at rates which totally prevent the production of plants all Critical Control Points during subculturing of plants are classified as Critical Control Points 2. However, losses due to laboratory contaminants can make the production of 'sensitive' plants (which have no active defence mechanisms and get contaminated at a high rate), impossible. High losses due to laboratory contamination can reduce or abolish the profitability of commercial micropropagation [33]. Under such circumstances some or even all of the Critical Control Points become a Critical Control Point 1.

Contamination at this stage can only be reduced by monitoring every Critical Control Point continuously. The number of organisms in the laboratory air can be tested by air samplers or particle counters (Table 2). To reduce the number of contaminants in the laboratory air attempts should be made to (1) protect the inside of the laboratory from the outside air and (2) to keep the laboratory as dust free as possible. This can only be achieved by alterations to the laboratory design. Windows should not be opened, hepa filter units should be fitted to the air-conditioning outlets and any other openings to the outside and the laboratory should only be entered through air locks. Any dust within the laboratory should be removed ideally by dry cleaning using a vacuum cleaner with an attached hepa-filter (cleaning in general must be carried out to the standards of hospital operating theatres).

Plant losses as a result of contamination should be monitored to pinpoint faulty laminar flow cabinets or operators with inefficient aseptic technique. Introduction of contaminants during handling of plant material can only be reduced by a very detailed training of operators in aseptic technique and continuous monitoring of standards and by routine servicing and testing of the laminar flow cabinets [33]. Protective clothing such as laboratory coats with closed sleeves, hair nets, surgical gloves and masks have also been shown to reduce the rate at which contaminants are introduced during subculturing [Leifert unpublished].

Many attempts have been made to prevent the introduction of, or eliminate, contaminants by subculturing plants on media containing antibiotics [29, 30], fungicides [9, 16] or even low concentrations of hypochlorite [11]. When plant cultures are uniformly and exclusively infected with one contaminant and when the contaminants are highly sensitive to the biocides used, some contaminants can be successfully eliminated [29, 30]. However, elimination of unidentified mixed populations of contaminants are usually unsuccessful or even more harmful to the plant then the contaminant alone [29, 30]. Prophylactic incorporation of antibiotics into the growth media can select for antibiotic resistant bacterial species such as *Pseudomonas fluorescens* and *Acinetobacter* in cultures [Leifert unpublished] and increases the risk of producing antibiotic resistant strains of bacteria. The treatment of unidentified mixed populations of microorganisms and the prophylactic use of antibiotics should therefore be avoided.

3.6 Incubation and storage of sterile culture vessels, media and plant cultures (stage 5)

A variety of severe hazards are found to be associated with the storage of media and plant cultures. Up to 7% of irradiated culture vessels became contaminated with *Candida* yeasts and *Penicillium* fungi during inappropriate storage [28, 33].

Fungal growth on growth room floors and walls and in drip trays of cooling systems increases the number of fungal spores in the air and on the outside surfaces of the culture vessels in the growthroom and resulted in increased losses due to fungal contamination. Fungi can enter the vessels either actively by germination and hyphal growth in the film of condensation water between the tissue culture container and container lid or when the containers are opened during subculturing [9]. Mite and thrip infestations of growth rooms have caused devastating effects in various commercial and research laboratories because they actively transferred laboratory fungi into plant tissue cultures [3, 23, 29]. The danger of large losses due to mite and thrip infestation at this stage suggest that the storage of plants and media should be classified as a Critical Control Point 1.

Contamination during storage and incubation of plant tissue cultures can only be prevented by introducing the same high level of hygiene described for the laboratory. Introduction of mites and thrips with the explants can be prevented by incubating cultures in separate 'quarantine' growth rooms for up to 1 year after initiation of tissue cultures and by microscopic examination for the presence of mites and thrips on the plants. Mite and thrip infestations are treated by (1) fumigation of growth rooms with insecticides such as pirimiphos-ethyl (ICI), (2) the cleaning of all shelves, floors walls and culture vessels with biocides and sealing of all culture vessels with plastic film to prevent migrating stages of mites from spreading through the growth room [3, 29]. However, eradication of mites or thrips from growth rooms is an expensive and time-consuming exercise which can easily be avoided.

3.7 Rooting and weaning of plants (stage 6)

Since many plant pathogenic bacteria and viruses stay latent *in vitro*, plants should be assessed for disease symptoms and again tested for the presence of plant diseases before and/or after they are weaned into soil. This should be done to confirm that they are free of known pathogens of the plant species. If plants have been sufficiently tested prior to tissue culture and during *in vitro* growth, the testing at weaning is mainly a safety measure. This stage therefore represents a Critical Control Point 2. The importance of testing at this stage could, however, increase if new restrictions on the international trade of plant tissue cultures require additional tests to be carried out before phyto-sanitary certificates are issued.

4. Conclusions

It is now generally accepted that large scale scientific and commercial plant tissue culture requires 'microbiological production control' at all tissue culture stages.

Methods for stock plant treatment, surface sterilisation and shoot tip culture have been optimised for a wide range of plant species. However, methods for detection of microorganisms, viruses and viroids in plants are still very laborious, expensive and far from perfect. We also understand very little about the resistance mechanisms *in vitro* which at best eliminate contaminants and at worse suppress growth and allow the 'latent' persistence of contaminants. In our opinion future research should concentrated in these areas.

Methods for reducing laboratory contamination by monitoring the different sources in the laboratory have developed to an extent which should allow large scale aseptic production of most plant species. However, success relies on a combination of good management, detailed training of employees and permanent production control using a system such as HACCP.

References

1. Anonymous (1982) The Oxoid Manual, 5th Ed. Basingstoke: Turnergraphic Ltd
2. Attree SM and Sheffield E (1986) An evaluation of Ficoll density gradient centrifugation as a method for eliminating microbial contamination and purifying plant protoplasts. Plant Cell Rep 5:288–291
3. Blake J (1988) Mites and thrips as bacterial and fungal vectors between plant tissue cultures. Acta Hort 225:163–166
4. Boxus PH and Terzi J-M (1988) Control of accidental contamination during mass propagation. Acta Hort 225:189–193
5. Cassells AC, Harmey MA, Carney BF, Mc Carthy E and Mc High A (1988) Problems posed by cultivable bacterial endophytes in the establishment of axenic cultures of *Pelargonium x domesticum*: the use of *Xanthomonas pelargonii*-specific ELISA, DNA probes and culture indexing in the screening of antibiotic treated and untreated donor plants. Acta Hort 225: 153–161
6. Collins CH, Allwood MC, Bloomfield SF and Fox A (1981) Disinfectants: Their use and Evaluation of Effectiveness. London/New York: Academic Press
7. Cooke DL, Waites WM and Leifert C (1992) Effect of *Agrobacterium tumefaciens*, *Erwinia carotovora*, *Pseudomonas syringae* and *Xanthomonas campestris* on plant tissue cultures of *Aster*, *Cheiranthus*, *Delphinium*, *Iris* and *Rosa*; disease development *in vivo* as a result of latent infection *in vitro*. J Plant Dis Protection 99:469–481
8. Cooke DL, Epton HAS, Waites WM and Leifert C (1993) Elimination of plant pathogenic bacteria by plant tissue culture; the problem of latent persistance *in vitro*. Proceedings of the 8th International Conference on Plant Pathogenic Bacteria, 1992, Paris, France (In press)
9. Danby S and Leifert C (1993) Fungal contaminants of *Coffea*, *Musa*, *Primula*, and *Iris* shoot cultures. Plant Growth Reg (In press)
10. Debergh P and Maene L (1981) A scheme for commercial propagation of ornamental plants by tissue culture. Sci Hort 14:335–345
11. De Fossard RA (1990) Micropropagation. Queensland: Xarma Pty Ltd
12. Dickinson CH and Preece TF (1976) Microbiology of aerial plant surfaces. London/New York: Academic Press

13. Duhem K, Le Mercier N and Boxus PH (1988) Difficulties in the establishment of axenic '*in vitro* cultures of field collected coffee and cacao germplasm. Acta Hort 225:67–77

14. Enjalric F, Carron MP and Lardet L (1988) Contamination of primary cultures in tropical areas: The case of *Hevea brasiliensis*. Acta Hort 225: 57–65

15. Fedotina VL and Krylova NV (1976) Ridding tobaccos of the mycoplasmic infection big bud by the method of tissue culturing. Dokl Bot Sci 228: 49–51

16. George EF and Sherrington (1984) Plant Propagation by Tissue Culture. Eversley, Basingstoke: Exegetics Ltd

17. Gregory PH (1973) The microbiology of the Atmosphere, pp 127–145. Aylesbury: Leonard Hill Books

18. Hill SA (1984) Methods in Plant Virology. Oxford/London: Blackwell Scientific Publications

19. Hoffman PN, Death JE and Coates D (1981) The stability of sodium hypochlorite solutions. In: CH Collins, MC Allwood, SF Bloomfield and A Fox, eds. Disinfectants: Their use and evaluation of effectiveness, pp 77–83. London/New York: Academic Press

20. Hoffmann GM and Schmutterer H (1983) Parasitäre Krankheiten und Schädlinge an landwirtschaftlichen Kulturpflanzen. Stuttgart: Eugen Ulmer Verlag

21. Hugo WB (1971) Inhibition and Destruction of the Microbial Cell. New York/London: Academic Press

22. Jacoli GG (1978) Sequential degeneration of mycoplasma like bodies in plant tissue cultures infected with *Aster* yellows. Can J Bot 56:133–140

23. Klocke JA and Myers (1984) Chemical control of thrips on cultured *Simmondsia chinensis* (Jojoba) shoots. Hort Sci 19:400

24. Krieg NR and Holt JG (1984) Bergey's Manual of Systematic Bacteriology Baltimore/London: Williams and Wilkins

25. Leifert C and Waites WM (1992) Bacterial growth in plant tissue cultures. J Appl Bacteriol 72:460–466

26. Leifert C, Waites WM and Nicholas JR (1989) Bacterial Contaminants of micropropagated plant cultures. J Appl Bacteriol 67:353–361

27. Leifert C, Waites, WM, Camotta H and Nicholas JR (1989) *Lactobacillus plantarum*; a deleterious contaminant of plant tissue culture. J Appl Bacteriol 67:363–370

28. Leifert C, Waites WM, Nicholas JR and Keetley JW (1990) Yeast contaminants of micropropagated plant cultures. J Appl Bacteriol 69:471–476

29. Leifert C, Ritchie J and Waites WM (1991) Contaminants of plant tissue and cell cultures. World J Microbiol Biotechnol 7:452–469

30. Leifert C, Camotta H, Wright S, Waites B Cheyne VA and Waites WM (1991) Elimination of bacteria from micropropagated plant cultures using antibiotics. J Appl Bacteriol 71:307–330

31. Leifert C, Pryce S, Lumsden PJ and Waites WM (1992) Effect of medium acidity on growth and rooting of different plant species growing *in vitro*. Plant Cell Tiss Org Cult 30:171–179

32. Leifert C, Camotta H and Waites WM (1992) Effect of antibiotics on micropropagated plants. Plant Cell Tiss Org Cult 29:153–160

33. Leifert C, Morris, C.E. and Waites WM (1993) Ecology of microbial saprophytes and pathogens in field grown and tissue cultured plants. CRC Crit Rev Plant Sci (In press)

34. Leifert C, Clark E and Rothery CA (1993) Micropropagation; the propagation of plants by tissue culture. Biological Sciences Reviews 5:31–35

35. Lelliott RA and Stead DE (1987) Methods for the Diagnosis of Bacterial Diseases in Plants. Oxford: Blackwell Scientific Publications

36. Matthews REF (1991) Plant Virology 5th Ed. New York: Academic Press

37. Mederos S and López C (1991) Control of organogenesis '*in vitro*' of *Pistacia atlantica* Desf rootstock. Acta Hort 289:135–136

38. Mellor FC and Stace-Smith R (1977) Virus-free potatoes by tissue culture. In: J Reinert and YPS Bajaj, eds. Applied and Fundamental Aspects of Plant Cell, Tissue, and Organ Culture, pp 599–615. Berlin: Springer Verlag

39. Mersch-Sundermann V (1989) Medizinische Mikrobiologie für MTA. Stuttgart/New York: Georg Thieme Verlag

40. Möllers C and Sarkar S (1989) Regeneration of healthy plants from *Cataranthus roseus* infected with mycoplasma-like organisms through callus culture. Plant Sci 60:83–89
41. Murashige T and Jones JB (1972) Cell and organ culture methods in virus disease therapy. Acta Hort 36:223–228
42. Omura T and Wakimoto S (1978) Effect of plant hormones on tobacco mosaic virus concentration in tobacco tissue cultures. J Fac Agric, Kyushu Univ 22:211–219
43. Paduch-Cichal and Kryczynski S (1987) A low temperature therapy and meristem-tip culture for elimination of four viroids from infected plants. J Phytopath 118:341–346
44. Russell AD, Hugo WB and Ayliffe GAJ (1982) Principles and Practice of Disinfection, Preservation and Sterilisation. Oxford: Blackwell Scientific Publications
45. Smith IM, Dunez J, Lelliott RA, Phillips DH and Archer SA (1986) European Handbook of Plant Diseases. Oxford/London: Blackwell Scientific Publications
46. Sykes G (1965) Disinfection and Sterilisation. London: E and FN Spon Ltd
47. Toyoda H, Oishi Y, Matsuda Y, Chatani K and Hirai T (1985) Resistance of cultured plant cells to tobacco mosaic virus. Phytopath Z 114:126–133
48. Ulrychova M and Petru E (1975) Elimination of mycoplasma in tobacco callus tissues (*Nicotiana glauca* GRAH) cultured in the presence of kinetin and IAA in the nutrient medium. Biol Plant 17:352–356
49. Waites WM (1988) Hazardous microorganisms and the hazard analysis critical control point system. Food Sci Technol Today 212:49–51
50. Walkey DGA (1978) *in vitro* methods for virus elimination. In: TA Thorpe (ed) Frontiers of Plant Tissue culture, pp. 245–254. Calgory: University of Calgory Press
51. Walkey DGA (1985) Applied Plant Virology. London: William Heinemann Ltd
52. Wang PJ and Hu CY (1980) Regeneration of virus-free plants through *in vitro* culture. In: Fiechter A, ed. Advances in Biochemical Engineering 18, pp 61–99 Berlin: Springer Verlag
53. White RF, Dumas E, Shaw P and Antoniew JF (1986) The chemical induction of PR (b) proteins and resistance to TMV infection in tobacco. Antiviral Research 6:177–185

Latent bacterial infections: epiphytes and endophytes as contaminants of micropropagated plants

H.E. GUNSON and P.T.N. SPENCER-PHILLIPS
Department of Biological Sciences, Faculty of Applied Sciences, University of the West of England, Coldharbour Lane, Bristol BS16 1QY, UK

1. Introduction

Chronic microbial contamination is probably the major problem faced by all involved in the micropropagation of plants, and has particularly serious economic consequences for commercial micropropagation companies [7]. Indeed, it is not unusual for whole batches of plant cultures to become contaminated synchronously, after appearing to be axenic for many subcultures (Dr W. Morgan, personal communication). It is perhaps surprising, therefore, that one perceived advantage of micropropagated plants is their disease free and axenic status [6]. This apparent conflict of evidence and belief results from what has been described as a natural reluctance to admit to contamination and a lack of fundamental understanding of the sources and biology of contaminating microorganisms [7]. The present paper aims to redress this unsatisfactory situation by focusing on latent bacterial infections, which include endophytic contaminants.

The main sources of contamination are considered to be: normal airborne and human associated microorganisms [25] that are introduced via poor aseptic technique, inefficient sterilization of instruments or failure of equipment [16, 37]; epiphytic microorganisms that have survived surface sterilization procedures; non-cultivatable fastidious pathogens and endophytic microorganisms that are protected from normal surface sterilization techniques [1, 14, 25, 38].

This last source is of particular concern because endophytic microorganisms cannot be eliminated by improving the efficacy of surface sterilization techniques. Control therefore relies on the use of antibiotics, which often do not provide satisfactory levels of control and have inherent problems that may preclude their use [11, 13, 15].

A critical appraisal of endophytes as sources of contamination is required before alternative control strategies can be founded on a sound scientific basis. An important, preliminary step is to adopt a definition of endophytes appropriate to micropropagation (note that the term 'endogenous' is misleading and should not be used to describe endophytic sources of contamination).

P.J. Lumsden, J.R. Nicholas and W.J. Davies (eds.), Physiology, Growth and Development of Plants in Culture, 379–396, 1994.
© 1994 *Kluwer Academic Publishers.*

Endophytes were defined originally by DeBary in 1866 [30] to describe fungi living inside plant tissues, whilst fungi living on the surface of their host were termed epiphytes. DeBary's definition, therefore, encompasses pathogens, mutualistic symbionts and saprotrophs, and can be somewhat ambiguous because individual species of these groups of organisms do not normally inhabit either the epiphytic or endophytic environment exclusively. Thus, the definition has been modified and the term endophyte is presently understood to describe a fungus [4] or bacterium [28] which causes inapparent, asymptomatic but parasitic infections within the tissues of plants [5].

It is recommended that the definition of endophytes is further modified to take into account two distinct groups of organisms. The first group consists of 'obligate endophytes', which exclusively occupy the endophytic niche in nature and are unable to grow and reproduce as epiphytes or away from their host. 'Facultative endophytes' include those organisms showing a relatively faithful association with a particular host, but which are able to complete their lifecycle away from living host tissue. They are sub-divided into 'constitutive' and 'inducible' categories. The former includes microorganisms which naturally occupy both epiphytic and endophytic habitats. In contrast, inducible endophytes are able to survive if transferred artificially to an endophytic environment, but are not able to multiply significantly until the physiological environment becomes more favourable to growth.

Evidence from the present study indicates that inducible endophytes are probably the most frequently encountered group associated with micro-propagated plants, and originate as epiphytes which are transferred to endophytic niches during excision of explants and subsequent manipulations *in vitro*. This conclusion is based upon experiments which focus on the ability of bacteria isolated from glasshouse grown stock plants to grow endophytically within explant and micropropagated tissues. Physiological and environmental factors which may mediate the transition of bacteria from cryptic to blatant contaminant are identified in a model system of *Zantedeschia aethiopica* and four representative bacteria. The role of methanol, detected in vessel head space gases by gas chromatography, is discussed in relation to contamination by pink pigmented facultative methylotrophs. Lysigenous formation of aerenchyma is related to nutrient availability and rapid growth of endophytes.

2. Materials and methods

2.1 Plant material and propagation

Plants of *Ficus benjamina*, *Dieffenbachia* sp. cvs Camilla and Compacta, *Syngonium podophyllum* cv. Noak and *Zantedeschia aethiopica* cv. Nina (provided by Twyford Plant Laboratories Ltd, Baltonsborough, U.K.) were selected because they were consistently associated with contamination during micropropagation (Dr W. Morgan, personal communication). Stock plants were maintained in a glasshouse under natural daylight, supplemented to give a minimum day length of 12 h, at 18 to 25 °C.

Plants grown *in vitro* were maintained in Magenta GA7 culture boxes (Sigma, U.K.) on multiplication medium which comprised Murashige and Skoog minimal organic medium (MSMO; Flow Laboratories, U.K.) with the addition of benzylaminopurine (BAP; Sigma, U.K.) at between 1–3 mg dm^{-3} to stimulate shoot multiplication. All species formed roots in culture without application of exogenous growth regulators, so rooted plants were obtained by transferring to MSMO with activated charcoal (1 g dm^{-3}). During micropropagation, plants were grown with a 12 h daylength (17 μmol m^{-2} s^{-1} photosynthetically active radiation) at 25 °C in a Vindon Scientific Ltd (U.K.) incubator.

2.2 Isolation and identification of putative endophytes and cryptic contaminants

Putative endophytes were isolated from glasshouse grown stock plants using methods routinely employed to prepare explants for culture initiation. Tissue samples were excised, surface sterilized by dipping in 70% ethanol and shaking for 5 min in 1% (v/v) Domestos solution (these two steps repeated for some tissues), and rinsed in 3 changes of sterile distilled water (SDW). Subsequent manipulations of explant tissues and confirmation of surface sterilization efficacy are described elsewhere [17]. Explants were placed onto the following media: potato dextrose agar (PDA); 2% malt extract agar; tap water agar; cornmeal agar; nutrient agar with 1% v/v glucose; peptone yeast agar. They were incubated at room temperature for a maximum of 2 weeks and putative bacterial endophytes were sub-cultured by streak plating to give single colony isolates.

One hundred apparently healthy micropropagated *Z. aethiopica* plants, taken from 4 micropropagation vessels, were screened for cryptic contaminants by shaking (100 rpm) tissue samples in nutrient broth (overnight, 25 °C). In most cases, isolated bacteria were identified by two different techniques: the monothetic scheme for identification of phytopathogenic bacteria devised by Schaad [31], and the fatty acid profiling (FAP) technique described by Stead [35]. Polyclonal antiserum to one isolate (SA7.1, identified as *Xanthomonas maltophilia*) from *in vitro Z. aethiopica* plants was raised in rabbits by the method of Lyons and Taylor [27], and was used for identification by a slide agglutination method and immunocytochemistry [17].

2.3 Assessment of endophytic and pathogenic abilities

The ability of bacteria to survive as inducible endophytes was confirmed by reintroduction to detached leaf tissue from stock plants of the same species from which the bacteria were isolated. The pathogens *Erwinia carotovora* subsp. *carotovora*, *Erwinia chrysanthemi* pv. *dieffenbachiae* and *Pseudomonas marginalis* pv. *marginalis* and SDW were used as controls. Inoculum was prepared from 24 h PDA slope cultures to give a series of 10 fold dilutions over the range 10^2 to 10^8 colony forming units (cfu) cm^{-3}. Precise cfu values were determined by a micro-plating method [22].

Leaves of approximately the same age were detached from plants and floated on water in a penicillin assay dish (area, 230 × 230 mm; depth, 30 mm) under a halogen lamp for 1–2 h before infiltration to ensure that stomata were open. Inoculum was introduced by pressing the tip of a syringe (without hypodermic needle attached) gently against their abaxial surface and infiltrating until the tissue became watersoaked. Each leaf was inoculated at 4–6 different sites with the same inoculum, suspended on glass rods in a penicillin assay dish lined with damp filter paper, and incubated at 25 °C under constant light (2.5 μmol m^{-2} s^{-1} photosynthetically active radiation) for 10 days. Bacteria were reisolated from leaf discs by a standard plating method [17].

The pathogenicity of selected bacterial isolates was determined by the tobacco hypersensitivity test [12]. *E. chyrsanthemi* pv. *dieffenbachiae, Pseudomonas syringae* pv. *morsprunorum* race 1 and *Pseudomonas syringae* pv. *phaseolicola* race 1A were used as positive controls, and *P. marginalis* pv. *marginalis* [3] and SDW as negative controls.

2.4 Localization of bacteria within Z. aethiopica tissue

Cryo-scanning electron microscopy was inappropriate for detecting bacteria associated with *Z. aethiopica* tissue, due to the preservation of extracellular polysaccharide which masked the bacterial cells. Bacteria were observed, however, by conventional scanning electron microscopy (SEM) on and in tissue taken aseptically from apparently healthy, axenic plants in micropropagation culture. Internal surfaces were exposed by freeze fracture [17], then specimens were critical point dried, gold coated and viewed with an Hitachi S-450 scanning electron microscope at an accelerating voltage of 10 Kv.

The polyclonal antiserum against isolate SA7.1 was used with a goat anti-rabbit fluorescein isothiocyanate (FITC) conjugate (Sigma, U.K.) to localize bacteria associated with *Z. aethiopica* roots. Segments of fixed roots and sections (approximately 1 μm thick) of LR White embedded tissues, prepared as controls or treated with antiserum [17], were viewed with a Nikon Optiphot microscope with epifluorescence optics (filter block B2-A, peak excitation 450–490 nm, transmission > 520 nm).

2.5 Facultative methylotrophy and methanol in culture vessels

A simple assay was developed to demonstrate facultative methylotrophy by bacterial isolates. A small, sterile glass vial was half filled with absolute methanol, the top plugged with cotton wool and then partially embedded by laying on its side in semi-set nutrient agar (NA) in a Petri dish. The set agar was streaked with one bacterial isolate; each was tested in triplicate, with plates of NA minus methanol as controls. Petri dishes were incubated at room temperature and facultative methylotrophy was confirmed when growth was sparse in control plates but profuse when methanol was present.

Methanol concentration in head space gases was determined in culture vessels containing *Z. aethiopica* plants at various ages and both multiplication and rooting stages. Head space gas (0.5 cm^3) was removed from each vessel and methanol detected by gas chromatography [17].

3. Results

3.1 Isolation and identification of putative endophytes from stock plants

Bacteria were isolated from all 5 stock plants examined (Table 1). The greatest number was from *F. benjamina* and the least from *Z. aethiopica* and *S. podophyllum*. *Pseudomonas* spp. comprised over 50% of the isolates, with approximately equal numbers identified as fluorescent and non-fluorescent by the Schaad scheme. Other bacteria were identified as *Agrobacterium* (16.2%), Coryneform (9.1%), non-pectolytic *Erwinia* (10.1%) and *Streptomyces* (10.1%) isolates.

The identification of organisms as determined by FAP is given in Table 2, along with the identification by the Schaad scheme, where available, for comparison. It should be noted that FAP is a single diagnostic test and so it should not be assumed that the first choice identification is correct. Interpretation of the results varies with the bacteria in question; in general, similarity index values greater than 0.7 are taken to represent excellent matches, whilst a single match of 0.5 has a strong likelihood of being correct and even values at the 0.1 level quite often represent closely related species (Dr D. Stead, personal communication).

Table 1. Number of separate isolations of bacterial genera and groups (identified by Schaad's (1988) scheme) from tissues of stock plants

Bacterial genus or grouping	Host plant					Total %
	Ficus benjamina	Dieffen- bachia cv. Camilla	Dieffen- bachia cv. Compacta	Syngonium podo- phyllum cv. Noak	Zantedeschia aethiopica cv. Nina	
Agrobacterium	0	5	5	4	4	16.2
Coryneform	4	2	1	2	0	9.1
Erwinia	4	3	3	0	0	10.1
Pseudomonas:						
Fluorescent	12	4	5	3	6	30.3
Non-fluorescent	3	2	10	4	3	24.2
Streptomyces	10	0	0	0	0	10.1
Total	33	16	24	13	13	100

Table 2. Identification of putative endophytes by the scheme divised by Schaad (1988) and fatty acid profiling (FAP) technique (Stead, 1988). The most probable identification selected by FAP is compared to the first choice determinations for each library of profiles consulted (1 = TSBA 3.0[a]; 2 = TSBA 3.2[b]; 3 = NCPPB[c]); NT = not tested

Isolate Code and host	Schaad identification	Fatty acid profile identification		
		Identification	Library	1st choice (similarity index)
Zantedeschia aethiopica				
B20.2	Fluorescent pseudomonad	*P. fluorescens* maybe *P. tolaasii*	2	*P. tolaasii* (0.807)
B20.3	Fluorescent psuedomonad	*Pseudomonas* rRNA group 1	1 2	*P. aureofaciens* (0.292) *P. marginalis* (0.336)
B30.2	Fluorescent pseudomonad	*P. fluorescens* complex	1 2	*P. chlororaphis* (0.659) *P. fluorescens* (0.722)
RB20.2	Fluorescent pseudomonad	*P. fluorescens* complex	1 2 3	*P. aureofaciens* (0.669) *P. tolaasii* (0.654) *P. aureofaciens* (0.501)
B5.1	Fluorescent pseudomonad	*P. fluorescens* complex maybe LOPAT grp 4	1 2	*P. aureofaciens* (0.590) *P. marginalis* (0.741)
B10.3	Fluorescent pseudomonad	*P. fluorescens* complex	1 2	*P. aureofaciens* (0.852) *P. gingeri* (0.847)
B10.2	*Agrobacterium*	*Bacillus* sp.	1 2	*B. megaterium* (0.493) *Xanthomonas jasminii* (0.447)
B20.1	*Agrobacterium*	*Bacillus* sp.	1 2	*B. megaterium* (0.333) *X. jasminii* (0.015)
170.A	*Agrobacterium*	*Agrobacterium* bv 1	1 2	*A. tumefaciens* bt 1 (0.865) *Agrobacterium* bv 1 (0.823)
170.B	Non-fluorescent pseudomad	Unidentified, some similarities with *Pseudomonas* rRNA grp 2	1	No match
171.A	Non-fluorescent pseudomonad	*Xanthomonas* sp. does not belong to any known taxa	1 2	*X. maltophilia* (0.118) *X. campestris* (0.146)
171.B	Non-fluorescent pseudomonad	,,	1 2	*X. maltophilia* (0.220) *X. campestris* (0.342)
173.A	Pseudomonad	*Comamonas* sp.	1 2	*C. testosteroni* (0.121) *C. testosteroni* (0.142)
SA7.1	NT	*Xanthomonas maltophilia*	1 2 3	*X. maltophilia* (0.195) *X. maltophilia* (0.626) *X. c. pv. malvacearum* (0.200)
C	NT	Unidentified, similarities to *P. cepacia*. Same as 170.B		

Table 2. (continued)

Isolate Code and host	Schaad identification	Fatty acid profile identification		
		Identification	Library	1st choice (similarity index)
Dieffenbachia var Camilla				
42.C	*Erwinia*	*Comamonas* sp.	1	*C. acidovorans* (0.631)
		similar to *C. testosteroni*	2	*P. putida* (0.463)
42.D	*Erwinia*	,,	1	*C. acidovorans* (0.497)
			2	*P. putida* (0.309)
41.B	*Agrobacterium*	Probably an *Acinetobacter*	1	*A. calcoaceticus* (0.502)
42.A	*Agrobacterium*	*Pseudomonas* rRNA grp 1	1	*C. acidovorans* (0.154)
			2	*C. acidovorans* (0.124)
89.B	Fluorescent pseudomonad	*Pseudomonas* rRNA grp 1	1	*P. aeruginosa* (0.449)
			2	*P. aeruginosa* (0.565)
90.A	*Agrobacterium*	*Xanthomonas* sp.	1	*X. campestris* (0.366)
			2	*X. campestris* (0.404)
113.A	Coryneform	Enterobacteriaceae	1	*Kluyvera* (0.277)
114.B	Coryneform	*Enterobacter* sp.	1	*E. cloacae* (0.810)
152.A	Pseudomonad	*R. rubrisubalbicans*	1	*P. rubrisubalbicans* (0.868)
			2	*P. rubrisubalbicans* (0.751)
152.B	*Agrobacterium*	Possibly *Bordetella* sp.	1	*Bordetella* (0.334)
152.D	Non-fluorescent pseudomonad	Unidentified		
Dieffenbachia var Compacta				
22.A	Pseudomonad	*P. rubrisubalbicans*	1	*P. rubrisubalbicans* (0.877)
			2	*P. rubrisubalbicans* (0.956)

[a,b] TSBA profile libraries from Hewlett Packard computer assisted fatty acid profile package
[c] NCPPB profiles from isolates in the NCPPB at ADAS Harpenden, U.K.

In all but two cases the bacteria were typical plant or soil associated isolates, with the genus *Pseudomonas* occurring the most frequently. It is particularly interesting to note that isolates 22.A and 152.A, from *Dieffenbachia* cv. Compacta and Camilla respectively, were both identified as *Pseudomonas rubrisubalbicans*, the causal organism of mottled stripe of sugar-cane. Bacteria belonging to the genera *Pseudomonas*, *Agrobacterium*, *Xanthomonas*, *Bacillus* and *Comamonas* were among the isolates from *Z. aethiopica*, although isolates of the last two genera were identified with similarity indices of less than 0.5.

Bacteria from the genera *Pseudomonas*, *Comamonas* and *Xanthomonas* were also isolated from *Dieffenbachia* sp. along with bacteria from the genera *Acinetobacter* and *Enterobacter*, all with similarity indices greater than 0.5. Only three bacterial isolates were unidentified to genus level by FAP; one of these (isolate C) was identified subsequently as a facultative methylotroph [17]. The identity of eleven other isolates, with similarity indices below 0.5, is uncertain.

3.2 Isolation and identification of contaminants from Z. aethiopica *plants in vitro*

All the broths containing the one hundred apparently healthy micropropagated *Z. aethiopica* plants were cloudy, after overnight incubation, due to bacterial growth. When broth samples were streaked onto NA only one colony type grew indicating a single contaminating bacterium, which was confirmed by subsequent identification of 5 isolates (including SA7.1, Table 2) by FAP. All were reported to be identical, and probably represent an isolate of the hetero-geneous species *Xanthomonas maltophilia*.

Although the *Z. aethiopica* cultures showed no obvious bacterial contamin-ation when screened, similar cultures left undisturbed for more than 8 weeks had macroscopically visible, yellow to orange coloured colonies of bacteria growing on the medium and at the base of plants. Apical meristems cultured on MSMO amended with BAP or coconut milk also became contaminated one week after initiation by the same bacterium [17], which was identified as *X. maltophilia* by the slide agglutination test using antiserum raised against SA7.1.

3.3 Assessment of endophytic and pathogenic abilities

The causal agent of stem rot in *Dieffenbachia* spp., *E. carotovora* pv. *dieffenbachiae*, inoculated at a density of 10^8 cfu cm^{-3}, caused a transient green island effect followed by rapid and total disintegration of the host tissue. In contrast, no significant symptoms developed over the 10 day incubation period with this bacterium at inoculation densities of 10^2 and 10^5 cfu cm^{-3}. None of the other bacteria caused disease symptoms in any host, except for three coryneform isolates from *F. benjamina* and a pseudomonad from *Dieffenbachia* cv. Camilla which caused necrosis limited to the point of contact with the syringe used for inoculation.

Some other isolates induced chlorosis, but this was due to mechanical damage as demonstrated by similar effects observed in the SDW control treatments. Whilst there was some indication that natural senescence was accelerated by the presence of the inoculum, four isolates (a fluorescent and non-fluorescent pseudomonad from *Dieffenbachia* and isolates 170.B and 171.A from *Z. aethiopica*) produced green island effects when reinoculated into their host. All bacteria were reisolated from inoculated tissue after 10 days, and from each inoculum density in the majority of cases, indicating that most were able to survive endophytically.

The pathogens *E. chrysanthemi* pv. *dieffenbachiae*, *P. syringae* pv. *morsprunorum* and *P. syringae* pv. *phaseolicola* were the only isolates which produced hypersensitive reactions in tobacco. The other isolates tested were 170.B, 171.A, B30.2, SA7.1 and C from *Z. aethiopica*, *P. rubrisubalbicans* (isolates 22.A and 152.A) from *Dieffenbachia*, and *P. marginalis* pv. *marginalis*.

3.4 Localization of bacteria associated with Z. aethiopica tissue

The external surfaces of *Z. aethiopica* plants from apparently axenic multiplication cultures viewed by SEM were found to be colonized by bacteria (Plate 2 A). On some surfaces bacteria were so numerous that few areas were left uncolonized, whereas in other areas colonization was sparse with cells arranged as discrete microcolonies. The bacteria were always rod shaped and when viewed on external surfaces were not associated with any extracellular matrix.

External protected surfaces, defined as those surfaces that are contiguous with the external surfaces of the plant but protected by other tissue, were also examined. Leaf primordia and apical meristems were found to be colonized by bacteria over most of the available surfaces, again with only a few areas left uncolonized. Freeze fractured petiole base tissue revealed external protected surfaces formed by petioles of older leaves closely appressed to form an outer layer (Plate 1 A & B). These surfaces were less densely colonized than the meristem region, with bacteria arranged in microcolonies. In one sample, large intercellular spaces were shown to contain particulate matter (Plate 1 A & C) which comprised an heterogeneous mix of reticulate matrix and rod shaped bacteria (Plate 1 D).

Freeze fractured roots showed an outer epidermal layer with root hairs enclosing the root cortex, which comprised parenchyma cells with large intercellular spaces, and in some samples large lacunae of aerenchyma tissue (Plate 2 B). Bacteria colonized the root surface including root hairs, whilst endophytic bacteria were observed within the cortex region where they colonized intercellular spaces, aerenchyma and parenchyma cell walls (Plate 2 B, C & D). Many of these bacteria were associated with a reticulate matrix of unknown origin.

The FITC conjugated antiserum applied to root segments from micropropagated *Z. aethiopica* showed bacteria present over the entire surface, often in discrete microcolonies. All labelled bacteria exhibited the high intensity of fluorescence shown by the homologous antigen. Epiphytic bacteria were located similarly in sections from LR White embedded roots (Plate 3 A). Additionally, labelled endophytic bacteria were located at host cell junctions and intercellular spaces within the cortical parenchyma (Plate 3 B & C). No fluorescence was observed in control samples.

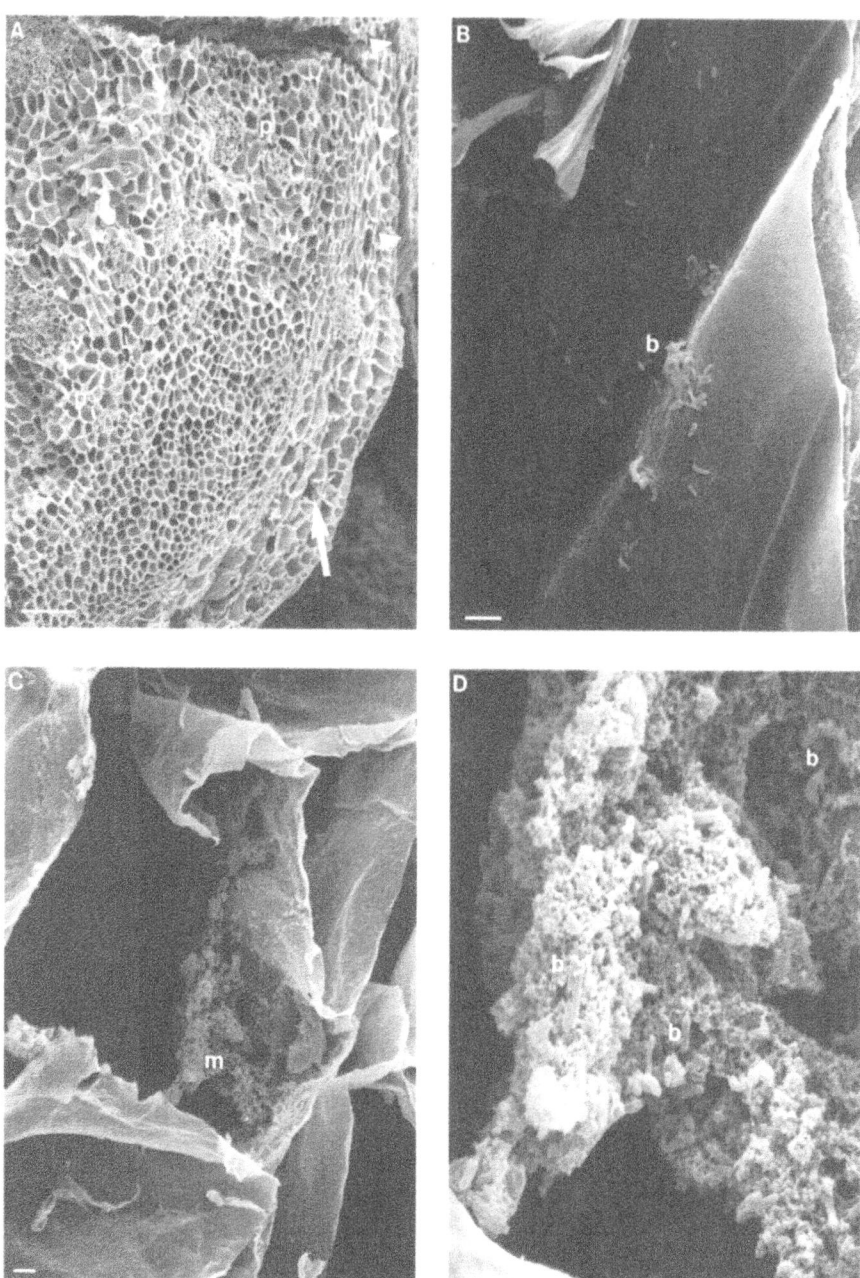

Plate 1. Scanning electron micrographs of a freeze fractured petiole base from *Z. aethiopica* plant growing in apparently axenic multiplication culture. A. Cortical parenchyma (p) of petiole base with particularly large intercellular space (arrowed) in outer layers beneath epidermis and protected external surface (arrowheads) provided by epidermal cells of leaf sheath. B. Microcolonies of bacteria (b) on protected surface (arrowheads in A). C & D. Intercellular space arrowed in A. Particulate material (m) adhering to plant cell walls comprises rod shaped bacteria (b) embedded in a reticulate matrix. Scale bars: A = 500 μm; B, C & D = 5 μm.

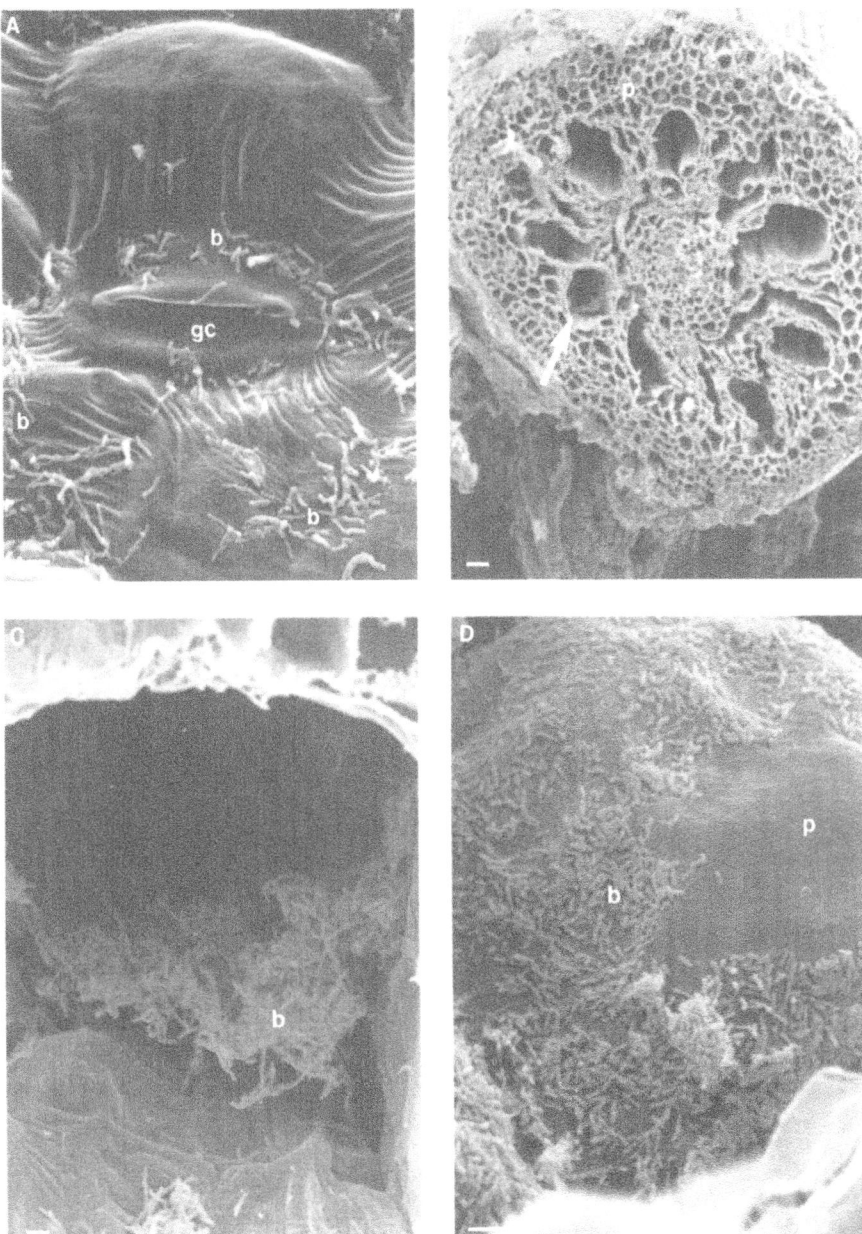

Plate 2. Scanning electron micrographs of *Z. aethiopica* plants taken from apparently axenic micropropagation culture. A. Epiphytic bacteria (b) colonizing leaf epidermis. Stomatal guard cell (gc). B & C. Freeze fractured root with aerenchyma arranged radially within cortical parenchyma (p). Lacuna arrowed in B shown in C to be colonized by bacteria (b). D. Bacteria (b) in intimate contact with root cortical parenchyma cell (p), exposed by freeze fracture. Scale bars: A, C & D = 5 μm; B = 50 μm.

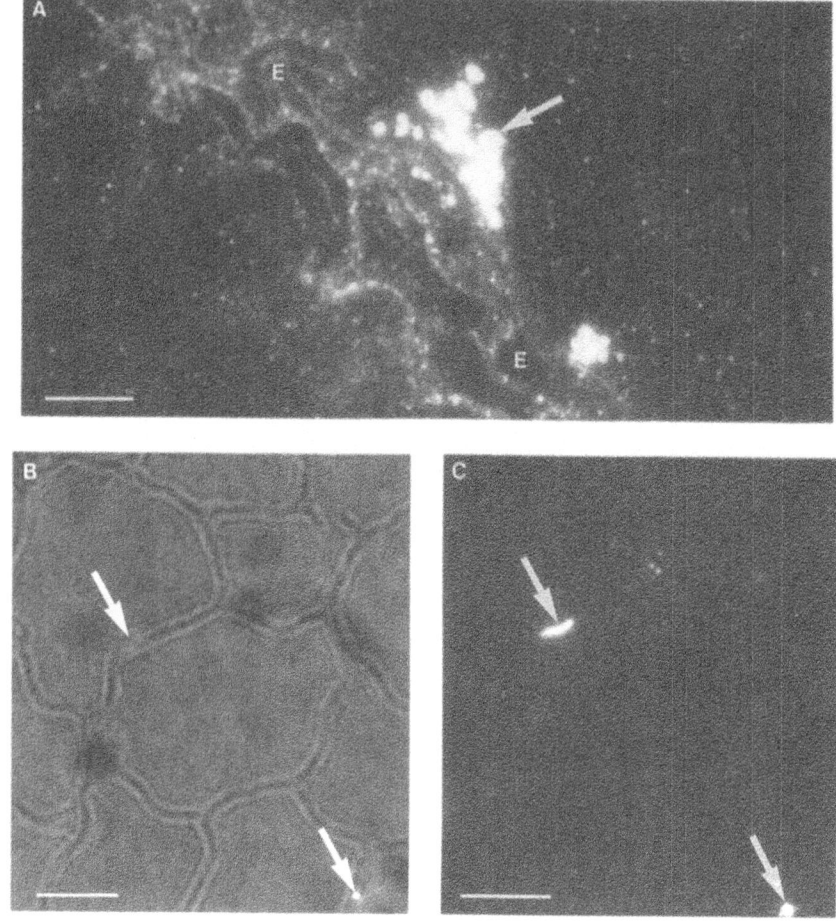

Plate 3. Roots from apparently axenic cultures of *Z. aethiopica*, embedded in LR White resin, sectioned and bacteria located by immunofluorescence (A & C, viewed with U.V.; B, bright field and U.V. illumination). A. Epiphytic colony attached to root epidermis (E). B & C. Endophytic bacteria at cell junctions and intercellular spaces within cortical parenchyma. Scale bar = 10 μm.

3.5 Facultative methylotrophy and methanol in culture vessels

Fluorescent pseudomonad B30.2, *Xanthomonas* sp. 171.A, unidentified isolate 170.B, *X. maltophilia* SA7.1 and isolate C, which appeared as a pink pigmented contaminant of micropropagated *Z. aethiopica* plants, were assayed for facultative methylotrophy. Isolate C alone showed sparse growth on NA minus methanol and profuse growth on NA in the presence of methanol, and is therefore a facultative methylotroph. Methanol was detected in the head space gas from *Z. aethiopica* cultures 7, 13, 14 and 15 weeks old, at concentrations ranging from 13.6 to 44 μg dm^{-3} (Fig. 1).

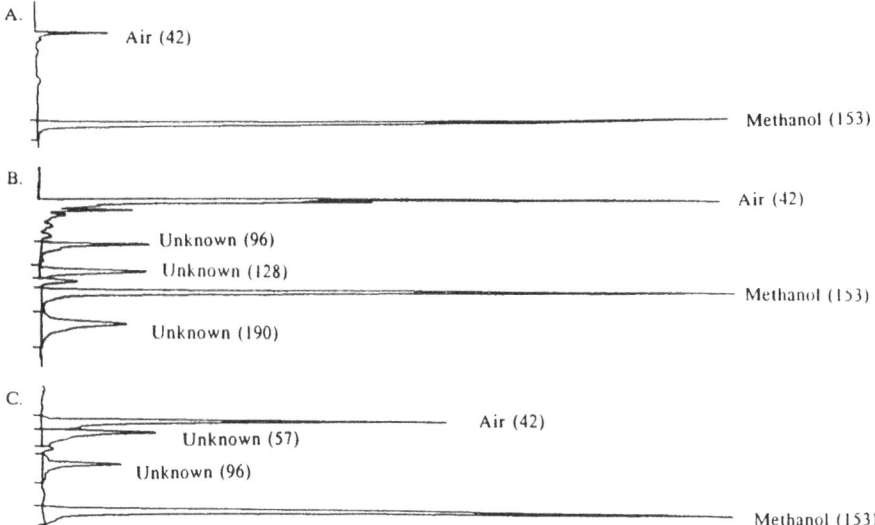

Fig. 1. Gas chromatography traces of standard sample of methanol (A) and headspace gas from plant tissue culture vessels (B & C). The latter are 0.5 cm³ samples from 6 (B) and 14 (C) week old *Zantedeschia aethiopica* multiplication cultures grown in MSMO with 3 mg dm⁻³ BAP. Peaks are labelled with the gas they represent. The retention times, in seconds, are given in brackets.

4. Discussion

4.1 Isolation and identitiy of putative endophytes

The greatest number of bacteria was isolated from bud tissue of *F. benjamina*, but the nature of this explant suggests that they were mostly epiphytes. The buds consisted of tightly furled vestigial leaves which provide a large surface area for epiphytic colonization. Their arrangement also makes it unlikely that the surface sterilization solutions penetrated beyond the outer layers. Indeed buds are recognized as a source of a proportion of phylloplane bacteria and fungi, which colonize leaves as they unfurl [24]. Although the numbers of bacteria isolated were less from *Dieffenbachia* cv. Camilla and Compacta, *S. podophyllum* and *Z. aethiopica*, bacteria isolated from bud tissue of these hosts are also more likely to be of epiphytic origin. Bacteria isolated from petiole or rhizome base tissue are, conversely, more probably of endophytic origin; indeed tropical Araceae are reported to be heavily colonized by an endophytic microflora [32].

Thus a range of important considerations should be taken into account when attempting to isolate endophytes, including choice of explant material and surface sterilization procedure. The explant should have no external surfaces protected from the sterilizing agent unlike, for example, bud tissue of *F. benjamina*. Surface sterilization protocols should be rigorous enough to eliminate epiphytic microorganisms. This may require special treatments to ensure all external protected surfaces are wetted, and a second immersion in sterilant following excision of outer layers of tissue. Effectiveness of the surface sterilization protocol can be confirmed by monitoring the disappearance of marked bacterial populations artificially applied to the surface of explant tissues. For example, the protocol used here had been tested with rifampicin resistant mutants [17].

Identification of isolated microorganisms enables reference to background information, such as the typical habitats and pathogenicity, which facilitates elimination of obvious airborne or human contaminants from the range of putative endophytes isolated. Published data on the bacteria isolated from the glasshouse grown plants in the present study indicate that the majority are typically plant associated [2, 31, 33]. It should be remembered, however, that little is known about the identity and biology of endophytic bacteria; thus it is important that any isolate is not dismissed *a priori*.

4.2 Endophytic ability and growth

The lack of disease symptoms and hypersensitive responses when isolated bacteria were reintroduced to host tissues (a modified form of Koch's postulates) and tobacco suggests that none were pathogenic. All, however, were shown to survive endophytically. The ability of four isolates to produce green island effects when reinoculated into their host is significant with respect to biotrophy. A green island is defined as 'a portion of green infected tissue surrounded by chlorotic tissue' [18], and is typically associated with lesions caused by biotrophic pathogens such as rust, downy and powdery mildew fungi [29]. Green islands surrounding infection sites are considered to represent tissue which remains in a juvenile condition, ensuring the pathogen is surrounded by metabolically active cells from which it can obtain nutrients. Thus induction of green islands by bacterial isolates may represent a useful adaptation to the endophytic environment, which would be expected for endophytes that are in the constitutive category.

The other bacterial isolates may be constitutive or inducible endophytes, and are likely to remain viable in explant material used for micropropagation. Epiphytic bacteria which evade surface sterilization, for example as micro-colonies held together and protected from sterilants by extracellular polysaccharide, are potential inducible endophytes if introduced to cut surfaces during manipulation of plant materials. Assuming the bacteria are not affected by the plant defence response to mechanical wounding, and that they remain viable, then a number of different scenarios may be postulated for both categories of endophyte.

For example, the bacteria could multiply at a very slow rate and not become apparent until they reached a critical density. Alternatively, they may remain viable but unable to multiply due to nutrient limitation. In both situations, bacterial cells would be transferred on sub-culturing to the progeny of micropropagated plants. Thus the presence of any endophytic microorganisms, constitutive or inducible, in micropropagated plant material will act as a source of macroscopic contamination if environmental or physiological changes occur to favour endophyte multiplication.

4.3 Xanthomonas maltophilia *in micropropagated* Zantedeschia aethiopica

The isolation of contaminating bacteria from apparently axenic cultures of *Z. aethiopica* plants in micropropagation highlights the problems of establishing and maintaining axenic cultures. All the plants screened were contaminated by the same bacterium, identified as *X. maltophilia* by FAP.

Bacteria located by immunofluorescence at parenchyma cell junctions in root tissue provide the source from which the large numbers of bacteria associated with aerenchyma tissue are derived. Formation of lysigenous aerenchyma involves dissolution of parenchyma cell walls as a result of cellulase activity [21]. Nutrients are released from the plant cell wall and cytoplasm, and the relatively large lacunae facilitate aeration. Increased oxygen and nutrient concentrations could have a chemotactic effect, encouraging movement of bacteria into the root. Thus aerenchyma formation may encourage establishment of endophytic associations, stimulate increased growth of existing endophytic populations, and explain the ability of a benign endophyte to switch to blatant contaminant. Further experiments are now required to test this hypothesis.

X. maltophilia does not appear to have any detrimental effect on the host under the normal 4 week sub-culture regime employed for shoot multiplication, and only became visible in plant cultures left for more than 8 weeks. Although Stead [34] suggests that all contaminating bacteria could cause the death of a plant at some stage in micropropagation, no obvious adverse effects were noted even at this level of contamination by *X. maltophilia*. No firm conclusions, however, can be drawn about detrimental or beneficial effects on *Z. aethiopica* because axenic cultures were not available for comparative experiments.

Attempts to initiate contaminant free cultures of *Z. aethiopica* from meristem tips were unsuccessful due to their contamination by *X. maltophilia*; scanning electron microscopy revealed that the apical meristem region was colonized by bacteria [17]. In these circumstances, meristem tip culture is clearly inappropriate for obtaining axenic plant cultures. Indeed, a similar problem has been observed when attempting to establish contaminant free cultures of field collected coffee and cacao germplasm [11].

4.4 *Influence of gaseous environment on contaminant growth*

Head space gas composition may, directly and indirectly, have an important role in the appearance of macroscopically visible contamination of micropropagated plants. For example, the formation of aerenchyma in *Z. aethiopica* plants in culture may be influenced by the permeability of the culture vessel seal. Aerenchyma develops in roots when deprived of oxygen in waterlogged soils due to ethylene accumulation [10, 21, 23], which also occurs in tissue culture vessels [19, 20]. Jackson and co-workers [19] demonstrated that this caused various deleterious affects on explant development, which may indirectly provide environments more favourable to proliferation by endophytic bacteria.

A further effect of the gaseous environment is illustrated by the presence of methanol in head space gases and identification of a pink pigmented, facultative methylotroph (PPFM) which appeared as a contaminant in multiplication cultures of *Z. aethiopica.*

Methanol accumulation is not surprising, as plant tissues are known to produce methanol as part of the normal process of growth [8]. Furthermore, methanol trapped by the plant cuticle is thought to provide an important nutrient for phylloplane microorganisms [8]. Cryptic contamination of plant cultures by PPFMs will be spread mechanically during sub-culturing, which facilitates colonization of the host plant and progeny. Eventually, accumulation of methanol in the tissue culture vessel is likely to combine with a critical PPFM population to favour profuse growth. This effect would be characterized by the appearance of macroscopically visible contamination of cultures, as the bacterium changed status from quiescent to blatant contaminant. Further work is required to explore the occurrence of methanol in tissue culture vessels, as well as effects on plant and microbial growth. Meanwhile, the simple assay for detecting facultative methylotrophs should prove invaluable to micro-propagators when similar bacteria contaminate their cultures.

Avoidance of these contamination problems could be effected by more frequent venting of tissue culture vessels, or the inclusion of a protected aperture to allow free diffusion of gases. Alternatively, the internal atmosphere could be monitored for the accumulation of critical gases with special detectors. Biosensors for the detection of methanol are already produced for the food and drink industries [26], whilst sensors that respond by a colour change are being developed for incorporation into commercial food packaging [9]. Similar systems would be ideal for monitoring the gaseous environment of plant tissue culture vessels: they do not require skilled labour or expensive equipment, and early detection of deleterious gas accumulation would facilitate management of culture hygiene.

5. Conclusions

It is clear that commercial tissue culture companies are well advised to appreciate the significance of identifying contaminants of micropropagated plants and should initiate research into endophytic sources of contamination. Whilst it is possible that any contaminant could cause the death of a plant at some stage of micropropagation [34], other organisms may cause more specific problems. For example, results presented here have shown *Dieffenbachia* stock plants acting as symptomless, alternate hosts for *P. rubrisubalbicans*, a pathogen non-indigenous to this country. This illustrates how inadvertent contravention of legislation could lead to the introduction of new pathogens with potentially devastating effects. The components that make a plant disease highly threatening are the ability of the causal agent to spread rapidly, cause serious losses and be difficult to control [36]. Latent infections of plant tissue cultures by bacterial endophytes satisfy all three of these criteria and so deserve greater consideration than they receive at present.

Acknowledgements

We are grateful to Twyford Plant Laboratories Ltd, Glastonbury for financial support and for supplying the plant material used in this project, and to John Clement for Plate 2 B, C & D.

References

1. Bastiaens L (1983) Endogenous bacteria in plants and their implications in tissue culture — a review. Med Fac Landbouww Rijksuniv Gent 48:1–11
2. Billing E (1976) The taxonomy of bacteria on the aerial parts of plants. In: CH Dickinson and TF Preece, eds. Microbiology of Aerial Plant Surfaces, pp 223–292. London: Academic Press
3. Bradbury JF (1986) Guide to Plant Pathogenic Bacteria. Slough: CAB International
4. Carroll G (1986) The biology of endophytism in plants with particular reference to woody perennials. In: NJ Fokkema and J Van Den Heuvel, eds. Microbiology of the Phyllosphere, pp 205–222. Cambridge: Cambridge University Press
5. Carroll G (1988) Fungal endophytes in stems and leaves: from latent pathogen to mutualistic symbiont. Ecol 69:2–9
6. Cassells AC (1992) Screening for pathogens and contaminating microorganisms in micropropagation. In: JM Duncan and L Torrance L, eds. Techniques for the Rapid Detection of Plant Pathogens, pp 179–192. Oxford: Blackwell Scientific Publications
7. Constantine DR (1986) Micropropagation in the commercial environment. In: LA Withers and PG Alderson, eds. Plant Tissue Culture and its Agricultural Applications, pp 175–186. London: Butterworths
8. Corpe WA and Rheem S (1989) Ecology of the methylotrophic bacteria on living leaf surfaces. FEMS Microb Ecol 62:243–250
9. Dambrot SM (1992) Biosensors galore. Bio/technol 10:129
10. Drew MC, Jackson MB and Giffard S (1979) Ethylene induced adventitious rooting and development of cortical air spaces (aerenchyma) in roots may be adaptive responses to flooding in *Zea mays* L. Planta 147:83–88
11. Duhem KLE, Mercier N and Boxus PH (1988) Difficulties in the establishment of axenic *in vitro* cultures of field collected coffee and cacao germplasm. Acta Hort 225:67–75
12. Fahy PC and Hayward AC (1983) Media and methods for isolation and diagnostic tests. In: PC Fahy and GJ Persley, eds. Plant Bacterial Diseases. A diagnostic guide, pp 337–378. London: Academic Press
13. Falkiner FR (1988) Strategy for the selection of antibiotics for use against common bacterial pathogens and endophytes of plants. Acta Hort 225:53–56
14. Fisse J, Batlle A and Pera J (1987) Endogenous bacteria elimination in ornamental explants. Acta Hort 212:87–90
15. Gilbert JE, Shohct S and Caligari PDS (1991) The use of antibiotics to eliminate latent bacterial contamination in potato tissue cultures. Ann App Biol 119:113–120

16. Giles KL and Morgan WM (1987) Industrial scale micropropagation. Trends Biotechnol 5:35–39
17. Gunson HE (1992) Endophytes and microbial contaminants of micropropagated plant species. PhD Thesis, University of the West of England, Bristol, UK
18. Habeshaw D (1984) Effects of pathogens on photosynthesis. In: RKS Wood and GJ Jellis, eds. Plant Diseases: Infection, Damage and Loss, pp 63–72. Oxford: Blackwell Scientific Publications
19. Jackson MB, Abbot AJ, Belcher AR, Hall KC, Butler R and Cameron J (1991) Ventilation in plant tissue cultures and effects of poor aeration on ethylene and carbon dioxide accumulation, oxygen depletion and explant development. Ann Bot 67:229–237
20. Jona R, Gribaudo I and Vigliocco R (1984) Effects of naturally produced ethylene in tissue culture jars. In: Y Fuchs and E Chalutz, eds. Ethylene. Biochemical, Physiological and Applied Aspects, pp161–162. The Hague: Martinus Nijhoff / Dr W Junk Publishers
21. Kawase M (1979) Role of cellulase in aerenchyma development in sunflower. Am J Bot 66:183–190
22. Keen NT, Ersek T, Long M, Bruegger B and Holliday M (1981) Inhibition of the hypersensitive reaction of soybean leaves to incompatible *Pseudomonas* spp. by blasticidin S, streptomycin, or elevated temperature. Physiol Plant Pathol 18:325–337
23. Konings H (1982) Ethylene-promoted formation of aerenchyma in seedling roots of *Zea mays* L. under aerated and non-aerated conditions. Physiol Plant 54:119–124
24. Leben C (1971) The bud in relation to the epiphytic microflora. In: TF Preece and CH Dickinson, eds. Ecology of Leaf Surface Microorganisms, pp 117–127. London: Academic Press
25. Leifert C, Waites WM and Nicholas JR (1989) Bacterial contaminants of micropropagated plant cultures. J Appl Bact 67:353–361
26. Luong JHT, Groom CA and Male KB (1991) The potential role of biosensors in the food and drink industries. Biosensors and Bioelectronics 6:547–554
27. Lyons NF and Taylor JD (1990) Serological detection and identification of bacteria from plants by the conjugated *Staphylococcus aureus* slide agglutination test. Plant Path 39:584–590
28. Mishagi IJ and Donndelinger CR (1990) Endophytic bacteria in symptom-free cotton plants. Phytopath 80:808–811
29. Pegg GF (1981) Involvement of growth regulators in the diseased plant. In: PG Ayres, ed. Effects of Disease on the Physiology of the Growing Plant, pp 149–177. Cambridge: Cambridge University Press
30. Petrini O (1986) Taxonomy of endophytic fungi of aerial tissues. In: NJ Fokkema and J Van Den Heuvel, eds. Microbiology of the Phyllosphere, pp 176–187. Cambridge: Cambridge University Press
31. Schaad NW (1988) Initial identification of common genera. In: NW Schaad, ed. Laboratory Guide for the Identification of Plant Pathogenic Bacteria, pp 1–15. 2nd edition. Minnesota: American Phytopathological Society
32. Staritsky G, Dekkers AJ, Louwaars NP and Zandvoort EA (1986) *In vitro* conservation of aroid germplasm at reduced temperatures and under osmotic stress. In: Withers LA and Alderson PG (eds) Plant Tissue Culture and its Agricultural Applications, pp 277–283. London: Butterworths
33. Starr MP (1981) Phytopathogenic bacteria. Berlin: Springer-Verlag
34. Stead DE (1988) Identification of bacteria by fatty acid profiling. Acta Hort 225:39–46
35. Stead DE (1992) Techniques for detecting and identifying plant pathogenic bacteria. In: JM Duncan and L Torrance, eds. Techniques for the Rapid Detection of Plant Pathogens, pp 76–114. Oxford: Blackwell Scientific Publications
36. Thurston DH (1973) Threatening plant diseases. Annu Rev Phytopathol 11:27–52
37. Trick I and Lingens F (1985) Aerobic spore-forming bacteria as detrimental infectants in plant tissue cultures. App Microbiol Biotechnol 21:245–249
38. Wilson ZA and Power JB (1989) Elimination of systemic contamination in explant and protoplast cultures of rubber (*Hevea brasiliensis* Muell Arg). Plant Cell Rep 7:622–625

Fungal contaminants of *Primula*, *Coffea*, *Musa* and *Iris* tissue cultures

S. DANBY[1], F. BERGER[2], D.J. HOWITT[1], A.R. WILSON[1], S. DAWSON[3] and C. LEIFERT[4]*

[1] *Department of Cell & Structural Biology, Stopford Building, Manchester, University, Manchester M13 9PT, UK*
[2] *Institute für Pflanzenbau und Tierhygiene in den Tropen und Subtropen, Universität Göttingen, 3400 Göttingen, Germany*
[3] *Institute of Environmental and Biological Sciences, Lancaster University, Lancaster LA1 4YQ, UK*
[4] *Department of Plant & Soil Science, University of Aberdeen, Aberdeen AB9 2UE, UK*
* *corresponding author*

1. Introduction

Successful micropropagation relies on successful disinfection of plant tissues taken from field or glasshouse grown plants and aseptic techniques and laboratory conditions which allow aseptic conditions in plant tissue cultures to be maintained for long periods of time [6, 8]. Although aseptic conditions were often achieved by surface disinfection, many plants cultures did not remain free of contaminants during *in vitro* culture [8]. Such recontamination of aseptic tissue cultures in the laboratory is now considered the main reason for losses during commercial micropropagation [8].

The list of organisms described as contaminants in plant tissue culture includes viruses, bacteria, yeasts, fungi, mites and thrips [6]. Contamination with bacteria has been described extensively in the literature and many sources of bacterial contamination have been characterized [8, 10, 14]. On the other hand, very few yeasts, and filamentous fungi found in plant tissue cultures have been identified and the sources of fungal contaminants remained unclear [3, 7]. However, losses due to fungal and yeast contamination, especially when connected to mite or thrip infestations have repeatedly resulted in large commercial losses [6, 8].

In this study we have therefore isolated and identified fungal contaminants from plant tissue cultures which had been growing apparently aseptically *in vitro* for longer than 1 year and compared these to fungi isolated from the laboratory air.

P.J. Lumsden, J.R. Nicholas and W.J. Davies (eds.), Physiology, Growth and Development of Plants in Culture, 397–403, 1994.

2. Materials and methods

2.1 Isolation of fungi from contaminated plant tissue cultures

Fifty eight culture vessels with fungal or yeast contaminated plant tissue cultures were taken at random from the growth rooms in a commercial micropropagation laboratory. Thirty five *Primula*, 12 *Coffea*, 7 *Musa* and 4 *Iris* cultures were used. Except in two instances, each culture vessel appeared to be contaminated with a single organism.

Contaminants were subcultured from the contaminated culture vessels onto isolation media using a number 2 cork-borer (6 mm diameter) for the filamentous fungi and a bacteriological loop for colonies resembling yeasts growth. Three types of isolation media were used: Murashige and Skoog [11] plant tissue culture medium (MS, Sigma) with 20 gl^{-1} sucrose and solidified with 12 gl^{-1} of bacteriological agar (Oxoid) adjusted to pH 5.7 (with NaOH), malt extract agar (MEA, Oxoid) and nutrient agar (NA, Oxoid; suspected yeast colonies only). At this stage 4 of the isolates were found to be bacteria by microscopic examination and discarded. Plates were incubated at 20 °C.

2.2 Isolation of fungi from laboratory air

Air samples were taken from the media preparation room, cutting or subculturing area, storage areas, the growth rooms and outside the laboratory building using a Casella 253 air sampler (Casella London Ltd., Kempston U.K.). Three replicate samples were taken at each site onto plates with malt extract agar (MEA) a medium favouring fungi. MEA-plates were then incubated for 2 days at 20 °C in the dark before colony counts were made. Plates were assessed again after 3 days to see if any further colonies had developed. Two sets of readings were taken, the first in October 1991 and a second in January 1992. This enabled a comparison to be made between spore counts in Autumn and Winter. Fifty four fungi and yeasts were isolated at random from air sample plates taken in October. These were subsequently identified to enable a comparison to be made between fungi found in the laboratory air and those isolated from contaminated plant tissue cultures.

2.3 Identification of filamentous fungi and yeasts

Filamentous fungi from both contaminated plant cultures and air samples were identified to family and, where possible, to species level, on the basis of hyphal, spore and sporophore morphology according to the descriptions given by Domsch *et al.* [2]. Fungal growth taken from MEA plates was stained with Lactophenol Blue and examined under a light microscope. Fungi which had not sporulated were placed under UV light (Philips TL40W/08; maximum radiant power at 350 nm) for up to 2 weeks to encourage spore formation. Yeasts were identified using API ID 32C yeast identification strips (BioMerieux, 69280 Marcy-l'Etoile, France).

3. Results and discussion

Of the 58 isolates from plants which had been *in vitro* for longer than 1 year 49% were *Penicillium* spp., 9% *Aspergillus* spp., 8% *Cladosporium* spp., 5% *Botrytis cinerea*, 5% yeast, 3% *Phoma* spp., 2% *Zygorrhynchus* spp., 2% *Trichophyton* spp. and 15% unidentified (Fig. 1). Fungal genera similar to those isolated at Microplants have also been found in other laboratories. When we identified cultures from 10 different laboratories at the Conference microbiology workshop we found that 35% of filamentous fungi were *Penicillium*, 17% *Aspergillus*, 8% *Alternaria* 4% *Botrytis* and 8% *Fusarium* spp. (Fig. 1). In contrast with many bacterial species [5] fungi isolated from contaminated cultures grew well on plant tissue culture media in the absence of plants. This is not surprising since plant tissue culture media have very similar nutrient compositions to common fungal media [15].

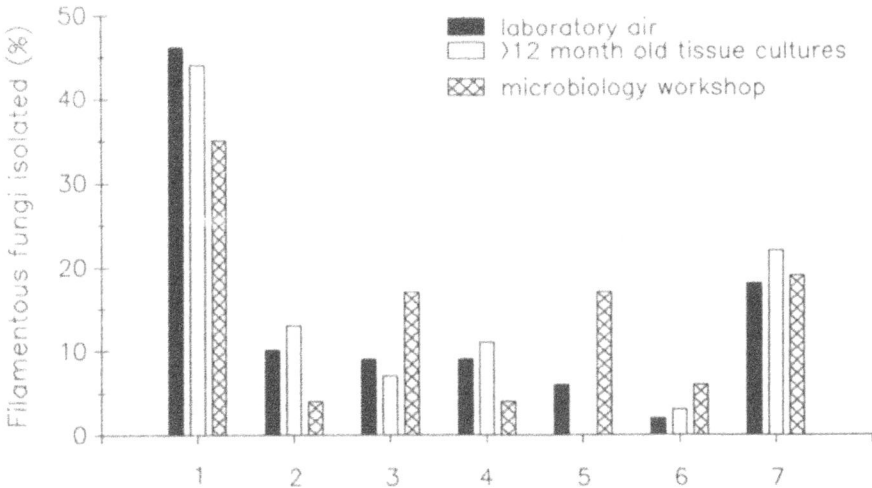

Fig. 1. Fungal genera (%) isolated from contaminated plant tissue cultures and the laboratory air in September 1991 at Microplants and contaminated plant tissue cultures from 10 different tissue culture laboratories at the microbiology workshop at the SEB/IAPTC conference in September 1992 in Lancaster. 1, *Penicillium*; 2, *Cladosporium*; 3, *Aspergillus*; 4, *Botrytis*; 5, *Alternaria*; 6, *Fusarium*; 7, other filamentous fungi.

Some fungal genera were only associated with certain plant species, for example only *Penicillium* spp. were isolated from *Iris* and *Musa*; whilst *Coffea* was also contaminated with *Cladosporium* and *Botrytis* and *Primula* were found to be also infected with *Aspergillus, Phoma, Botrytis, Cladosporium* and yeasts.

Of the 54 isolates taken from air samples 44% were *Penicillium* spp., 13% *Cladosporium* spp., 6% *Aspergillus* spp., 7% yeasts, 6% *Botrytis cinerea*, 4% *Trichocladium opacum*, 4% *Fusarium* spp., 2% *Alternaria tenuissima* and 13% unidentified (Fig 1.).

The 7 yeast isolates were identified as *Rhodotorula* spp., *Cryptococcus albidus, Debaryomyces polymorphus, Trichosporon cutaneum* and *Candida* spp. (Table 1).

Table 1. Yeasts isolated from contaminated plant tissue cultures and the laboratory air at Microplants Ltd. in September 1991

Yeast strain	API computer identification	% probability
From contaminated tissue cultures		
P21	*Candida* spp.	*
P27	*Debaryomyces polymorphus*	99.6
P29	*Trichosporum cutaneum*	95.2
	(*Candida humicola*	4.6)
From the laboratory air		
G4/1	*Rhodotorula rubra*	92.4
	(*Rhodotorula glutinis*	7.6)
G4/6	*Cryptococcus albidus*	99.0 (&)
G4/7	*Candida* spp.	*
H/1	not-identifiable	

* identification to species level not possible using API32C test strips only
(&) Identification was made on the basis of the aesculin test to separate closely related species.

Fungal populations found in contaminated plant cultures and the laboratory air were similar (Fig. 1) indicating the laboratory air as a main source of contamination. Since there was no means of filtration of air coming in to the laboratory it can be assumed that at least a proportion of the contaminants enter the laboratory from the outside. To test this hypothesis we compared air counts outside and inside the laboratory in autumn when spore counts in the air are high due to the grass harvest around the laboratory and in winter when fewer fungal spores are found in the outside air (Fig. 2). Fungal concentrations in the air of the various rooms inside the laboratory building were found to correlate well with the concentrations in the outside air (Fig. 2). In autumn, when the outside air contained high numbers of fungal spores, high counts were also found inside most laboratory rooms, whereas counts in winter were lower both inside and outside the laboratory (Fig. 2). There were also differences between different rooms in the laboratory. In September the highest spore densities were found in growth room 3, the cutting room (the room containing the laminar flow cabinets) and one of the hallways.

Fungal spores produced in contaminated plant cultures can be passed to other vessels by mite and thrip vectors; and the mite species *Sideroptes graminum* had been isolated from cultures containing *Botrytis cinerea* in the same laboratory (Leifert, unpublished). This indicates another potential source of fungal contamination in the laboratory since mites and thrips have been shown to be responsible for sudden and extensive outbreaks of fungal contamination [8].

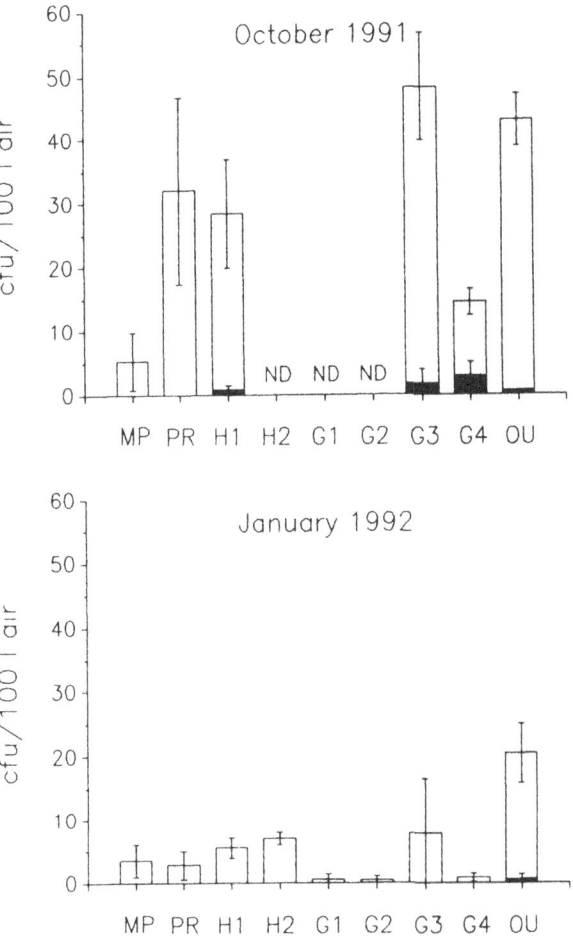

Fig. 2. Numbers of cfu isolated on malt extract agar from the laboratory air in September 1991 and January 1992. □, colonies of filamentous fungi; ■, bacteria and yeasts colonies. MP, media preparation; PR, cutting room with laminar flow cabinets; H1, hallway outside PR; H2, hallway outside G1 &G2; G1 & G2, indoor growth rooms; G3 & G4, outside growth rooms; OU, outside air.

Fungal contamination can only be avoided if the air coming in to the laboratory is filtered and by a high level of hygiene in the laboratory. Culture vessels should, for example, be wiped with disinfectants prior to subculture to kill spores which have settled on the containers and so avoid them being introduced into cultures when these are opened. By working deep inside the laminar flow cabinets the introduction of spores from the laboratory air can be reduced greatly [6, 8]. Fungal growth in plant tissue cultures can also be suppressed by incorporation of fungicides such as benomyl, iprodione or penconazole into the growth media [1, 4, 12, 13, 16]. However, we found that like many antibiotics [8, 9] fungicides reduce plant growth and that some stop plants from rooting [Danby & Leifert, unpublished].

Spread of fungal contaminants in the growth room can be avoided by the quarantine of all newly introduced tissue cultures in a separate growth cabinet or room, by regular microscopic examination of cultures for mites and thrips, by removing all fungal contaminated cultures from growth rooms as soon as contamination becomes visible, by sealing of uninfected cultures with plastic film, by wiping all shelves, floors and walls with disinfectants and by fumigating growth rooms with insecticides/acaricides such as pirimiphos-ethyl [6, 8].

Some of the fungal genera found including *Alternaria, Botrytis, Phoma* and *Cladosporium* are known plant pathogens. Weaning of plants which become infected with fungi in the laboratory (which is common practise in many laboratories) might therefore result in plant disease and losses during weaning of plants. *Botrytis cinerea* is especially known to cause significant losses during acclimatization of tissue cultured plants. Plants contaminated with such fungi should therefore be discarded rather than weaned.

Acknowledgements

We would like to thank Dr. P. C. Harper and Mr. S. Kerr-Liddell and others at Microplants Ltd. (Longnor, Derbyshire, UK.) for their help and cooperation during this project.

References

1. Clarke JH, Norman JA & Lavery E (1989) Some observations on contamination of animal cell cultures by the fungus *Aspergillus fumigatus*, and suggested control measures. Cell Biol Int Rep 13:773–779
2. Domsch KH, Gams W & Anderson T (1980) Compendium of Soil Fungi, Vol. I. London/New York: Academic Press
3. Enjalric E, Carron MP & Lardet L (1988) Contamination of primary cultures in tropical areas: the case of *Hevea brasiliensis*. Acta Hort 225:57–65
4. Hauptmann RM, Wildholm JM & Paxton JD (1985) Benomyl a broad spectrum fungicide for use in plant cell and protoplast cultures. Plant Cell Rep 4:129–132
5. Leifert C & Waites WM (1992) Bacterial growth in plant tissue culture media. J Appl Bacteriol 72:460–466
6. Leifert C & Waites WM (1993) Dealing with contaminants in plant tissue and cell culture; Hazard analysis, Critical control points. Plant Growth Reg (In press)
7. Leifert C, Waites WM, Nicholas JR & Keetley JW (1990) Yeast contaminants of micropropagated plant cultures. J Appl Bacteriol 69:471–476
8. Leifert C, Ritchie JY & Waites WM (1991) Contaminants of plant-tissue and cell cultures. World J Microbiol Biotechnol 7:452–469
9. Leifert C, Camotta H, Wright S, Waites B, Cheyne VA & Waites WM (1991) Elimination of *Lactobacillus plantarum, Corynebacterium* spp., *Staphylococcus saprophyticus, Pseudomonas paucimobilis* from micropropagated *Hemerocallis, Choisya* and *Delphinium* cultures using antibiotics. J Appl Bacteriol 71:307–330
10. Knauss JF & Miller JM (1978) A contaminant *Erwinia carotovora*, affecting commercial plant tissue cultures. In Vitro 14:754–756

11. Murashige T & Skoog F (1962) A revised medium for rapid growth and bioassays with tobacco tissue cultures. Physiologia Plantarum 15:473–497
12. Shields R, Robinson SJ & Anslow PA (1984) Use of fungicides in plant tissue culture. Plant Cell Rep 3:33–36
13. Thurston KC, Spencer SJ & Arditti J (1979) Phytotoxicity of fungicides and bacteriocides in orchid culture media. Am J of Bot 66:825–835
14. Trick I & Lingens F (1985) Aerobic spore-forming bacteria as detrimental infectants in plant tissue cultures App Microbiol and Biotech 21:245–249
15. Vogel HJ (1956) Complete growth medium for *Neurospora* (Medium N). Microbiology and Genetical Bulletin 13, 42
16. Wyler RA, Murleach A & Mohl H (1979). An imidazole derivative (penconazole) as an antifungal agent in cell culture systems. In Vitro 15:745–750

Activity of antibiotics produced by *Bacillus subtilis* and *Bacillus pumilus* against common fungal contaminants of plant tissue cultures

SHARON DANBY, S.P. HAMPSON, S. JOSHI, D.C. SIGEE, H.A.S. EPTON and C. LEIFERT[1]*

Department of Cell and Structural Biology, University of Manchester, Stopford Building, Manchester M13 9PT, UK
[1] *Department of Plant & Soil Science, University of Aberdeen, Aberdeen AB9 2UE, UK*
* corresponding author

1. Introduction

Various *Bacillus* species are currently under investigation as bio-control agents for fungal diseases of horticultural and agricultural crop plants [4]. Their activity is thought to be due to the production of antifungal peptide antibiotics [5]. Such compounds have been partially purified and have been suggested as preservatives for plant tissue culture [2].

The aim of this investigation was to test the antifungal activity of crude extracts from fermentation broths from two *Bacillus* strains against a range of common fungal contaminants isolated from plant tissue cultures. Since fungi and yeasts have also been shown to reduce the medium pH of plant tissue culture media [1,3] we also tested the activity of antibiotics at different pH values using *Botrytis cinerea* as the test organism.

2. Materials and methods

2.1 Preparation of bacterial fermention broth extracts

The activity of crude fermentation browth extracts from batch cultures of *Bacillus subtilis* strain CL27 or *Bacillus pumilus* strain CL45 in cabbage broth 5 (CB5; see [4] for medium composition) or nutrient broth (NA; Oxoid) was determined against a number of strains of the most frequently isolated fungal contaminants (see Danby *et al.* [1] this volume for details of the strains used). Bacteria were grown on the media for 2 or 4 days (2d,4d) before crude extracts were prepared (see [4] for method). In the Figures and Tables the media and fermentation times are abbreviated (eg. NA/2d = grown for 2 days on nutrient agar).

P.J. Lumsden, J.R. Nicholas and W.J. Davies (eds.), Physiology, Growth and Development of Plants in Culture, 404–408, 1994.

2.2 Testing of extracts against fungal isolates

Spore suspensions of all fungi used were prepared as described previously [4]. Twenty ml of molten Murashige & Skoog medium [6], containing 12 g/l agar (pH 5.7) was cooled to 40 °C and inoculated with 0.2 ml of an appropriate dilution step of fungal suspension containing $4x10^5$ spores/ml. For investigation of the influence of media pH the pH of different batches of media were adjusted to pH 5.6 (with NaOH) before autoclaving and then altered to values of between 8.5 and 2.6 after autoclaving using NaOH or HCl prior to inoculation with fungal spores. The medium was then poured into either Petri-dishes or 7 Micro-detection plates (Proteus Molecular Design Ltd., Marple, Cheshire, UK.) and allowed to set. Wells were made 2 cm from the edge of Petri-dishes or in the centre of each channel in a Microdetection-plate using a 4 mm diameter cork borer. Thirty μl of crude fermentation broth extracts were then added to each well, leaving 2 wells as controls. Plates were incubated at 20 °C and diameters of zones of inhibition measured after 2–4 days as soon as fungal growth was clearly visible.

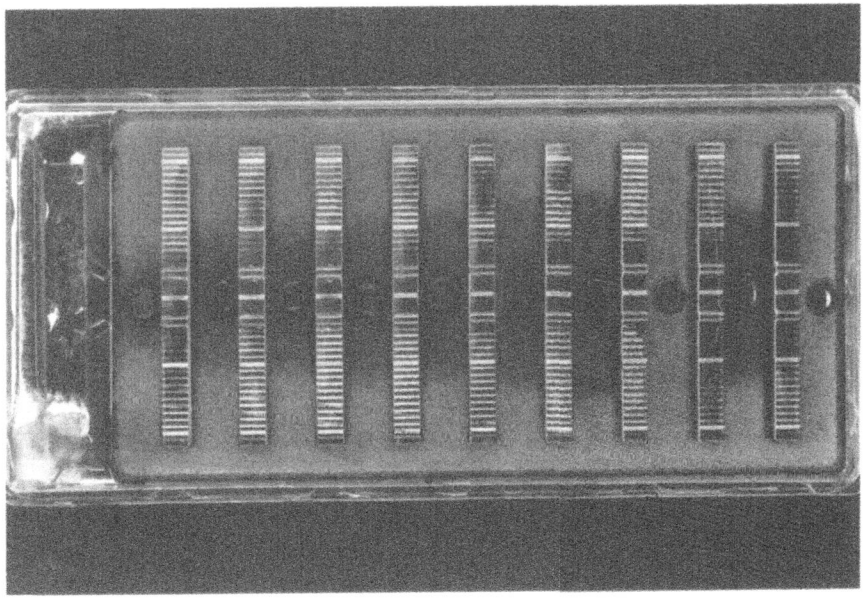

Plate 1. Inhibition zones formed on *Botrytis cinerea* seeded MS medium around wells with dilutions of crude extract CB5/2d from *Bacillus subtilis* strain CL27 batch cultures media on Micro-detection plate (original dimensions of the plate are 13.5 x 6 cm). Well 8 and 10 were inoculated with medium only.

3. Results and discussion

3.1 Sensitivity of petri dish and sens-acute plates for detection of antifungal activity in crude fermentation extracts

When crude extracts were diluted with CB5 broth to determine the minimal concentrations of antibiotics that can be detected on Petri- and Microdetection-plates much higher dilution steps of the crude extracts produced inhibition zones on *Botrytis* seeded media on sens-acute plates than conventional petri dishes (Fig. 1). Because Microdetection plates were found to be more sensitive in detecting the activity of *Bacillus* antibiotics in crude extracts they were used in all subsequent trials.

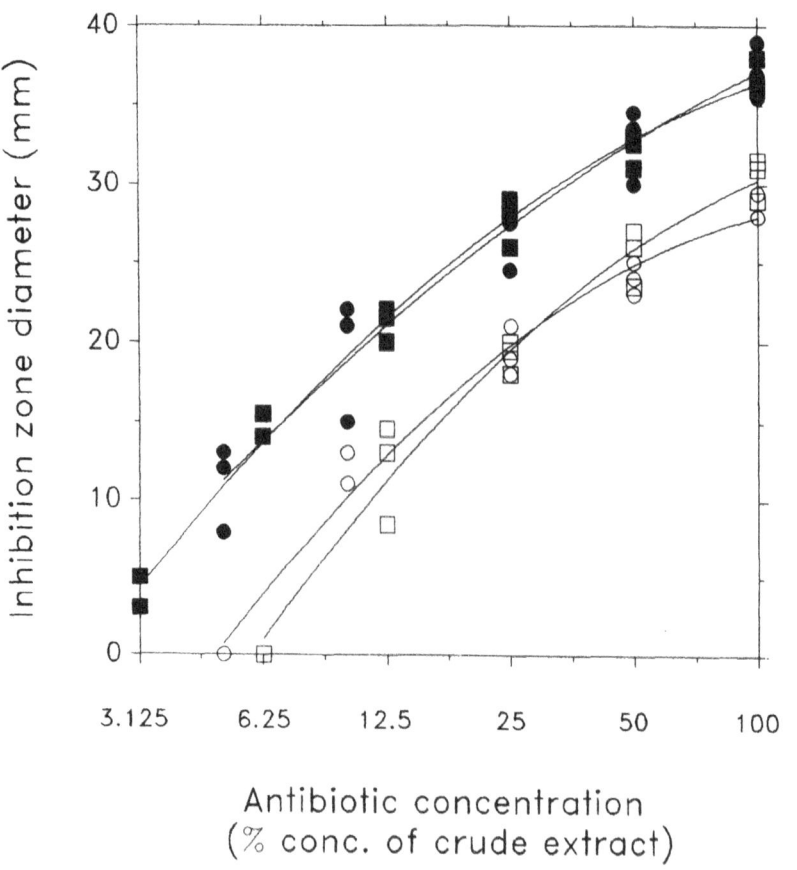

Fig. 1. Comparison of inhibition zones formed on *Botrytis cinerea* seeded MS medium around wells containing different concentrations of crude extracts CB5/4d on Petri-dishes (□, *Bacillus subtilis* CL27; ○, *Bacillus pumilus* CL45) and Microdetection plates (■, *B. subtilis* CL27; ●, *B. pumilus* CL45). Second order polynomial curves were fitted to the data (horizontal axis is $\log_2$ scale).

3.2 Range of antifungal activity

Of the four crude extracts only one, the crude extract from *Bacillus subtilis* grown for 4 days on 5% cabbage broth, was active against a wide range of different fungi and yeasts. However, some of the other extracts of the same *Bacillus subtilis* strain showed higher activity against individual fungal strains (Table 1.) This strongly suggests that different antibiotics are produced by the same strain on different fermentation media. Different strains of the same fungal genus showed similar inhibition patterns (for example, all *Botrytis* strains tested show similar responses, being inhibited by all 4 crude extracts; Table 1).

Table 1. Inhibition zones (mm) formed in fungal seeded media around wells containing crude extracts of *Bacillus subtilis* strain CL27 and *Bacillus pumilus* strain CL45 fermentation broths

| Fungal genera | Strain | Bacillus subtilis | | | Bacillus pumilus |
		Ca5/ 2d	Ca5/ 4d	NB/ 4d	Ca5/ 4d
Aspergillus	P1	0	8	8	0
	H9	0	4	4	0
Alternaria	OUT3	0	17	17	0
Botrytis	P19	27	15	10	34
	C1	21	14	9	24
	MP3	30	19	12	28
	MP5	12	8	12	26
	G3/2	22	14	9	26
	ST1	10	10	9	30
Debaryomyces	P27	0	0	6	0
Penicillium	P5	0	0	8	0
	I4	0	0	11	0
	M4	0	0	0	0
	PR5	0	0	7	0
Phoma	P13	0	5	6	5
Trichocladium	H7	6	5	14	0
Trichophyton	C3	0	0	11	0
Trichosporum	P29	31	12	0	46
Zygorrhynchus	P23	0	0	23	20
Yeasts					
Rhodotorula	G4/1	0	0	0	0
Cryptococcus	G4/6	0	0	8	0

Ca5 cabbage extract agar [4]
NB Nutrient broth

3.3 Effect of pH on antifungal activity

Crude extracts (CB5/4d) from *Bacillus subtillis* were found to produce inhibition zones of between 30 and 35 mm diameter when the pH was adjusted to between 3 and 6. However above pH 6.5 inhibition zones were below 20 mm (individual results not shown). The high activity at lower pH values could be benificial for treatment of fungal contaminants in plant tissue cultures, since most fungal contaminants were shown to reduce the medium pH of plant tissue culture media to around 3 [1, 3].

However, a method for large scale production of the antibiotics needs to be developed and the compounds have to be tested for their human and phytotoxicity before they can be realistically used as biocides in plant tissue culture.

References

1. Danby S, Sigee DC, Epton HAS and Leifert C (1992) Fungal contaminants of *Coffea*, *Musa*, *Primula* and *Iris* shoot tissue cultures. Plant Growth Regulation (In press)
2. Hussain S, Lane SD and Price DN (1990) Control of contamination problems in plant tissue culture systems. Abstract VIIth International Congress on Plant Tissue Culture, Amsterdam, The Netherlands, 24–29 June 1990: p 235
3. Leifert C, Waites WM, Nicholas JR and Keetley JW (1990) Yeast contaminants of micropropagated plant cultures. J Appl Bacteriol 69:471–476
4. Leifert C, Sigee DC, Stanley R, Knight C and Epton HAS (1992) Biocontrol of *Botrytis cinerea* and *Alternaria brassicicola* on Dutch white cabbage by bacterial antagonists at cold store temperatures. Plant Pathol (In press)
5. McKeen CD, Reiley CC and Pusey PL (1986) Production and partial characterisation of antifungal substances antagonistic to *Monilinia fructicola* from *Bacillus subtilis*. Phytopathol 76:136–139
6. Murashige T and Skoog F (1962) A revised medium for rapid growth and bioassays with tobacco tissue cultures. Physiol Plant 15:473–497
7. Quesnel LB (1992) Apparatus for Microbiology Testing. British Patent No. 2233760

Physiology, growth and development of plants and cells in culture – the way ahead

RICHARD D. FIRN, NAVIN SHARMA* and JOHN DIGBY

Department of Biology, University of York, York, YO1 5DD, UK
* Current address: Hindustan Lever Research Centre, Bombay, India

1. Establishing the needs and capacities of plant cells and plant organs – the original objective of plant cell culture

One of the strange aspects of the subject area sometime termed 'plant tissue culture' is that it is rarely concerned with the culture of plant tissues and is more usually concerned with the culture of cells or organs [2]. This imprecise terminology has not only given a somewhat spurious unity to an area but it also seems to have obscured the fact that the culture of cells and the culture of organs are actually somewhat different topics. For instance the cultural needs of each system depends on its own inherent biosynthetic capacity and developmental potential, both of which vary very considerably depending on intrinsic and extrinsic factors. A consideration of these factors should form a basis for a rational means of culturing cells, organs or indeed plants. Indeed one of the early strands connecting work on the culture of cells and organs was the aim of understanding the degree of nutritional autonomy which cells of various types possessed. For instance, Haberland's [4] early attempts to culture cells in isolation were conducted with the objective of determining the degree of nutritional autonomy [7]. Haberland understood that photosynthetic plants have very simple nutritional requirements:

- CO_2
- H_2O
- O_2
- a relatively small number of simple ions to provide the remaining essential elements

However, Haberland wanted to know how far this autonomy extended to tissues or individual cells, consequently he adopted the methodology that had been so successfully applied to determining nutritional autonomy in microorganisms — the culturing of cells on defined media. Although Haberland did not achieve his goals with his cell culture studies, the workers who followed had more success and in general reached the general conclusions:

- organs are more autonomous than tissues
- tissues are more autonomous than cells

P.J. Lumsden, J.R. Nicholas and W.J. Davies (eds.), Physiology, Growth and Development of Plants in Culture, 409–421, 1994.

- nutritional autonomy is found at lower organisational levels in simpler organisms.

The most notable early successes in culturing organs were attempts to culture roots [12] and callus cultures, both systems where a carbon substrate obviously had to be supplied. This need for a carbon source necessitated the adoption of aseptic techniques in order to avoid microbial contamination. This one aspect of plant tissue culture, the aseptic technique, necessary because the culture medium was able to support rich microbial growth, was then to become a central feature of all approaches to the culture of cells and organs. Indeed the basic techniques, equipment and concepts for plant cell culture were obviously very greatly influenced by similar ideas being used so successfully by microbiologists and plant pathologists. Using these concepts, plant physiologists and plant biochemists devised culture media and cultural methods (both still remarkable similar to those used by microbiologists) to facilitate the suspension culture of cells and culture of cell masses on agar surfaces. The ability of such cultures to grow and differentiate into organs or indeed into whole plantlets excited a generation of physiologists. However, as noted by Street [9] the culture of organs seemed to become an end in itself:

> The pioneer workers seemingly found the successful aseptic cultures of plant organs and tissues so rewarding that there followed many descriptive papers concerned only with the announcing of similar cultures but adding little to the solution of major problems in plant physiology

Some of those interested in tissue culture turned their attention away from asking physiological questions and explored the possible practical importance of these new skills. The clonal propagation of commercially important species seemed a valid goal. This 'high-tech' approach to plant propagation was seen as a fine example of the appliance of science, even though such work was initially largely confined to research laboratories. Only in the 1960s was real commercialisation to begin with the extension of these cultural practices to plant organ culture. Eventually the micropropagation industry did indeed take these techniques out of the research laboratory into the commercial laboratory, often on a semi-industrial scale.

So it can be seen that the ideas about the ways and means of culturing plant cells and organs arose from a number of sources and the objectives of the various studies were equally varied. However, because the subject was thought as having a basic conceptual unity, ideas were sometimes transferred without good reason from one system to another and sometimes to the detriment of the system under study. Most crucially, concepts about the nutrition of cultures were to become very vague and all too often culture media were chosen with the belief that the more complete the medium the better. This vague and misleading thinking is well illustrated by a quotation from a recent chapter on nutrition in a book about plant tissue culture:

> The basic nutritional requirement of cultured plant cells are very similar to

those of whole plants ... macronutrients, micronutrients, vitamins, amino-acid or other N-containing supplements, sugars, other undefined organic supplements, solidifying agents and growth regulators" K. C. Torres, 1989 [10].

The author, instead of emphasising the crucial differences between the nutritional requirements of cells and plants, actually gives the impression that the differences are minor and then provides a list of ingredients only two of which are actually needed by most whole plants.

2. A rational basis to the nutritional requirements of cultured cells and organs

The whole plant is essentially a community of cells with different specialisations; some are optimised for nutrient uptake, some for transport, others have certain biosynthetic capacities, etc.. However, for many basic biosynthetic and ionic regulation capacities, specialisation is not extreme and within any individual plant, regions of considerable autonomy can exist. In lower plants (such as some alga and certain stages of moss or fern development) this autonomy seems to operate at the cell level but in most higher plants autonomy is evident only in some of the organs. For example, individual organs (e.g. some leaves) may have qualitative requirements which do not differ greatly from the whole plant. In such cases, the rest of the plant may be considered to act more as an efficient supply mechanism for simple compounds rather than as a well controlled source of metabolites. Hence, when considering the nutritional requirements for the culture of cells or organs, a consideration of the inherent biosynthetic capacities is a sensible starting place.

The next aspect to consider is the developmental potential inherent in the system. Developmental plasticity is a common feature of some plant cells and most plant organs and these capacities may be used to produce new abilities or new structures to compensate for any imbalance that might exist in the supply of the materials to the plant tissue or plant organ.

Thus three factors
• the simple basic nutritional needs of most plant cells
• considerable regional autonomy of most biosynthetic functions within plants
• the capacity for developmental plasticity

should be key elements in devising rational ways of culturing plant, plant parts or even plant cells. These three elements are those which are so strikingly different from the key elements to be taken into consideration when culturing animal cells or organs.

3. Propagation from parts of plants

Given these key features, it is not surprising that vegetative propagation of plants is often simple and has been practiced so successfully for centuries by gardeners and horticulturist. The removal of a 'cutting' from a plant simply removes a relatively autonomous part of the plant from the main structure and brings into play available developmental plasticity. The newly isolated organ develops new capacities or structures to replace the functions previously provided by other parts of the plant to which the organ was originally attached. The material to sustain this production of new structures or the expansion of existing ones comes from three possible sources:
• current production
• current uptake
• stored or mobilisable resources

The exact contribution that each of these three sources will provide will vary depending on the circumstances and gardeners have learned over the years how to manipulate the physiology of the parent plant and the explant to optimise the success of propagation. For instance, the time of year the cutting is taken is crucial to the process in some cases presumably because the physiological state of the parent plant must be optimal if the explant is to survive and develop. Likewise, some cuttings benefit from the presence of photosynthetic structures, others do not. In general terms, the conditions of the explant must be manipulated in order to
• cause the production of some new cell types
• stimulate the expansion and development of some existing structures
• stimulate the provision of materials to sustain each of these processes
• avoid fungal and bacterial invasion

The first two requirements will usually be met if the inherent developmental plasticity is expressed. The production of new structures will require the production of new cell types and may involve cycles of cell dedifferentiation, cell proliferation and organogenesis. Although these processes can be influenced by the provision of substances (such as plant growth regulators) externally to the explant, the opportunities for anything more than minor manipulation, possibly by potentiating some inherent capacity, is limited. An example of such an effect of an exogenous chemical intervention would be the stimulation of rooting in cuttings or explants by supplying high concentrations of auxins to the basal ends of such explants. The second type of developmental response, the expression of an existing developmental fate, is also important. For example, if the axilliary buds can be induced to grow in an explant, new photosynthetic organs soon become available to sustain future development. Such development may require that a developmental block that existed in the structure before excision is removed. Unfortunately, the control of such developmental stages is often poorly understood and this hinders some attempts to manipulate organ development. For instance, the poorly understood physiology of apical dominance is obviously important in this respect and the ability to control this phenomenon by the application of exogenous chemicals is still very limited. Hence it must be concluded that currently, rather limited opportunities exist to usefully and predictably alter the development of explants in any very specific manner. Although many cultural regimes for plant organs and cells involve quite specific additions of chemicals supposedly to manipulate the developmental pattern, the rationale behind such additions is usually based on simplistic thinking about hormonal control that is now very debatable. At the time when such knowledge was being accumulated by what now seem very empirical methods (1940s–1970s), very simplistic ideas about the role of plant growth substances were prevalent, ideas that are no longer widely accepted. The practices are still retained in plant tissue culture because they 'work' but it is dangerous to assume that such practices are rational and that they demonstrate a mastery and understanding of the processes involved.

The third factor involved in the culture of explants is the provision and supply of materials needed for cell division, expansion and differentiation. As outlined earlier, depending on the degree of biosynthetic autonomy of the structure being cultured, the needs will vary very greatly. Unfortunately while the simple needs of whole plants have been determined very precisely and the complex needs of cell cultures have been subject to very extensive studies, the needs of various types of explants has received less attention.

4. Leaving behind the life support mentality

The smallest plant to flower regularly in the UK is apparently Chaffweed (*Cetunculus minimus*) which has a seed weight of only 0.03mg. The fact that such a very small seed, when placed in soil teaming with a rich microbiological flora can soon give rise to a vigorous seedling suggests that very little material is needed to produce a seedling and that the seedling itself can usually avoid being overwhelmed by microorganisms. Hence an explant weighing many orders of magnitude more than such a seed and which contains a meristem should contain sufficient stored material to sustain the growth and development needed until self sufficiency is reached again as a result of growth and development. Certainly the experience of gardeners suggests that this is true in many cases and also suggests that conditions for the propagation from cuttings can be chosen so that microbial infection is minimised. Yet those attempting micropropagation adopt conditions that are very different from the traditional horticulturist and for microproagation it is usually assumed that the explant has rather little capacity to sustain or defend itself. The provision of a nutritionally rich medium (traditionally M&S salts with added sugars and vitamins, a medium originally developed to sustain non-photosynthetic plant cells) on which to grow the explant may help to accelerate the growth and development in the short term but at what cost?

- the medium is so nutritionally rich that it sustains the growth of microorganisms hence the need to sterilise the explants and grow them in closed containers to avoid further contamination (interestingly it is now known that many contaminating organisms usually harm the explants indirectly, either by altering the pH of the medium or by secreting toxins, hence sterility is required to avoid the harmful consequences of the microbes not the actual microbes themselves)

- the growth of the explants in closed sterile containers results in shoots developing abnormally because of the high humidity, such plantlets then being difficult to acclimatise to life outside their containers when they encounter a normal humidity.

- the labour involved manipulating the cultures under sterile conditions means that it is impractical to change the medium on which the explants are growing. Therefore the medium used may have supra optimum levels of ions initially and possibly suffer later from ion depletion. There may also be an accumulation of toxic substances from the cut surface of the explant. It is also difficult to adjust the medium composition to be optimal as the explant progresses through various developmental stages

- the closed containers necessary for sterility make it difficult to introduce beneficial gases (CO_2) or remove detrimental gases (C_2H_4)

- the high osmotic potential of the medium due to the high concentration of added nutrients may cause physiological stress

In other words, the growth of explants on supplemented M&S medium is a bit like placing **all** hospital patients (even those suffering from non life-threatening conditions) on life-support systems. In the short term it might be beneficial but it might actually create a physiological dependence that is harmful.

5. Low-tech approaches

With these considerations in mind, we have been studying explant culture methods in an attempt to:

- eliminate all chemicals other than mineral ions in the culture medium, hence reducing the need for strict sterility

- eliminate the use of a solidified medium as a means of supporting the plant, both as a means of reducing costs and also providing a better environment

- enable the adoption of larger, simple culture containers in which the nutrient and gaseous atmosphere can be easily and simple controlled

- ease the means of changing medium composition, so as to make it possible to change the medium composition to match the changing requirements of the culture as it develops and also in order to avoid nutrient or growth substance depletion

- reduce the problems of microbial contamination by growing explants under conditions where the medium pH could be controlled and toxin build-up prevented

Three methods have been evaluated.

5.1 Mist micropropagation

The provision of a liquid culture medium to plantlets/explants could be achieved in a number of ways and after consideration we tried misting for the following reasons:

(a) It provides a flexible approach in which the nutrient medium can be varied easily.
(b) Misting can be achieved using simple equipment which is easy to set up.
(c) Large numbers of plantlets can be served by a single misting unit.
(d) When mist is used, a large surface area of the explant is exposed to nutrients hence uptake of the nutrient may be increased.
(e) The supply of water can be regulated easily to provide a preweaning phase.

   We designed and successfully tested such an 'aeroponic' system (Fig. 1). In this system, a spray of autoclaved Murashige and Skoog's salts containing 1.5mg/l kinetin and 0.5mg/l IAA was applied, from top, to *Chrysanthemum* (var Xantha) explants. The mist was generated by a spinning disc technique using a Turbair X-J ultra low volume sprayer (Turbair Ltd., Waltham Cross, Hartfordshire, U.K.). For trial purposes, a single mist unit was fitted to a plastic chamber (approx. 24 x 24 x 10 cm) and approximately 200 explants were accommodated in each chamber. Liquid was supplied to the disc by a low cost peristaltic pump. Both peristaltic pump and sprayer were connected to a timer unit, which in most trials turned on every 20 min for 10 seconds. However the growth of the explants was not greatly influenced by the frequency of spraying (Fig. 2).

   The explants were supported in the wells of disposable plastic microtitre plates. A small hole was drilled in the bottom of the wells to allow excess

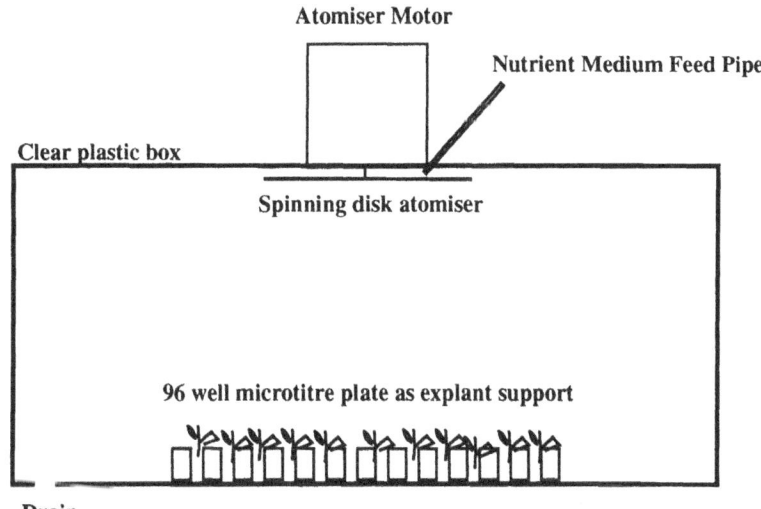

Fig. 1. Design of an aeroponic system.

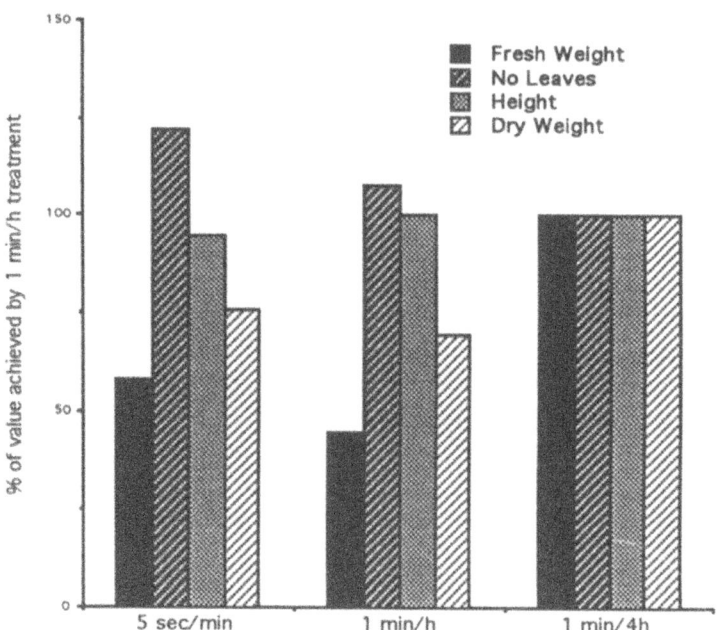

Fig. 2. Effect of the frequency of spraying on Chrysanthemum explant growth in the aeroponic unit.

medium to drain. The drained medium was collected in another chamber fitted at the bottom of the aeroponic unit, and this spent medium could be reutilised after autoclaving or could be discarded.

The performance of explants grown in aeroponics was compared with a conventional micropropagation system. Explants were taken from glasshouse-grown Chrysanthemum plants. They were surface sterilised in 10% sodium hypochlorite with 0.01% (approximately) Tween 20 for 10 min. They were rinsed three times in sterilised distilled water before being placed inside the jar for conventional micropropagation or in the misting apparatus. Weekly measurements were made of fresh weight, number of leaves and dry weight. The results (Fig. 3) showed that:

1. initially the fresh weight, number of leaves and dry weight of explants in aeroponics did not match with those grown on solidified or liquid medium supplemented with sucrose.
2. by the end of 4 weeks, the plants grown in aeroponics matched conventionally micropropagated plants.

To find out the significance of stored reserves, an experiment on effect of size of explant was carried out. Three different sizes of explants viz. 1 cm, 1.5 cm and 3 cm, were compared in the mist system. After 4 weeks in culture the observations were made on height of the developed shoots, fresh weight, number of leaves and dry weight. There were clear-cut differences, the largest explants formed taller shoots with maximum number of leaves and had higher

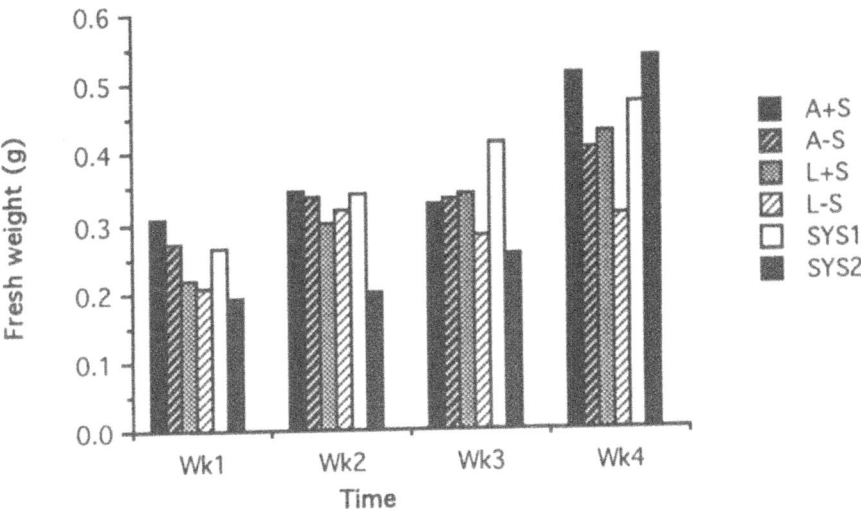

Fig. 3. A comparison of growth (fresh weight) of Chrysanthemum plants grown in aeroponically and plants grown *in vitro* on agar-solidified media and on liquid media.
A = agar grown; S = sucrose; L = liquid culture grown; SYS = aeroponic culture

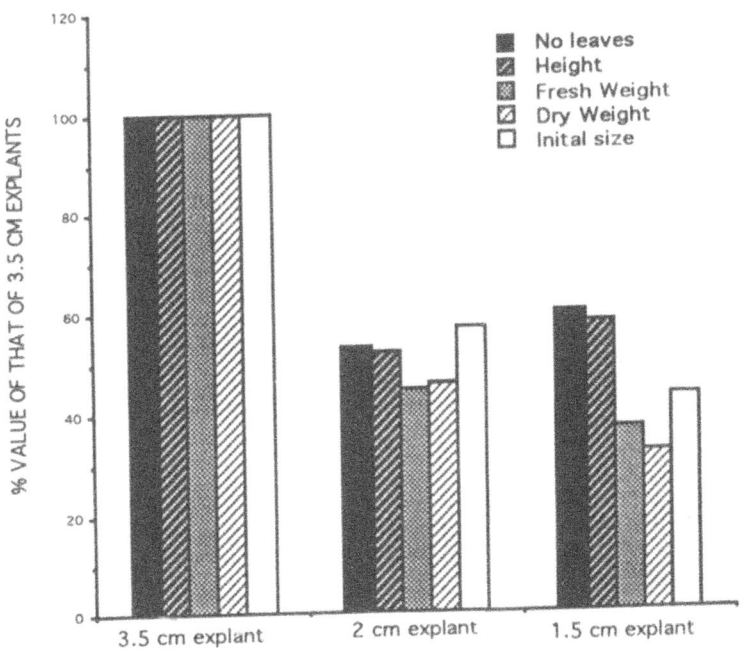

Fig. 4. The effect of initial explant size on subsequent growth of Chrysanthemum plants grown in the aeroponic system. All values normalised against the largest initial explants.

fresh and dry weight as compared to plants developed on 1 and 1.5 cm explants (Fig. 4). Other experiments confirmed this finding.

An indication that photosynthesis was a significant factor in the good performance during mist propagation was the observation that explants grown in darkness performed very poorly. This led to preliminary studies of the effects of CO_2 enrichment, both in the aeroponic and flow through systems (see later).

When leaves from conventionally *in vitro* grown explants, mist grown explants and greenhouse grown plants were compared, it was apparent that mist grown plants are physiologically similar to greenhouse grown plants and quite different from conventionally micropropagated plantlets. This comparison was valid whether the rates of transpiration , photosynthesis or stomatal apertures were observed.

5.2 Liquid flow system

In this system the explants were placed in the wells of the microtitre plates inside a sterile clear polystyrene chamber as used for the misting experiments (Fig. 5). The microtitre plates floated on liquid medium so that only the lower portion (base) of explants came in contact with the solution. The culture medium could be allowed to run to waste or recirculated through the chamber via a peristaltic pump after passing through two filters, one cleared the solution (0.45μm Millipore filter) and other strerilised it (0.22μm Millipore filter).

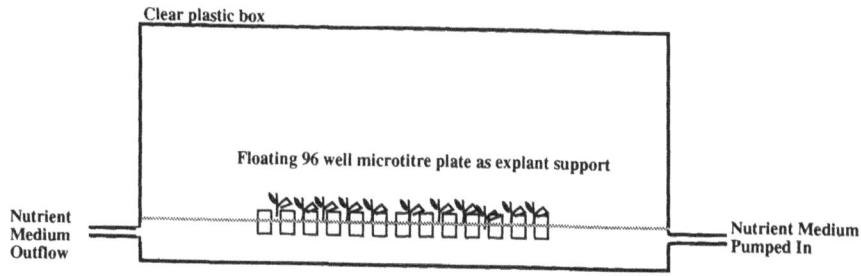

Fig. 5. Design of a liquid flow system for explant culture.

The attractions of the liquid flow system are:
- the simplicity and the fact that it would be possible to scale up the process
- the composition of the medium could be monitored and changed easily and pulse treatments to the basal end could be achieved (for example a short pulse of an auxin to stimulate the initiation of roots followed by an auxin-free medium to encourage root growth)
- it was possible to control the humidity of the atmosphere around the shoots
- preliminary experiments (Fig. 6) revealed that growing the plants in enriched carbon dioxide atmospheres (600–1000ppm) in this system resulted in a considerable promotion of growth

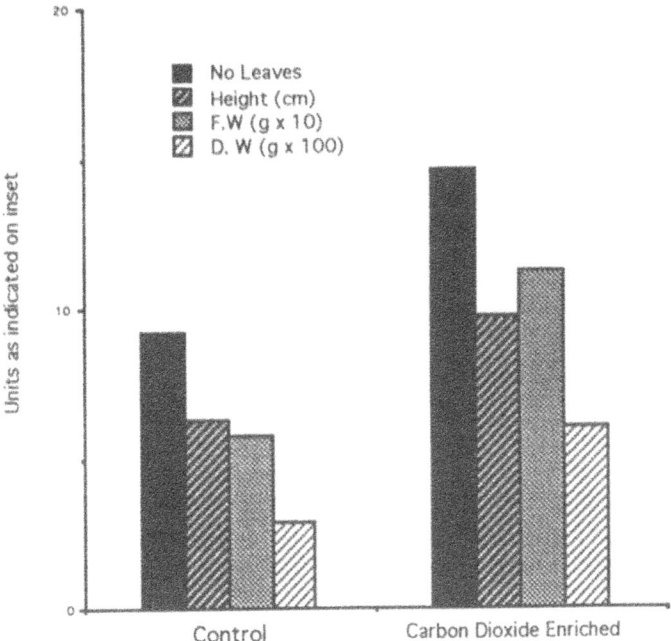

Fig. 6. The effect of CO_2 enrichment on the growth of Chrysanthemum explants in the liquid flow system.

These simple trials seem to support concepts outlined earlier — that explants should be able to sustain their initial own growth and development from endogenous sources and should be able to soon supplement these resources with current production. By adopting an approach which aims only to provide simple mineral salts to the explants, one is liberated from the need for strict aseptic techniques. This in turn allows one to use larger containers in which it is practical to maintain a more normal humidity and in which one can facilitate carbon dioxide enrichment. The former factor helps such systems to produce plants which are physiologically more normal than conventionally micropropagated plants and the latter allows helps compensate for the slower growth that such systems sustain initially.

6. The ways ahead?

There is no single way ahead because there are many different objectives depending on whether one is interested in culturing cells or organs, or whether one is operating in a research or commercial environment. If there is any message that we would like people to take from this article it is that each piece of work must have its own aims and while it is valuable to borrow ideas from related work, there are dangers in assuming that any one methodology is universally useful.

Cell culture? A question that must be asked is whether there will be quite such an emphasis on cell culture in future. Two aspects of cell culture which seemed exciting in previous times now seem less so. First, the use of cell cultures as a means of studying the physiology and biochemistry of cells is no longer as attractive as it once seemed because many of the interesting, unique aspects of plant growth and development simply do not occur in cell cultures and the remaining more basic aspects of cell functioning are often as well studied in other systems. Second, the concept of cell culture on an industrial scale as a means of producing products of interest may well be overtaken by the currently more attractive idea of genetically engineering appropriate plants to make such products in easily grown and harvested organs. Cell culture may of course play a vital role in the transformation process for some species but it will be a facilitating role rather than a final objective.

Organ culture and micropropagation. Micropropagation as currently operated is labour intensive and requires some technical skills, hence can rarely compete with simpler horticultural, silvicultural or agricultural practices able to produce the same plants. Three options are available to the micropropagation industry:
• relocation to areas of low cost labour
• the adoption of robotics [1]
• the development of lower cost, more efficient systems
    The first alternative has we gather already begun and seems inevitable because the level of technology needed is not high. The second alternative seems attractive until it is recognised that the manipulations required in micro-propagation are subtle and complex due to the biological variation inherent in the units being manipulated and the unit value is very low. Both these factors mitigate against the widespread adoption of robotics in the near future and the automation of some of the tasks which do not involve the plants themselves seem likely to be more cost effective in the short term. The final alternative, the fundamental rethink of the concept of micropropagation seems to us the most attractive alternative. The essence of this approach is that one does not try to adopt a low-tech version of a process actually devised for a research laboratory. Instead one devises new, simple techniques, based on sound physiological principles. In particular, it has already been shown by previous workers that carbon provision for micropropagated explants can be provided by photosynthesis and can replace sugars in the medium [3, 4, 5, 6, 8, 11]. Our own trial have shown that simple photoautotrophic systems have much to commend them and deserve much further study.

Acknowledgements

We would like to thank the Ministry of Agriculture, Food and Fisheries for financial support. The advice of Dr Philip Orton, Askham Bryan Agricultural College, was invaluable.

References

1. Aitken-Christie J, Davies HE, Siviter J and Mairn B (1989) Automation in micropropagation of *Radiata pine*. In Vitro Cell Dev Biol 25(A):22
2. Bailey IW (1943) Some misleading terminologies in the literature of 'plant tissue culture'. Science 93:539
3. Grout BWW and Aston MJ (1978) Transplanting of cauliflower plants regenerated from meristem culture. II. Carbon dioxide fixation and the development of photosynthetic ability. Hort Res 17:65–71
4. Haberland G (1902 but reprinted and translated in 1969) Experiments on the culture of isolated plant cell. Bot Rev 35:68–88
5. Kozai T, Koyama Y and Watanabe I (1988) Development of a photoautotrophic tissue culture system for shoots and or/plantlets at rooting and acclimatisation stages. Acta Hort 230:153–158
6. Kozai T and Ivanami Y (1988) Effects of CO_2 enrichment and sucrose concentration under high photon fluxes on plantlet growth of Carnation (*Dianthus caryophyllus*) in tissue culture during the preparation stage. J Jap Hort Sci 57:279–288
7. Krikorian AD and Berquam DL (1969) Plant cell and tissue cultures: the role of Haberland. Bot Rev 35:59–67
8. Lasko AN, Reisch BI, Mortensen J and Roberts M (1986) Carbon dioxide enrichment for stimulation of growth of *in vitro*-propagated grape vines after transfer from culture. J Amer Soc Hort Sci 111:634–638
9. Street HE (1969) Growth in organised and unorganised system. In: Steward FC (ed) Plant Physiology A treatise, Vol Vb, pp 3–224. Academic Press
10. Torres TC (1989) Tissue Culture Techniques for Horticultural Crops. New York: Van Nostrand Reinhold
11. Walker S, Simpkins J and Roberts AV (1990) Autotrophic growth of *Rosa hybrida* cv Mountbatten on sucrose free medium. Abstracts Int Congress on Plant Tissue Culture. p 41
12. White PR (1943) The Handbook of Plant Tissue Culture Lancaster, PA, USA: The Jacques Press 277pp

Index

P.J. Lumsden, J.R. Nicholas and W.J. Davies (eds.), Physiology, Growth and Development of Plants in Culture, 422–427, 1994.
© 1994 *Kluwer Academic Publishers.*

The manufacturer's authorised representative in the EU is Springer
Nature Customer Service Centre GmbH, Europaplatz 3, 69115 Heidelberg,
Germany. If you have any concerns regarding our products, please
contact ProductSafety@springernature.com

Printed and bound by CPI Group (UK) Ltd, Croydon, CR0 4YY

24/04/2026

02096348-0010